THE OXFORD HANDBOOK OF
FUNCTIONAL DATA ANALYSIS

THE OXFORD HANDBOOK OF

FUNCTIONAL DATA ANALYSIS

Edited by
FRÉDÉRIC FERRATY
and
YVES ROMAIN

OXFORD
UNIVERSITY PRESS

Great Clarendon Street, Oxford OX2 6DP

Oxford University Press is a department of the University of Oxford.
It furthers the University's objective of excellence in research, scholarship,
and education by publishing worldwide in

Oxford New York

Auckland Cape Town Dar es Salaam Hong Kong Karachi
Kuala Lumpur Madrid Melbourne Mexico City Nairobi
New Delhi Shanghai Taipei Toronto

With offices in

Argentina Austria Brazil Chile Czech Republic France Greece
Guatemala Hungary Italy Japan Poland Portugal Singapore
South Korea Switzerland Thailand Turkey Ukraine Vietnam

Oxford is a registered trade mark of Oxford University Press
in the UK and in certain other countries

Published in the United States
by Oxford University Press Inc., New York

© Oxford University Press 2011

The moral rights of the authors have been asserted
Database right Oxford University Press (maker)

First published 2011

All rights reserved. No part of this publication may be reproduced,
stored in a retrieval system, or transmitted, in any form or by any means,
without the prior permission in writing of Oxford University Press,
or as expressly permitted by law, or under terms agreed with the appropriate
reprographics rights organization. Enquiries concerning reproduction
outside the scope of the above should be sent to the Rights Department,
Oxford University Press, at the address above

You must not circulate this book in any other binding or cover
and you must impose the same condition on any acquirer

British Library Cataloguing in Publication Data

Data available

Library of Congress Cataloging in Publication Data

Data available

Typeset by SPI Publisher Services, Pondicherry, India
Printed in Great Britain
on acid-free paper by
CPI Antony Rowe, Chippenham, Wiltshire

ISBN 978–0–19–956844–4

1 3 5 7 9 10 8 6 4 2

Preface

What is functional data?

Technological progress in computational tools and memory capacity is allowing us to handle larger and larger datasets. At the same time, monitoring devices and electronic equipment like sensors (for recording images, temperature, gas composition, etc.) are becoming more and more sophisticated. This high-tech revolution is offering us the opportunity to observe phenomena in a more and more accurate way by producing statistical information sampled over finer and finer grids. A standard one-dimensional grid is that for time; such time-dependent data occur when one observes a continuous process such as in biomechanics (e.g. human movements), econometrics (e.g. the stock market index), geophysics (e.g. spatio-temporal events like El Niño or a time series of satellite images), or medicine (e.g. electrocardiograms and electroencephalograms). Numerous other examples may be found in chemometrics where the one-dimensional grid corresponds to wavelengths (spectrometric curves). Such datasets become extremely large as soon as one considers a 2-D or 3-D grid; this is the case when one is studying a collection of surfaces or hyperspectral images. Furthermore, such statistical information can be made much more sophisticated if one adds other kinds of grid (for instance a time grid).

Of course, this is not an exhaustive description of the kinds of high-dimensional data that the statistician may encounter, but it gives a good idea of the complexity of their structure and also of their great potential for application. As already noted, each kind of information in such datasets is sampled over a grid which is simply a discrete approximation of a "continuous" set (e.g. the positive real line in the case of time or wavelength). In this sense, the information in such high-tech datasets varies over some continuum and this is why, from a mathematical point of view, one can consider such data as a collection of functional mathematical objects (for example, as a collection of curves or of surfaces). Finally, the terminology "functional data" has been adopted by the statistics community to designate this kind of high-dimensional data, and so "functional data analysis" (FDA) may refer to the many statistical methods used to deal with functional data, the trajectories of random function, stochastic processes, and so on.

Motivated by numerous fields of application, functional data analysis has become more and more popular in the statistics community; the extensive and dense literature now available on this topic gives a good idea of its important role in modern statistics. Pioneering statistical work in the fields of chemometrics and climatology was published in the 1960s. However, the study of functional data was really popularized at the end of the 1990s with the publication of the benchmark book by Ramsay and Silverman entitled *Functional Data Analysis*. Since that time, numerous areas of statistics have been extended to cover functional data: bootstraps, classifications, factor and multivariate analysis, preprocessing, regression, resampling, tests, time series, etc.

Why a handbook on functional data?

The idea of this volume goes back to 2007. At that time, our working group STAPH (a research group on functional and operatorial statistics, founded in 1999 at the Institute of Mathematics in Toulouse, France), based on long experience in these areas, had developed much collaboration and many contacts with the international statistical community. These rather young areas of statistics were by then sufficiently mature for STAPH to decide to organize the 1st International Workshop on Functional and Operatorial Statistics (IWFOS), which took place in Toulouse (France) in June 2008. This was an opportunity to gather together a large number of international experts in the field of high-dimensional data analysis, with a special focus on functional data and on the mathematical environment used to deal with such data. At the same time, we were contacted by Oxford University Press with a request to edit a contributed volume dealing with functional data. The coinciding of this request with the organization of IWFOS, combined with our experience in this area, argued in favor of editing such a handbook on functional data.

Guidelines and intended audience

The aim of this handbook is to provide a benchmark volume for anyone interested in this high-tech field of statistics. Its scientific purpose is to present the state-of-the-art of the statistics dealing with functional data (and its mathematical background) by gathering together most of the major advances in this area. International experts have been invited to contribute to this volume, under the guideline that each chapter has to give priority to bibliographical information and to the main original ideas. In addition, the particularity of this book is that it offers a wide range of the most important statistical topics (classification, inference, factor-based analysis, regression modeling, resampling methods, time series, random processes) while at the same time considering practical, methodological, and theoretical aspects of the problems. It is worth noting that, because of the continuous nature of such functional data, it becomes possible to carry out on it various kinds of infinitesimal calculations such as differentiation, integration, approximations, and many other more sophisticated mathematical tools.

Therefore, two challenges clearly appear in the setting of functional data: the need to implement new statistical methodologies (and here there are evidently practical issues) as well as to enlarge the mathematical background and toolbox. Lasting and relevant advances in this fascinating modern field of statistics will require developments in both the practical and theoretical aspects of the subject. This is why we have decided in this volume to provide a combination of practical, methodological, and theoretical chapters, and some which combine these various points of view. As a result, this book can be read on various levels which allows us to accommodate a broad audience: engineers, practitioners, and graduate students as well as academic researchers, not only in statistics and probability but also in the numerous related domains of application.

CONTENT

Functional data analysis is a very attractive field of research because it broadens all the main topics of statistics: classification (un/supervised), factor-based analysis, inference, modeling, time series, resampling, random functions, etc. So this book will deal with all these important topics in statistics (and probability), but of course in the setting of functional data. The first part of the book is devoted to regression modeling, which has received a great deal of attention from the statistics community. Linear, semi- and nonparametric approaches with independent or dependent observations are presented. This allows us to cover a wide range of statistical methods for functional data: linear/nonparametric functional regression, varying coefficient models, and linear/nonparametric functional processes (i.e. functional time series). In the second part, the focus is on various related benchmark methods/tools for functional data analysis: preprocessing functional data using curve registration methods, functional principal component analysis as a useful factor-based exploratory tool, sparseness and functional data, resampling/bootstrap methods and functional data, and classifying functional data. Finally, the third part is oriented towards some of the fundamental mathematical aspects of the infinite-dimensional setting, with a focus on the stochastic background and operatorial statistics: vector-valued function integration, operator geometry, inequalities in Banach spaces, spectral and random measures linked to stationary processes, and operatorial statistics linked to quantum statistics.

CHALLENGES FOR THE FUTURE

This handbook is the first to give such a broad overview of functional statistics and its mathematical environment. As a consequence, this volume also provides a good resource for identifying which topics seem to be underdeveloped at present, in order to set out new directions for future research. For instance, semiparametric methodologies (varying coefficient models, random effect models, partial linear models, etc.), Bayesian approaches, spatial functional statistics and differential equation models (dynamic systems) provide a sample of very attractive topics for research in the context of functional data that have not been much investigated up to now. We can be sure that all these various subjects will receive a great deal of attention in the near future. From a more general viewpoint, it is essentially the nonlinear (nonparametric) aspect of these methods which makes their development more difficult in the context of high-dimensional data such as functional data. There is a need to develop and use suitable mathematical tools to explore nonlinear structure in high-dimensional spaces. Advances in the theory of nonlinear operators in general spaces should play a major role in enabling significant progress to be made in differential geometry, numerical approximation methods, and infinitesimal calculus in such high-dimensional settings.

Another crucial problem in infinite-dimensional spaces is the nonexistence of a benchmark/universal measure such as the Lebesgue measure in the finite-dimensional case. Indeed, in the infinite-dimensional situation, whenever one deals with any probability density function, one has to assume the existence of a relevant measure, which makes the statistician's work harder. Therefore one way to make progress in the nonparametric distributional aspects of function spaces might be to propose statistical methods that are valid for a large class of measures, which will certainly also require advances in probability theory. Finally, one can expect that datasets with much more sophisticated structures will be produced in the future. So constructing statistical methods for data with values in abstract spaces is not just about doing mathematics, but it also allows us to anticipate new kinds of datasets and hence the new technologies which will produce them. This highlights the constant need for feedback between statistics and mathematics, especially in high-dimensional settings (the link between statistics and other disciplines such as biology, computer science, geophysics, economics, and medicine is already obvious because of the numerous applications that exist). This argues in favor of giving priority not only to the practical and methodological aspects of statistics, but also to deep theoretical developments (see the third part of this volume for a large spectrum of open problems in mathematical research that are directly linked with high-dimensional statistics).

Acknowledgements

We would like to express our grateful thanks to all the outstanding experts for their valuable contributions to this volume. All of them immediately agreed to take part in this great scientific adventure. The high quality of their work, as well as their helpful and careful involvement in the peer-reviewing process, has ensured the high scientific level of this handbook. One can easily predict that it will become a benchmark in the field of statistics for functional data analysis. All the authors are again warmly thanked. Finally, we would like to express our special thanks to the STAPH working group (http://www.math.univ-toulouse.fr/staph), which is a dynamic research team at the Institute of Mathematics in Toulouse (France). This volume could not have been produced without the ongoing support of the STAPH group; this handbook has without doubt come out of the work of this group and all its members and collaborators are gratefully acknowledged.

Frédéric Ferraty
Yves Romain

September 2010

Contents

List of Contributors — xiii
List of Figures — xv
List of Datasets — xvii

PART I REGRESSION MODELING FOR FDA

1. A unifying classification for functional regression modeling — 3
   FRÉDÉRIC FERRATY AND PHILIPPE VIEU

2. Functional linear regression — 21
   HERVÉ CARDOT AND PASCAL SARDA

3. Linear processes for functional data — 47
   ANDRÉ MAS AND BESNIK PUMO

4. Kernel regression estimation for functional data — 72
   FRÉDÉRIC FERRATY AND PHILIPPE VIEU

5. Nonparametric methods for α-mixing functional random variables — 130
   LAURENT DELSOL

6. Functional coefficient models for economic and financial data — 166
   ZONGWU CAI

PART II BENCHMARK METHODS FOR FDA

7. Resampling methods for functional data — 189
   TIMOTHY MCMURRY AND DIMITRIS POLITIS

8. Principal component analysis for functional data: methodology, theory, and discussion — 210
   PETER HALL

9. Curve registration — 235
   JAMES RAMSAY

10. Classification methods for functional data — 259
    AMPARO BAÍLLO, ANTONIO CUEVAS, AND RICARDO FRAIMAN

11. Sparseness and functional data analysis — 298
    GARETH JAMES

PART III TOWARDS A STOCHASTIC BACKGROUND IN INFINITE-DIMENSIONAL SPACES

12. Vector integration and stochastic integration in Banach spaces — 327
    NICOLAE DINCULEANU

13. Operator geometry in statistics — 355
    KARL GUSTAFSON

14. On Bernstein type and maximal inequalities for dependent Banach-valued random vectors and applications — 383
    NOUREDDINE RHOMARI

15. On product measures associated with stationary processes — 423
    ALAIN BOUDOU AND YVES ROMAIN

16. An invitation to operator-based statistics — 452
    YVES ROMAIN

Index — 483

List of Contributors

Amparo Baíllo Department of Mathematics, Faculty of Sciences, Universidad Autónoma de Madrid, Campus de Cantoblance, 28049 Madrid, Spain, amparo.baillo@uam.es

Alain Boudou Institut de Mathématiques de Toulouse, UMR CNRS 5219, Equipe de Laboratoire de Statistique et Probabilités, 118 route de Narbonne, 31062 Toulouse Cedex 9, France, boudou@math.univ-toulouse.fr

Zongwu Cai Department of Mathematics & Statistics, University of North Carolina at Charlotte, Charlotte, NC 28223, USA, The Wang Yanan Institute for Studies in Economics, Xiamen University, Xiamen, Fujian 361005, China, and China Ministry of Education (MOE) key Laboratory of Econometrics at Xiamen University, Xiamen, Fujian 361005, China, zcai@uncc.edu

Hervé Cardot Institut de Mathématiques de Bourgogne, UMR CNRS 5584, Université de Bourgogne, 9 avenue Savary, BP 47870, 21078 Dijon, France, Herve.Cardot@u-bourgogne.fr

Antonio Cuevas Department of Mathematics, Faculty of Sciences, Universidad Autónoma de Madrid, Campus de Cantoblanco, 28049 Madrid, Spain, antonio.cuevas@uam.es

Laurent Delsol Institut de Statistique, Université Catholique de Louvain, 20 voie du Roman Pays, 1348 Louvain-la-Neuve, Belgium, laurent.delsol@uclouvain.be

Nicolae Dinculeanu Department of Mathematics, University of Florida, 475 Little Hall, Gainesville, FL 32611-8105, USA, nd@math.ufl.edu

Frédéric Ferraty Institut de Mathématiques de Toulouse, Université Paul Sabatier, 118 route de Narbonne, 31062 Toulouse Cedex 9, France, frederic.ferraty@math.univ-toulouse.fr

Ricardo Fraiman Departmento de Matemática, Universidad de San Andrés, Vito Dumas 284, Buenos Aires, Argentina and Universidad de la República, Montevideo, Uruguay, rfraiman@udesa.edu.ar

Karl Gustafson Department of Mathematics, University of Colorado at Boulder, Campus Box 395, Boulder, CO 80309-0395, USA, gustafs@euclid.colorado.edu

Peter Hall Department of Mathematics and Statistics, The University of Melbourne, Parkville, VIC, 3010, Australia and Department of Statistics, University of California, Davis, CA-95616 USA, halpstat@ms.unimelb.edu.au

Gareth James Marshall School of Business, University of Southern California, Los Angeles, CA 90089, USA, gareth@usc.edu

André Mas I3M, Université Montpellier 2, Place Eugène Bataillon, 34095 Montpellier Cedex 5, France, mas@math.univ-montp2.fr

Timothy McMurry Department of Mathematical Sciences, DePaul University, 2320 N. Kenmore Avenue, Chicago, IL 60614, USA, tmcmurry@depaul.edu

Dimitris Politis Department of Mathematics, University of California, San Diego, La Jolla, CA 92093-0112, USA, politis@math.ucsd.edu

Besnik Pumo Agrocampus Ouest, Centre d'Angers, 2 rue Le Nôtre, 49045 Angers, France, Besnik.Pumo@agrocampus-ouest.fr

James Ramsay Department of Psychology, McGill University, 1205 Dr Penfield Avenue, Montreal, Quebec, Canada, H4A 1B1, ramsay@psych.mcgill.ca

Noureddine Rhomari Département de Mathématique et Informatique, Université Mohammed 1er, Faculté des Sciences, Boulevard Mohammed VI, BP 717, 60 000 Oujda, Morocco, rhomari@fso.ump.ma

Yves Romain Institute de Mathématiques de Toulouse UMR CNRS 5219, Equipe de Laboratoire de Statistique et Probabilités, 118, route de Narbonne, 31062 Toulouse Cedex 9, France, yves.romain@math.univ-toulouse.fr

Pascal Sarda Institut de Mathématiques de Toulouse, Université Paul Sabatier, 118 route de Narbonne, 31062 Toulouse Cedex 9, France, pascal.sarda@math.univ-toulouse.fr

Philippe Vieu Institut de Mathématiques de Toulouse, Université Paul Sabatier, 118 route de Narbonne, 31062 Toulouse Cedex 9, France, philippe.vieu@math.univ-toulouse.fr

List of Figures

Figure 3.1	Ornstein–Uhlenbeck and Wong processes	57
Figure 3.2	Monthly mean El Niño sea surface temperatures	67
Figure 5.1	Monthly sea temperature evolution	133
Figure 5.2	Yearly El Niño sea surface temperature curves	135
Figure 6.1	Estimated functional coefficients in time-varying coefficient models	178
Figure 6.2	Nonparametric estimation of time-varying beta coefficients	182
Figure 9.1	Human growth acceleration curves and their corresponding landmark-registered curves	237
Figure 9.2	Amplitude/phase variation	238
Figure 9.3	Growth acceleration curves and corresponding time-warping functions	240
Figure 9.4	De-registered sine curves with phase variation by one-parameter warping functions	242
Figure 9.5	Unregistered/registered sine curves with amplitude and phase variation	245
Figure 9.6	Registered/unregistered mean temperature curves	246
Figure 9.7	Continuously registered sine curves and estimated warping functions	250
Figure 9.8	Continuous registration of the landmark-registered height acceleration curves	250
Figure 9.9	Mean of the continuously registered acceleration curves and landmark-registered acceleration curves	251
Figure 9.10	Harmonics for the growth acceleration curves resulting from combining registration and principal component analysis	253
Figure 9.11	Deformation functions associated with the phase variation in growth acceleration not accounted for by the two principal components	255
Figure 9.12	Principal component scores for acceleration curves registered to two principal components	255

Figure 11.1	Spinal bone and nephropathy longitudinal data	299
Figure 11.2	Confidence intervals for the mean function and both principal components for the spinal bone data	306
Figure 11.3	Linear discriminant analysis for the spinal bone data	309
Figure 11.4	Cluster mean curves and linear discriminant plots for the membranous nephropathy data	312
Figure 11.5	Estimated covariance and correlation functions for the spinal bone data	313
Figure 11.6	Discriminant curves for the spinal bone and nephropathy datasets	314
Figure 11.7	Predicted curves for two subjects from the spinal bone data	315
Figure 11.8	Linear discriminants with respect to ethnicity for spinal bone data	317

List of Datasets

Dataset 1	Monthly El Niño sea surface temperature (http://www.cpc.ncep.noaa.gov/data)	Chapters 3, 5
Dataset 2	Wages and education	Chapter 6
Dataset 3	Common stock price of Microsoft stock	Chapter 6
Dataset 4	Functional Magnetic Resonance Imaging (fMRI) data	Chapter 7
Dataset 5	Sulfur dioxide concentrations	Chapter 7
Dataset 6	Language classification	Chapter 7
Dataset 7	Berkeley growth data (http://ego.psych.mcgill.ca/faculty/ramsay/datasets.html)	Chapters 9, 10
Dataset 8	Canadian weather dataset (http://ego.psych.mcgill.ca/faculty/ramsay/datasets.html)	Chapter 9
Dataset 9	Electrocardiagram (ECG) data (http://www.physionet.org)	Chapter 10
Dataset 10	Cell data	Chapter 10
Dataset 11	Tecator data—NIR spectra (http://lib.stat.cmu.edu/datasets/tecator)	Chapter 10
Dataset 12	Phoneme data for speech recognition (http://www-stat.stanford.edu/~tibs/ElemStatLearn)	Chapter 10
Dataset 13	Medflies data (http://anson.ucdavis.edu/~mueller)	Chapter 10
Dataset 14	Gene expression profiles (http://www.blackwellpublishing.com/rss/Volumes/Bv69p4.htm)	Chapter 10
Dataset 15	Food industry data	Chapter 10
Dataset 16	Spinal bone mineral density dataset	Chapter 11
Dataset 17	Membranous nephropathy dataset	Chapter 11

PART I
REGRESSION MODELING FOR FDA

CHAPTER 1

A UNIFYING CLASSIFICATION FOR FUNCTIONAL REGRESSION MODELING

FRÉDÉRIC FERRATY

PHILIPPE VIEU

1.1 INTRODUCTION

TECHNOLOGICAL progress, both in collecting and storing data, is now providing datasets of a functional nature (curves, images, etc.) more and more often; for the last few years it has been a truly exciting challenge for the statistical community to propose models and methods for dealing with them. The book by Ramsay and Silverman (1997) has popularized functional data analysis as a specific field of statistics. Other recent monographs on this topic include those by Ramsay and Silverman (2002, 2005) and Ferraty and Vieu (2006), while a very good overview of the literature can be found in the numerous special issues that various statistical journals have recently devoted to functional data analysis or to related topics (Davidian *et al.* 2004; González Manteiga and Vieu 2007; Valderrama 2007; Ferraty 2010).

Our aim here focuses on the regression framework. Our objective is to model the link between two variables X and Y, when the explanatory variable (X) is of a functional nature. The literature on functional regression began and quickly grew in the early 1990s with linear modeling. The limited flexibility of such "parametric" modeling also suffers, in infinite-dimensional settings, from a lack of graphical tools for checking the accuracy of the model. To avoid these drawbacks, during the last decade "nonparametric" ideas have been adapted to the functional variable setting, providing much more flexible models. The success of these nonparametric functional models/methods, as much for insuring good asymptotic behaviour as for practical issues, relies on a suitable topological structure on the functional space that allows the dimensionality effects of the variable X to be controlled for. Even more recently, intermediate models have been developed in order to control these dimensionality effects without falling back into the disadvantages of a strong linearity assumption. These models are based on suitable functional adaptations of the standard multivariate models (additive modeling, single index modeling, partial linear modeling, etc.) and are particularly interesting when the variable X is multifunctional.

This chapter is organized as follows. Remaining in the functional setting (that is, when the covariate X takes values in some infinite-dimensional space), in Section 1.2 we define what can be called parametric, semiparametric, or nonparametric models, and we discuss how semiparametric modeling can be interpreted in terms of dimension reduction. Then a few specific examples are developed in Section 1.2.3. These examples have already been studied in the functional regression literature and they include functional linear regression, smooth nonparametric functional modeling, additive functional regression, and single functional index modeling. In Section 1.2.4 we define some new models, directly adapted to functional variables from the existing standard multivariate literature. These new models have not yet been studied, but this situation will certainly be rectified in the next few years.

1.2 Classification of Regression Models

1.2.1 The regression problem

Let X and Y be two random variables, defined on the same probability space, and taking respectively values in E and $\mathbb{R}$. The regression relation between X and Y can be written in the following way:

$$Y = r(X) + \varepsilon = \mathbb{E}(Y|X) + \varepsilon, \tag{1.1}$$

where ε is a real random variable such that $E(\varepsilon|X) = 0$. In what follows, we will consider such kinds of problems in a functional setting, that is, when the space E of possible values for the explanatory variable X is of infinite dimension.

To keep things as simple as possible, our classification is proposed only for regression problems with real response. The same kinds of definitions as are proposed below could be directly extended at least in two directions: on the one hand they could be extended to more general regression settings when Y takes values in higher- (and possibly also infinite-) dimensional spaces, and on the other hand they could also be similarly rewritten for modeling any other operator that depends on the probability distribution of (X, Y), and not only for the regression operator (this last point might involve conditional distribution, density, the hazard function, ...).

Definition 1.1 *The regression problem described in (1.1) is called* functional *if and only if E is an infinite-dimensional space.*

In this situation r is called the functional regression operator. Note that at this stage we do not need any linearity assumption on this operator, which is just an element of the space $\mathcal{O}(E, \mathbb{R})$ of all the correspondances between E and $\mathbb{R}$.

Basically, a regression model is a set of a priori assumptions made on the joint distribution of the pair (X, Y) that allow us to restrict the regression operator to belong to some specific class of operators. Formally, a regression model can be summarized by an assumption of the form

$$r \in \mathcal{M} \subset \mathcal{O}(E, \mathbb{R}). \tag{1.2}$$

Obviously, the wider the class $\mathcal{M}$ and the more flexible the model, the smaller will be the modeling error. But considering a large class of models leads to two main disadvantages. From a practical point of view, one can lose the ability to interpret the results. From a theoretical point of view, the larger the class $\mathcal{M}$, the slower the rate of convergence of the estimator. Therefore the trade-off between too general models and too restrictive ones is a central problem in statistics (and not only in regression), and the literature usually classifies the various models according to the dimensionality of the class $\mathcal{M}$. Roughly speaking, the dividing line between parametric and nonparametric modeling comes from the size of $\mathcal{M}$ (i.e. of finite or infinite dimension). This is specified in the next definition which is stated in a general framework, and not only for the regression operator r.

Definition 1.2 *Let $g \in \mathcal{O}(F, \mathbb{R})$ be some functional operator from some space F into $\mathbb{R}$. A class C_g is called a* parametric *class for the operator g if it is indexed by a finite number of elements of F. Otherwise, the class C_g is called a* nonparametric *class for the operator g.*

1.2.2 Model classification

The general Definition 1.2 induces a first natural splitting of models into parametric and nonparametric regression models. This is the aim of the next definition.

Definition 1.3 *The regression model defined by (1.1) and (2.2) is called parametric (respectively, nonparametric) if and only if $\mathcal{M}$ is a parametric (respectively, nonparametric) class for the operator r.*

Such a definition has been explicitly stated in Ferraty and Vieu (2006). Note that, in finite-dimensional cases (that is, when $E = \mathbb{R}^p$), other definitions of nonparametric models can be proposed by looking at the joint distribution of (X, Y) rather than at the regression operator r. The reason why we have stated Definition 1.3 in this way is because of its easy extension to abstract infinite-dimensional spaces E. In order to make this definition clearer, various examples of models will be described later in Section 1.2.3. The main advantage of Definition 1.3 is that it provides a precise dividing line between parametric and nonparametric, which is quite natural in the sense that it is based on a dimensionality consideration with regard to the space $\mathcal{M}$ where the statistical target (namely the regression operator r) lives.

However, the simple dichotomy induced by Definition 1.3 does not allow us to distinguish finely the wide variety of nonparametric models. Indeed, as is already well known in standard unfunctional regression analysis, the main problem in nonparametric statistics is to control the dimensionality effects of X and the sparseness of the data which follows from it (see Section 1.2.5 for extensive references on this point). Keeping once again in mind what has been done in standard unfunctional multivariate nonparametric statistics, a natural idea consists in introducing some intermediary models which tend to reduce the dimensionality of the problem without falling into the drawbacks of a too unflexible parametric model. To make this idea precise, we will consider that the regression operator r can be written by means of various components, each of these components acting on a space of "lower" dimension. In other words, let $E_1, \ldots, E_{j_0}, G_1, \ldots, G_{j_0}$, and $\mathcal{E}$ be spaces, and assume that there is a one-to-one correpondance, denoted by Ψ, between the class $\mathcal{M}$ and the product space:

$$\mathcal{M} \xleftrightarrow{\Psi} \mathcal{M}_1 \times \cdots \times \mathcal{M}_{j_0} \times \mathcal{E}, \tag{1.3}$$

where the $\mathcal{M}_j$'s are nonempty subspaces of the operator spaces $\mathcal{O}(E_j, G_j)$. For any $r \in \mathcal{M}$ we use the notation

$$\Psi(r) = (g_1, \ldots, g_{j_0}, f).$$

Definition 1.4 *The model defined by (1.1), (2.2), and (1.3) is called a* dimensionality-reduction nonparametric functional model *if and only if each $\mathcal{M}_j$ is a nonparametric class for the operator g_j.*

Such a model offers the particular advantage of changing the above statistical problem (i.e. the estimation of r) into several problems (i.e. the estimation of $g_1, \ldots, g_{j_0}, f$), each of these new problems being concerned with the estimation of a target living in a space of lower dimensionality than r. The nonparametric feature is retained by means of the classes $\mathcal{M}_j$ introduced to model each operatorial component g_j. This definition will be clarified in Section 1.2.3 by means of a few examples. At this stage, let us just mention that there is no need for further assumptions on the space $\mathcal{E}$, which is introduced simply to allow for the modeling of a complementary component f which is not of operatorial nature.

A particular subclass of dimensionality-reduction nonparametric functional models consists of those where we let the complementary component f be of parametric form. To be precise, one can assume that

$$\mathcal{M} \xleftrightarrow{\Psi} \mathcal{M}_1 \times \cdots \times \mathcal{M}_{j_0} \times \mathcal{E}_1, \times \cdots \times \mathcal{E}_{k_0}, \tag{1.4}$$

where the $\mathcal{M}_j$'s are defined as before and where each $\mathcal{E}_k$ is a nonempty subset of $\mathcal{E}$. In this case for any $r \in \mathcal{M}$ we use the notation

$$\Psi(r) = (g_1, \ldots, g_{j_0}, \theta_1, \ldots, \theta_{k_0}).$$

Definition 1.5 *The model defined by* (1.1), (1.2), *and* (1.4) *is called a* semiparametric functional regression model *if and only if each $\mathcal{M}_j$ is a nonparametric class for g_j and if each $\mathcal{E}_k$ is a parametric class for θ_k.*

It clearly appears from Definitions 1.4 and 1.5 that semiparametric functional models are special cases of dimensionality-reduction nonparametric functional regression models. The next section will highlight this definition through a simple example.

1.2.3 Some important functional regression models

The aim of this section is to provide four examples of functional regression models. Each example corresponds to one among the various families of model introduced before. The first example is the standard functional linear model which illustrates the family of parametric functional regression models. Then we will also introduce a pure nonparametric model, making regularity assumptions only on the regression operator. The third example is a functional version of additive regression that will be seen to be a dimensionality-reduction model, while the single functional index model will be seen as a typical semiparametric functional regression approach.

Example (a) A parametric example: the functional linear model

The linear regression model is the first one to have been adapted to functional variables. Basically, it consists in assuming that

$$\mathcal{M} = \{\text{linear and continuous operators from } E \text{ into } \mathbb{R}\}. \qquad (1.5)$$

It should be emphasized that such a model is usually investigated in the literature under the constraint that the functional space E in which X takes its values is a Hilbert space with inner product $\langle \cdot, \cdot \rangle$. This allows us, by means of Riesz's theorem, to say that to each $r \in \mathcal{M}$ there can be associated one single element $r^* \in E$ in the following way:

$$\forall x \in E, r(x) = \langle x, r^* \rangle. \qquad (1.6)$$

This model was introduced by Ramsay and Dalzell (1991) and also presented by Hastie and Mallows (1993) in a discussion of Frank and Friedman's (1993) paper (see Ramsay and Silverman (2005) for a general bibliography). It is not our aim to give details of this model; further contributions in this book will be devoted to it (see especially Chapter 2). At this stage, we just wish to point out that the one-to-one correspondence between $\mathcal{M}$ and E makes the following result obvious:

Remark 1.1 *The functional linear regression model (FLR) defined by (1.1), (2.2), and (1.5) is a parametric functional model according to Definition 1.3.*

Example (b) A nonparametric example: the continuity-type functional model

The main idea in nonparametric statistical modeling, in either the functional or standard multivariate situations, consists in suppressing any restriction on the shape of the object to be estimated. In the functional regression framework, this leads us to avoid linearity assumptions and to make only regularity assumptions on the target operator r. This is done by introducing a topological structure on the space E in order to define what we call the regularity of the operator r. A usual way of doing this consists in assuming the existence of some semi-metric d on the space E. Note that this semi-metric is simply introduced for topological reasons in order to have some measure of proximity between functional elements, and it is not necessarily linked with the σ-field endowing the space E. For instance, the following continuity-type model has been introduced in Ferraty and Vieu (2000):

$$\mathcal{M} = \{\text{continuous operators from } E \text{ into } \mathbb{R}\}, \qquad (1.7)$$

where the continuity is derived from the semi-metric d. This model has now been much studied (see Ferraty and Vieu (2006) for an extensive discussion of it), and it will be the subject of other contributions in this book (see Chapters 4 and 5). Here, we wish to point out the following obvious fact.

Remark 1.2 *The continuity-type model defined by (1.1), (2.2), and (1.7) is a nonparametric functional regression model according to Definition 1.3.*

Of course many other regularity models can be defined by means of additional regularity assumptions on r (such as, for instance, Hölder continuity or differentiability-type hypotheses), and all of them are still contained in the family of nonparametric models.

Example (c) A dimensionality reduction example: the additive functional model

In standard multivariate nonparametric statistics, additive modeling is a popular way for constructing dimensionality-reduction models (see Stone (1985) as an example of an older reference and Mammen and Park (2006) for one of the most recent one). Moreover, this model can be easily extended to the functional setting. The idea consists in writing the target operator r in some additive way as a function of various other operators r_j, $j = 1, \ldots, r_{j_0}$, each acting on some specific part of the variable X. It is a natural dimension-reduction model since each operator r_j acts on a space of lower dimensionality than E. Formally, the space E is written as a product

$$E = E_1 \times \cdots \times E_{j_0},$$

and the functional variable is decomposed into j_0 variables, each of them taking values in some of the spaces E_j:

$$X = (X^1, \ldots, X^{j_0}).$$

The additive hypothesis amounts to assuming that the regression target operator can be written as

$$r(X) = \mu + \sum_{j=1}^{j_0} r_j(X^j), \tag{1.8}$$

with, for identifiability reasons, $\mu = \mathbb{E}Y$ and $\mathbb{E}(r_j(X^j)) = 0$ for all j. Note that there is no restriction on the spaces E_j being functional or not. This allows us to take into account a wide variety of situations in which some of the components of X can be either functional or multivariate. The nonparametric feature of the model appears in the fact that only regularity assumptions are made on each operatorial component r_j. As before, such regularity assumptions can take various forms: continuity, Hölder continuity, differentiability,... To fix our ideas we will just remain with simple continuity-type modeling that consists in assuming that $\forall j = 1, \ldots, j_0$,

$$r_j \in \mu_j = \{\text{continuous operators from } E_j \text{ into } \mathbb{R}\}. \tag{1.9}$$

The extra real parameter μ does not need any special assumptions. In other words, and to make a connection with the general presentation stated in (1.3), we just have to assume that

$$\mu \in \mathcal{E} = \mathbb{R}.$$

Until now, this model has only been used for a few applied problems with data curves (see Aneiros *et al.* (2004)) without any deep theoretical advances. See, however, Chapter 4 in this book for more discussion. Here we just want to point out how such an additive approach appears in the general classification of models described earlier.

Remark 1.3 *The additive functional model (AFM) defined by (1.1), (1.8), and (1.9) is a dimensionality-reduction nonparametric functional model according to Definition 1.4.*

Example (d) A semiparametric example: the single functional index model

Another very simple way of reducing the dimensionality of a regression problem is to assume that the variable X acts on the response Y only through its projection onto some one-dimensional subspace. This idea was introduced by Härdle *et al.* (1993) in the standard unfunctional multivariate situation and it leads to the so-called single index model (see, for example, Delecroix *et al.* (2006) for recent advances). The same procedure can be followed in the functional setting, leading to the single functional index regression model described just below. This model was introduced by Ferraty *et al.* (2003), and it consists in assuming that the space E is Hilbertian, with inner product denoted by $\langle \cdot, \cdot \rangle$ and associated norm $\| \cdot \|$, and in writing the regression target operator as:

$$r(X) = g(\langle X, \theta \rangle). \tag{1.10}$$

The operatorial component is modeled by means of the nonparametric hypothesis

$$g \in \mathcal{M}_1 = \{\text{differentiable operators from } \mathbb{R} \text{ into } \mathbb{R}\}, \tag{1.11}$$

while the extra unfunctional component θ is simply such that

$$\theta \in E \text{ and } \langle \theta, e_1 \rangle = 1,$$

where e_1 is the first element of an orthonormal basis of E. The last condition acting on θ ensures identifiability. This model can obviously be written in the general form described in (1.4) with a single operatorial nonparametric component (that is, with the general notation stated before, $j_0 = 1$, $g_1 = g$, and $E_1 = \mathbb{R}$) and with a single functional parametric component/index (that is, with $k_0 = 1$, $\mathcal{E}_1 = E$, and $\theta_1 = \theta$). Finally, we get the following result:

Remark 1.4 *The functional index model (FIM) defined by (1.1), (1.10), and (1.11) is a semiparametric functional regression model according to Definition 1.5.*

It is not the purpose of this chapter to discuss this model in detail (see Ait-Saidi *et al.* (2008) for the most recent advances here). Note that this model will also be discussed in Chapter 4 of this book. However, it is worth noting here that the particular semiparametric form introduced in (1.10) clearly appears, on the one hand, as a natural extension of the standard linear regression model (1.6) and, on the other hand, as a special case of the nonparametric model (1.7) with the new explanatory variable $\langle X, \theta \rangle$ in place of X. The dimensionality-reduction effect also appears clearly since the variable $\langle X, \theta \rangle$ lives in the space $\mathbb{R}$ which is of lower dimension than the previous functional space E.

1.2.4 A few generalizations of the functional models

In Section 1.2.3 we have given details of the functional models which have already received some attention in recent literature. However, many other models known to be of interest in standard multivariate nonparametric regression can be expected to be adapted to the functional setting. The aim of this section is to define precisely some of these new functional models, leaving their statistical studies as open problems. Given their great importance in functional data analysis, we are sure that most of these open problems will be successfully investigated in the near future by the scientific community.

Model (a) Direct extensions of the additive functional model

The first natural extension of additive models consists in introducing some known link function inside the additive relationship between Y and X. This idea has been widely developed in multivariate situations (see Hastie and Tibshirani (1986) and Stone (1986)) and can be adapted easily for functional variables, as indicated below. As discussed before for the additive model, suppose that $E = E_1 \times \cdots \times E_{j_0}$, and write $X = (X^1, \ldots, X^{j_0})$. The general additive model consists in assuming that the regression target operator can be written as

$$r(X) = L\left(\mu + \sum_{j=1}^{j_0} r_j(X^j)\right), \qquad (1.12)$$

with, again for identifiability reasons, $\mu = \mathbb{E}Y$ and $\forall j, \mathbb{E}(r_j(X^j)) = 0$. The link function L is supposed to be known. Conditional on suitable nonparametric modeling for the unknown operatorial components r_j, this model belongs to the class of dimension-reduction functional regression models. To be precise, we have:

Remark 1.5 *The generalized additive functional model (GAFM) defined by (1.1), (1.9), and (1.12) is a dimensionality-reduction nonparametric functional model according to Definition 1.4.*

Another natural extension consists in allowing possible interaction terms (still modeled in an additive way) between the various components X^j of X. This idea, well developed in multivariate analysis (Chen 1991; Stone 1994), can be extended to functional variables as follows. Let X^S be some subset of the whole family of components $\{X^1, \ldots, X^{j_0}\}$, that is,

$$\forall S \subset \{1, \ldots, j_0\}, \ X^S = (X_j, j \in S).$$

Let us also denote by E_S the space in which X^S takes its values:

$$E_S = \prod_{j \in S} E_j.$$

The interactive model consists in modeling the effect of X on Y, in an additive way, from the effects of a fixed number of variables X^S. Precisely, let $\mathcal{S}$ be the set of subsets of $\{1, \ldots, j_0\}$

$$r(X) = L\left(\mu + \sum_{S \in \mathcal{S}} r_S(X^S)\right), \quad (1.13)$$

where $\mu \in \mathbb{R}$, the link function L is known, and

$$\forall S \in \mathcal{S}, \ \mathbb{E}(r_S(X^S)) = 0.$$

The nonparametric feature of the model appears in regularity assumptions on each operatorial component r_S. For instance, one can introduce a simple continuity-type assumption:

$$\forall S \in \mathcal{S}, \ r_S \in \mathcal{M}_S = \{\text{continuous operators from } E_S \text{ into } \mathbb{R}\}. \quad (1.14)$$

In summary, we have the following result:

Remark 1.6 *The interactive additive functional model (IAFM) defined by* (1.1), (1.13), *and* (1.14) *is a dimensionality-reduction nonparametric functional model according to Definition 1.4.*

Model (b) Direct extensions of the single functional index model

The single functional index model defined by (1.10) can be naturally extended by assuming that X does not act on Y through one projection but through various projections. This idea has been used in a multivariate setting (see Stoker (1986) for first advances and Bickel et al. (2006) for the most recent ones), and we propose to adapt it to functional variables as follows. Assuming that the space E is Hilbertian, with inner product denoted by $\langle \cdot, \cdot \rangle$ and associated norm $\|\cdot\|$, the so-called multiple functional index regression model consists in writing the regression operator as

$$r(X) = g\left(\langle X, \theta_1 \rangle, \ldots, \langle X, \theta_{k_0} \rangle\right). \quad (1.15)$$

The operatorial component is modeled by means of the nonparametric hypothesis

$$g \in \mathcal{M}_1 = \{\text{differentiable operators from } \mathbb{R}^{k_0} \text{ into } \mathbb{R}\}, \quad (1.16)$$

while the extra unfunctional components θ_k are such that

$$\forall k = 1, \ldots, k_0, \; \langle \theta_k, e_1 \rangle = 1,$$

e_1 being the first element of an orthonormal basis of E. This model can obviously be written in the general form described in (1.4) with a single operatorial nonparametric component (that is, with the general notation stated above, $j_0 = 1$, $g_1 = g$, and $E_1 = \mathbb{R}^{k_0}$) and with multiple functional parametric components $\theta_1, \ldots, \theta_{k_0}$. In other words, we have the following result:

Remark 1.7 *The multiple functional index model (MFIM) defined by (1.1), (1.15), and (1.16) is a semiparametric functional regression model according to Definition 1.5.*

This is clearly a dimensionality-reduction model, since the operatorial component of the model (that is, g) acts on a k_0-dimensional space: the effect of the infinite-dimensional variable X on Y can be expressed through the new explanatory variable $(\langle X, \theta_1 \rangle, \ldots, \langle X, \theta_{k_0} \rangle)$ which lives in the space $\mathbb{R}^{k_0}$.

In the same spirit, another model can be defined by letting each projection $\langle X, \theta_k \rangle$ have its own operatorial influence on the response Y. This projection-pursuit idea has been widely developed in the multivariate setting (Friedman and Stuetzle 1981; Huber 1985; Hall 1989) and extends naturally to functional variables as follows. Assume that

$$r(X) = \mu + \sum_{k=1}^{k_0} r_k \left(\langle X, \theta_k \rangle \right), \tag{1.17}$$

with identifiability conditions of the form

$$\forall k = 1, \ldots, k_0, \; \langle \theta_k, e_1 \rangle = 1 \text{ and } \mathbb{E}\left(r_k(\langle X, \theta_k \rangle)\right) = 0.$$

The operatorial components are modeled by means of a nonparametric hypothesis, such as, for example $\forall k = 1, \ldots, k_0$,

$$r_k \in \mathcal{M}_k = \{\text{continuous operators from } \mathbb{R} \text{ into } \mathbb{R}\}, \tag{1.18}$$

while

$$\mu \in \mathbb{R} \text{ and } \forall k = 1, \ldots, k_0, \; \theta_k \in E.$$

This model can obviously be written in the general form described in (1.4) with operatorial nonparametric components (that is, with the general notation stated above, $j_0 = k_0$, $g_j = r_j$, and $E_j = \mathbb{R}$) and with multiple functional parametric components $\theta_1, \ldots, \theta_{k_0}$. Thus we have the following result:

Remark 1.8 *The projection-pursuit functional model (PPFM) defined by (1.1), (1.17), and (1.18) is a semiparametric functional regression model according to Definition 1.5.*

Model (c) Partial linear functional model

Another way of reducing the dimensionality of a problem (classically used in multivariate settings) consists in mixing the nature of the effects of the various components X^k of X: some of them may act in a nonparametric way while some others may act linearly (see Eubank and Whitney (1989) for previous work on this model and Härdle *et al.* (2000) for a broad presentation of recent advances in this field). Suppose that

$$E = E_1 \times \cdots \times E_{j_0} \times E_{j_0+1} \times \cdots \times E_{j_0+k_0},$$

and write $X = (X^1, \ldots, X^{j_0+k_0})$, and assume that each of the spaces E_{j_0+j}, $j = 1, \ldots, k_0$ is Hilbertian with its own inner product $\langle \cdot, \cdot \rangle_{j_0+j}$. The partial linear functional approach consists in modeling the effect of X on Y by some combination (once again additive) of nonparametric effects for the first components X_j, $j = 1, \ldots, j_0$ and of linear effects for the others. More precisely, the model can be written as

$$r(X) = \mu + \sum_{j=1}^{j_0} r_j(X^j) + \sum_{j=j_0+1}^{j_0+k_0} \langle X^j, \theta_j \rangle_j, \qquad (1.19)$$

with $\mu = \mathbb{E}Y$. For identifiability reasons, it is also assumed that

$$\forall j \le j_0, \ \mathbb{E}(r_j(X^j)) = 0 \text{ and } \forall j > j_0, \ \mathbb{E}(\langle X^j, \theta_j \rangle_j) = 0.$$

The reader will find early developments of this model in Aneiros and Vieu (2006). A broader presentation will be made in Chapter 4. Our wish here is simply to note that, under suitable general nonparametric conditions on the operatorial components r_j, this model obviously enters into the family of semiparametric models. For instance, under continuity-type models, we have the following result:

Remark 1.9 *The partial linear functional model (PLFM) defined by (1.1), (1.19), and (1.9) is a semiparametric functional regression model according to Definition 1.4.*

Model (d) Varying-coefficient functional model

In a similar way, the varying-coefficient regression models widely studied in multivariate situations (see Hastie and Tibshirani (1993)) can easily be extended to the functional setting. Suppose that

$$E = E_1 \times \cdots \times E_{j_0} \times E_{j_0+1} \times \cdots \times E_{2j_0},$$

and write $X = (X^1, \ldots, X^{2j_0})$, and assume that each of the spaces E_{j_0+j}, $j = 1, \ldots, j_0$ is Hilbertian with its own inner product $\langle \cdot, \cdot \rangle_{j_0+j}$. The model is defined by assuming that

$$r(X) = \mu + \sum_{j=1}^{j_0} \langle g_j(X^j), X^{j+j_0} \rangle_{j_0+j}, \qquad (1.20)$$

with $\mu = EY$. For identifiability reasons, it is also assumed that

$$\forall j = 1, \ldots, j_0, \ \mathbb{E}\left(\langle g_j(X^j), X^{j+j_0}\rangle_{j_0+j}\right) = 0. \tag{1.21}$$

Here the g_j are operators defined on E_j and taking values on E_{j+j_0}. Under suitable general nonparametric conditions on these operatorial components, this model obviously enters into the family of dimensionality-reduction regression models. For instance, under a continuity-type model such as

$$\forall j = 1, \ldots, j_0, g_j \in \mathcal{M}_j = \{\text{continuous operators from } E_j \text{ to } E_{j+j_0}\},$$

we have the following result:

Remark 1.10 *The varying-coefficient functional model (VCFM) defined by (1.1), (1.20), and (1.21) is a dimensionality-reduction functional model according to Definition 1.4.*

More details can be found in Chapter 6 of this book, which is devoted to varying-coefficient models.

Model (e) A few others

Of course, many other possibilities could be introduced and most of them will doubtless be investigated over the next few years. Indeed, various combinations of the above models can be introduced. For instance, the introduction of linear relations between some of the operators g_j appearing in (1.20) would lead to the so-called partial linear varying-coefficient functional model (PLVCFM), which is known to have interesting features in some multivariate situations. Also, as was done in (1.12) when generalizing the additive model, all the models defined below could be extended in a natural way by introducing some known link function L. Finally, rather than modeling the regression operator r, one might wish to model the regression of some transformed variable $T(Y)$ on the functional covariate X:

$$T(Y) = r_T(X) + \varepsilon_T = \mathbb{E}(T(Y)|X) + \varepsilon_T, \ \mathbb{E}(\varepsilon_T|X) = 0, \tag{1.22}$$

where T is an unknown transformation. In the context of this transformation regression model, one could imagine introducing a specific constraint (additive, semi-linear, varying-coefficient, ...) on this new transformed problem. This idea has already been introduced in the multivariate context with some additive models for the transformed regression r_T (see Breiman and Friedman (1985) and Burman (1991)).

1.2.5 Complementary bibliography

Let us first note that it was not our goal to give an exhaustive survey of the standard literature on regression modeling for multivariate (unfunctional) variables. Our purpose has been to describe some of the most popular models, with two main aims in mind. On the one hand, we propose a general unifying classification of all these models and, on the other hand, we define various functional versions of them. Here are some general references which will allow the reader to have a nice overview of the extensive recent methodological literature on semiparametric and dimension-reduction modelling: Hastie and Tibshirani (1990), Bickel *et al.* (1993), Pelegrina *et al.* (1996), Horowitz (1998), Schimek (2000), Ruppert *et al.* (2003), Sperlich *et al.* (2003), and Härdle *et al.* (2004). Similarly, a selected set of general references on nonparametric regression modeling includes: Collomb (1981 and 1985), Härdle (1990), Green and Silverman (1994), Sarda and Vieu (2000), and Györfi *et al.* (2002). Without any doubt, most of the models and methodologies described in this extensive multivariate literature will, over the next few years, have their corresponding functional data adaptation.

It is also worth noting that regression problems also arise when one tries to predict future values of some time series. In this context as well, the same kind of classification appears when modeling the operatorial link with the future of the process and with a continuous path of its past. While the modeling considerations are the same as in standard functional regression, the statistical challenge is much more difficult because of the dependence structure in the data that needs to be controlled for, and the literature is much less extensive than in the general regression setting. A key reference for the linear modeling of the autoregression relation between the paths of the process is the monograph by Bosq (2000), while recent nonparametric advances are discussed in Ferraty and Vieu (2006) and Delsol (2008). The only papers dealing with semiparametric functional time-series modeling are that by Ait-Saidi *et al.* (2005) on the single functional index time-series model, and that of Aneiros and Vieu (2008) for partial linear functional time-series models. Other contributions in this book (see Chapters 3 and 5) are devoted to this question.

1.3 CONCLUSIONS

This chapter has proposed a general method for classifying the various statistical models that one can introduce for functional regression, and has discussed how various examples of models are assigned to different families under this classification. As with any mathematical definition, this classification has an obvious subjective element that could be summarized in both of the following ideas. Firstly, the word *functional* is used to describe situations in which the observed statistical variable is functional (that is, when it takes values in an infinite-dimensional space). Secondly, the word *parametric* is used when an operator is modeled under conditions that allow us to summarize it by a finite number of *parameters*; the word parameter being used for elements living in the same space as the one on which the modeled operator is defined. As a consequence, the word *nonparametric* is used in other cases. As a second consequence, the word *semiparametric* is used in intermediary settings when some features of the operator are modeled nonparametrically and some others are modeled parametrically. This classification is of interest for two main reasons. Firstly, it provides precise definitions of what are parametric and nonparametric models; these definitions join the standard ones for multivariate/unfunctional standard regression modeling. Secondly, the classification also provides a precise definition of what can be called semiparametric modeling (usable both in functional settings, as is our main wish here, and in standard multivariate regression).

Finally, we would like to reject any naive point of view that wants to say that any functional data analysis problem contains some smoothing feature (because the data are of an infinite-dimensional nature) and therefore should be called nonparametric. It is worth stressing again that our classification (nonparametric, semiparametric, parametric) is carried out according to various different kinds of models for the regression operators. In addition, it is true that when dealing with the functional explanatory variable X (for instance, with a random function $X = \{X(t), t \in [0, T]\}$), there is an existing data-registration question that could include, among other things, a preliminary smoothing treatment. When this is the case, this smoothing is made nonparametrically in the sense that the link between $X(t)$ and t is nonparametric, but the operatorial link between the response Y and the curve X can be modeled in any way (parametric, nonparametric, or semiparametric).

We wish to conclude by emphasizing the fact that these thoughts on functional regression modeling have grown up around the numerous discussions we have had with the working group STAPH in Toulouse. This group intends to develop all the functional features of modern statistics, and its activities have obviously played an important role in our own activities and therefore have had a great impact on this contribution. All the activities of this group are available online (see Staph (2009)).

References

Aït-Saidi, A., Ferraty, F., Kassa, R. (2005). Single functional index model for a time series. *Rev. Roumaine Math. Pures Appl.*, **50**, 321–30.

Aït-Saidi, A., Ferraty, F., Kassa, R., Vieu, P. (2008). Cross validated estimations in the single functional index model. *Statistics*, **42**, 475–94.

Aneiros-Perez, G., Cardot, H., Estevez Perez, G., Vieu, P. (2004). Maximum ozone forecasting by functional nonparametric approaches. *Environmetrics*, 15, 675–85.

Aneiros-Perez, G., Vieu, P. (2006). Semi-functional partial linear regression. *Statist. Probab. Lett.*, 11, 1102–10.

Aneiros-Perez, G., Vieu, P. (2008). Time series prediction: a semi-functional partial linear model. *Journal of Multivariate Analysis*, 99, 834–57.

Bickel, P.J., Klaassen, C.A., Ritov, Y., Wellner, J.A. (1993). *Efficient and Adaptive Estimation for Semiparametric Models.* Johns Hopkins University Press, Baltimore.

Bickel, P.J., Ritov, Y., Stoker, T. (2006). Tailor-made tests for goodness of fit to semiparametric hypotheses. *Ann. Statist.*, 34, 721–41.

Bosq, D. (2000). *Linear Processes in Functional Spaces. Theory and Applications.* Lecture Notes in Statistics, 149. Springer, New York.

Breiman, L. and Friedman, J.H. (1985). Estimating optimal transformations for multivariate regression and correlation. *J. Amer. Statist. Assoc.*, 80, 580–619.

Burman, P. (1991). Rates of convergence for the estimate of the optimal transformations of variables. *Ann. Statist.*, 19, 702–23.

Chen, Z. (1991). Interaction spline models and their convergence rates. *Ann. Statist.*, 19, 1855–68.

Collomb, G. (1981). Estimation non-paramétrique de la régression: revue bibliographique. *Internat. Statist. Rev.*, 49, 75–93.

Collomb, G. (1985). Nonparametric regression: an up-to-date bibliography. *Statistics*, 16, 309–24.

Davidian, M., Lin, X., Wang, J.L. (2004). Introduction to: Emerging issues in longitudinal and functional data analysis. *Statist. Sinica*, 2004, 613–14.

Delecroix, M., Hristache, M., Patilea, V. (2006). Semiparametric M-estimation in single-index regression. *J. Statist. Plann. Inference*, 136, 730–69.

Delsol, L. (2008). Advances on asymptotic normality in nonparametric functional time series analysis. *Statistics*, 43, 13–33.

Eubank, R.L., Whitney, P. (1989). Convergence rates for estimation in certain partially linear models. *J. Statist. Plann. Inference*, 23, 33–43.

Ferraty, F., Peuch, A., Vieu, P. (2003). Modèle à indice fonctionnel simple. *Comptes Rendus Math. Académie Sciences Paris*, 336, 1025–8.

Ferraty, F., Vieu, P. (2000). Dimension fractale et estimation de la régression dans des espaces vectoriels semi-normés. *Comptes Rendus Académie Sciences Paris*, 330, 139–42.

Ferraty, F., Vieu, P. (2006). *Nonparametric Functional Data Analysis. Theory and Practice.* Springer, New York.

Ferraty, F. (2010). Special Issue: Statistical Methods and Problems in Infiniter-dimensional Spaces (Ed.). *J. Mult. Anal.*, 101, 305–490.

Frank, I.E., Friedman, J.H. (1993). A statistical view of some chemometrics regression tools (with discussion). *Technometrics*, 35, 109–48.

Friedman, J.H., Stuetzle, W. (1981). Projection pursuit regression. *J. Amer. Statist. Assoc.*, 76, 817–23.

González Manteiga, W., Vieu, P. (2007). Statistics for functional data. *Comput. Statist. Data Anal.*, 51, 4788–92.

Green, P.J., Silverman, B.W. (1994). *Nonparametric Regression and Generalized Linear Models. A Roughness Penalty Approach.* Monographs on Statistics and Applied Probability, 58. Chapman & Hall, London.

Györfi, L., Kohler, M., Krzyzak, A., Walk, H. (2002). *A Distribution-free Theory of Nonparametric Regression.* Springer, New York.

Härdle, W. (1990). *Applied Nonparametric Regression.* Oxford University Press, Oxford.

Härdle, W., Hall, P., Ichimura, H. (1993). Optimal smoothing in single index models. *Ann. Statist.*, 21, 157–78.

Härdle, W., Liang, H., Gao, J. (2000). *Partially Linear Models.* Physica, Heidelberg.

Härdle, W., Müller, M., Sperlich, S., Werwatz, A. (2004). *Nonparametric and Semiparametric Models.* Springer, New York.

Hall, P. (1989). On projection pursuit regression. *Ann. Statist.*, 17, 573–88.

Hastie, T., Mallows, C. (1993). Discussion of Frank and Friedman's paper: A statistical view of some chemometrics regression tools. *Technometrics*, 35, 140–8.

Hastie, T., Tibshirani, R. (1986). Generalized additive models (with discussion). *Statist. Sci.*, 1, 297–318.

Hastie, T., Tibshirani, R. (1990). *Generalized Additive Models.* Chapman and Hall, London.

Hastie, T., Tibshirani, R. (1993). Varying-coefficients models. *J. Roy. Statist. Soc.*, 55, 757–96.

Horowitz, J. (1998). *Semiparametric Methods in Econometrics.* Lecture Notes in Statistics, 131. Springer, New York.

Huber, P.J. (1985). Projection pursuit. *Ann. Statist.*, 13, 435–75.

Mammen, E., Park, B. (2006). A simple smooth backfitting method for additive models. *Ann. Statist.*, 34, 2252–71.

Pelegrina, L., Sarda, P., Vieu, P. (1996). On multidimensional nonparametric regression. In *Proceedings in Computational Statistics, COMPSTAT 1996* (A. Prat, ed.), 149–60. Physica, Heidelberg.

Ramsay, J.O., Dalzell, C.J. (1991). Some tools for functional data analysis. With discussion and a reply by the authors. *J. Roy. Statist. Soc. Ser. B*, 53(3), 539–72.

Ramsay, J.O., Silverman, B.W. (1997). *Functional Data Analysis.* Springer, New York.

Ramsay, J.O., Silverman, B.W. (2002). *Applied Functional Data Analysis: Methods and Case Studies.* Springer, New York.

Ramsay, J.O., Silverman, B.W. (2005). *Functional Data Analysis* (second edition). Springer, New York.

Ruppert, D., Wand, M., Carroll, R. (2003). *Semiparametric Regression.* Cambridge Series in Statistical and Probabilistic Mathematics, 12. Cambridge University Press, Cambridge.

Sarda, P., Vieu, P. (2000). Kernel regression. In *Smoothing and Regression. Approaches, Computation, and Application* (M. Schimek, ed.), 43–70. John Wiley & Sons, New York.

Schimek, M. (2000). *Smoothing and Regression. Approaches, Computation, and Application.* John Wiley & Sons, New York.

Sperlich, S., Härdle, W., Gökhan, A. (2003). *The Art of Semiparametrics. Selected Papers from the Conference Held in Berlin, 2003.* Physica/Springer, Heidelberg.

Staph, (2009). Working group on functional and operatorial statistics, Toulouse, France. Activities online at *http://www.math.univ-toulouse.fr/staph/*

Stoker, T. (1986). Consistent estimation of scaled coefficients. *Econometrica*, 54 1461–81.

STONE, C. (1985). Additive regression and other nonparametric models. *Ann. Statist.*, **13**, 689–705.

STONE, C. (1986). The dimensionality reduction principle for generalized additive models. *Ann. Statist.*, **14**, 590–606.

STONE, C. (1994). The use of polynomial splines and their tensor products in multivariate function estimation. *Ann. Statist.*, **22**, 118–84.

VALDERRAMA, M. (2007). An overview to modelling functional data. *Computational Statistics*, **22**, 331–5.

CHAPTER 2

FUNCTIONAL LINEAR REGRESSION

HERVÉ CARDOT

PASCAL SARDA

2.1 INTRODUCTION

FUNCTIONAL linear regression (FLR) is an area of functional data analysis that has been intensively researched, motivated by many real-world case studies. For instance, in numerous situations the aim is to explain the variations in, or to predict future values of, some scalar response by using information from a high-dimensional set of predictors. In this general situation, there exists a class of methodologies that attempt to reduce the number of predictors in some appropriate way and/or postulate that only a small number of regression coefficients are significant. The latter methods are based on an assumption of sparsity of the parameter vector and, moreover, it is assumed that the predictors are almost uncorrelated: see, for instance, the recent paper of Candès and Tao (2007). The procedures deduced in such cases are most often intended for statistical applications in genomics (gene expression studies, for instance). On the other hand, they are not (at least directly) suitable in cases of high correlation and especially not when the predictor is functional, i.e. when it emerges from the discretizations of curves. It is, however, this latter case that we consider in this chapter. For example, Frank and

Friedman (1993) are interested in statistical applications in chemometrics where typical predictors are spectrometric curves digitized at more than 100 equispaced frequencies. In this particular example, ordinary least squares (OLS) fails since the covariance matrix that must be inverted is ill-conditioned. A first strategy therefore consists in adapting OLS in order to circumvent the inversion of the badly-conditioned design matrix, i.e. in using methods intended for the case of multicollinearity. See here Frank and Friedman (1993), who propose a unifying view of ridge regression, principal component regression, and partial least squares regression; see also Hocking (1976) who gives an overview of methods for variable selection in linear regression models and considers the particular problem of multicollinearity; finally see the recent paper of Cook (2007), with a discussion on dimension reduction in regression, based in particular on principal components and related methodologies.

Another approach consists in describing the link between the response and the functional predictor through a continuous version of linear regression, such as that proposed in Ramsay and Dalzell (1991) and Cardot et al. (1999a). Thus, the functional linear regression (FLR) models that follow integrate the discretization error and allow us, in principle, also to analyze the effect of the unobserved points in between the discretizations (while the multivariate methods mentioned above are not intended to be used for these objectives). A large literature has emerged that aims at investigating this type of functional modelization. This also corresponds to the more general taste for FDA that has appeared in the last decade. Now, functional linear regression allows us to describe the relationship between the predictor and the response in a functional way, and above all allows us to describe correlations between the predictors in a general and elegant way. To this end, the so-called Karhunen–Loève expansion plays a central role: it requires that the norm of the underlying random predictor curve has a finite second moment (i.e. is of second order) and is strongly connected to functional principal components. More details on this issue can be found in Chapter 8 of this book.

In this chapter we give a selected bibliography on functional linear regression, highlighting the main contributions from applied and in particular from theoretical points of view. In Section 2.2, we first define FLR in the case of a scalar response and show how its modelization can also be extended to the case of a functional response. This latter case has been less investigated in the literature and this explains why we have chosen to concentrate mainly on a scalar response. In any case, our main aim is to make inferences on the slope (functional) parameter, most often with the objective of prediction. The different estimation procedures of this slope parameter are classified into two categories: estimators based on projections and estimators obtained by considering a penalized least squares minimization problem. A third group of methods contains approaches derived from one of the above classes with specific adaptation in order to take into account the discretization of the curves or the sparsity of the discretization points of the observed trajectories. More details on how to deal with sparse data are given in Chapter 11 of this book. We then given an overview of the main asymptotic properties separating results on mean square prediction error and results on L^2 estimation error. The final section presents some related models, such as generalized functional linear models, FLR on quantiles, and finally a complementary bibliography as well as some open problems. Finally, we note at this point that the chapter does not include the estimation of multivariate linear models with a functional response: for this, see Faraway (1997) or Chiou et al. (2004).

2.2 FUNCTIONAL LINEAR REGRESSION

As mentioned in the introduction, functional linear regression has been introduced for the cases where the predictors are curves, whereas the responses are either scalars or curves. We begin our presentation of the model with the most frequently-investigated situation in the literature, i.e. when the responses are scalars. Thus, the set of data consists of a sequence (X_i, Y_i), $i = 1, \ldots, n$, where X_i is a real function defined on an interval I of $\mathbb{R}$ and $Y_i \in \mathbb{R}$. The most popular modelization in this case is based on random designs, assuming that one observes pairs (X_i, Y_i) that have the same distribution as a generic (X, Y). The case of deterministic controlled curves X_i is seldom considered. We mention in this connection the work of Cuevas et al. (2002) where the authors study functional linear regression for a fixed design. These authors consider a controlled input $x_i(t)$, $i = 1, \ldots, n$ and functional random responses, and study a functional version of the linear regression model. They provide motivation for this setting with different examples in which a fixed design could be more relevant, and discuss the choice of the design, which is crucial in order to get consistent estimators. In what follows we will only focus on random designs.

The minimal assumption on the curves X_i is that they are square integrable over I. We note at this point that one usually takes the interval $[0, 1]$ for I, in order to simplify notation and developments further. We thus consider the set $L^2([0, 1])$ of real square-integrable functions defined on $[0, 1]$. This set, equipped with its natural inner product defined as $\langle f, g \rangle = \int_0^1 f(t) g(t) \, dt$ and the induced norm $\| \cdot \|$, is a Hilbert space. Once again this is the most commonly adopted modelization of the curves X_i, although sometimes more restrictive Hilbert spaces of functions are considered such as Sobolev spaces or reproducing-kernel Hilbert Spaces (see, for instance, Smale and Zhou (2007)).

The random curves X_i are assumed to have a finite second moment

$$\mathbb{E}(\|X_i\|^2) < +\infty,$$

from which follows the existence of a mean curve $\mathbb{E}(X_i) = \{\mathbb{E}(X_i(t)), t \in [0, 1]\}$, and especially the existence of the covariance operator Γ_X, denoted as Γ when there is no ambiguity. The operator Γ is defined on $L^2([0, 1])$ and takes values in $L^2([0, 1])$ as follows:

$$\forall u \in L^2([0, 1]), \; \Gamma u = \mathbb{E}(\langle X_i - \mathbb{E}(X_i), u \rangle (X_i - \mathbb{E}(X_i))).$$

It can be easily seen that the operator Γ is an integral operator whose kernel is the covariance function of the random function X_i:

$$\forall u \in L^2([0, 1]), \forall t \in [0, 1], \; \Gamma u(t) = \int_0^1 \text{Cov}(X_i(s), X_i(t)) u(s) \, ds.$$

The operator Γ is non-negative, symmetric, Hilbert–Schmidt, and thus compact. Without giving more details, we simply recall that the eigenvalues of Γ, λ_j, $j = 1, 2, \ldots$, are all positive, and zero is the only accumulation point of the spectrum. Denoting by v_j, $j = 1, 2, \ldots$, the sequence of orthonormal eigenfunctions associated with the eigenvalues λ_j, sorted in decreasing order, we have the so-called Karhunen–Loève decomposition of the curve X_i, which plays a central role here. This decomposition is written

$$\forall t \in [0, 1], \; X_i(t) - \mathbb{E}(X_i(t)) = \sum_{j=1}^{\infty} \xi_j v_j(t), \tag{2.1}$$

where ξ_j, $j = 1, 2, \ldots$, are centered uncorrelated random variables with variance equal to λ_j: $\mathbb{E}(\xi_j) = 0$, $\mathbb{E}(\xi_j \xi_{j'}) = \lambda_j \delta_{jj'}$. Note that equality (2.1) holds in quadratic mean, as well as almost surely (a.s.) when X is Gaussian.

We come back now to the definition of the functional linear regression model presented in Section 2.3 of Chapter 1. This model can be thought of as a continuous version of multivariate linear regression: the link between the predictors and the responses is analyzed through the following relation:

$$Y_i = \alpha + \int_0^1 \beta(t) X_i(t) \, dt + \epsilon_i, \; i = 1, \ldots, n, \tag{2.2}$$

where α is the intercept and $\beta \in L^2([0, 1])$ is the slope function, and ϵ_i, $i = 1, \ldots, n$ is a sequence of independent and identically distributed (i.i.d.) centered random variables uncorrelated with X_i: $\mathbb{E}(\epsilon_i) = 0$ and $\mathbb{E}(X_i(t) \epsilon_i) = 0$ for t a.e. in $[0, 1]$. The main point of interest consists in the estimation of the slope function β, from which estimation of the intercept follows.

Leaving aside for the moment the question of identifiability in the relation (2.2) (existence and uniqueness of β), we first note that the function β can be defined equivalently in two other ways. Indeed, equation (2.2) leads to

$$Y_i - \mathbb{E}(Y_i) = \int_0^1 \beta(t) (X_i(t) - \mathbb{E}(X_i)(t)) \, dt + \epsilon_i, \; i = 1, \ldots, n. \tag{2.3}$$

It is easy to see that β satisfying relation (2.3) is equivalent to

$$\beta = \underset{\psi \in L^2([0,1])}{\arg\min} E\left[\left(Y_i - \mathbb{E}(Y_i) - \int_0^1 \psi(t)(X_i(t) - \mathbb{E}(X_i)(t))dt\right)^2\right]. \quad (2.4)$$

Now, multiplying both sides of relation (2.3) by $X_i - \mathbb{E}(X_i)$ and taking expectation leads to

$$\mathbb{E}((X_i - \mathbb{E}(X_i))(Y_i - \mathbb{E}(Y_i)))$$
$$= \mathbb{E}\left((X_i - \mathbb{E}(X_i))\int_0^1 \beta(t)(X_i(t) - \mathbb{E}(X_i)(t))\,dt\right)$$
$$= \Gamma\beta.$$

Identifying $\mathbb{E}((X_i - \mathbb{E}(X_i))(Y_i - \mathbb{E}(Y_i)))$ and the cross-covariance operator Δ, i.e. the functional defined as $\Delta\psi = \mathbb{E}(\langle\psi, (X_i - \mathbb{E}(X_i))\rangle(Y_i - \mathbb{E}(Y_i)))$, we have

$$\Delta = \Gamma\beta. \quad (2.5)$$

Equation (2.5) is a continuous version of the normal equations that the vector of regression coefficients of the usual multivariate linear regression satisfies. Thus, relations (2.3), (2.4), and (2.5) are three different equivalent definitions of the functional parameter β.

Relation (2.5) makes clear at this point that the existence of β is not as simple as in the multivariate case where the design matrix is just assumed to be a full-rank matrix. Since the operator Γ is a Hilbert–Schmidt operator, it does not admit a continuous inverse as long as the range of Γ is an infinite-dimensional space. Consequently, a solution β cannot be directly derived from (2.5). First of all, let us note that the functional parameter can be identified only in the orthogonal space of the kernel of Γ since if some function β satisfies (2.5) and if β_1 is in the kernel of Γ, i.e. $\Gamma\beta_1 = 0$, then $\beta + \beta_1$ also satisfies (2.5).

Taking $v_j, j = 1, 2, \ldots$, to be a complete orthonormal system of eigenfunctions, we can write $\beta = \sum_{j=1}^{\infty} \langle \beta, v_j \rangle v_j$. By (2.5),

$$\langle \mathbb{E}((X_i - \mathbb{E}(X_i))(Y_i - \mathbb{E}(Y_i))), v_j \rangle = \lambda_j \langle \beta, v_j \rangle, \quad j = 1, 2, \ldots \quad (2.6)$$

which allows us to obtain the coordinate of β on the functions v_j.

From now on we look for a solution in the closure of $\mathcal{I}m(\Gamma) = \{\Gamma x, x \in L^2([0, 1])\}$ or we assume without loss of generality that the kernel of Γ, $\text{Ker}(\Gamma)$, is reduced to zero. Now, inverting (2.6), we get the expansion for β:

$$\beta = \sum_{j=1}^{\infty} \frac{\langle \mathbb{E}((X_i - \mathbb{E}(X_i))(Y_i - \mathbb{E}(Y_i))), v_j \rangle}{\lambda_j} v_j$$
$$= \sum_{j=1}^{\infty} \frac{\mathbb{E}(\xi_j(Y_i - \mathbb{E}(Y_i)))}{\lambda_j} v_j, \quad (2.7)$$

and the function β will belong to $L^2([0, 1])$ if and only if the following condition is satisfied:

Condition 2.1 The random variables X and Y satisfy

$$\sum_{j=1}^{\infty} \frac{(\mathbb{E}[\xi_j(Y_i - \mathbb{E}(Y_i))])^2}{\lambda_j^2} < \infty.$$

Condition 2.1 ensures the existence and uniqueness of a solution β of the optimization problem (2.4) in the closure of $\mathcal{I}m(\Gamma)$. This condition is also called the Picard condition in the field of linear inverse problems (see, for example, Kress (1989)). Let us note that this condition is automatically fulfilled when the regression function of Y given X is a continuous linear functional, i.e. $\mathbb{E}(Y|X) = a + \langle \beta, X \rangle$.

Finally, let us note that (2.7) tells us that the estimation of β is a hard task since the eigenvalues λ_j decrease rapidly towards zero: as a matter of fact, $\sum_{j=1}^{\infty} \lambda_j < +\infty$.

The definitions above can be extended to the case where the response is functional. Assuming now that Y_i is a random function defined on an interval J and having a finite second moment, $\mathbb{E}(\|Y_i\|^2) < +\infty$, functional linear regression is defined through the relation

$$\forall t \in J, \ Y_i(t) = \int_I \beta(s, t) X_i(s) \, ds + \epsilon_i(t), \ i = 1 \ldots, n, \qquad (2.8)$$

where $\epsilon_i, i = 1, \ldots, n$, are i.i.d. random functions defined on $L^2([0, 1])$, with $\mathbb{E}(\epsilon(t)) = 0$ for t a.e., having a finite second moment, and ϵ_i and X_i are uncorrelated in the sense that $\mathbb{E}(X_i(t)\epsilon_i(s)) = 0$ for s and t a.e. The function parameter β is an element of $L^2(I \times J)$. Using the same developments as above one may show that a necessary and sufficient condition for identifiability of the model is:

Condition 2.1' The random variables X and Y satisfy

$$\sum_{j,j'=1}^{\infty} \frac{\mathbb{E}^2(\xi_{X,j}\xi_{Y,j'})}{\lambda_{X,j}^2} < \infty,$$

where $\xi_{X,j}$ (respectively, $\xi_{Y,j'}$) are the coefficients in the Karhunen–Loève decomposition of X (respectively, Y).

Under condition 2.1', a unique solution β exists in the orthogonal space of the kernel of the covariance operator of X, Γ_X, and this solution is written

$$\beta = \sum_{j=1}^{\infty} \frac{\langle \mathbb{E}((X - \mathbb{E}(X))(Y - \mathbb{E}(Y))), v_{X,j}v_{Y,j} \rangle}{\lambda_{X,j}} v_{X,j} v_{Y,j}.$$

Note that the case of a functional response has most often been investigated for functional autoregressive models (see, for instance, Mas (2007)). This model is described in detail in Chapter 3 of this book.

2.3 Estimation

As seen in the previous section, the estimation of the function β can be related to an ill-posed inverse problem. The specificity of functional linear regression is that the covariance operator Γ is itself unknown, which adds extra difficulty in some sense. As usual in the inverse problem literature, the estimation of β is performed using a regularization method combined here with an estimation of Γ. This has led to two kinds of estimation procedures. The first type of estimators relies on projection onto embedded finite-dimensional spaces whose dimensions grow along with the sample size; the most popular method is functional principal component regression. This consists in considering an empirical version of (2.7) where the eigenelements of Γ are replaced by the eigenelements of the empirical covariance operator, and where the sum is taken over a finite-dimensionalal space. This method is sometimes combined with some pre-smoothing of the curves or with a post-smoothing procedure for the functional parameter.

The other class of estimators is based on an expansion of the functional parameter in some basis of functions of $L^2([0, 1])$ combined with the minimization of the empirical version of (2.5), with the addition of a penalization term in order to regularize the solution. The most popular basis is probably that of spline functions, which have some well-known good features of approximation and stability. Finally, let us also mention another type of estimation procedure based on an adaption of partial least squares to the functional setting and developed by Preda and Saporta (2005) or Reiss and Ogden (2007).

2.3.1 Projection-based estimators

We first review the main projection-based estimators in which regularization is performed through dimension reduction.

Functional principal component regression (FPCR)
Generally speaking, the class of estimators of β based on functional PCA consists of an ordinary least squares (OLS) regression of the responses Y_i on the vectors of principal components, $(\langle \widehat{\varphi}_1, X_i \rangle, \ldots, \langle \widehat{\varphi}_K, X_i \rangle)^\tau$, where $\widehat{\varphi}_1, \ldots, \widehat{\varphi}_K$ are the eigenfunctions associated to the K largest eigenvalues of the empirical covariance operator Γ_n based on the sample $X_1, \ldots, X_n$. In this definition K is some positive integer defining the dimension of the projecting space, which plays the role of a smoothing parameter. This definition corresponds to the more usual version of the estimator based on functional PCA.

The basic FPCR estimator $\widehat{\beta}$ of β is defined as

$$\widehat{\beta} = \sum_{j=1}^{K} \frac{\Delta_n \widehat{\varphi}_j}{\widehat{\lambda}_j} \widehat{\varphi}_j,$$

where Γ_n is the empirical covariance operator,

$$\Gamma_n = \frac{1}{n} \sum_{i=1}^{n} \langle X_i, \cdot \rangle X_i,$$

and Δ_n is the cross-covariance operator,

$$\Delta_n = \frac{1}{n} \sum_{i=1}^{n} X_i Y_i.$$

Several authors have studied the estimator $\widehat{\beta}$ or defined alternative versions of it. The latter combine FPCR with a smoothing step, either of the curves X_i or of the estimator β obtained as above. Indeed, it has been pointed out in a simulation study that this estimator of the function β can be too rough in some situations even for large n (see Cardot et al. (1999a)). Among contributions in this area, we note Cardot et al. (1999a, 2003b), Cai and Hall (2006), Hall et al. (2006), Hall and Hosseini-Nasab (2006, 2009), Cardot et al. (2007), Hall and Horowitz (2007), and Reiss and Ogden (2007).

Projection estimators

Projection estimators, which consist in approximating the regression function by projections onto finite-dimensional spaces, are very similar to principal component regression. These estimators are easy to implement and are built by projecting the functional data onto a finite sequence of functions of $L^2[0,1]$. Then one only needs to estimate the coordinates of the slope function in this finite-dimensional subspace. Ramsay and Dalzell (1991) have built estimators using projection onto a B-spline basis, while Li and Hsing (2007) consider a complex Fourier basis. More recently Cardot and Johannes (2010) have proposed a general framework that relies on an orthonormal basis and the introduction of a thresholding term, controlled by a tuning parameter γ, that prevents the estimator from being too unstable. Let us consider a sequence $\psi_1, \ldots, \psi_k, \ldots$ of orthonormal functions in $L^2[0,1]$; their estimator can be defined as follows:

$$\widehat{\beta}(t) = \sum_{\ell=1}^{k} \widehat{\beta}_\ell \mathbb{1} \left\{ \left\| [\Gamma_{n,k}]^{-1} \right\| \leq \gamma \right\} \psi_\ell(t), \quad t \in [0,1], \qquad (2.9)$$

where the $\widehat{\beta}_\ell$ are the generic elements of the vector of coordinates obtained by least squares projection, $\mathbb{1}$ is the indicator function, and $\Gamma_{n,k}$ is the empirical covariance operator restricted to the space generated by $\{\psi_1, \ldots, \psi_k\}$. The new thresholding step can be seen as an improvement of the estimator proposed by Ramsay and Dalzell (1991). From an inverse-problem perspective this approach is similar to the linear Galerkin procedure (Engl et al. 2000). It has the advantage (compared to principal component regression) that it does not require the estimation of the eigenfunctions of the empirical covariance operator.

2.3.2 Penalized least squares estimators

The second main class of estimators of the parameter β consist in expanding the functional parameter β in some given deterministic basis of functions in $L^2([0, 1])$ and in minimizing a penalized least squares criterion. Several candidates for such a basis have been proposed in the literature; the most popular are probably spline functions. Several versions of estimators based on spline functions have been studied: they are of course very similar.

We define below the spline estimators proposed by Cardot et al. (2003b), which combine ideas from Marx and Eilers (1996) and Hastie and Mallows (1993). Suppose that q and k are integers and let S_{qk} be the space of *splines* defined on $[0, 1]$ with degree q and with $k - 1$ equispaced interior knots. The space S_{qk} has dimension $q + k$ and one can derive a basis by means of normalized B-splines $\{B_{k,j}, j = 1, \ldots, k + q\}$ (see de Boor (1978)). In what follows we denote by $\mathbf{B}_k$ the vector of all the B-splines and by $\mathbf{B}_k^{(m)}$ the vector of derivatives of order m of all the B-splines for some integer m ($m < q$).

The penalized B-spline estimator of β is thus defined as

$$\widehat{\beta}_{PS} = \sum_{j=1}^{q+k} \widehat{\theta}_j B_{k,j} = \mathbf{B}_k' \widehat{\theta}, \qquad (2.10)$$

where $\widehat{\theta}$ is a solution of the minimization problem

$$\min_{\theta \in \mathbb{R}^{q+k}} \frac{1}{n} \sum_{i=1}^n \left(Y_i - \sum_{j=1}^{q+k} \langle \theta_j B_{k,j}, X_i \rangle \right)^2 + \rho \left\| \mathbf{B}_k^{(m)'} \theta \right\|^2, \qquad (2.11)$$

with smoothing parameter $\rho > 0$. The solution $\widehat{\theta}$ of the minimization problem (2.11) is given by

$$\widehat{\theta} = \widehat{\mathbf{C}}_\rho^{-1} \widehat{\mathbf{b}} = \left(\widehat{\mathbf{C}} + \rho \mathbf{G}_k\right)^{-1} \widehat{\mathbf{b}}, \qquad (2.12)$$

where $\widehat{\mathbf{C}}$ is the $(q + k) \times (q + k)$ matrix with elements

$$n^{-1} \sum_{i=1}^n \langle B_{k,j}, X_i \rangle \langle B_{k,l}, X_i \rangle,$$

$\widehat{\mathbf{b}}$ is the vector in $\mathbb{R}^{q+k}$ with elements $n^{-1}\sum_{i=1}^{n}\langle B_{k,j}, X_i\rangle Y_i$, and where $\mathbf{G}_k$ is the $(q+k) \times (q+k)$ matrix with elements $\langle B_{k,j}^{(m)}, B_{k,l}^{(m)}\rangle$. In the special case $m = 0$, the minimization criterion (2.11) becomes

$$\frac{1}{n}\sum_{i=1}^{n}(Y_i - \langle \mathbf{B}_k'\boldsymbol{\theta}, X_i\rangle)^2 + \rho\|\mathbf{B}_k'\boldsymbol{\theta}\|^2,$$

which is a functional generalization of the ridge regression criterion. The difference between this estimator and those proposed by Marx and Eilers (1996) is only in the penalization term, which is slightly different but cannot be interpreted as the $L^2[0, 1]$ norm of a fixed-order derivative.

2.3.3 Discretization and sparse data

We have supposed until now that the trajectories X_i have been observed for all possible time points. With real datasets one generally gets functional data that are discretized at equidistant design points of time, $0 \le t_1 < t_2 < \cdots < t_p \le 1$, and maybe corrupted by noise

$$Z_{ij} = X_i(t_j) + \epsilon_{ij} \quad i = 1, \ldots, n, \ j = 1, \ldots, p,$$

where ϵ_{ij} is a white noise, with finite variance. In the absence of noise, the empirical covariance operators Γ_n and Δ_n can be approximated using quadrature rules, replacing inner products in $L^2[0, 1]$ by summations (see Ramsay and Silverman (2005) for a more detailed presentation). When observations are corrupted by noise and/or the discretization points differ from one curve to another using a sufficiently fine grid, a common approach consists in expanding the trajectories in basis functions such as B-splines or Fourier bases in order to get functional denoised data.

Note that smoothing spline estimators are also natural estimators when the discretization points have the same location for all curves. This approach has been proposed in the literature by Goutis (1998) and Crambes et al. (2009). In this case, one can build the estimator by considering a decomposition in a B-spline basis with knots located at the discretization points and a similar penalized criterion as in (2.11).

Sparse data

The case of sparse data is more delicate to deal with and generally needs much greater computational effort. We say that we have sparse data when the number of discretization points is relatively small for each curve. Then one observes

$$Z_{ij} = X_i(t_{ij}) + \epsilon_{ij}, \quad i = 1, \ldots, n, \ j = 1, \ldots, p_i,$$

with p_i relatively small.

Two different estimation procedures are proposed in the literature. The first consists in maximizing the log likelihood using the EM algorithm by a finite-dimensional basis expansion of both the slope function and the functional covariates. Let us note that this approach relies heavily on the assumption that the data are Gaussian when the model under consideration is the FLM, but can be extended to generalized linear models for probability distributions belonging to the exponential family (see James (2002) for generalized linear models and James et al. (2000) for PCA).

The latter approach, developed more recently in a series of articles (Yao et al. 2005a, 2005b; Hall et al. 2006, etc.), is based on scatterplot smoothing such as that using kernels or local polynomials. It first consists in estimating the mean function by passing a nonparametric smoother through the data (t_{ij}, Z_{ij}). Then the covariance function of the X_i's and the cross-covariance function between the X_i and the Y_i can be estimated in a similar way, once the estimated mean trajectory has been subtracted. For example, the covariance function, which is a bivariate function of time, is estimated by applying a bivariate smoother to the data $((t_{ij}, t_{ij'}), \gamma_{ijj'})$, where

$$\gamma_{ijj'} = (Z_{ij} - \widehat{\mu}(t_{ij}))(Z_{ij'} - \widehat{\mu}(t_{ij'}))$$

and $\widehat{\mu}(t_{ij})$ is the estimated mean function of the trajectories at point t_{ij}. Note that this approach is only valid when the location of the measurement points are random and independent of the trajectories, which is often the case for medical longitudinal studies.

Note finally that for both approaches, prediction of a new value of Y can be made for the FLM, even with a functional covariate X that is sparsely observed, thanks to the BLUP (Best Linear Unbiased Prediction) formula.

2.4 ASYMPTOTIC RESULTS

While there has been a prolific literature on the development of estimation algorithms for functional linear regression over the last fifteen years, there are only a few papers dealing with the asymptotic behavior of the estimators of the slope function β and of the intercept α. However over the past two years, several authors have carried out deeper studies on the properties of these estimators and one can now find a substantial literature on this topic that will probably deserve further investigation in the future. Below we give the main results and try to discuss and compare the different asymptotic results when the response is scalar. Results for a functional response have not yet been thoroughly investigated and we will just mention below the few works on this topic.

First, we would like to distinguish two types of asymptotics. On the one had, one can be interested in evaluating the error between the estimator and the true curve β based on a functional distance such as, for instance, the L^2 distance in the space of square-integrable functions, i.e. $\|\beta - \widehat{\beta}\|^2$. At the other extreme one may be interested in evaluating the error in the prediction of a new value of Y. A consideration of the mean square error of prediction leads us to consider the distance between the estimator and the true curve with respect to a seminorm constructed from the covariance operator Γ. Indeed, when the predictive curves are centered, it is easy to see that

$$\mathbb{E}\left(\langle \beta - \widehat{\beta}, X_{n+1}\rangle^2 \big| \widehat{\beta}\right) = \|\beta - \widehat{\beta}\|_\Gamma^2 =: \langle \Gamma(\beta - \widehat{\beta}), \beta - \widehat{\beta}\rangle,$$

where X_{n+1} is a random function having the same distribution as X and independent of $X_1, \ldots, X_n$. In the general case, it can be shown that $\|\widehat{\beta} - \beta\|_\Gamma^2$ determines the mean square error of prediction up to a term with a rate no greater than n^{-1} (see Crambes et al. (2009)). In some settings, one considers the mean square error of prediction for a fixed curve x which is written $\langle \beta - \widehat{\beta}, x\rangle$.

Contrary to what happens in the case of multivariate linear regression, the two types of errors are very different in the context of functional linear regression. As a matter of fact, Cai and Hall (2006) or Crambes et al. (2009) emphasize that an optimal estimator of β (i.e. that minimizes the L^2 error) will be over-smoothed for prediction. Thus the estimation of β is sensitive and depends on the goal one wants to attain. The majority of the literature concerns asymptotics for prediction.

We may distinguish between three types of assumptions needed for asymptotic studies. The first group deals with smoothness conditions on the slope function β, as happens in general functional (nonparametric) estimation. However for functional linear regression, despite the traditional meaning of smoothness (i.e. the existence of derivatives up to a given order and Lipschitz conditions) some authors measure smoothness in terms of the spectral decomposition of β in the basis of eigenfunctions as, for example, in Cai and Hall (2006) (see the developments below). Asymptotic properties also depend on the regularity of the predictive curve X. In functional (nonparametric) estimation, regularity of the distribution of the predictive vector is most often associated with the existence of a density function which is bounded from below on some compact set. When the predictor is a curve, and thus the existence of a density no longer holds, regularity is measured by means of conditions on the covariance operator of X. Roughly speaking, these types of assumptions deal with the rate of decay towards zero of the sequence of eigenvalues of Γ.

In all the work dealing with asymptotics in functional linear regression, the Karhunen–Loève decomposition plays a central role.

2.4.1 Mean square prediction error

Cai and Hall (2006) study the asymptotic behavior of an estimator with respect to the error of prediction for a fixed curve x, i.e. $\langle \beta - \hat{\beta}, x \rangle$. The estimator of β is of the same kind as the one based on functional principal regression with smoothing parameter (dimension projection) $K = K_n$. In fact, they propose a slightly modified version of the estimator described in the previous section in order to ensure that all the moments of the estimator are finite: the modification consists in truncating the estimator when its L^2 norm is greater than some "threshold" value depending on n.

The asymptotic behavior of this estimator is described in terms of the spectral decomposition of the curves and when the eigenvalues λ_j of the covariance operator Γ of the predictors have a polynomial rate of decay. In the following, β_j, $j \geq 1$, represent the coefficients in the spectral decomposition of β: $\beta = \sum_{j \geq 1} \beta_j v_j$. The authors consider a fixed curve x with

$$x = \sum_{j \geq 1} x_j v_j.$$

Hence the rate of convergence of the mean square prediction error depends on the speed of decrease of the coefficients β_j and x_j: it is assumed that $|\beta_j| < C_1 j^{-b}$ while $x_j < C_2 j^{-\gamma}$, $\gamma > 1/2$. The eigenvalues λ_j are assumed to satisfy, for some $a > 1$,

$$C^{-1} j^{-a} \leq \lambda_j \leq C j^{-a}, \quad \lambda_j - \lambda_{j+1} \geq C^{-1} j^{-a-1}, \quad \text{for } j \geq 1. \tag{2.13}$$

Finally, the authors show that one has the following result provided that $b \geq a + 2$:

$$\sup_{\beta \in \mathcal{B}} \mathbb{E}\left(\langle \beta - \hat{\beta}, x \rangle \right)^2 = O(\tau), \tag{2.14}$$

where

$$\tau = \begin{cases} n^{-1}, & \text{if } a + 1 < 2\gamma, \\ n^{-1} \log n, & \text{if } a + 1 = 2\gamma, \\ n^{-2(b+\gamma-1)/(a+2b-1)}, & \text{if } a + 1 > 2\gamma. \end{cases}$$

The result (2.14) is proved for predictive curves and error variables having moments which are all finite. The choice of the smoothing parameter is of course particularly important. It satisfies

$$K = \begin{cases} n^{1/2(b+\gamma-1)}, & \text{if } a + 1 < 2\gamma, \\ (n/\log n)^{1/(a+2b-1)}, & \text{if } a + 1 = 2\gamma, \\ n^{1/(a+2b-1)}, & \text{if } a + 1 > 2\gamma. \end{cases}$$

As mentioned above, the value of the smoothing parameter should be different when optimizing the L^2 error instead of the mean square prediction error (see also Section 2.4.2). Moreover, Cai and Hall (2006) show that the rates derived in (2.14) are minimax optimal over the class of functions β belonging to the set of functions such that $|\beta_j| < C_1 j^{-b}$, and when the random curve X_i is a Gaussian process and the errors are normally distributed.

Thus it appears that the rate of convergence of the error of prediction for a fixed curve x, as derived in Cai and Hall (2006), depends heavily on the smoothness of the predictor curve in terms of its spectral decomposition. In particular, the optimal rate n^{-1} can only be obtained for very smooth curves. We will discuss these results below on the basis of their comparison with results on prediction for a random curve X_{n+1}.

Crambes et al. (2009) study the mean square prediction error for an out-of-sample random curve X_{n+1}. They estimate β by means of the smoothing splines estimator described in the previous section. Unlike in the work of Cai and Hall (2006), the smoothness of β is measured in terms of the existence of derivatives up to a given order: for a positive integer $m \in \{1, 2, \ldots\}$, it is assumed that

$$\beta \text{ is } m \text{ times differentiable and } \beta^{(m)} \text{ belongs to } L^2([0, 1]). \qquad (2.15)$$

As in Cai and Hall (2006), they assume a certain polynomial rate of decrease for the eigenvalues of Γ. More precisely, they suppose that the predictive curves X_i can be approximated by some arbitrary low-dimensional linear function space with dimension $k \geq 1$, i.e. the distance between X_i and this space is of order less than Ck^{-2q}, $q \geq 0$. It turns out that if the predicting curves X_i are sufficiently smooth, i.e. if they have q_1 derivatives with $X^{(q_1)}$ Lipschitz continuous, then the above condition holds. However, this condition may also be satisfied for non-smooth curves X_i such as Brownian motion trajectories (note that a minimal Lipschitz condition on the curves X_i is assumed in order to control the discretization error term). In any case, the condition implies that the eigenvalues are such that $\sum_{j=k+1}^{+\infty} \lambda_j = O(k^{-2q})$ and, since the sequence of eigenvalues λ_j is a decreasing sequence, this implies that

$$\lambda_j = O(j^{-2q-1}). \qquad (2.16)$$

Defining the estimator of the intercept α as $\widehat{\alpha} = \overline{Y} - \langle \widehat{\beta}, \overline{X} \rangle$, Crambes et al. (2009) obtain the following convergence rate under an additional moment condition on the predictive curves and for $q \geq 1$:

$$\mathbb{E}\left(\left(\widehat{\alpha} - \alpha + \langle \beta - \widehat{\beta}, X_{n+1}\rangle\right)^2 \mid \widehat{\beta}, \widehat{\alpha}\right) = O_P(n^{-(2m+2q+1)/(2m+2q+2)}). \qquad (2.17)$$

The rate (2.17) holds as soon as the smoothing parameter ρ is of order $n^{-(2m+2q+1)/(2m+2q+2)}$. The authors show, moreover, that the rate of convergence is lower in a large class of possible slope functions $\mathcal{C}_{m,D}$ with m continuous derivatives and having the integral of their square bounded by D, and a set of distributions of the predictive curves $\mathcal{P}_{q,C}$ such that (2.16) holds:

$$\liminf_{n \to \infty} \sup_{\widehat{\beta}} \sup_{P \in \mathcal{P}_{q,C}, \beta \in \mathcal{C}_{m,D}} P\left(\|\beta - \widehat{\beta}\|_\Gamma \geq c_n \cdot n^{-(2m+2q+1)/(2m+2q+2)}\right) = 1, \qquad (2.18)$$

where c_n denotes an arbitrary sequence of positive integers tending to zero as n tends to infinity. This shows that the rate in (2.17) is minimax optimal over the classes $\mathcal{C}_{m,D}$ and $\mathcal{P}_{q,C}$.

Cardot and Johannes (2010) consider the thresholded projection estimator $\widehat{\beta}$ described in Section 2.3. They measure the error of this estimator with respect to the criterion

$$\mathbb{E}\left\|\widehat{\beta} - \beta\right\|_\omega^2,$$

where the weighted norm is defined by $\|f\|_\omega^2 = \sum_{j\geq 1} w_j |\langle f, \psi_j \rangle|^2$, $(\omega_j)_{j\in\mathbb{N}}$ being a positive sequence of weights and $\{\psi_1, \ldots\}$ forming a basis of $L^2([0,1])$ as defined in previous section. They first state, under general conditions on the link between the covariance operator, the weighted norm, and moment assumptions, lower and upper bounds for the error with respect to this general criterion. It turns out that such a weighted norm allows us to consider a wide range of estimation errors which can measure, for particular sequences of basis functions and particular sequences of weights, either prediction error or estimation error of a given-order derivative. If, for example, one supposes that the slope function β belongs to the space of periodic square-integrable functions $\mathcal{W}_m$ having m derivatives with $\beta^{(m)}$ absolutely continuous and $f^{(m)} \in L^2([0,1])$, it is possible to derive explicitly the rate of convergence for the mean square prediction error. Under some moment conditions on the predictive curves and on the error terms, Cardot and Johannes (2010) show, in this periodic context, that

$$\sup_{\beta \in \mathcal{W}_m^p} \left\{ \mathbb{E}\left\|\widehat{\beta} - \beta\right\|_\Gamma^2 \right\} = O(n^{-2(p+2a)/(2p+2a+1)}), \tag{2.19}$$

when the decay of the eigenvalues of Γ is of order $|j|^{-2a}$ for some $a > 1/2$, while

$$\sup_{\beta \in \mathcal{W}_m^p} \left\{ \mathbb{E}\left\|\widehat{\beta} - \beta\right\|_\Gamma^2 \right\} = O(n^{-1}(\log n)^{1/2a}), \tag{2.20}$$

when $\lambda_j = \exp(-|j|^{2a})$. In the expressions above, $\mathcal{W}_m^p$ represents some ellipsoid of the space $\mathcal{W}_m$ having an ad hoc weighted norm bounded by p. Concerning the dimension of the projecting space k, the upper bounds above are stated for $k \sim n^{1/(2m+2a+1)}$ for a polynomial rate of decay of the eigenvalues, while $k \sim (\log n)^{1/(2a)}$ in the exponential case. In both cases, the threshold γ satisfies $\gamma \sim n$. It is shown, moreover, that the rates of convergence are optimal in the sense that for any estimator $\widetilde{\beta}$ of β

$$\sup_{\beta \in \mathcal{W}_m^p} \left\{ \mathbb{E}\left\|\widetilde{\beta} - \beta\right\|_\Gamma^2 \right\} \geq C\delta_n, \tag{2.21}$$

where δ_n is either the "polynomial" or "exponential" rate given above.

For the purpose of comparison of the different results described above, we may emphasize that in the case of eigenvalues λ_j with a polynomial rate of decay,

Crambes et al. (2009) and Cardot and Johannes (2010) find the same mean square prediction rate of convergence: the connection between the parameters q and a being $q = a - 1/2$. The results are obtained from slightly different conditions on the distributions of the predictive curves and on the slope function β. However, they consider the same smoothness conditions on the slope parameter β, measured in terms of the number of derivatives. Cardot and Johannes (2010) have the additional condition of the periodicity of this function. In both cases, smoothness of the predictive curves is measured in terms of the rate of decay of the eigenvalues of Γ. For a polynomial rate, the optimal rate of convergence is that derived above: it appears that this rate improves when the values of the parameters m and a increase. Cardot and Johannes (2010) additionally study the case of an exponential rate of decay of the eigenvalues. They find in this case that the optimal rate of convergence does not depend on the value of m (the number of derivatives of β), provided that m is some integer greater or equal to 1 (see above). We refer to this paper for a deeper discussion on this point.

Things are different in the approach proposed by Cai and Hall (2006). Here, the smoothness of a fixed curve x is measured in terms of its spectral decomposition. The essential condition on the structure of the coefficients x_j may be re-expressed in the following form: there exist some $\nu \in \mathbb{R}$ and $0 < D_0 < \infty$ such that $D_0^{-1} j^\nu \leq x_j^2/\lambda_j \leq D_0 j^\nu$ for all $j = 1, 2, \ldots$ Rates of convergence then follow from the magnitude of ν, and it is shown that parametric rates n^{-1} (or $n^{-1}(\log n)^2$) are achieved if $\nu \leq -1$. However, Crambes et al. (2009) show that such a condition cannot be satisfied for a predictive random curve having a Gaussian distribution, in the sense that if $\nu \leq 0$, then

$$P\left(D_0^{-1} j^\nu \leq \frac{x_{n+1,j}^2}{\lambda_j} \leq D_0 j^\nu \text{ for all } j = 1, 2, \ldots\right) = 0$$

for all $0 < D_0 < \infty$. This clearly demonstrates that the parametric rates n^{-1} cannot be achieved for the prediction error $\langle \beta - \widehat{\beta}, X_{n+1} \rangle^2$. Finally, this proves that the rates of convergence are very different for a fixed curve x than for a random one. This fact has also been highlighted in Cardot et al. (2007).

Some authors have also been interested in getting a central limit theorem for the prediction that would allow us to build asymptotic confidence bands. Cardot et al. (2007) study the Central Limit Theorem (CLT) for the same kind of estimators as those in Cai and Hall (2006). In fact, they introduce a general regularization procedure leading to an estimator of the form

$$\widehat{\beta} = \Gamma_n^\dagger \Delta_n,$$

where $\Gamma_n^\dagger = \sum_{j=1}^n f_n(\widehat{\lambda}_j) \widehat{v}_j \otimes \widehat{v}_j$ with f_n a decreasing function defined on $[c_n, +\infty)$, c_n being a threshold. For $f_n(x) = (1/x) I_{[c_n, +\infty)}(x)$, the estimator is exactly the functional principal component regression estimator, while other choices of f_n lead to ridge-type estimators, estimators based on Tikhonov regularization of Γ_n, etc. At first they show that it is not possible to obtain a CLT for the topology of the functional space, but that a CLT can be obtained for the weak topology or, in other words, for the error of prediction. They study both the fixed and the random case, i.e. they consider the quantities $\langle \widehat{\beta} - \beta, X_{n+1} \rangle$ where X_{n+1} is an out-of-sample random curve, and $\langle \widehat{\beta} - \widehat{\Pi}_k \beta, x \rangle$ where x is a fixed curve and $\widehat{\Pi}_k$ is the orthogonal projection of the space spanned by the k first eigenfunctions of the empirical covariance operator. They show that the results are very different for the two cases. Whereas the optimal n^{-1} rate can be obtained for the fixed design, this is not possible for the random design framework. Indeed, in the random case the rates of convergence in the CLT are given by $\sqrt{n/\text{tr}(\Gamma^\dagger \Gamma)}$, while the rates for the fixed design are $\sqrt{n}/\|\Gamma^{1/2}\Gamma^\dagger x\|$. While the denominator for the rate of convergence for the fixed design may be finite and thus the parametric rate $\sqrt{n}$ can be reached (at least for a fixed curve x such that $\|\Gamma^{-1/2}x\|$ is finite), this is not the case for the random design since the denominator in the rate of convergence always tends to infinity.

This again illustrates the difference between the random and the fixed designs, as has been pointed out before. The results in Cardot et al. (2007) are stated for a specific choice of smoothing parameter (which is, roughly speaking, similar to the dimension of the projecting space in an FPC regression): its rate of decrease as the sample size grows to infinity must be rather slow. The assumptions on the predictive curves and on the slope parameter β are quite general. In particular, almost no smoothness assumptions are required for β and the form of the eigenvalues is not specified, but only mild conditions are assumed which are satisfied for both polynomially or exponentially decreasing sequences of eigenvalues. Finally the only condition on the predictive curve is that it has a finite fourth moment. More specific assumptions would certainly lead to more precise rates of convergence, such as those obtained by Cai and Hall (2006).

2.4.2 The L^2 estimation error

Hall and Horowitz (2007) derive optimal rates of convergence for the estimation error $\|\widehat{\beta} - \beta\|^2 = \int_0^1 (\widehat{\beta}(t) - \beta(t))^2 dt$. The estimator being considered here is the one based on FPCR described in Section 2.3.1 (see also Cai and Hall (2006)). Under similar assumptions as in Cai and Hall (2006) on the slope function, i.e. when the spectral coefficients b_j are bounded up to a constant by j^{-b}, they obtain

$$\lim_{D \to \infty} \limsup_{n \to \infty} \sup_{F \in \mathcal{F}} P_F \left\{ \int_0^1 (\widehat{\beta} - \beta)^2 > Dn^{-(2b-1)/(a+2b)} \right\} = 0, \qquad (2.22)$$

where $\mathcal{F}$ is a set of distributions of the pair (X_i, Y_i) satisfying several conditions such as: X_i has a finite fourth moment, the eigenvalues of the covariance operator Γ have a polynomial rate of decay with exponent equal to a (see condition (2.3) above), and the slope function β satisfies the boundary condition described above. Hall and Horowitz (2007) also show that $n^{-(2b-1)/(a+2b)}$ is a lower bound for the rates of convergence and consequently that the estimator $\widehat{\beta}$ is a minimax optimal estimator of β. Optimality is reached for a tuning parameter of order $k \sim n^{1/(a+2b)}$. They also obtain minimax bounds for the expected L^2 error and for an alternative estimator $\widetilde{\beta}$ based on a ridge procedure with ridge tuning parameter ρ. This latter result is obtained under weaker conditions on the eigenvalues.

Cardot and Johannes (2010) also consider the estimation of derivatives of the slope function β. They generalize the estimation procedure defined in the previous section to the estimation of a derivative $\beta^{(s)}$ of order $s \geq 0$ of β. The assumptions are of the same nature as for the mean square error of prediction. In the periodic setting, they obtain the following upper bound for the rates of convergence with respect to the L^2 norm, when the eigenvalues of the covariance operator Γ have a polynomial rate of decay, the dimension of the projecting space k is such that $k \sim n^{1/(2m+2a+1)}$, and the threshold $\gamma = n$,

$$\sup_{\beta \in \mathcal{W}_p^\rho} \left\{ \mathbb{E} \left\| \widehat{\beta}^{(s)} - \beta^{(s)} \right\|^2 \right\} = O(n^{-(2m-2s)/(2m+2a+1)}). \tag{2.23}$$

When the eigenvalues of Γ have an exponential rate of decay, the upper bound is given by

$$\sup_{\beta \in \mathcal{W}_p^\rho} \left\{ \mathbb{E} \left\| \widehat{\beta}^{(s)} - \beta^{(s)} \right\|^2 \right\} = O((\log n)^{-(m-s)/a}), \tag{2.24}$$

provided that the dimension k satisfies $k \sim (\log n)^{1/(2a)}$ and the threshold $\gamma = n$. They also show that the rates given above are minimax optimal over the functional space $\mathcal{W}_m^\rho$, as defined in the previous section.

2.5 CONNECTED MODELS AND EXTENSIONS

2.5.1 The generalized functional linear model

As in the case of multivariate linear regression, a natural extension of functional linear regression is the generalized functional linear model. This model is of interest when the response has discrete positive values and the predictor is functional. This includes discrete responses with binomial, multinomial, or Poisson distributions and thus leads to functional versions of logistic, binomial, or Poisson regression. The dataset consists here of random pairs (X_i, Y_i), $i = 1, \ldots, n$, where the X_i's are random curves with the same features as in the previous sections whereas Y_i is a real-valued random variable which may be discrete.

We may distinguish two points of view, as in the multivariate case: the functional quasi-likelihood models studied by Müller and Stadtmüller (2005) and the generalized functional linear model which relies on the exponential family of distributions (Marx and Eilers (1999); James (2002); Cardot and Sarda (2005)). In both cases, given a monotone and twice continuously differentiable function g, the generalized functional linear model is written

$$Y_i = g\left(\alpha + \int_0^1 \beta(t) X_i(t) \, dt\right) + \epsilon_i, \quad i = 1, \ldots, n, \tag{2.25}$$

where $(\epsilon_i)_i$ are i.i.d. random variables such that

$$\mathbb{E}(\epsilon_i | X_i) = 0, \quad i = 1, \ldots, n.$$

For quasi-likelihood models, one additionally assumes that the conditional variance is a function of the conditional mean:

$$\text{Var}(\epsilon_i | X_i) = \sigma^2(\mu) = \tilde{\sigma}^2(\eta),$$

where μ is the conditional mean and $\eta = \alpha + \int_0^1 \beta(t) X_i(t) \, dt$. Thus, an exponential family of distributions is not necessarily required in the work of Müller and Stadtmüller (2005). Conversely, Marx and Eilers (1999), James (2002), and Cardot and Sarda (2005) consider such types of families of distributions. Several parametrizations can be found in the literature; they are equivalent. For instance, following Stone (1986), Cardot and Sarda (2005) assume that the conditional distribution of Y given $X = x$ belongs to the exponential family described by

$$\exp\{b_1(\eta) y + b_2(\eta)\} \nu(dy), \tag{2.26}$$

where ν is a nonzero measure on $\mathbb{R}$ which is not concentrated at a single point, the function b_1 is twice continuously differentiable, and b_1' is a strictly positive function on $\mathbb{R}$. Then the function b_1 is strictly increasing and b_2 is twice continuously differentiable on $\mathbb{R}$. The mean μ of the distribution is

$$\mu = g(\eta) = -\frac{b_2'(\eta)}{b_1'(\eta)}.$$

To get identifiability of the model, it is assumed that the eigenvalues of the covariance operator of the X_i are nonzero and that, for some interval S in $\mathbb{R}$ on which ν is concentrated, we have $b_1''(\eta) y + b_2''(\eta) < 0$, for all $\eta \in R$ and all $y \in S$. This condition is satisfied for classical distributions such as the Bernoulli, the Poisson, or the Gamma distribution.

As before, two main types of estimators of the slope function β are proposed in the literature: projection estimators and penalized likelihood estimators. Müller and Stadtmüller (2005) propose an estimator based on a truncated Karhunen–Loève expansion. The k-dimensional vector of the coefficients of the estimator $\widehat{\beta}$ of β in the basis formed by the (estimated) eigenfunctions of the covariance operator is defined as the solution of a score equation. In fact, the estimator is even defined in a more general way as other orthogonal expansions of the functional predictors are considered in order to reduce dimensionality. Müller and Stadtmüller (2005) prove the asymptotic normality of the error between this k-dimensional vector and the vector of coefficients of β in the same basis with respect to a L^2 metric defined through a generalized covariance operator. The same type of estimators, i.e. based on a truncated Karhunen–Loève expansion, are also considered in Cardot et al. (2003a) in a multinomial model with functional predictors. The aim of this work is the prediction of land use using a multilogit functional linear model due to the temporal evolution of coarse resolution SPOT4/VEGETATION satellite images. Escabias et al. (2004) compare, by means of simulation studies, two versions of estimators based on functional principal component analysis for functional logistic regression. We also mention the functional partial least squares logit regression model proposed by Escabias et al. (2007) as an alternative to functional principal component logistic regression.

The other type of estimators have been considered in Marx and Eilers (1999) and Cardot and Sarda (2005). This type consists in expanding the slope coefficient in a basis of functions, such as B-splines, and maximizing a penalized likelihood criterion. The estimators defined in these two papers only differ by the penalty term: Cardot and Sarda (2005) consider a penalty proportional to the square L^2 norm of the derivative of order m, while Marx and Eilers (1999) use a discrete version of this norm. In both cases the estimator β is a linear combination of B-splines of order q with k equispaced knots on $[0, 1]$: $\widehat{\beta} = \mathbf{B}'_k \widehat{\boldsymbol{\theta}}$, where $\widehat{\boldsymbol{\theta}}$ is a solution of some maximization problem. With the parametrization of the exponential family introduced in Cardot and Sarda (2005), this can be written, for some smoothing (penalty) parameter $\rho > 0$,

$$\max_{\boldsymbol{\theta} \in R^{q+k}} \frac{1}{n} \sum_{i=1}^{n} \left(b_1\left(\langle \mathbf{B}'_k \boldsymbol{\theta}, X_i \rangle\right) Y_i + b_2\left(\langle \mathbf{B}'_k \boldsymbol{\theta}, X_i \rangle\right) \right) - \frac{1}{2} \rho \boldsymbol{\theta}^\tau \mathbf{G}_m \boldsymbol{\theta}, \qquad (2.27)$$

where $\mathbf{G}_m$ is a positive symmetric matrix obtained from difference operations of order m in Marx and Eilers (1999), or with a general term equal to the inner product of the derivative of order m of B-splines in Cardot and Sarda (2005). The solution of this optimization problem is generally obtained in a few iterations of the Newton–Raphson or Fisher scoring algorithm. Cardot and Sarda (2005) find an upper bound for the mean square error of prediction which depends on the smoothness of the slope function β (in terms of derivatives). The conditions on the distribution of the predictive curves are very mild and, as a matter of fact, there are no specific assumptions on the rates of decay towards zero of the eigenvalues λ_j. Considering the recent results obtained for the functional linear model, we may conjecture that it should be possible to obtain better rates with additional conditions on the eigenvalues of the operator Γ.

2.5.2 Functional linear regression on quantiles

It is also of interest to estimate the conditional median or quantile of a real covariate, given a functional predictor value. Cardot et al. (2005) consider this problem and generalize the linear quantile regression model of Koenker and Bassett (1978). Given i.i.d. random pairs (X_i, Y_i), $i = 1, \ldots, n$, where Y_i is a real random variable and X_i is a random curve with the same properties as before, the aim is to estimate a conditional quantile of Y_i given $X_i = x$, where x is some fixed curve in $L^2([0, 1])$. For some $\delta \in (0, 1)$, the conditional quantile of Y of order δ is defined as the function g_δ such that

$$P\left(Y_i \leq g_\delta(x) | X_i = x\right) = \delta. \tag{2.28}$$

As shown in Koenker and Bassett (1978), one may define g_δ as

$$g_\delta(x) = \operatorname{argmin} E\left(l_\delta(Y_i - a) | X_i = x\right), \quad a \in \mathbb{R} \tag{2.29}$$

with

$$l_\delta(u) = |u| + (2\delta - 1)u.$$

Note that the case $\delta = 1/2$ corresponds to the conditional median.

Cardot et al. (2005) consider a linear model which is the functional version of the quantile regression model of Koenker and Bassett (1978). The functional g_δ is assumed to be linear and continuous, and thus may be written as

$$g_\delta(x) = \int_0^1 \beta(t) x(t) \, dt, \tag{2.30}$$

where β is the slope function belonging to $L^2([0, 1])$, or equivalently

$$Y_i = \int_0^1 \beta(t) X_i(t) \, dt + \epsilon_i, \tag{2.31}$$

where ϵ_i are i.i.d. random variables such that $P(\epsilon_i \leq 0) = \delta$. The variable ϵ_i is also assumed to be independent of X_i. Cardot et al. (2005) show that identifiability of the model follows from the condition $f_\epsilon(0) > 0$, where f_ϵ is the distribution of

ϵ_i. An estimator of β is constructed by expanding the slope function in a finite-dimensional function space and minimizing a penalized loss criterion. Taking the space of spline functions with degree q with k equispaced interior knots in $[0, 1]$, this estimator can be written

$$\widehat{\beta} = \sum_{\ell=1}^{k+q} \widehat{\theta}_\ell B_\ell = {}^t\mathbf{B}_{k,q}\widehat{\boldsymbol{\theta}},$$

with $\widehat{\boldsymbol{\theta}}$ the solution of the minimization problem

$$\min\left\{\frac{1}{n}\sum_{i=1}^{n} l_\beta\left(Y_i - \langle {}^t\mathbf{B}_{k,q}\boldsymbol{\theta}, X_i\rangle\right) + \rho \left\|\left({}^t\mathbf{B}_{k,q}\boldsymbol{\theta}\right)^{(m)}\right\|^2 \,\Big|\, \boldsymbol{\theta} \in \mathbb{R}^{k+q}\right\}. \qquad (2.32)$$

The authors propose an iterative algorithm in order to compute solutions of the minimization problem. Assuming smoothness of the slope function β (existence of derivatives) and some regularity of the distribution of the errors, Cardot et al. (2005) obtain upper bounds for the rates of convergence of the mean square prediction error.

This work, which is to our knowledge all that has been done in a linear setting, deserves further investigation. From an asymptotic point of view, we believe that optimal rates of convergence could be derived in the same way as in the case of linear regression. Note also that quantile regression is often linked to M-estimation and the same features could be developed in a functional setting.

2.5.3 Complementary bibliography

In this chapter, we have made a tour of recent advances in functional linear regression. Being exhaustive is quite impossible nowadays because of the huge number of publications related to this topic. We would like to conclude by discussing briefly some complementary work and addressing some open questions.

In the case of multiple linear regression, the interpretation of the slope coefficient may be of great importance in applications. For instance, the question of no effect of the functional predictor on the real covariate Y is particularly relevant. As a matter of fact Cardot et al. (2003c) address this problem and propose a test of no effect, i.e. of the nullity of the slope function β or, more generally, of the null hypothesis $H_0 : \beta = \beta_0$, where β_0 is some given function (possibly equal to zero). They propose two test statistics based on the norm of the empirical normalized cross-covariance operator Δ_n. While the distribution of the first statistic is approximated by a χ^2 distribution, they state the asymptotic normality of the second statistic. In each case, asymptotic p-values can be obtained. Cardot et al. (2003c) consider the question of testing for the nullity of β as well as the "partial" nullity of β, i.e. its nullity on some subintervals of [0, 1]. For testing global nullity, they introduce a fully automatic permutation test based on the operator Δ_n. Another pseudo-likelihood ratio test procedure is introduced to test nullity of one slope coefficient when there are several functional predictors. In addition, this procedure can be applied to test the nullity of β on subintervals of [0, 1].

Regression diagnostics have also been proposed in this functional context by Chiou and Müller (2007), based on an analysis of the residuals of the regression. Among the various techniques proposed in their work, the analog of Cook's distance is derived, allowing one to determine high-leverage functional predictors.

Looking further into the complexity of the modelization, some authors have recently proposed the study of extensions of the functional linear model. Mas and Pumo (2009) consider a new type of functional linear model which can take into account both the trajectories and their first-order derivatives. Another way to get more flexible models is to transpose nonparametric ideas into this functional context. Cardot and Sarda (2008) consider a varying-coefficient model in which the slope function depends nonparametrically on a real covariate. Antoniadis *et al.* (2006) and, more recently, Ait Saidi *et al.* (2008), look for informative directions in the functional space and estimate nonparametrically the link between the projected functional variables in this reduced-dimension space and the response variable Y. Using similar ideas James and Silverman (2005) study functional additive models, adapting projection pursuit and generalized additive model ideas to a functional context. Their approach is very general; both functional and real covariates can enter the model and effective algorithms are given.

In practice, a crucial problem is the choice of the smoothing parameters that come into the definition of the estimator. We have seen in Section 2.4 that the values of these parameters determine the asymptotic behavior of the estimator. In practical situations, however, one needs to have some data-driven rules for the choice of the value of this tuning parameter. In several applied studies, authors have proposed using generalized cross-validation (GCV) to choose either the dimension k of the space of lower dimension in functional principal component-type estimators, or the penalty parameter ρ for the penalized least squares estimators such as splines. In the latter case the procedure adapts, in the functional context, the generalized cross validation introduced by Wahba (1990). The asymptotic theory for this procedure is, in most cases, an open problem. Nevertheless, Crambes *et al.* (2009) show that the average squared error (ASE) of their smoothing spline estimator, constructed with a smoothing parameter selected by GCV, is asymptotically first-order equivalent to the ASE obtained with the (theoretical) optimal smoothing parameter ρ_{opt}. The latter is defined as the minimizer of the mean square error of prediction. They additionally show that a similar result is also valid when the GCV criterion is used to select the order m of the smoothing splines.

References

Ait Saidi, A., Ferraty, F., Kassa, R., Vieu P. (2008). Cross-validated estimation in the single functional index model. *Statistics*, 42, 475–94.

Antoniadis, A., Amato, U., De Feiss, I. (2006). Dimension reduction in functional regression with applications. *Computational Statistics & Data Analysis*, 50, 2422–46.

Cai, T.T., Hall, P. (2006). Prediction in functional linear regression. *Ann. Statist.*, 34, 2159–79.

Candès, E., Tao, T. (2007). The Dantzig selector: statistical estimation when p is much larger than n. *Ann. Statist.*, 35, 2313–51.

Cardot, H., Crambes, C., Sarda, P. (1999a). Quantile regression when the covariates are functions. *J. Nonparam. Statist.*, 17, 841–56.

Cardot, H., Faivre, R., Goulard, M. (2003a). Functional approaches for predicting land use with the temporal evolution of coarse resolution remote sensing data. *Journal of Applied Statistics*, 30, 1185–99.

Cardot, H., Ferraty, F., Mas, A., Sarda, P. (2003c). Testing hypothesis in the functional linear model. *Scand. J. Statist.*, 30, 241–55.

Cardot, H., Ferraty, F., Sarda, P. (1999b). Functional linear model. *Statist. & Prob. Letters*, 45, 11–22.

Cardot, H., Ferraty, F., Sarda, P. (2003b). Spline estimators for the functional linear model. *Statistica Sinica*, 13, 571–91.

Cardot, H., Johannes, J. (2010). Thresholding projection estimators in functional linear models. *J. Multivariate Analysis*, 101, 395–408.

Cardot, H., Mas, A., Sarda, P. (2007). CLT in functional linear regression models. *Probab. Theory Related Fields*, 138, 325–61.

Cardot, H., Sarda, P. (2005). Estimation in generalized linear models for functional data via penalized likelihood. *J. Multivariate Anal.*, 92, 24–41.

Cardot, H., Sarda, P. (2008). Varying-coefficient functional linear regression models. *Communications in Statistics: Theory and Methods*, 37, 3186–203.

Chiou, J., Müller, H.G. (2007). Diagnostics for functional regression via residual processes. *Computational Statistics & Data Analysis*, 51, 4849–63.

Chiou, J., Müller, H.G., Wang, J.-L. (2004). Functional response models. *Statistica Sinica*, 14, 659–77.

Cook, R.D. (2007). Dimension reduction in regression (with discussion). *Statistical Science*, 22, 1–43.

Crambes, C., Kneip, A., Sarda, P. (2009). Smoothing splines estimators in functional linear regression. *Ann. Statist.*, 37, 35–72.

Cuevas, A., Febrero, M., Fraiman, R. (2002). Linear functional regression: the case of fixed design and functional response. *Canad. J. Statist.*, 30, 285–300.

de Boor, C. (1978). *A Practical Guide to Splines*. Springer, New York.

Engl, H.W., Hanke, M., Neubauer, A. (2000). *Regularization of Inverse Problems*. Kluwer Academic, Dordrecht.

Escabias, M., Aguilera, A.M., Valderrama, M.J. (2004). Principal component estimation of functional logistic regression: discussion of two different approaches. *Nonparametric Statistics*, 16, 365–84.

ESCABIAS, M., AGUILERA, A.M., VALDERRAMA, M.J. (2007). Functional PLS logit regression model. *Comp. Statist. Data Analysis*, 51, 4891–902.

FARAWAY, J. (1997). Regression analysis for a functional response. *Technometrics*, 39, 254–61.

FRANK, I.E., FRIEDMAN, J.H. (1993). A statistical view of some chemometrics regression tools. *Technometrics*, 35, 109–48.

GOUTIS, C. (1998). Second-derivative functional regression with applications to near infrared spectroscopy. *Journal of the Royal Statistical Society, B*, 60, 103–14.

HALL, P., HOROWITZ, J.L. (2007). Methodology and convergence rates for functional linear regression. *Ann. Statist.*, 35, 70–91.

HALL, P., HOSSEINI-NASAB, M. (2006). On properties of functional principal component analysis. *J. R. Statist. Soc. B*, 68, 109–26.

HALL, P., HOSSEINI-NASAB, M. (2009). Theory for high-order bounds in functional principal components analysis. *Math. Proc. Comb. Phil. Soc.*, 146, 225–56.

HALL P., MÜLLER H.G., WANG, J.L. (2006). Properties of principal component methods for functional and longitudinal data analysis. *Annals of Statistics*, 34, 1493–517.

HASTIE, T.J., MALLOWS, C. (1993). A discussion of "A statistical view of some chemometrics regression tools" by I.E. Frank and J.H. Friedman. *Technometrics*, 35, 140–3.

HOCKING, R.R. (1976). The analysis and selection of variables in linear regression. *Biometrics*, 32, 1–49.

JAMES, G. (2002). Generalized linear models with functional predictor variables. *Journal of the Royal Statistical Society, B.*, 64, 411–32.

JAMES, G., HASTIE, T., SUGAR, C. (2000). Principal component models for sparse functional data. *Biometrika*, 87, 587–602.

JAMES, G., SILVERMAN, B.W. (2005). Functional adaptive model estimation. *J. Amer. Statist. Assoc.*, 100, 565–76.

KOENKER, R., BASSETT, G. (1978). Regression quantiles. *Econometrica*, 46, 33–50.

KRESS, R. (1989). *Linear Integral Equations*. Springer, New York.

LI, Y., HSING, T. (2007). On rates of convergence in functional linear regression. *J. Multivariate Anal.*, 98, 1782–804.

MARX, B.D., EILERS, P.H. (1996). Generalized linear regression on sampled signals with penalized likelihood. In *Statistical Modelling, Proceedings of the Eleventh International Workshop on Statistical Modelling* (Forcina, A., Marchetti, G.M., Hatzinger, R., Galmacci, G., eds). Orvietto.

MARX, B.D., EILERS, P.H. (1999). Generalized linear regression on sampled signals and curves: a P-spline approach. *Technometrics*, 41, 1–13.

MAS, A. (2007). Weak convergence in the functional autoregressive mode. *J. Multivariate Anal.*, 98, 1231–61.

MAS A., PUMO, B. (2009). Functional linear regression with derivatives. *Journal of Nonparametric Statistics*, 21, 19–40.

MÜLLER, H.-G., STADTMÜLLER, U. (2005). Generalized functional linear models. *Ann. Statist.*, 33, 774–805.

PREDA, C., SAPORTA, G. (2005). PLS regression on a stochastic process. *Computational Statistics & Data Analysis*, 48, 149–58.

RAMSAY, J.O., DALZELL, C.J. (1991). Some tools for functional data analysis (with discussion). *Journal of the Royal Statistical Society, B*, 53, 539–72.

RAMSAY, J.O., SILVERMAN, B.W. (2002). *Applied Functional Data Analysis: Methods and Case Studies.* Springer, New York.

RAMSAY, J.O., SILVERMAN, B.W. (2005). *Functional Data Analysis.* Springer, New York.

REISS, P.T., OGDEN, R.T. (2007). Functional principal component regression and functional partial least squares. *J. Amer. Statist. Assoc.*, **102**, 984–96.

SMALE, S., ZHOU, D. (2007). Learning theory estimates via integral operators and their approximations. *Constructive Approximation*, **26**, 153–72.

STONE, C.J. (1986). The dimensionality reduction principle for generalized additive models. *Ann. Statist.* **14**, 590–606.

WAHBA, G. (1990). *Spline Models for Observational Data.* SIAM, Philadelphia.

YAO, F., MÜLLER, H.-G., WANG, J.-L. (2005a). Functional data analysis for sparse longitudinal data. *J. Amer. Statist. Assoc.*, **100**, 577–90.

YAO, F., MÜLLER, H.-G., WANG, J.-L. (2005b). Functional linear regression analysis for longitudinal data. *Ann. Statist.*, **37**, 2873–903.

CHAPTER 3

LINEAR PROCESSES FOR FUNCTIONAL DATA

ANDRÉ MAS

BESNIK PUMO

3.1 INTRODUCTION

LINEAR processes on functional spaces were born about fifteen years ago, and this topic had the same fast development as other areas of functional data modeling such as PCA or regression. These linear processes aim at generalizing to random curves the classical ARMA (auto regressive moving average) models widely known in time-series analysis. They offer a wide spectrum of models suited to statistical inference for continuous-time stochastic processes within the paradigm of functional data. Essentially designed to improve the quality and the range of prediction, they give birth to challenging theoretical and applied problems.

The aim of this chapter is twofold. First of all, we want to provide the reader with the basic theory and applications of linear processes for functional data. The second goal consists in providing an overview of the state-of-the-art to complement the monograph by Bosq (2000). Many crucial theorems are given in this book, to which we will frequently refer. Consequently, even if this chapter is self-contained, we will pay special attention to recent results published from 2000 to 2008, and try to draw the lines of future and promising research in this area.

It is worth recalling now the approach that leads to modeling and inference from curve data. We begin with a continuous-time stochastic process $(\xi_t)_{t\geq 0}$. The paths of ξ are divided into equally-spaced pieces of trajectory. Each of these pieces is then viewed as a random curve. With mathematical symbols we write:

$$X_k(t) = \xi_{kT+t}, \quad 0 \leq t \leq T,$$

where T is fixed. The function $X_k(\cdot)$ maps $[0, T]$ to $\mathbb{R}$ and is random. Observing ξ over $[0, nT]$ produces an n-sample $X_1, \ldots, X_n$. Obviously the choice of T is crucial; it is usually left to the practitioner and may be linked with seasonality (with period T). Dependence along the paths of ξ will create dependence between the X_i's. However, this approach is not restricted to the whole path of the stochastic process. One could just as well imagine modeling whole curves observed at discrete intervals: the interest-rate curve at day k, $I_k(\delta)$, is for instance a function linking duration δ (as an input) to the associated interest rates (as outputs). Observing these curves, whose random variations will depend on the financial markets, for n days produces a sample similar in nature to the one described above, although there is no underlying continuous-time process in this situation, rather a surface $(k, \delta, I_k(\delta))$. We refer, for example, to Kargin and Onatski (2008) for an illustration.

We will then propose some statistical models here that mimic or adapt the scalar or finite-dimensional approaches for time series (see Brockwell and Davis (1991)). Each of these (random or not) functions will be viewed as a vector in a vector space of functions. This paradigm has been adopted for a long time in probability theory as can be seen in, for instance, Ledoux and Talagrand (1991) and references therein. But the first book entirely dedicated to the formal and applied aspects of statistical inference in this setting is certainly that of Ramsay and Silverman (1997), followed by Bosq (2000), Ramsay and Silverman again (2002), then Ferraty and Vieu (2006).

In what follows, we will consider centered processes with values in a Hilbert space of functions denoted by H with inner product $\langle \cdot, \cdot \rangle$ and norm $\|\cdot\|$. The Banach setting, although more general, has several drawbacks. The reasons for focusing on Hilbert spaces are both theoretic and practical. First, many fundamental asymptotic theorems may be stated under simple assumptions in this setting. The central limit theorem (CLT) is a good example. Considering random variables with values in $C([0, 1])$ or in Hölder spaces, for instance, leads to very specific assumptions in order to obtain the CLT and computations are often difficult, whereas in a Hilbert space, moment conditions are both necessary and sufficient. The nice geometric features of Hilbert spaces allow us to consider denumerable bases, projections, etc. in a framework that generalizes the Euclidean space with few drawbacks. Besides, in practice, recovering curves from discretized observations by interpolation or smoothing techniques such as splines or wavelets yields functions in Sobolev spaces, say $W^{m,2}$ (here m is an order of differentiation connected with the desired smoothness of the output), that are all Hilbert spaces. We refer to Ziemer (1989) or to Adams and Fournier (2003) for monographs on Sobolev spaces.

In statistical models, unknown parameters will be functions or linear operators (the counterpart of matrices in Euclidean space), the latter being of particular interest. We now give some basic facts about operators which will be of great use in what follows.

Several monographs are dedicated to operator theory, which is a major theme within mathematical science. Classical references here are Dunford and Schwartz (1988) and Gohberg *et al.* (1991). The adjoint of the operator T is classically denoted T^*. The Banach space of compact operators $\mathcal{C}$ on a Hilbert space H is separable when endowed with the classical operator norm $\|\cdot\|_\infty$:

$$\|T\|_\infty = \sup_{x \in \mathcal{B}_1} \|Tx\|,$$

where $\mathcal{B}_1$ denotes the unit ball of the Hilbert space H. The space $\mathcal{C}$ contains the set of Hilbert–Schmidt operators, which is a Hilbert space and is denoted by $\mathcal{S}$. Let T and S belong to $\mathcal{S}$. The inner product of T and S, and the norm of T, are respectively defined by:

$$\langle S, T \rangle_\mathcal{S} = \sum_p \langle Se_p, Te_p \rangle,$$

$$\|T\|_\mathcal{S}^2 = \sum_p \|Te_p\|^2,$$

where $(e_p)_{p \in \mathbb{N}}$ is a complete orthonormal system (c.o.n.s.) in H. The inner product and the norm defined above do not depend on the choice of the c.o.n.s. $(e_p)_{p \in \mathbb{N}}$. Nuclear (or trace-class) operators are another important family of operators for which:

$$\sum_p \|Te_p\| < +\infty.$$

It is evident that a trace-class operator is also Hilbert–Schmidt. Many of the asymptotic results mentioned from now on that involve random operators are usually obtained for the Hilbert–Schmidt norm, unless explicitly stated otherwise. It should be noted as well that this norm is thinner than the usual operator norm.

The next section is devoted to general linear processes. Then we focus on the autoregressive model and its recent advances, which are developed in the third section. We conclude with some issues for future work.

3.2 GENERAL LINEAR PROCESSES

Linear processes on function spaces generalize the classical scalar or vector linear processes to random elements which are curves or functions and are more generally valued in an infinite-dimensional separable Hilbert space H.

Definition 3.1 *Let $(\epsilon_k)_{k\in\mathbb{N}}$ be a sequence of i.i.d. centered random elements in H, let $(a_k)_{k\in\mathbb{N}}$ be a sequence of bounded linear operators from H to H such that $a_0 = I$, and let $\mu \in H$ be a fixed vector. If*

$$X_n = \mu + \sum_{j=0}^{+\infty} a_j \left(\epsilon_{n-j}\right), \tag{3.1}$$

$(X_n)_{n\in\mathbb{N}}$ is a linear process on H (written from now on an "H-linear process") with mean μ.

Unless explicitly mentioned, the mean function μ will always be assumed to be null (and the process X to be centered). It seems that, following a series of paper on this model dating back to the late 1990s-early 2000s, this model now creates less inspiration in the research community. We guess that the recent paper by Bosq and Blanke (2007) and the book by Bosq and Blanke (2007) may introduce some fresh ideas here. We now state some basic facts: the invertibility and the convergence of the estimated moments.

3.2.1 Invertibility

When the sequence ϵ is a strong H-white noise, that is, a sequence of i.i.d. random elements such that $\mathbb{E} \|\epsilon\|^2 < +\infty$, and whenever

$$\sum_{j=0}^{+\infty} \|a_j\|_\infty^2 < +\infty, \tag{3.2}$$

the series defining the process $(X_n)_{n\in\mathbb{N}}$ given in (3.1) converges under the square norm and almost surely by the a-1 law. The strict stationarity of X_n is also ensured. The problem of invertibility is addressed in Merlevède (1995).

Theorem 3.1 *If $(X_n)_{n\in\mathbb{N}}$ is 0 linear process with values in H defined by (3.1) and such that*

$$1 - \sum_{j=1}^{+\infty} z^j \|a_j\|_\infty \neq 0 \quad \text{for } |z| < 1, \tag{3.3}$$

then $(X_n)_{n \in \mathbb{N}}$ is invertible:

$$X_n = \epsilon_n + \sum_{j=1}^{+\infty} \rho_j (X_{n-j}),$$

where all the ρ_j's are bounded linear operators in H with $\sum_{j=1}^{+\infty} \|\rho_j\|_\infty < +\infty$ and the series converges in mean square and almost surely.

Remark 3.1 *We deduce from this that ϵ_n is the innovation of the process X and that (3.1) coincides with the Wold decomposition of X.*

We now give some convergence theorems for the mean and the covariance of Hilbert-valued linear processes. These probabilistic results are well known but essential for statistical inference.

3.2.2 Asymptotics

It is worth mentioning a general method for proving asymptotic results for linear processes. If several approaches are possible, it turns out that (in the authors' opinion) one of the most fruitful relies on approximating the process X_n by truncated versions such as:

$$X_{n,m} = \sum_{j=0}^{m} a_j (\epsilon_{n-j}),$$

where $m \in \mathbb{N}$. The sequence $X_{n,m}$ is for fixed m, blockwise independent: $X_{n+m+1,m}$ is indeed stochastically independent of $X_{n,m}$ if the ϵ_j's are. The outline of the proofs usually consists in proving asymptotic results for the m-dependent sequence $X_{n,m}$, then letting m tend to infinity with an accurate control of the residual $X_n - X_{n,m} = \sum_{j=m+1}^{+\infty} a_j (\epsilon_{n-j})$.

The mean

Asymptotic results for the mean of a linear process may be found in Merlevède (1996) and Merlevède et al. (1997). Even if the first of these articles is in a way more general, we focus here on the second since it deals directly with the mean of the non-causal process indexed by $\mathbb{Z}$:

$$X_k = \sum_{j=-\infty}^{+\infty} a_j (\epsilon_{k-j}).$$

The authors obtain sharp conditions for the CLT of $S_n = \sum_{k=1}^{n} X_k$.

Theorem 3.2 *Let* $(a_j)_{j \in \mathbb{Z}}$ *be a sequence of operators such that:*

$$\sum_{j=-\infty}^{+\infty} \|a_j\|_\infty^2 < +\infty.$$

Then

$$\frac{S_n}{\sqrt{n}} \to_w N(0, A\Gamma_\epsilon A^*),$$

where $N(0, A\Gamma_\epsilon A^*)$ *is the H-valued centered Gaussian random element with covariance operator* $A\Gamma_\epsilon A^*$, *where* $\Gamma_\epsilon = \mathbb{E}(\epsilon_0 \otimes \epsilon_0)$ *is the covariance operator of* ϵ_0 *and* $A = \sum_{j=-\infty}^{+\infty} a_j$.

We recall that if u and v are two vectors in H then the notation $u \otimes v$ stands for the rank-one linear operator from H to H defined by $(u \otimes v)(x) = \langle v, x \rangle u$.

This result is extended, under additional assumptions, to the case of strongly mixing ϵ_k's. Note that the problem of weak convergence for the mean of stationary Hilbertian processes under mixing conditions has been addressed in Maltsev and Ostrovskii (1982). A standard equi-integrability argument and classical techniques provide the following rates of convergence for S_n. Nazarova (2000) has proved the same sort of theorem when X is a linear random field with values in a Hilbert space. Now we turn to the rate of convergence of the empirical mean in quadratic mean and almost surely (a.s.). The following theorem may be found in Bosq (2000):

Proposition 3.1 *Let* $X_k = \sum_{k=0}^{+\infty} a_j(\epsilon_{k-j})$ *and* $S_n = \sum_{k=1}^n X_k$, *then*

$$n\mathbb{E}\left\|\frac{S_n}{n}\right\|^2 \to \sum_{k=-\infty}^{+\infty} \mathbb{E}\langle X_0, X_k \rangle,$$

$$\frac{n^{1/4}}{(\log n)^{1/2+\epsilon}} \left\|\frac{S_n}{n}\right\| \to 0 \quad a.s.$$

for all $\epsilon > 0$.

We now turn to covariance operators.

Covariance operators

The situation is slightly more complicated than that for the mean, due to the presence of tensor product.

Definition 3.2 *The theoretical covariance operator at lag* $h \in \mathbb{N}$ *of a process X is defined by:*

$$\Gamma_h = \mathbb{E}(X_h \otimes X_0).$$

The linear operator Γ_h *is nuclear on H when the second-order strong moments of X are convergent. Its empirical counterpart based on the sample is:*

$$\Gamma_{n,h} = \frac{1}{n} \sum_{t=1}^{n} X_{t+h} \otimes X_t.$$

The covariance operator of the process $\Gamma_0 = \Gamma$ is selfadjoint, positive, and nuclear. Consequently Γ is Hilbert–Schmidt and compact.

It should be noted that Γ_h is not in general a symmetric operator, unlike in the case of the classical covariance operator Γ_0. Weak convergence of covariance operators for H-linear processes has been addressed by Mas (2002). It is assumed that:

$$\mathbb{E} \|\epsilon_0\|^4 < +\infty$$

$$\sum_{k=-\infty}^{+\infty} \|a_k\|_\infty < +\infty,$$

then the vector of the h covariance operators up to any fixed lag h is asymptotically Gaussian under the Hilbert–Schmidt norm.

Theorem 3.3 *Let us consider the following linear and Hilbert-space-valued process:*

$$X_t = \sum_{j=-\infty}^{+\infty} a_j \left(\epsilon_{t-j}\right),$$

then

$$\sqrt{n} \begin{pmatrix} \Gamma_{n,0} - \Gamma_0 \\ \Gamma_{n,1} - \Gamma_1 \\ \vdots \\ \Gamma_{n,h} - \Gamma_h \end{pmatrix} \xrightarrow[n \to +\infty]{w} G_\Gamma,$$

where $G_\Gamma = \left(G_\Gamma^{(0)}, \ldots, G_\Gamma^{(h)}\right)$ is a Gaussian centered random element with values in $\mathcal{S}^{h+1}$. Its covariance operator is

$$\Theta_\Gamma = \left(\Theta_\Gamma^{(p,q)}\right)_{0 \leq p,q \leq h},$$

which is a nuclear operator in $\mathcal{S}^{h+1}$ defined blockwise for all T in $\mathcal{S}$ by

$$\Theta_\Gamma^{(p,q)} (T) = \sum_h \Gamma_{h+p-q} T \Gamma_h + \sum_h \Gamma_{h+q} T \Gamma_{h-p} + A_q \left(\Lambda - \Phi\right) A_p (T), \quad (3.4)$$

where Λ, Φ, and A_p are linear operators from $\mathcal{S}$ to $\mathcal{S}$ defined respectively by

$$\Lambda (T) = \mathbb{E} \left((\epsilon_0 \otimes \epsilon_0) \tilde{\otimes} (\epsilon_0 \otimes \epsilon_0)\right) (T)$$

$$\Phi (T) = C \left(T + T^*\right) C + \left(C \tilde{\otimes} C\right) (T)$$

$$A_p (T) = \sum_i a_{i+p} T a_i^*.$$

As by-products, weak convergence results for the eigenelements of $\Gamma_{n,0} - \Gamma_0$, that is, for the PCA of the stationary process X, can be derived. The reader interested in these developments should refer to Mas and Menneteau (2003a), which sets out a general method for deriving asymptotics for the eigenvalues and eigenvectors of $\Gamma_{n,0}$ (the by-products of the functional PCA) from the covariance sequence itself. Perturbation theory is the main tool, using a modified delta-method.

3.2.3 Perspectives and trends: towards generalized linear processes?

It turns out that the literature on inference methods for general linear processes is rather meager. Obviously, simultaneously estimating many a_j's would seem to be complicated and it is not necessarily needed since the functional AR process, which will be described in the next section, is quite successful and easier to handle. However, general linear processes provide the starting points for very interesting theoretical problems where dependence plays a key role. We finally mention the paper by Dedecker and Merlevède (2003), especially Section 2.4 which is dedicated to proving a conditional central limit theorem for linear processes under mild assumptions, and also Dedecker and Merlevède (2007), whose Section 3.3 deals with rates in the law of large numbers. This research may provide theoretical material which will enable us to go deeper into the asymptotic study of these processes.

In a recent article, Bosq (2007) introduces the notion of linear process in a broad sense. The definition remains essentially the same as in (3.1), but the operators $(a_j)_{j \in \mathbb{N}}$ may now be unbounded, which finally generalizes the notion. A key role is played by linear closed spaces (LCS) which were introduced by Fortet (1995). A LCS $\mathcal{G}$ is a subspace of L_H^2, the space of random variables with values in H and finite strong second moment, such that:

(i) $\mathcal{G}$ is closed in H;
(ii) if $X \in \mathcal{G}, l(X) \in \mathcal{G}$ for all bounded linear operators l.

This theory, involving projection onto LCS, weak and strong orthogonality, and dominance of operators, allows Bosq to revisit and extend the notions of linear process, Wold decomposition, and Markovian process when the bounded $(a_j)_{j \in \mathbb{N}}$ may be replaced with measurable mappings $(l_j)_{j \in \mathbb{N}}$. Several examples are given: derivatives of functional processes such as in the MAH $X_n = \epsilon_n + c\epsilon'_n$, arrays of linear processes, truncated Ornstein–Uhlenbeck processes, ... The personal communication Bosq (2010) discusses these extensions to tensor products of linear processes and will certainly shed new light on their covariance structure. We also refer to Chapters 10 and 11 in the book by Bosq and Blanke (2007) for an explanation of these concepts.

3.3 AUTOREGRESSIVE PROCESSES

3.3.1 Introduction

Here the model generalizes the classical autoregressive process AR(1) for scalar or multivariate time series to functional data. It was introduced for the first time by Bosq (1991). Let $X_1, \ldots, X_n$ be a sample of random curves for which a stochastic dependence is suspected (for instance, the curve of temperature observed over n days at a given place). We assume that all the X_i's take values in a Hilbert space H and set:

$$X_n = \rho(X_{n-1}) + \epsilon_n, \qquad (3.5)$$

where ρ is a linear operator from H to H and $(\epsilon_n)_{n \in \mathbb{N}}$ is a sequence of H-valued centered random elements, usually with a common covariance operator. The model is simple, with a single unknown operator; however, it allows us to retain the possibility of getting rid of various assumptions either on the operator ρ (linear, compact, Hilbert–Schmidt, symmetric or not, etc.) or on the dependence between the ϵ_n's. The latter are quite often independent and identically distributed (i.i.d.) but alternatives are possible (such as mixing or, more naturally, martingale differences). Bosq (2000) has proved that assumption (3.2) actually comes down to the existence of $a > 0$, $b \in [0, 1]$ such that, for all $p \in \mathbb{N}$:

$$\|\rho^p\|_\infty \leq a b^p,$$

which ensures that (3.5) admits a unique stationary solution. The process $(X_n)_{n \in \mathbb{N}}$ is Markov as soon as $\mathbb{E}(\epsilon_n | X_{n-1}, \ldots, X_1) = 0$ and is called an ARH (autoregressive Hilbertian) process of order 1. As often noted, the interest of the model lies in its predictive power. The estimation of ρ is usually the first and necessary step before the derivation of the statistical predictor, given the new input $X_{n+1}: \widehat{\rho}(X_n)$. The prediction is often compared with an ARMA model or with nonparametric smoothing techniques. The global treatment of the trajectory as a function often ensures better long-term prediction, but this comes with a cost: more tedious numerical procedures.

Remark 3.2 *Obviously this model falls within the range of linear models for functional data and hence may be considered as a parametric model in the paradigm of Chapter 1.*

Yet it differs in several aspects from the linear regression models described in Chapter 2: the data are obviously dependent and the unknown parameter is an operator and not a function. These are not minor changes.

Representation of stochastic processes by functional AR

Various real-valued processes allow for the ARH representation. In Figure 3.1 we have plotted the graphs of two simulated processes, the Ornstein–Uhlenbeck (O–U) process and the Wong process. The O–U process ($\eta_t, t \in R$) is a real stationary Gaussian process:

$$\eta_t = \int_{-\infty}^{t} e^{-a(t-u)} dw_u, \ t \in R,$$

where $(w_t)_{t \in R}$ is a bilateral standard Wiener process and a is a positive constant. Bosq (1996) gives the ARH representation $X_n = \rho(X_{n-1}) + \epsilon_n$ with values in $L^2 := L^2[0,1]$, where $X_n(t) = \eta_{n+t}, t \in [0,1], n \in Z$, and ρ is a degenerate linear operator

$$\rho(x)(t) = e^{-at} x(1), \ t \in [0,1], x \in L^2$$

and

$$\epsilon_n(t) = \int_{n}^{n+t} e^{-a(n+t-s)} dw_s, \ t \in [0,1], n \in Z.$$

The Wong process is a mean-square differentiable stationary Gaussian process which is zero-mean and is defined for $t \in R$ by:

$$\xi_t = \sqrt{3} \exp\left(-\sqrt{3}\, t\right) \int_0^{\exp(2t/\sqrt{3})} w_u\, du.$$

Cutting R into intervals of length 1 and defining $X_n(t) = \xi_{n+t}$ for $t \in [0,1]$, Mas and Pumo (2007) obtain an ARH representation, $X_n = A(X_{n-1}) + \epsilon_n$, of this process with values in the Sobolev space $W := W^{2,1} = \{u \in L^2, u' \in L^2\}$ and

$$\epsilon_n(t) = \sqrt{3}\, \exp[-\sqrt{3}\,(n-1+t)] \int_{\exp[2(n-1)/\sqrt{3}]}^{\exp[2(n-1+t)/\sqrt{3}]} \left[w_u - w_{\exp[2(n-1)/\sqrt{3}]}\right] du$$

for $t \in [0,1]$. The linear and degenerate operator A is given by $\phi + \Psi(D)$, where D is the ordinary differential operator and

$$[\phi(f)](t) = [\exp(-\sqrt{3}t) + \sqrt{3}c(t)] f(1), \quad [\Psi(D)(f)](t) = c(t) f'(1),$$

with $c(t) = \frac{\sqrt{3}}{2} \exp(-\sqrt{3}t) \cdot \{\exp(2t/\sqrt{3}) - 1\}$.

Other examples are given in the paper by Bosq (1996) or in the classic book Bosq (2000).

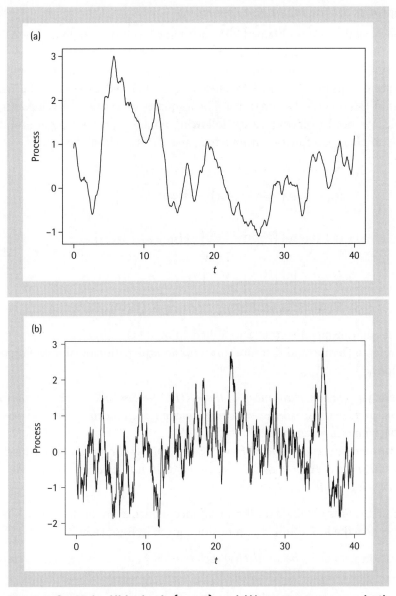

Fig. 3.1 Ornstein–Uhlenbeck ($a = 1$) and Wong processes are both evaluated at instants $t_i = 0.02 * i$.

A major area for development here is the investigation of deeper links between autoregressive functional processes and diffusion processes or stochastic differential equations with more general autocorrelation operators. But this remains an open question; the work by Ramsay (2000) on this topic certainly deserves more attention in order to extend it.

Asymptotics for the mean and covariance

Obviously all the results obtained for general linear processes hold for the ARH(1): namely, for the mean and the covariance operator. Some new results are stated below (they are new essentially with respect to Bosq (2000)) and are related to moderate deviations (Mas and Menneteau (2003b)) and to laws of the iterated logarithm (Menneteau (2003a)). Let η be a square-integrable real-valued random variable. We need to introduce the following notation (I_X and J_X are functions from H to H, J_Γ is a function from S to S, and K_Γ is a subset of S):

$$u_1 = \rho(X_0) \otimes \epsilon_1 + \epsilon_1 \otimes \rho(X_0) + \epsilon_1 \otimes \epsilon_1 - \Gamma_\epsilon, \tag{3.6}$$

$$I_X(x) = \sup_{h \in H} \{\langle h, x - \rho(x)\rangle - \mathbb{E}\exp\langle h, \epsilon_1\rangle\},$$

$$J_X(x) = \frac{1}{2}\inf\{\mathbb{E}\eta^2 : x = \mathbb{E}[\eta(I_H - \rho)^{-1}(\epsilon_1)]\},$$

$$J_\Gamma(s) = \frac{1}{2}\inf\{\mathbb{E}\eta^2 : s = \mathbb{E}[\eta(I_S - R)^{-1}(u_1)]\},$$

$$K_\Gamma = \{\mathbb{E}[\eta(I_S - R)^{-1}(u_1)] : \eta \in L^2(P), \mathbb{E}\eta^2 \leq 1\},$$

and R is a linear operator from S to S defined by $R(s) = \rho s \rho^*$.

We refer to Dembo and Zeitouni (1993) for an explanation of large and moderate deviations.

Theorem 3.4 *The empirical mean of the ARH(1) process follows the large deviation principle in H with speed n^{-1} and rate function I_X, and the moderate deviation principle with rate function J_X.*
The covariance sequence of the ARH(1), $\Gamma_n - \Gamma$, follows the moderate deviation principle in the space of Hilbert–Schmidt operators with rate function J_Γ, and the law of the iterated logarithm with limit set K_Γ.

The first results obtained on the covariance sequence are in Bosq (1991), but we mention here the interesting decomposition given in Bosq (1999).

Proposition 3.2 *Let X_n be an ARH(1) such that $\mathbb{E}\|X_0\|^4 < +\infty$, then the tensorized process*

$$Z_i = X_i \otimes X_i - \Gamma$$

is an autoregressive process with values in S such that:

$$Z_i = R(Z_{i-1}) + u_i,$$

where $R(S) = \rho S \rho^*$ and u_1 was defined at (3.6). The sequence u_i is a martingale difference sequence with respect to the filtration $\sigma(\epsilon_i, \epsilon_{i-1}, \ldots)$.

Remark 3.3 *The methods of proofs for the mean or covariance of linear functional processes in general (either ARH or linear) used above are based on truncation. The* truncated process is then applied to general asymptotics for m-dependent sequences. The residual term is controlled by assumptions on the norm of the generating operators. These techniques are free from mixing conditions (see Chapter 5 for inference methods for dependent sequences).

3.3.2 Two issues related to the general estimation problem

Identifiability

The moment method provides the following normal equation:

$$\Delta = \rho\Gamma, \tag{3.7}$$

where

$$\Gamma = \mathbb{E}\left(X_1 \otimes X_1\right),$$
$$\Delta = \mathbb{E}\left(X_2 \otimes X_1\right)$$

are the covariance operator (respectively, the cross-covariance operator of order one) of the process $(X_n)_{n\in\mathbb{Z}}$.

The first step here consists in checking that the Yule–Walker equation (3.7) correctly defines the unknown parameter ρ.

Proposition 3.3 *When inference on ρ is based on the moment equation (3.7), identifiability holds if* $\ker \Gamma = \{0\}$.

The proof of this proposition is clear since, taking $\tilde{\rho} = \rho + u \otimes v$ where v belongs to the kernel of Γ, whenever this set is nonempty we see that

$$\tilde{\rho}\Gamma = \rho\Gamma + u \otimes \Gamma v = \rho\Gamma$$

and hence that (3.7) holds for $\tilde{\rho} \neq \rho$.

Consequently, the injectivity of Γ is a basic assumption which would be difficult to remove and which requires that the eigenvalues of Γ are infinite and strictly positive. These eigenvalues will be denoted $(\lambda_i)_{i\in\mathbb{N}}$, where we assume once and for all that the λ_i's are arranged in decreasing order with $\sum_{i\in\mathbb{N}} \lambda_i$ finite. The corresponding eigenvectors (respectively, eigenprojectors) will be denoted $(e_i)_{i\in\mathbb{N}}$ (respectively, $(\pi_i)_{i\in\mathbb{N}}$ where $\pi_i = e_i \otimes e_i$). Heuristically, with (3.7) at hand we should expect that Γ^{-1} exists in order to estimate ρ; this inverse will not be defined if Γ is not one-to-one.

The inverse problem

Even if identifiability is ensured, estimating ρ is a difficult task because of an underlying inverse problem which stems from equation (3.7). The notion of inverse (or ill-posed) problem is classical in mathematical analysis (see, for instance Tikhonov and Arsenin (1977) or Groetsch (1993)). In our framework this problem can be explained by claiming that equation (3.7) implies that any attempt to estimate ρ will result in a highly unstable estimate. In simple terms, which will be developed below, this results from the inversion of Γ. A canonical example of an inverse problem is the numerical inversion of an ill-conditioned matrix (that is, a matrix with eigenvalues close to zero).

The first stumbling block comes from the fact that we cannot deduce from (3.7) that $\Delta \Gamma^{-1} = \rho$. We know that a sufficient condition for Γ^{-1} to be defined as a linear mapping is that $\ker \Gamma = \{0\}$. Then Γ^{-1} is an unbounded symmetric operator on H. Some consequences of this are gathered together in the next proposition:

Proposition 3.4 *When Γ is injective, Γ^{-1} may be defined. It is a measurable linear mapping defined on a domain in H, denoted $\mathcal{D}\left(\Gamma^{-1}\right)$ and defined by:*

$$\mathcal{D}\left(\Gamma^{-1}\right) = \operatorname{Im} \Gamma = \left\{ x = \sum_{p=1}^{+\infty} x_p e_p \in H, \sum_{p=1}^{+\infty} \frac{x_p^2}{\lambda_p^2} < +\infty \right\}.$$

This domain is dense in H. It is not an open set for the norm topology of H. The operator is unbounded, which means that it is continuous at no point of $\mathcal{D}\left(\Gamma^{-1}\right)$. In addition, $\Gamma^{-1}\Gamma = I_H$ but $\Gamma\Gamma^{-1} = I_{|\mathcal{D}(\Gamma^{-1})}$ and $\Gamma\Gamma^{-1}$, which is not defined on the whole of H, may be continuously extended to H.

For similar reasons (3.7) implies $\Delta \Gamma^{-1} = \rho_{|\operatorname{Im}\Gamma} \neq \rho$ and $\Delta \Gamma^{-1}$ may be continuously and formally extended to the whole of H. In fact Γ, and hence Γ^{-1}, are unknown. However if Γ were totally accessible, we could find a way to regularize the odd mathematical object that is Γ^{-1}. In the literature on inverse problems (see, for instance, Groetsch (1993)) one often replaces Γ^{-1} by a linear operator "close" to it but endowed with additional regularity (continuity/boundedness) properties, say $\Gamma^{\dagger}$. The Moore–Penrose pseudo-inverse is an example of such an operator but many other techniques exist. Indeed, starting from

$$\Gamma^{-1}(x) = \sum_{l \in \mathbb{N}} \frac{1}{\lambda_l} \pi_l(x)$$

for all x in $\mathcal{D}\left(\Gamma^{-1}\right)$ one may set, for instance:

$$\Gamma^{\dagger}(x) = \sum_{l \leq k_n} \frac{1}{\lambda_l} \pi_l(x) \tag{3.8}$$

$$\Gamma^{\dagger}(x) = \sum_{l} \frac{1}{\lambda_l + a_n} \pi_l(x) \tag{3.9}$$

$$\Gamma^{\dagger}(x) = \sum_{l} \frac{\lambda_l}{\lambda_l^2 + a_n} \pi_l(x), \tag{3.10}$$

where k_n is an increasing and unbounded sequence of integers and a_n is a sequence of positive real numbers decreasing to 0. The three operators in the display above are indexed by n, and they are all bounded with increasing norm and are known as the spectral cut-off, penalized, and Tikhonov regularized inverses of Γ. They share the following pointwise convergence property:

$$\Gamma^{\dagger} x \to \Gamma^{-1} x$$

for all x in $\mathcal{D}\left(\Gamma^{-1}\right)$.

In practice if Γ_n is a convergent estimator of Γ, the regularizing methods introduced below can be applied to Γ_n which is usually not invertible (see below for an example). It should be noted at this point that the regularization for the inverse of the covariance operator appears in the linear regression model for functional variables:

$$y = \langle X, \varphi \rangle + \epsilon$$

when estimating the unknown φ (see Cardot et al. (2007) and also Chapter 2 of this volume).

Finally a general method for estimating ρ may be proposed when we have with estimates of Γ and Δ at hand, say Γ_n and Δ_n: compute $\Gamma_n^{\dagger}$ and take, for the estimate and the predictor based on the new input X_{n+1}, respectively:

$$\widehat{\rho}_n = \Delta_n \Gamma_n^{\dagger} \quad \text{and} \quad \widehat{\rho}_n(X_{n+1}).$$

Obviously examples of such estimates are the empirical covariance and cross-covariance operators:

$$\Gamma_n = \frac{1}{n} \sum_{k=1}^{n} X_k \otimes X_k,$$

$$\Delta_n = \frac{1}{n-1} \sum_{k=1}^{n-1} X_{k+1} \otimes X_k,$$

where the X_k's have been reconstructed by interpolation techniques.

For instance the spectral cut-off version for Γ_n is

$$\Gamma_n^{\dagger} = \sum_{l \leq k_n} \frac{1}{\widehat{\lambda}_l} \widehat{\pi}_l,$$

where the eigenvalues $\widehat{\lambda}_l$ and the eigenprojectors $\widehat{\pi}_l$ are by-products of the functional PCA of the sample $X_1, \ldots, X_n$ (i.e. eigenelements of Γ_n; see Chapter 8 for an overview of functional PCA).

Remark 3.4 *This inverse problem indeed turns out to be the main serious abstract concern when making inferences on the ARH model.*

3.3.3 Convergence results for the autocorrelation operator and the predictor

As the data under consideration are of a functional nature, inference on ρ cannot be based on likelihood. The Lebesgue measure cannot be defined on infinite-dimensional spaces. However, it must be mentioned that Mourid and Bensmain (2006) have suggested the adaptation of Grenander's theory of sieves (Grenander (1981) and Geman and Hwang (1982)) to this issue. They prove consistency in two very important cases: when ρ is a kernel operator and when ρ is Hilbert–Schmidt. In the former case ρ is identified with the associated kernel K, developed on a basis of trigonometric functions along the sieve:

$$\Theta_m = \left\{ K \in L^2 : K(t) = c_0 + \sum_{k=1}^{m} \sqrt{2} c_k \cos(2\pi k t), \ t \in [0,1], \ \sum_{k=1}^{m} k^2 c_k^2 \leq m \right\}.$$

This approach is truly original within the literature on functional data and could certainly be extended to other problems of linear or nonlinear regression.

The seminal paper dealing with the estimation of the operator ρ dates back to 1991 and is due to Bosq (1991). Several consistency results were obtained immediately afterwards in Pumo's (1992) and Mourid's (1995) PhD theses. Then Pumo (1998) focused on random functions with values in $C([0,1])$ using specific techniques. Besse and Cardot (1996), and later Besse et al. (2000), implement spline and kernel methodologies with applications to climatic variations. These studies may be viewed as pioneering in terms of kernel methods for functional data (see Chapter 4 of this volume). Amongst several interesting ideas, they introduce a local covariance estimate:

$$\widehat{\Gamma}_{h_n} = \frac{\sum_{i=1}^{n} [X_i \otimes X_i] K(\|X_i - X_n\|/h)}{\sum_{i=1}^{n-1} K(\|X_i - X_n\|/h)}$$

and a local cross-covariance estimate which emphasize data close to the last observation. This method makes it possible to consider data in which there are departures from the assumption of stationarity. This issue of the estimation of ρ is also treated in Guillas (2001) and Mas (2004).

In a recent paper, Antoniadis and Sapatinas (2003) carry out wavelet estimation and prediction in the ARH(1) model. The inverse problem is highlighted through a class of estimates stemming from the deterministic literature on this topic. This class of estimates is compatible with wavelet techniques and leads to consistency of the predictor. The method is applied to the "El Nino" dataset, which has tended to be used as a benchmark for comparing the performances of the various predictions.

Ruiz-Medina et al. (2007) consider the functional principal oscillation pattern (POP) decomposition of the operator ρ as an alternative to functional PCA decomposition. They implement a Kalman filter on the state-space equation obtained at the preceding step, and thereby derive the optimal predictor. This original approach, illustrated by various simulations, seems to be suited to spatial functional data as well.

Kargin and Onatski (2008) introduce the notion of "predictive factor," which seems to be more suitable than the PCA basis for projecting the data if one really focuses on the predictor (and not on the operator itself). A double regularization (penalization and projection) provides them with a rate of $O\left(n^{-1/6} \log^\beta n\right)$ (where $\beta > 0$) for the prediction mean square error.

In Mas (2007), the problem of weak convergence is addressed. The main results are given in the theorem below:

Theorem 3.5 *It is impossible for $\widehat{\rho}_n - \rho$ to converge in distribution for the classical norm topology of operators. But under moment assumptions, if $\left\|\Gamma^{-1/2}\rho\right\|_\infty < +\infty$ and if the spectrum of Γ is convex then, when $k_n = o\left(n^{1/4}/\log n\right)$,*

$$\sqrt{\frac{n}{k_n}} \left(\widehat{\rho}_n\left(X_{n+1}\right) - \rho\widehat{\Pi}_{k_n}\left(X_{n+1}\right)\right) \xrightarrow{w} \mathcal{G},$$

where $\mathcal{G}$ is an H-valued Gaussian centered random variable with covariance operator Γ_ϵ and $\widehat{\Pi}_{k_n}$ is the projector onto the first k_n eigenvectors of Γ_n.

Remark 3.5 *The first sentence of Theorem 3.5 is quite surprising but is a direct consequence of the underlying inverse problem. Finally considering the predictor weakens the topology and has a smoothing effect on $\widehat{\rho}_n$. This phenomenon (which was exploited in Antoniadis and Sapatinas (2003)) appears additionally in the linear regression model for functional data (see Cardot et al. (2007)).*

It should be noted that rates of convergence are difficult to obtain (see Guillas (2001) or Kargin and Onatski (2008, Theorem 3)) and are rather slow with respect to those obtained in the regression model. An exponential inequality is given in Theorem 8.8 of Bosq (2000), but it seems that a more systematic study of the mean square prediction error has not yet been carried out and that optimal bounds are not available.

Hypothesis testing

A very recent article by Horvath et al. (2010) focuses on the stability of the autocorrelation operator in comparison with change-point alternatives. In fact model (3.5), based on the sample $X_1, \ldots, X_n$, is slightly modified to become:

$$X_n = \rho_n\left(X_{n-1}\right) + \epsilon_n$$

and the authors test

$$H_0 : \rho_1 = \cdots = \rho_n$$

against the alternative:

$$H_A : \text{there exists } k^* \in \{1, \ldots, n\} : \rho_1 = \cdots = \rho_{k^*} \neq \rho_{k^*+1} = \cdots = \rho_n.$$

The test is based on the projection of the process X_n onto the first p eigenvectors of functional PCA and on an accurate approximation of the long-run covariance matrix. The asymptotic distribution is derived by means of empirical process techniques. The consistency of the test is obtained and a simulation/real case study dealing with credit card transaction time-series is treated.

It turns out that Laukaïtis and Rackauskas (2002) had considered the same sort of problem a few years earlier. They introduce a functional version of the partial sum process of estimated residuals:

$$S(t) = \sum_{k=2}^{\lfloor t \rfloor} [X_k - \widehat{\rho}(X_{k-1})]$$

and obtain weak convergence results for its normalized version to an H-valued Wiener process. This formal theorem yields different strategies (dyadic increments of partial sums or moving residual sums) for deriving a test.

It seems, however, that the topic of hypothesis testing has rarely been addressed, and yet quite promising (even though serious) theoretical and technical problems appear here, once again in connection with the inverse problem mentioned earlier in this chapter.

3.3.4 Extensions of the ARH model

Various extensions have been proposed for the ARH(1) model in order to improve its prediction performance. The first is the natural extension to an autoregressive process of order p with $p > 1$, denoted ARH(p), defined by

$$X_n = \rho_1 X_{n-1} + \cdots + \rho_p X_{n-p} + \epsilon_n.$$

Using the Markov representation $Y_n = \rho' Y_{n-1} + \epsilon'_n$ where

$$\rho' = \begin{bmatrix} \rho_1 & \rho_2 & \cdots & \rho_n \\ I & 0 & \cdots & 0 \\ \vdots & & & \vdots \\ 0 & & & 1 \end{bmatrix}, \quad Y_n = \begin{bmatrix} X_n \\ X_{n-1} \\ \vdots \\ X_{n-p+1} \end{bmatrix}, \quad \text{and } \epsilon'_n = \begin{bmatrix} \epsilon_n \\ 0 \\ \vdots \\ 0 \end{bmatrix},$$

and I denotes the identity operator, Mourid (2002) obtains asymptotic results for projector estimators and predictors.

Damon and Guillas (2002) introduce an autoregressive Hilbertian process model with exogenous variables, denoted ARHX(1), which is intended to take into account the dependence structure of random curves under the influence of explanatory variables. The model is defined by the equation

$$X_n = \rho(X_{n-1}) + a_1(Z_{n,1}) + \cdots + a_q(Z_{n,q}) + \epsilon_n, \ n \in Z,$$

where $a_1, \ldots, a_q$ are bounded linear operators in H and $Z_{n,1}, \ldots, Z_{n,q}$ are ARH(1) exogenous variables; it is assumed that the noises of the $q+1$ H-valued autoregressive processes are independent. Damon and Guillas obtain some limit theorems, derive consistent estimators, and present a simulation study in order to illustrate the accuracy of the estimation and compare the forecasts with other functional models.

Guillas (2002) considers an H-valued autoregressive stochastic sequence (X_n) under several regimes, such that the underlying process (I_n) is stationary. Under some dependence assumptions on (I_n) he proves the existence of a unique stationary solution; he also states a law of large numbers and the consistency of the covariance estimator. Following the same idea, Mourid (2004) introduces and studies the autoregressive process with random operators $X_n = \rho_n X_{n-1} + \epsilon_n$, where $(\rho_n, n \in Z)$ is stationary and independent of (ϵ_n). Results similar to those for the classical ARH(1) are obtained.

A new model, called the ARHD process, which considers the derivative curves of an ARH(1) model, has been introduced by Marion and Pumo (2004). In a recent paper, Mas and Pumo (2007) introduce and study a slightly new model:

$$X_n = \phi X_{n-1} + \Psi\left(X'_{n-1}\right) + \epsilon_n,$$

where X_n are random functions with values in the Sobolev space

$$W^{2,1} = \{u \in L^2[0,1], u' \in L^2[0,1]\},$$

ϕ is a compact operator from W to W, Ψ is a compact operator from $L^2[0,1]$ to $W^{2,1}$, and $\|\phi h + \Psi h'\| \leq \|h\|$ for $h \in W^{2,1}$. Convergent estimates are obtained through an original double-penalization method. Simulations on real data show that the predictions are comparable to those obtained by other classical methods based on ARH(1) modelization. Tests on the derivative part and models with higher derivatives may be interesting from both a theoretical and practical point of view.

3.3.5 Numerical considerations

In this section we present some numerical aspects of prediction when the data are curves observed at discrete points. To our knowledge, the prediction methods based on linear processes are limited to the application of the ARH(1) model, since tractable algorithms using general linear processes in Hilbert spaces do not exist (see Merlevède (1997)). However, some partial results are available for moving-average processes in Hilbert spaces; these will be briefly discussed in the next section.

The literature in which the ARH(1) model is used to make predictions is varied and rich, and deals with different domains:

- Environment: Besse *et al.* (2000); Antoniadis and Sapatinas (2003); Mas and Pumo (2007); Fernández de Castro *et al.* (2005); Damon and Guillas (2002);
- Economy and finance: Kargin and Onatski (2008);
- Electricity consumption: Cavallini *et al.* (1994);
- Medical sciences: Marion and Pumo (2004); Glendinning and Fleet (2007).

From a technical point of view, the different approaches for implementing an ARH proceed in two steps. The first step consists in decomposing data into some functional basis in order to reconstruct them on the whole observed interval. Most of the methods use a spline or wavelet basis and assume that the curves belong to the Sobolev $W^{2,k}$ space of functions such that the kth derivative is square integrable. We invite the reader to refer to the papers by Besse and Cardot (1996), Pumo (1998), and Antoniadis and Sapatinas (2003), among others, for detailed discussions on the use of splines and wavelets for numerical estimation and prediction using an ARH(1) model and for the numerical results presented hereafter. The second step consists in choosing the tuning parameters required by these methods, for example, the dimension of the projection subspace for the projection estimators. A general method used by the above authors is based on a cross-validation approach, which gives satisfactory results in applications. Finally we note that alternative approaches to prediction based on ARH(1) modelization are proposed by Mokhtari and Mourid (2002) and Mourid and Bensmain (2006). In Mokhtari and Mourid (2002), the authors use a Parzen approximation in a reproducing-kernel spaces framework. Some simulation studies are presented in Mokhtari and Mourid (2008).

In order to compare the methods described above, we consider a climatological time series describing the El Niño Southern Oscillation (see, for example Besse *et al.* (2000) or Smith *et al.* (1996) for a description of the data[1]). The series gives the monthly mean El Niño sea surface temperature index from January 1950 to December 1986 and is presented in Fig. 3.2. We will compare the ARHD predictor with various functional prediction methods.

We compare the predictors of monthly temperature during 1986, given the data up until 1985, by two criteria: mean square error (MSE) and relative mean absolute error (RMAE), defined by:

$$\text{MSE} = \frac{1}{12} \sum_{i=1}^{12} \left(X_n^i - \hat{X}_n^i \right)^2, \quad \text{RMAE} = \frac{1}{12} \sum_{i=1}^{12} \frac{\left| X_n^i - \hat{X}_n^i \right|}{X_n^i},$$

[1] The data is freely available from http://www.cpc.ncep.noaa.gov/data/indices/index.html

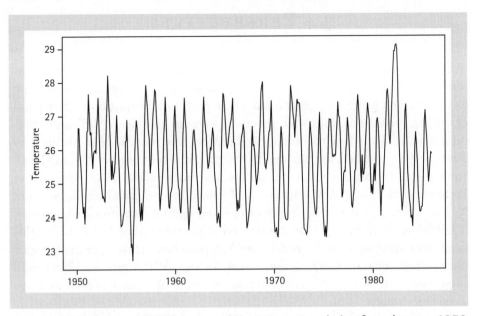

Fig. 3.2 Monthly mean El Niño sea surface temperature index from January 1950 to December 1986.

Table 3.1 Mean square error (MSE) and RMAE errors for the prediction of the El Niño index during 1986.

Prediction method	MSE	RMAE (%)
Wavelet II	0.063	0.89
Splines	0.065	0.89
ARHD	0.167	1.25
ARH(1): linear spline	0.278	2.4
SARIMA	1.457	3.72

where X_n^i (respectively, $\widehat{X}_n^i$) denotes the ith month's observation (respectively, prediction). The two criteria for the various functional predictors are given in Table 3.1. The results show that the best methods are Wavelet II (one of the wavelet approaches proposed in Antoniadis and Sapatinas (2003)) and spline smoothing ARH(1). Globally, the predictors obtained using the ARH(1) model are better and numerically faster than the classical SARIMA $(0, 1, 1) \times (1, 0, 1)_{12}$ model (the best SARIMA model based on classical criteria).

3.4 Perspectives

In the preceding sections we highlighted two important statistical problems concerning H-linear processes. The first was discussed in Section 3.2.3 in relation to inference on general or generalized linear processes. Estimation with the aim of making predictions using such processes seems to give rise to difficult technical problems. Some new results in this direction have recently been obtained by Bosq (2007) by introducing the moving average process of order $q \geq 1$, MAH(q). Some partial consistency results for the specific process MAH(1) are presented in Turbillon et al. (2008). A MAH(1) is an H-valued process satisfying the equation $X_t = \epsilon_t + \ell(\epsilon_{t-1})$, where ℓ is a compact operator and (ϵ_t) is a strong white noise. It follows easily from (3.3) that this process is invertible when $\|\ell\| < 1$. The difficulty in estimating ℓ, as for real-valued MA processes, stems from the fact that the moment equation is non-linear, unlike for the ARH(1) process. Under mild conditions, Turbillon et al. (2008) propose two types of estimators for ℓ and give consistency results.

The second problem concerns the ARH(1) model and its extensions. In Section 3.3.3 we recalled some serious theoretical and technical problems that arise for hypothesis testing. These problems are very important, particularly from a practical point of view. As an example, consider the ARHD model and the test addressing the significance of the derivative in the model. Another issue may prove to be the characterization of real-valued, or more generally linear, processes that allow an ARH representation. While some examples exist that admit an ARH or MAH representation (see Bosq (2000)), a general approach for recognising real processes which allow such a representation is a topic for future study.

The above questions are important from a theoretical point of view, in particular for research in statistics. For those who analyze data in the form of discretized curves, it has become more and more necessary to have analog description tools available, as for ARMA(p, q) real-valued processes. In this direction, we cite Hyndman and Shang (2008) as an example in the case of visualizing functional data and identifying functional outliers.

Acknowledgements

The authors thank Frédéric Ferraty, Yves Romain, and the whole group STAPH for initiating this work as well as for permanent and fruitful collaboration, and are grateful to Professor Denis Bosq for helpful discussions and pointing out recent articles on functional linear processes.

References

Adams, R.A., Fournier, J.J.F. (2003). *Sobolev Spaces*(second edition). Academic Press, New York.

Antoniadis A., Sapatinas, T. (2003). Wavelet methods for continuous-time prediction using representations of autoregressive processes in Hilbert spaces. *J. of Multivariate Analysis*, 87, 133–58.

Besse, P., Cardot, H. (1996). Approximation spline de la prévision d'un processus fonctionnel autorégressif d'ordre 1. *Canad. J. Statist*, 24, 467–87.

Besse, P., Cardot, H., Stephenson, D. (2000). Autoregressive forecasting of some climatic variations. *Scand. J. Statist*, 27, 673–87.

Bosq, D. (1991). Modelization, nonparametric estimation and prediction for continuous time processes. *Nato Asi Series C*, 335, 509–29.

Bosq, D. (1996). Limit theorems for Banach-valued autoregressive processes. Applications to real continuous time processes. *Bull. Belg. Math. S.c.*, 3, 537–55.

Bosq, D. (1999). Représentation autorégressive de l'opérateur de covariance empirique d'un ARH(1). Applications. *C.R. Acad. Sci.*, 329(I), 531–4.

Bosq, D. (2000). *Linear Processes in Function Spaces*. Springer, New York.

Bosq, D. (2007). General linear processes in Hilbert spaces and prediction. *J. Stat. Planning and Inference*, 137, 879–94.

Bosq, D. (2010). Tensorial products of functional ARMA processes. *J. of Multivariate Analysis* (to appear).

Bosq, D., Blanke, D. (2007). *Inference and Prediction in Large Dimensions*. Wiley–Dunod, Chichester.

Brockwell, P., Davis, A. (1991). *Time Series: Theory and Methods*. Springer, New York.

Cardot, H., Mas, A., Sarda, P. (2007). CLT in functional linear regression models. *Probability Theory and Related Fields*, 138, 325–61.

Cavallini, A., Montanari, G.C., Loggini, M., Lessi, O., Cacciari M. (1994). Nonparametric prediction of harmonic levels in electrical networks. In *Proceedings of IEEE ICHPS VI, Bologna*, Italy, 165–71.

Damon, J., Guillas, S. (2002). The inclusion of exogenous variables in functional autoregressive ozone forecasting. *Environmetrics*, 13(7), 759–74.

Damon, J., Guillas, S. (2005). Estimation and simulation of autoregressive Hilbertian processes with exogenous variables. *Stat. Infer. for Stoch. Proc.*, 8 (2), 185–204.

Dedecker, J., Merlevede, F. (2003). The conditional central limit theorem in Hilbert spaces. *Stochastic Process. Appl.*, 108, 229–62.

Dedecker, J., Merlevede, F. (2007). Convergence rates in the law of large numbers for Banach valued dependent variables. *Teor. Veroyatnost. i Primenen*, 52, 562–87.

Dembo A., Zeitouni, O. (1993). *Large Deviations Techniques and Applications*. Jones and Bartlett, London.

Dunford, N., Schwartz, J.T. (1988). *Linear Operators*, Vols I and II. Wiley, New York.

Fernández de Castro, B., Guillas, S., González Manteiga, W. (2005). Functional Samples and Bootstrap for Predicting Sulfur Dioxide Levels. *Technometrics*, 47(2), 212–22.

Ferraty, F., Vieu, P. (2006). *Nonparametric Functional Data Analysis. Theory and Practice*. Springer, New York.

Fortet, R. (1995). *Vecteurs, Fonctions et Distributions Aléatoires dans les Espaces de Hilbert*. Hermes, Paris.

GEMAN, S., HWANG, C.R. (1982). Nonparametric maximum likelihood estimation by the method of sieves. *Ann. Statist.*, **10**(2), 401–14.

GLENDINNING, R.H., FLEET, S.L. (2007). Classifying functional time series. *Signal Processing*, **87**(1), 79–100.

GOHBERG, I., GOLDBERG, S., KAASHOEK, M.A. (1991). *Classes of Linear Operators, Vols I & II. Operator Theory: Advances and Applications*. Birkhaüser, Basel.

GRENANDER, U. (1981). *Abstract Inference*. Wiley, New York.

GROETSCH, C. (1993). *Inverse Problems in the Mathematical Sciences*. Vieweg, Wiesbaden.

GUILLAS, S. (2001). Rates of convergence of autocorrelation estimates for autoregressive Hilbertian processes. *Statist. Probab. Lett.*, **55**(3), 281–91.

GUILLAS, S. (2002). Doubly stochastic Hilbertian processes. *J. Appl. Probab.*, **39**(3), 566–80.

HORVATH, L., HUSKOVA, M., KOKOSZKA, P. (2010). Testing the stability of the functional autoregressive process. *J. of Multivariate Analysis*, **101**(2), 352–67.

HYNDMAN, R., SHANG, H.L. (2008). Bagplots, boxplots and outlier detection for functional data. In *Functional and Operatorial Statistics* (Dabo-Niang, S. and Ferraty, F., eds), 201–8. Physica, Heidelberg.

KARGIN, V., ONATSKI, A. (2008). Curve forecasting by functional autoregression. *J. of Multivariate Analysis*, **99**, 2508–26.

LABBAS, A., MOURID, T. (2002). Estimation and prediction of a Banach valued autoregressive process. *C. R. Acad. Sci. Paris*, **335**(9), 767–72.

LAUKAITIS A., RACKAUSKAS, A. (2002). Functional data analysis of payment systems. *Nonlinear Analysis: Modeling and Control*, **7**, 53–68.

LEDOUX, M., TALAGRAND, M. (1991). *Probability in Banach spaces – Isoperimetry and Processes*. Springer, Berlin.

MALTSEV, V.V., OSTROVSKII, E.I. (1982). Central limit theorem for stationary processes in Hilbert space. *Theor. Prob. and its Applications*, **27**, 357–9.

MARION, J.M., PUMO, B. (2004). Comparaison des modèles ARH(1) et ARHD(1) sur des données physiologiques. *Annales de l'ISUP*, **48**(3), 29–38.

MAS, A. (2002). Weak convergence for the covariance operators of a Hilbertian linear process. *Stoch Process. App*, **99**, 117–35.

MAS, A. (2004). Consistance du prédicteur dans le modèle ARH(1): le cas compact. *Annales de l'Isup*, **48**, 39–48.

MAS, A. (2007). Weak convergence in the functional autoregressive model. *J. of Multivariate Analysis*, **98**, 1231–61.

MAS A., MENNETEAU, L. (2003a). Perturbation approach applied to the asymptotic study of random operators. *Progress in Probability*, **55**, 127–34.

MAS A., MENNETEAU, L. (2003b). Large and moderate deviations for infinite-dimensional autoregressive processes. *J. of Multivariate Analysis*, **87**, 241–60.

MAS, A., PUMO, B. (2007). The ARHD process. *J. of Statistical Planning and Inference*, **137**, 538–53.

MENNETEAU, L. (2005). Some laws of the iterated logarithm in Hilbertian autoregressive models. *J. of Multivariate Analysis*, **92**, 405–25.

MERLEVÈDE, F. (1995). Sur l'inversibilité des processus linéaires à valeurs dans un espace de Hilbert. *C. R. Acad. Sci. Paris*, **321**(I), 477–80.

MERLEVÈDE, F. (1996). Central limit theorem for linear processes with values in Hilbert space. *Stoch. Proc. and their Applications*, **65**, 103–14.

MERLEVÈDE, F. (1997). Résultats de convergence presque sûre pour l'estimation et la prévision des processus linéaires hilbertiens. *C. R. Acad. Sci. Paris*, **324**(I), 573–6.

MERLEVÈDE, F., PELIGRAD, M., UTEK, S. (1997). Sharp conditions for the CLT of linear processes in Hilbert space. *J. of Theoritical Probability*, **10**(3), 681–93.

MOKHTARI, F., MOURID, T. (2002). Prediction of autoregressive processes via the reproducing kernel spaces. *C. R. Acad. Sci. Paris*, **334**, 65–70.

MOKHTARI, F., MOURID, T. (2008). Prévision des processus ARC(1) via le prédicteur de Parzen. Exemples et simulations. *Ann. I.S.U.P.*, **52**, 1–2, 81–90.

MOURID, T. (1995). Contribution à la statistique des processus autorégressifs à temps continu. PhD Thesis, Univ. Paris VI.

MOURID, T. (2002). Estimation and prediction of functional autoregressive processes. *Statistics*, **36**(2), 125–38.

MOURID, T. (2004). The Hilbertian autoregressive process with random operator. (In french.) *Annales de l'ISUP*, **48**(3), 79–86.

MOURID, T., BENSMAIN, N. (2006). Sieves estimator of the operator of a functional autoregressive process. *Stat. and Probab. Letters*, **76**(1), 93–108.

NAZAROVA, A.N. (2000). Normal approximation for linear stochastic processes and random fields in Hilbert space. *Math. Notes*, **68**, 363–9.

PUMO, B. (1992). Estimation et prévision de processus autorégressifs fonctionnels. Applications aux processus à temps continu. PhD Thesis, Univ. Paris VI.

PUMO, B. (1998). Prediction of continuous time processes by $C[0, 1]$-valued autoregressive process. *Statist. Infer. for Stoch. Processes*, **3**(1), 297–309.

RAMSAY, J.O. (2000). Differential equation models for statistical functions. *Canad. J. Statist*, **28**, 225–40.

RAMSAY, J.O., SILVERMAN, B.W. (1997). *Functional Data Analysis*. Springer, New York.

RAMSAY, J.O., SILVERMAN, B.W. (2002). *Applied Functional Data Analysis*. Springer, New York.

RUIZ-MEDINA, M.D., SALMERON, R., ANGULO, J.M. (2007). Kalman filtering from POP-based diagonalization of ARH(1). *Comp. Statist. Data Anal.*, **51**, 4994–5008.

SMITH, T.M., REYNOLDS, R.W., LIVEZEY, R.E., STOKES, D.C. (1996). Reconstruction of historical sea surface temperatures using empirical orthogonal functions. *Journal of Climate*, **9**(6), 1403–20.

TIKHONOV, A.N., ARSENIN, V.Y. (1977). *Solutions of Ill-posed Problems*. V.H. Winstons and Sons, Washington.

TURBILLON, S., BOSQ, D., MARION, J.M., PUMO, B. (2008). Parameter estimation of moving averages in Hilbert spaces. *C. R. Acad. Sci. Paris*, **346**, 347–50.

WONG, E. (1966). Some results concerning the zero-crossings of Gaussian noise. *SIAM J. Appl. Math.*, **14**(6), 1246–54.

ZIEMER, W.P. (1989). *Weakly Differentiable Functions. Sobolev Spaces and Functions of Bounded Variations*. Graduate Text in Mathematics **120**. Springer, New York.

CHAPTER 4

KERNEL REGRESSION ESTIMATION FOR FUNCTIONAL DATA

FRÉDÉRIC FERRATY
PHILIPPE VIEU

4.1 INTRODUCTION

OVER the last decade there has been great interest in developing new models for regression problems involving functional data. The main aim of these models is to overcome the lack of flexibility of the standard linear functional modeling ideas, and recent research has naturally been oriented towards nonparametric (and more recently semiparametric) models. The general presentation done in Chapter 1, provides a detailed discussion of all these various ways of modeling functional regression problems. While Chapter 2 focuses on linear modeling, this chapter is concerned with recent nonparametric and semiparametric advances.

In the standard literature on nonparametric regression, convolution kernels play a major role (see Collomb (1981, 1985); Härdle (1990); Sarda and Vieu (2000); Schimek (2000); and Györfi *et al.* (2002) for bibliographical surveys and general discussions of standard multivariate nonparameric regression). They are many reasons for this. From a practical point of view, these estimates are quite appealing (and now widely used) because they are very easy to implement. This feature combines with various attractive theoretical properties, such as for instance the fact that they can reach Stone's optimal rates of convergence for multivariate nonparametric regression problems (see Stone (1982)). Finally, another nice feature of kernel ideas is the fact that they provide several different classes of estimators. In the nonparametric setting, this includes the standard Nadaraya–Watson estimator, but also the k-nearest neighbour (kNN) and local polynomial estimators. But this holds not only for nonparametric modeling, since kernel ideas can be used in most semiparameric models or in dimensionality-reduction models.

Therefore amongst the first developments linking nonparametric regression modeling and functional data analysis, kernel ideas have taken a determining role. The main goal of this chapter is to present the various statistical techniques based on kernel smoothing ideas that have recently been developed for functional regression estimation problems. It is worth noting that even if this field of statistics is rather young, since the first paper in this area is only ten years old (see Ferraty and Vieu (2000)), the literature is now sufficiently developed to make an exhaustive presentation of the results impossible. So our presentation of these methods is carried out using a necessary selective set of results. This selection has been motivated by the wish to make the presentation as simple as possible: that is why strong consistency type results have been preferentially chosen. Other kinds of results, such as L_2 consistency or asymptotic normality, are presented in a more synthetic way. The second guideline followed during the selection of material has been to emphasize the local specificities of functional data modeling: this is why special attention has been given to pointwise consistency results.

The first part of this contribution (Section 4.2) concerns nonparametric regression modelling and is organized as follows. A few asymptotic pointwise results are presented in Sections 4.2.2 and 4.2.3 for three popular functional regression estimates constructed by means of kernel ideas, namely the functional version of the Nadaraya–Watson estimate, a functional adaptation of the local linear method, and the kNN functional estimate. In each case the rates of convergence are specified and the influence of the local structure of the functional space is highlighted. In a second approach, Section 4.2.4 discusses uniform consistency type results. In this setting the existing literature is much less developed, and to our knowledge the only available result is for Nadaraya–Watson type estimates. The emphasis here is on how such a uniform theoretical result turns out to be a key tool for solving further problems of great applied impact, such as automatic bandwidth selection or bootstrapping. Finally, all the results presented so far are commented on in Section 4.2.5, which also presents a short survey of the state-of-the-art in nonparametric functional regression.

The second part of this chapter concerns recent advances in the area of alternative models which fall between parametric (linear) models and purely nonparametric ones. Special attention is paid to additive modeling (see Section 4.4), while semiparametric models are discussed in Section 4.3; here special attention is given to single functional index models (see Section 4.3.2) and to partial linear functional models (see Section 4.3.3). The main point all these models have in common is the need for a preliminary purely nonparametric functional stage, and it will be emphasized how the uniform consistency developments presented earlier in Section 4.2.4 are indispensable tools for their study. As a consequence, and because uniform consistency results are only known for Nadaraya–Watson type functional estimates, all the methods developed in Sections 4.4 and 4.3 will concern such kinds of estimates.

The large range of possible models for functional regression problems opens the door to a new kind of question: which model is to be used in practice? One way of looking at this problem is to construct goodness-of-fit testing procedures, and once again kernel methods turn out to be of great interest. This field has been very underdeveloped in the literature up to now; the few existing studies will be presented in Section 4.5. Once again, the uniform consistency developments described earlier in Section 4.2.4 are necessary tools, and consequently only Nadaraya–Watson type functional estimates will be considered.

It is worth noting that Sections 4.2–4.5 will offer the opportunity for presenting an exhaustive survey of the literature on kernel methods for functional regression. However, kernel ideas can be used in other settings than regression, and Section 4.6.1 will be devoted to a brief presentation of this literature. The main focus of this chapter is on theoretical considerations, but the final section will briefly discuss the impact of kernel methods on practical real-data analysis involving functional (curves) datasets.

4.2 KERNELS IN NONPARAMETRIC FUNCTIONAL REGRESSION

4.2.1 Models and estimates

Consider the following regression problem:

$$Y = r(X) + \varepsilon, \qquad (4.1)$$

where Y and ε are real random variables with $E(\varepsilon|X) = 0$, and where X is a covariate which is allowed to belong to some abstract topological space E which can be of infinite dimension. In what follows, P_X denotes the probability measure of the random variable X. This is called a functional regression model with scalar response. For the sake of generality, we do not restrict ourselves to Hilbert or Banach spaces and the topology endowing the space E is only assumed to be associated with a semi-metric d. The only restriction on the operator r is a Lipschitz-type regularity hypothesis:

$$\exists \beta > 0, \exists C < +\infty, \forall (x, x') \in E \times E, \ |r(x) - r(x')| \leq Cd(x, x')^{\beta}. \quad (4.2)$$

According to the general classification of regression models presented in Chapter 1 (see Definition 1.3 therein), this model is nonparametric in the sense that the general assumption (4.2) does not allow us to characterize the operator r by a finite number of elements of E. In other words, r is a nonlinear real-valued operator defined on the functional space E.

Given a sample of independent pairs (X_i, Y_i) distributed as (X, Y), the statistical problem consists in estimating the functional nonlinear operator r. In what follows, three different estimates of r will be considered: the Nadaraya–Watson functional estimate, the kNN functional estimate, and the local linear functional estimate, each of them being constructed as a locally-weighted average. More precisely, if x is an element of E, these estimates have the following form:

$$\hat{r}(x) = \frac{\sum_{i=1}^{n} W_n(x, X_i) Y_i}{\sum_{i=1}^{n} W_n(x, X_i)}. \quad (4.3)$$

4.2.2 Pointwise almost-sure consistency

Throughout this section, x is a fixed point of the functional space E; and strong pointwise consistency results (with rates) are presented for three different estimates constructed in the manner of (4.3). Proofs are presented in a very synthetic way, the main objective being to emphasize their main ideas. Precise references will be given for the reader interested in having all the technical details of these proofs. Comments on these results, as well as a complementary bibliography, are referred to Section 4.2.5.

General conditions

The weights $W_n(x, X_i)$ are constructed differently according to the type of estimate to be considered, but in each case they are based on some real-valued kernel function K satisfying the following set of standard conditions:

$$K \geq 0, \ \int_{\mathbb{R}} K = 1, \ K \text{ is Lipschitz on } [0, 1), \text{ and support}(K) \subset [0, 1]. \quad (4.4)$$

Basically, the weight $W_n(x, X_i)$ vanishes as soon as X_i is "too far" from x. This is controlled by means of a smoothing parameter $h = h_n$, called bandwidth, which determines the size of the neighbourhood around x (in the sense of the topology associated with d) outside of which the weights are zero. This means that the estimate is not (for a fixed point x) using the whole statistical sample, but only the pairs (X_i, Y_i) for which X_i falls inside of the ball

$$\mathcal{B}(x, h) = \{x' \in E, d(x, x') \leq h\},$$

and naturally its behavior will depend strongly on the number of such pairs. From a mathematical point of view, these local concentration effects can be controlled by means of the following small-ball probability function:

$$\phi_x(\epsilon) = P_X(\mathcal{B}(x, \epsilon)),$$

which is assumed to be such that $\phi_x(\epsilon) > 0$ for any $\epsilon > 0$. For each $x \in E$, this concentration function will be linked with the bandwidth h (depending on n) through the following set of standard assumptions:

$$\lim_{n \to \infty} h = 0 \text{ and } \lim_{n \to \infty} \frac{n\phi_x(h)}{\log n} = \infty. \tag{4.5}$$

There is also a need for a strong link between the kernel function K and the small-ball probability function ϕ_x. Here, this is stated through the following technical hypothesis. Assume that there exist three constants $(C_1, C_2, \epsilon_0) \in]0, +\infty[^3$ such that:

$$\forall \epsilon \leq \epsilon_0, \ C_1 \phi_x(\epsilon) \leq EK\left(\frac{d(x, X)}{\epsilon}\right) \leq C_2 \phi_x(\epsilon). \tag{4.6}$$

As stated in Ferraty and Vieu (2006, Chapter 4), condition (4.6) is not too restrictive in the sense that it is satisfied in most situations.

Finally, in a standard way, we need some condition to control the tails of the distribution of Y. This is achieved by using the following moment assumption:

$$\forall m \geq 2, \ E(|Y|^m | X = \cdot) \text{ exists and is continuous.} \tag{4.7}$$

Kernel estimation

The first estimate to be considered is a functional version of the Nadaraya–Watson convolution kernel estimate. It can be defined as follows:

Definition 4.1 *The kernel functional regression estimate, denoted from now on by $\hat{r}_{\text{ker}}$, is defined by the local weighted average equation (4.3) with the following weights:*

$$W_n(x, X_i) = W_{n,\text{ker}}(x, X_i) = K\left(\frac{d(x, X_i)}{h}\right).$$

This estimate was introduced for nonparametric functional regression problems (see Ferraty and Vieu (2000)), and various asymptotic results are now available in the literature. In the following theorem we present a simple result giving the rate of almost sure (a.s.) convergence, while more insights into the existing literature are referred to Section 4.2.5.

Theorem 4.1 *Under conditions* (4.2), (4.4), (4.5), (4.6), *and* (4.7), *we have*

$$\hat{r}_{\text{ker}}(x) - r(x) = O\left(h^{\beta}\right) + O\left(\sqrt{\frac{\log n}{n\phi_x(h)}}\right), \quad a.s. \tag{4.8}$$

Proof of Theorem 4.1 A complete proof of this result can be found in Ferraty and Vieu (2006a, Chapter 6). Here, we simply present its main lines of argument. The key step in the proof consists in the following decomposition:

$$\hat{r}_{\text{ker}}(x) = \frac{\hat{r}_{\text{ker},0}(x)}{\hat{r}_{\text{ker},1}(x)}, \tag{4.9}$$

where for $j = 0$ or 1:

$$\hat{r}_{\text{ker},j}(x) = \frac{1}{nE\,W_{n,\text{ker}}(x, X_i)} \sum_{i=1}^{n} Y_i^j\, W_{n,\text{ker}}(x, X_i).$$

This decomposition has been adapted to take into account the fact that no assumption about the existence of density for the functional variable X is made, and it is therefore slightly different from the classical decomposition performed in standard multivariate kernel estimation (see, for instance, Collomb (1976, 1984) for the earliest results in this area, and Sarda and Vieu (2000) for a bibliographical survey). The bias terms can be computed directly by using condition (4.2) defining nonparametric modeling, and we get

$$E\hat{r}_{\text{ker},0}(x) = 1 \quad \text{and} \quad E\hat{r}_{\text{ker},1}(x) = r(x) + O(h^{\beta}). \tag{4.10}$$

Both the other terms can be expressed as sums of independent and centered real random variables:

$$\Delta_{i,j} = Y_i^j\, \frac{W_{n,\text{ker}}(x, X_i)}{E\,W_{n,\text{ker}}(x, X_i)} - E\left(Y_i^j\, \frac{W_{n,\text{ker}}(x, X_i)}{E\,W_{n,\text{ker}}(x, X_i)}\right), \, i = 1, \ldots, n, \, j = 0, 1,$$

for which a Bernstein-type exponential inequality can be used. Conditions (4.6) and (4.7) allow us to bound the moments of these variables in such a way that:

$$\exists C > 0, \forall m \geq 2, \; E\Delta_{i,j}^m \leq C\phi_x(h)^{1-m}. \tag{4.11}$$

Once these bounds are obtained, one can use any kind of exponential-type inequality (for instance, Ferraty and Vieu (2006a), Corollary A.8-ii) to obtain, for some $\eta_0 > 0$:

$$\sum_{n=1}^{\infty} P\left(\left|\frac{1}{n}\sum_{i=1}^{n} \Delta_{i,j}\right| \geq \eta_0 \sqrt{\frac{\log n}{n\phi_x(h)}}\right) < +\infty, \tag{4.12}$$

Because of the Borel–Cantelli Lemma, this is sufficient to get that:

$$\hat{r}_{\text{ker},j}(x) - E\hat{r}_{\text{ker},j}(x) = O\left(\sqrt{\frac{\log n}{n\phi_x(h)}}\right), \text{ a.s., } j = 0, 1. \quad (4.13)$$

Finally, Theorem 4.1 follows directly from (4.9), (4.10), and (4.13). □

Local linear smoothing

A natural extension of the standard Nadaraya–Watson estimate consists in using local linear ideas. This leads to the following local linear regression estimate:

Definition 4.2 *The local linear functional regression estimate, denoted from now on by $\hat{r}_{\text{LL}}$, is defined by the local weighted average equation (4.3) when the weights are constructed from some known operator β from E^2 into $\mathbb{R}$ by:*

$$W_n(x, X_i) = W_{n,\text{LL}}(x, X_i)$$
$$= \sum_{j \neq i}^{n} \beta(X_i, x)(\beta(X_i, x) - \beta(X_j, x))K\left(\frac{d(x, X_i)}{h}\right) K\left(\frac{d(x, X_j)}{h}\right).$$

This estimate was introduced for nonparametric functional regression problems in Barrientos-Marin (2007). Note that the interest of this estimate is that it can be seen as the explicit solution in a of the minimization problem:

$$\min_{(a,b) \in \mathbb{R}^2} \sum_{i=1}^{n} (Y_i - a - b\,\beta(X_i, x))^2 K\left(\frac{d(x, X_i)}{h}\right).$$

In the literature there are fewer asymptotic results than for the estimate $\hat{r}_{\text{ker}}$. In the following theorem we present its rate of pointwise almost sure convergence, while more insights into the existing literature are referred to Section 4.2.5. The study of the estimate $\hat{r}_{\text{LL}}$ needs additional technical assumptions linking the operator β, the semi-metric d, and the kernel function K. To improve the readability of this chapter, we summarize these conditions by assuming that there exist two constants $(M_1, M_2) \in]0, +\infty[^2$ such that:

$$\forall (k, l) \in \mathbb{N}^* \times \mathbb{N}, \ E\left(|\beta(X, x)|^l K\left(\frac{d(x, X)}{h}\right)^k\right) \leq M_1 h^l \phi_x(h) \quad (4.14)$$

and

$$E\left(\beta(X, x)^2 K\left(\frac{d(x, X)}{h}\right)\right) \geq M_2 h^2 \phi_x(h). \quad (4.15)$$

It is not our purpose here to discuss these conditions, but their high degree of generality appears in Lemma A.1 of Barrientos-Marin et al. (2010).

Theorem 4.2 *Under conditions* (4.2), (4.4), (4.5), (4.6), (4.7), (4.14), *and* (4.15) *we have:*

$$\hat{r}_{LL}(x) - r(x) = O\left(h^\beta\right) + O\left(\sqrt{\frac{\log n}{n\phi_x(h)}}\right), \quad a.s. \qquad (4.16)$$

Proof of Theorem 4.2 A complete proof of this result can be found in Barrientos-Marin *et al.* (2009). While the technical details are rather heavy, the main ideas of the proof are quite simple and will be presented now. The proof follows the same general framework as for the kernel estimate $\hat{r}_{ker}$ in Theorem 4.1, and is based on the following decomposition

$$\hat{r}_{LL}(x) = \frac{\hat{r}_{LL,0}(x)}{\hat{r}_{LL,1}(x)}, \qquad (4.17)$$

where for $j = 0$ or 1:

$$\hat{r}_{LL,j}(x) = \frac{1}{nE W_{n,LL}(x, X_i)} \sum_{i=1}^{n} Y_i^j W_{n,LL}(x, X_i).$$

The bias terms can be computed by using condition (4.2), without any more difficulty than for the kernel estimate $\hat{r}_{ker}$ in Theorem 4.1, and one arrives at:

$$E\hat{r}_{LL,1}(x) = r(x)E\hat{r}_{LL,0}(x) + O(h^\beta). \qquad (4.18)$$

The treatment of the dispersion terms is more difficult since they cannot be written as simple sums of i.i.d. centered random variables. The idea here is to use the following decomposition, for $j = 0, 1$:

$$\hat{r}_{LL,j}(x) = A\left[B_1 B_2 - C_1 C_2\right],$$

where

$$A = \frac{n^2 h^2 \phi_x(h)^2}{E W_{n,LL}(x, X_1)},$$

$$B_1 = \frac{1}{n}\sum_{i=1}^{n} \frac{Y_i^j K\left(\frac{d(x,X_i)}{h}\right)}{\phi_x(h)}, \quad B_2 = \frac{1}{n}\sum_{i=1}^{n} \frac{\beta(X_i, x)^2 K\left(\frac{d(x,X_i)}{h}\right)}{h^2 \phi_x(h)},$$

and

$$C_1 = \frac{1}{n}\sum_{i=1}^{n} \frac{Y_i^j \beta(X_i, x) K\left(\frac{d(x,X_i)}{h}\right)}{h\phi_x(h)}, \quad C_2 = \frac{1}{n}\sum_{i=1}^{n} \frac{\beta(X_i, x) K\left(\frac{d(x,X_i)}{h}\right)}{h\phi_x(h)}.$$

In other words, and because A is bounded, we have

$$\hat{r}_{LL,j}(x) - E\hat{r}_{LL,j}(x) = O\left((B_1 B_2 - E B_1 B_2) - (C_1 C_2 - E C_1 C_2)\right).$$

Each of the terms B_1, B_2, C_1, and C_2 can be written as a sum of i.i.d. real random variables and can therefore be treated by standard tools. To fix our ideas, if we look at the term B_1 we can write

$$B_1 = \frac{1}{n} \sum_{i=1}^{n} W_i,$$

and we can bound the moments of the variables W_i, exactly as was done to obtain (4.11), simply by using conditions (4.7), (4.14), and (4.15). In this way, we arrive at

$$E\left|W_i^m\right| = O\left(\phi_x(h)^{1-m}\right).$$

Then, exactly as we obtained (4.13) for the estimate $\hat{r}_{\text{ker}}$, the use of a standard exponential inequality leads to:

$$B_1 - E\,B_1 = O\left(\sqrt{\frac{\log n}{n\phi_x(h)}}\right), \text{ a.s.}$$

The same kind of expression can similarly be obtained for the terms B_2, C_1, and C_2, and finally

$$\hat{r}_{\text{LL},j}(x) - E\hat{r}_{\text{LL},j}(x) = O\left(\sqrt{\frac{\log n}{n\phi_x(h)}}\right), \text{ a.s.} \tag{4.19}$$

Theorem 4.2 follows directly from (4.17)–(4.19).

kNN estimation

Another extension of the standard Nadaraya–Watson estimate consists in using locally adaptive bandwidth constructed by means of k-nearest neighbour (kNN) ideas. Such a kNN functional regression estimate can be defined as follows:

Definition 4.3 *The k-nearest neighbour functional estimate, denoted from now on by $\hat{r}_{\text{kNN}}$, is defined by the local weighted average equation (4.3) with the following weights:*

$$W_n(x, X_i) = W_{n,\text{kNN}}(x, X_i) = K\left(\frac{d(x, X_i)}{H_k(x)}\right),$$

the bandwidth $H_k(x)$ being defined from a sequence $k = k_n$ of integers by:

$$H_k(x) = \inf\{h > 0, \#\{i, X_i \in \mathcal{B}(x, h)\} = k\}.$$

This estimate was introduced for nonparametric functional regression problems by Burba et al. (2008). Its main appealing feature when compared with the estimate $\hat{r}_{\text{ker}}$ is that it uses a local adaptive bandwidth (that is, a bandwidth depending on x) which is controlled by a discrete parameter $k \in \mathbb{N}$ (rather than by a continous one h). On the other hand, its main drawback will be the additional high level of technicality needed to study its theoretical behavior, because of the randomness of the new parameter $H_k(x)$ which depends on the whole functional sample X_i, $i = 1, \ldots, n$. In the literature there are not as many asymptotic results as for the estimate $\hat{r}_{\text{ker}}$. In the following theorem we present its rate of pointwise almost sure convergence; more details of the existing literature are referred to Section 4.2.5. It should be natural to express the rates of convergence as functions of the number k of neighbours. However, we have decided to follow another approach in order to unify the presentation of the results for all the various estimates studied in this chapter. This is the reason why we make the following additional restriction:

$$\phi_x(\cdot) \text{ is continuous and strictly increasing around } 0, \qquad (4.20)$$

which allows us to define precisely the value h such that:

$$\phi_x(h) = \frac{k}{n}. \qquad (4.21)$$

Theorem 4.3 *Assume that conditions (4.2), (4.4), (4.6), (4.7), and (4.20) hold. Assume also that k is such that the bandwidth defined by (4.21) satisfies condition (4.5). Then we have*

$$\hat{r}_{\text{kNN}}(x) - r(x) = O\left(h^\beta\right) + O\left(\sqrt{\frac{\log n}{n\phi_x(h)}}\right), \quad a.s. \qquad (4.22)$$

Proof of Theorem 4.3 A complete proof of this result can be found in Burba *et al.* (2009). While technical details are rather heavy, the main ideas of the proof are quite simple and will be presented now. The key point consists in using condition (4.20) to construct, for any nonnegative increasing sequence such that $\lim_{n\to\infty} \beta_n = 1$, two deterministic bandwidths h^- and h^+ in the following way:

$$\phi_x(h^-) = \sqrt{\beta_n} \frac{k}{n} \text{ and } \phi_x(h^+) = \sqrt{\beta_n^{-1}} \frac{k}{n}.$$

Then, by a Chernoff-type inequality, one has that the random bandwidth $H_k(x)$ satisfies

$$1_{\{h^- \leq H_k(x) \leq h^+\}} \to 1, \quad a.s. \qquad (4.23)$$

We note that both bandwidths h^- and h^+ satisfy condition (4.5). Therefore both kernel estimates $\hat{r}_{\text{kNN}}$ constructed using these bandwidths are such that Theorem 4.1 holds. Finally, by a suitable functional adaptation of standard ideas previously developed in Collomb (1980) for real variables X, one can show that the result (4.23) is enough to insure that the kernel estimate using the random bandwidth $H_k(x)$ (that is, precisely, the kNN estimate $\hat{r}_{\text{kNN}}$) satisfies the same asymptotic property as both the kernel estimates $\hat{r}_{\text{kNN}}$ constructed using the bandwidths h^- and h^+. □

4.2.3 Other pointwise asymptotic results

Presentation

In addition to the pointwise almost sure consistency results stated below, it is possible to obtain asymptotic results for functional estimates with respect to other modes of convergence. However, this can be more difficult, and the existing literature (as far as we know) is restricted only to the Nadaraya–Watson kernel estimate $\hat{r}_{\mathrm{ker}}$. In what follows we will present a few additional pointwise results for this estimate, including an L_2 asymptotic expansion, general L_p consistency results, and asymptotic normality. Because the statements of these results need tedious conditions and notation, we have decided to present them in a very synthetic way just to give the main ideas, and we refer readers to the relevant literature for more insights. Recall that x is a fixed point of the functional space E.

L_2 expansion for the kernel estimate

When stating such results, one is not only interested in finding the rates of convergence but also in obtaining precise expressions for the constant terms involved in the leading terms of the asymptotic expansion. To make this possible, it is necessary to strengthen the nonparametric model. This is done by introducing the following real-valued function:

$$\forall s \in \mathbb{R}, \ \xi(s) = E\left[(r(X) - r(x)) | d(X, x) = s\right],$$

and by changing condition (4.2) into the following:

$$\xi(0) = 0 \text{ and } \xi'(0) \text{ exists.} \qquad (4.24)$$

The conditions on the parameters of the estimate $\hat{r}_{\mathrm{ker}}$ also have to be slightly modified into the following ones:

$$\forall s \in [0, 1), \ K'(s) \leq 0 \text{ and } K(1) > 0, \qquad (4.25)$$

$$\lim_{n \to \infty} h = 0 \text{ and } \lim_{n \to \infty} n\phi_x(h) = \infty, \qquad (4.26)$$

while the concentration function also has to be such that

$$\phi_x(0) = 0. \qquad (4.27)$$

The leading terms of the asymptotic expansion of the pointwise L_2 error of the estimate $\hat{r}_{\mathrm{ker}}$ can be expressed by means of the following function (which is assumed to exist):

$$\tau(s) = \lim_{\epsilon \to 0} \frac{\phi_x(\epsilon s)}{\phi_x(\epsilon)}.$$

We refer readers to Ferraty et al. (2007) for a discussion on the low restrictiveness of these assumptions (and, more specifically, for the links between (4.24) and the differentiability properties of the nonlinear operator r, as well as for a precise expression of the function τ in various specific situations).

Theorem 4.4 *Assume that the conditions of Theorem 4.1 are satisfied as well as (4.25), (4.26), and (4.27). Then we have:*

$$E\left[\hat{r}_{\text{ker}}(x) - r(x)\right]^2 = B^2 h^2 + V \frac{1}{n\phi_x(h)} + o\left(h^2 + \frac{1}{n\phi_x(h)}\right), \qquad (4.28)$$

where

$$B = \xi'(0) \frac{K(1) - \int_0^1 (sK(s))' \tau(s)\, ds}{K(1) - \int_0^1 K'(s) \tau(s)\, ds}$$

and

$$V = E\left[\varepsilon^2 | X = x\right] \frac{K^2(1) - \int_0^1 2K(s) K'(s) \tau(s)\, ds}{(K(1) - \int_0^1 K'(s) \tau(s)\, ds)^2}.$$

Proof of Theorem 4.4 This proof is omitted; it can be found in Ferraty et al. (2007), under conditions slightly weaker than those presented here. Let us just note that it is based on the decomposition (4.9) and on precise asymptotic expansions of the bias and the variance of both terms $\hat{r}_{\text{ker},0}(x)$ and $\hat{r}_{\text{ker},1}(x)$. □

Pointwise asymptotic normality for the kernel estimate

In the same spirit one can also specify the asymptotic distribution of the estimation error with the kernel estimate $\hat{r}_{\text{ker}}$. This is achieved in the next result, where the following notation is used:

$$U_{n,x} = \sqrt{\frac{n\phi_x(h)}{V}} \left[(\hat{r}_{\text{ker}}(x) - r(x)) - Bh\right].$$

Theorem 4.5 *Assume that the conditions of Theorem 4.4 are satisfied and that $B \neq 0$. Assume also that*

$$\lim_{n \to \infty} h \sqrt{n\phi_x(h)} < \infty. \qquad (4.29)$$

Then we have:

$$U_{n,x} \xrightarrow{\mathcal{L}} \mathcal{N}(0, 1). \qquad (4.30)$$

Proof of Theorem 4.5 This result was obtained by Ferraty et al. (2007), under conditions slightly weaker than those presented here. Its proof is rather short and it combines the asymptotic expansions of the bias and variance obtained before (see Theorem 4.4), together with a standard central limit theorem for an array of i.i.d. centered real random variables. Note that, in the above-mentioned paper, condition (4.29) was omitted but it is obviously necessary in order that the term $o(h)$ appearing in the bias part of (4.28) does not disrupt the asymptotic normality. □

Note that, because of Slutsky's Theorem, (4.30) remains true if we change the quantities $\phi(h)$, B, and V by some consistent estimators $\hat{\phi}(h)$, $\hat{B}$, and $\hat{V}$.

L_p expansion for the kernel estimate

The final result on pointwise convergence that will be presented concerns the L_p errors of the estimate $\hat{r}_{\text{ker}}$.

Theorem 4.6 *Under the conditions of Theorem 4.5, we have for any $p \in \mathbb{N}$:*

$$E U_{n,x}^p \to E W^p, \qquad (4.31)$$

where W is a standard $\mathcal{N}(0, 1)$ random variable.

Proof of Theorem 4.6 This result was obtained by Delsol (2007), under conditions slightly weaker than those presented here (including non-integer values for p). The main difficulty of the proof consists in checking that the random variable $U_{n,x}^p$ is uniformly integrable. Once this is done, the convergence in distribution of $U_{n,x}$ towards W (stated before in Theorem 4.5) directly implies that the result (4.31) holds. □

Note that, subject to suitable additional condition allowing us to neglect the bias term, one directly obtains asymptotic expansions of L_p errors. This result is presented in the following corollary, which was also stated by Delsol (2007) under slightly weaker assumptions.

Corollary 4.1 *Under the conditions of Theorem 4.5, and if the bandwidth satisfies*

$$\lim_{n \to \infty} h\sqrt{n\phi_x(h)} = 0,$$

then we have, for any $p \in \mathbb{N}$,

$$E\left[(\hat{r}_{\text{ker}}(x) - E\hat{r}_{\text{ker}}(x))^p\right] = \left(\frac{V}{n\phi_x(h)}\right)^{\frac{p}{2}} E W^p + o\left(\left(\frac{1}{n\phi_x(h)}\right)^{\frac{p}{2}}\right), \qquad (4.32)$$

where W is a $\mathcal{N}(0, 1)$ random variable and V is defined as in Theorem 4.4.

Note that the same kinds of results were also obtained in Delsol (2007) for the uncentered L_p errors

$$E\left[(\hat{r}_{\text{ker}}(x) - r(x))^p\right],$$

but we have decided not to present them here because the leading terms of their asymptotic expansions are rather complicated and would require the introduction of long and tedious notation.

4.2.4 Uniform asymptotic results

Presentation

The aim of this section is to obtain uniform almost sure consistency (with rates) over some subset S of the functional space E and then to discuss how these theoretical results have great practical impact because of their possible applications to further advances in automatic bandwidth choice, bootstrapping, as well as on various other topics.

It turns out that the gap between pointwise and uniform results is not so easy to bridge in a functional setting as it is for standard multivariate nonparametric statistics. The first consequence of this difficulty is the lack of a wide variety of results, and it turns out that uniform results are only known (as far as we are aware) in a functional setting for the Nadaraya–Watson estimate $\hat{r}_{\text{ker}}$ (while extensions to other estimates such as $\hat{r}_{\text{LL}}$ or $\hat{r}_{\text{kNN}}$ should be possible, but are still open questions). This is the reason why, from now on, we concentrate entirely on the kernel estimate $\hat{r}_{\text{ker}}$. Another direct consequence of the difficulty of passing from pointwise to uniform results is the deterioriation of the rates of convergence, which will be controlled by means of Kolmogorov entropy considerations.

Entropy: a tool for controlling uniform rates of convergence

While it is well known that in multivariate nonparametric statistics the rates of convergence remain the same for pointwise as for uniform consistency (at least when the set S is compact), this is not always the case when functional variables are involved (see the final section below for more comments on this subject). Once again, the topological structure on the semi-metric space (E, d) will play a prominent role, as well of course as the complexity of the specific subset S on which uniform results are stated. A natural tool that enables this to be clearly seen consists in linking the rates of convergence of the estimate $\hat{r}_{\text{ker}}$ with the Kolmogorov entropy (see Kolmogorov and Tikhomirov (1959)) of the set S, whose definition is recalled now.

Definition 4.4 *Let $\epsilon > 0$ be fixed. Let $N_\epsilon(S)$ be the minimal number of open balls in E of radius ϵ needed to cover S. Then the ϵ-entropy of the set S is defined by*

$$\psi_S(\epsilon) = \log(N_\epsilon(S)). \tag{4.33}$$

In the following section, the rates of uniform consistency of the kernel estimate $\hat{r}_{\text{ker}}$ over the set S will be presented. For that, we need a uniform version of the small-ball probability function. We recall that $\phi_x = P_X(\mathcal{B}(x, \epsilon))$ and assume that there exists a function ϕ_S and three constants a_1, a_2, and a_3 such that, for any $\epsilon > 0$:

$$0 < a_1 \phi_S(\epsilon) \leq \inf_{x \in S} \phi_x(\epsilon) \leq \sup_{x \in S} \phi_x(\epsilon) \leq a_2 \phi_S(\epsilon) < \infty \text{ and } \phi'_S(\epsilon) \leq a_3. \quad (4.34)$$

One also needs the following technical conditions on the entropy ψ_S and on the concentration function ϕ_S:

$$\exists n_0, \forall n > n_0, \frac{(\log n)^2}{n \phi_S(h)} < \psi_S\left(\frac{\log n}{n}\right) < \frac{n \phi_S(h)}{\log n}, \quad (4.35)$$

and

$$\exists \delta > 0, \sum_{i=1}^{n} e^{-\delta \psi_S\left(\frac{\log n}{n}\right)} < \infty. \quad (4.36)$$

It is not our purpose here to discuss these assumptions. Let us just note that they are not that restrictive, in the sense that they are satisfied in most cases of statistical interest (detailed comments can be found in Ferraty *et al.* (2009)).

Rates of uniform consistency of the kernel regression estimate

The following theorem states the rate of uniform consistency of the Nadaraya–Watson functional kernel estimate $\hat{r}_{\text{ker}}$ defined before. This result allows us to estimate the regression operator r at a random point (see Corollary 4.2), and this will be a key tool for further developments (see the following three sections).

Theorem 4.7 *Assume that conditions (4.2), (4.4), (4.7), (4.34), (4.35), and (4.36) hold. Also assume that (4.5) and (4.6) hold for any $x \in S$. Then we we have*

$$\sup_{x \in S} |\hat{r}_{\text{ker}}(x) - r(x)| = O\left(h^\beta\right) + O\left(\sqrt{\frac{\psi_S\left(\frac{\log n}{n}\right)}{n \phi_S(h)}}\right), \quad a.s. \quad (4.37)$$

Proof of Theorem 4.7 A complete proof of this result can be found in Ferraty *et al.* (2009). Here we just present its main lines of argument. As for the proof of the pointwise result (see Theorem 4.1 above), it is based on the decomposition (4.9). For the treatment of the bias terms there is no additional problem linked with uniformity, and from condition (4.2) we obtain that

$$\sup_{x \in S} |E\hat{r}_{\text{ker},0}(x) - 1| = 0 \text{ and } \sup_{x \in S} |E\hat{r}_{\text{ker},1}(x) - r(x)| = O(h^\beta). \quad (4.38)$$

The techninal difficulty linked with uniformity appears when computing the dispersion terms. To make the presentation simpler, let us introduce the notation $\epsilon_n = \log n / n$, $N = N_{\epsilon_n}(S)$, and for $k = 1, \ldots, N$ let us denote by x_k the centers of the N balls of radius ϵ_n which are needed to cover the set S (see Definition 4.4).

Moreover, for each $x \in S$, we denote by $x_{k(x)}$ the center of the ball which contains x. The idea is to use, for $j = 0, 1$, the following decomposition:

$$\sup_{x \in S} |\hat{r}_{\ker,j}(x) - E\hat{r}_{\ker,j}(x)| \leq A + B + C, \qquad (4.39)$$

with

$$A = \sup_{x \in S} |\hat{r}_{\ker,j}(x) - \hat{r}_{\ker,j}(x_{k(x)})|,$$
$$B = \sup_{x \in S} |\hat{r}_{\ker,j}(x_{k(x)}) - E\hat{r}_{\ker,j}(x_{k(x)})|,$$
$$C = \sup_{x \in S} |E\hat{r}_{\ker,j}(x_{k(x)}) - E\hat{r}_{\ker,j}(x)|.$$

The term A is dealt with by using the Lipschitz property of the kernel function K, which leads to

$$A \leq \frac{C}{n} \sum_{i=1}^{n} \frac{\epsilon_n Y_i^j}{h \phi_S(h)} \mathbf{1}_{X_i \in \mathcal{B}(x,h) \cup \mathcal{B}(x_{k(x)},h)},$$

and one can use any kind of exponential-type inequality (for instance Corollary A.8-ii in Ferraty and Vieu (2006)) to get:

$$A = O\left(\sqrt{\frac{\epsilon_n \log n}{n h \phi_S(h)}}\right), \quad a.s.$$

The same thing can be done for the term C, and finally the definition of ϵ_n and condition (4.35) allow us to get

$$A + C = O\left(\sqrt{\frac{\psi_S\left(\frac{\log n}{n}\right)}{n \phi_S(h)}}\right), \quad a.s. \qquad (4.40)$$

Regarding the term B, we can write, for some $\eta > 0$:

$$P\left(B > \eta \sqrt{\frac{\psi_S(\epsilon_n)}{n \phi_S(h)}}\right) \leq N \left(\max_k P\left(|\hat{r}_{\ker,j}(x_k) - E\hat{r}_{\ker,j}(x_k)| > \eta \sqrt{\frac{\psi_S(\epsilon_n)}{n \phi_S(h)}}\right)\right).$$

By proceeding in the same way as for (4.12) and by using the Borel–Cantelli Lemma, we obtain that:

$$B = O\left(\sqrt{\frac{\psi_S\left(\frac{\log n}{n}\right)}{n \phi_S(h)}}\right), \quad a.s. \qquad (4.41)$$

Finally, Theorem 4.7 follows directly from (4.9), (4.38), (4.39), (4.40), and (4.41). □

The next result is an obvious consequence of Theorem 4.7 that will be used repeatedly later.

Corollary 4.2 *Assume that the conditions of Theorem 4.7 are satisfied. Then, for any random variable Z taking values on the functional space E, we have:*

$$|\hat{r}_{\text{ker}}(Z) - r(Z)| 1_{Z \in S} = O\left(h^{\beta}\right) + O\left(\sqrt{\frac{\psi_S\left(\frac{\log n}{n}\right)}{n \phi_S(h)}}\right), \quad a.s. \tag{4.42}$$

Application to bandwidth choice

In this section we will present some automatic data-driven bandwidth selection procedures based on cross-validation ideas, for which an asymptotic optimality result can be stated. From a technical point of view, uniform consistency results play a major role in the statement of such optimality results and that is why the literature on bandwidth selection (at least in the infinite-dimensional setting) is restricted to only the kernel estimate $\hat{r}_{\text{ker}}$.

Again let x be a fixed functional element in E. The bandwidth-choice problem consists in finding some data-driven bandwidth $\hat{h}_x = \hat{h}_x(X_1, \ldots, X_n)$ that minimizes some error of estimation, such as for instance the L_2 error:

$$L_{2,x}(h) = E(\hat{r}_{\text{ker}}(x) - r(x))^2.$$

Taking their inspiration from ideas developed in standard multivariate problems (see Härdle and Marron (1985) or Vieu (1991)), the following functional version of a local cross-validation criterion has recently been proposed by Benhenni et al. (2007):

$$CV_x(h) = \frac{1}{n} \sum_{j=1}^{n} \left(Y_j - \hat{r}_{\text{ker}}^{-j}(X_j)\right)^2 f_{n,x}(X_j), \tag{4.43}$$

and the bandwidth is defined by minimizing this data-driven criterion over some set $H = H_n$ of possible ones:

$$\hat{h}_{x,CV} = \arg\min_{h \in H} CV_x(h). \tag{4.44}$$

In this definition, $\hat{r}_{\text{ker}}^{-j}$ is the leave-one-out kernel estimate constructed by deleting the jth pair (X_j, Y_j):

$$\forall y \in E, \; \hat{r}_{\text{ker}}^{-j}(y) = \frac{\sum_{i \neq j}^{n} W_{n,\text{ker}}(y, X_i) Y_i}{\sum_{i \neq j}^{n} W_{n,\text{ker}}(y, X_i)}, \tag{4.45}$$

and $f_{n,x}$ is a real-valued positive weight function concentrated around the functional element x of interest. To fix our ideas, in what follows we will simply consider the following natural choice for $f_{n,x}$:

$$f_{n,x}(y) = 1_{y \in \mathcal{B}(x,g)}, \tag{4.46}$$

the parameter g being a sequence of positive real numbers that tend to zero more slowly than the smoothing parameter h to be selected. Of course, more general forms of weighting are possible (including, for instance, smooth versions of $f_{n,x}$).

The following conditons are necessary in order to obtain the optimality property of the cross-validated bandwidth $\hat{h}_{x,CV}$. Assume that there is some subset S such that $\mathcal{B}(x, g) \subset S \subset E$ for which the following conditions are satisfied:

$$\exists \gamma_1 > 0, \ \#H = O(n^{\gamma_1}), \tag{4.47}$$

$$\exists \gamma_2 \in (0, 1), \ \forall h \in H, \ g = Ch^{\gamma_2}, \tag{4.48}$$

$$\exists \gamma_3 > 0, \ \forall h \in H, \ \phi_S(h) = O(n^{-\gamma_3}). \tag{4.49}$$

Theorem 4.8 *Assume that the conditions of Theorems 4.4 and 4.7 are satisfied for any $h \in H$. Assume in addition that (4.46), (4.47), (4.48), and (4.49) hold. Then we have:*

$$\lim_{n \to \infty} \frac{L_{2,x}(\hat{h}_{x,CV})}{\inf_{h \in H} L_{2,x}(h)} = 1, \ a.s. \tag{4.50}$$

Proof of Theorem 4.8 This result can be found in Benhenni et al. (2007) under slightly weaker assumptions. Here we just present the main ideas of the proof. The proof is based on the following decomposition:

$$CV_x(h) = A_1(h) + A_2(h) + A_3, \tag{4.51}$$

where

$$A_1(h) = \frac{1}{n} \sum_{j=1}^{n} \left(\hat{r}_{\text{ker}}^{-j}(X_j) - r(X_j) \right)^2 f_{n,x}(X_j)$$

and

$$A_2(h) = -\frac{2}{n} \sum_{j=1}^{n} \left(\hat{r}_{\text{ker}}^{-j}(X_j) - r(X_j) \right) \left(\hat{r}_{\text{ker}}^{-j}(X_j) - Y_j \right) f_{n,x}(X_j).$$

The proof is broken down into two steps. First of all, taking inspiration from the general results stated by Marron and Härdle (1986), one can show the asymptotic equivalence between $A_1(h)$ and the quadratic error $L_{2,x}(h)$. Then, because the term A_3 does not depend on h, it suffices to show that $A_2(h)$ is of lower order than $L_{2,x}(h)$. This last point requires technical calculations, but the main ideas are to use the asymptotic expansion for $L_{2,x}(h)$ given in Theorem 4.4 and then to show that $A_2(h)$ is of lower order than the leading terms in this expansion. This last point requires repeated use of Theorem 4.7 in order to control the terms $\left(\hat{r}_{\text{ker}}^{-j}(X_j) - r(X_j) \right)$ appearing in $A_2(h)$, and this is why the conditions assumed for Theorem 4.8 are basically those needed to obtain both Theorems 4.4 and 4.7. □

It is worth noting that the same kind of result can be obtained if the pointwise L_2 error is changed into an integrated one, subject to a suitable modification of the cross-validation criterion (based on the use of a weight function which no longer depends on n). This extension is not presented here (see Rachdi and Vieu (2007)) in order to avoid introducing additional long and tedious notation. Note, finally, that these results on cross-validation are of even more interest than in the case of finite-dimensional nonparametric statistics. This is because in the functional situation it is hard to figure out how alternative techniques could be developed. For instance, while plug-in techniques are a competitive alternative to bandwidth selection rules in multivariate settings, the high complexity of the constants appearing in the L_2 asymptotic expansions (see, for instance, Theorem 4.4) does not lead us to expect too much development of these ideas in the functional setting. Perhaps the bootstrap ideas to be developed in the next section could lead to an alternative bandwidth selection method, but this has still to be developed.

Applications to bootstrapping

Another important field for which the result of Theorem 4.5 has been of major interest is bootstrapping. Keeping in mind the results that exist in standard multivariate nonparametric regression (see Mammen (2000) for a discussion and extensive references), there are many ways of thinking about bootstrapping in regression problems. Extensions to functional variables are not very easy and have not been much developed; the literature is mainly devoted to residual-type bootstrapping. This is quite natural since the residuals that appear in the functional regression model (4.1) are real random variables, and therefore the gap between the multivariate and functional settings is narrower than it would be if one was resampling from the data (X_i, Y_i) themselves.

Because of the scarcity of literature in the relatively new field of functional bootstrapping, our goal is restricted to the kernel estimate $\hat{r}_{\text{ker}}$ and to a fixed functional element $x \in E$. The main goal of bootstrapping is to provide, by suitable random generation of new samples, an approximation of the distribution of the estimation error $(\hat{r}_{\text{ker}}(x) - r(x))$ in order to overcome the problems linked with the use of the standard normal asymptotic approximation, which requires preliminary estimates of bias and variance. Given the highly complex structure one observes in the results (4.28) and (4.30), it goes without saying that bootstrapping can be expected to be even more interesting with functional data than it is in standard nonparametric statistics. It is not our purpose here to present a precise statement of the theoretical behavior of bootstrapping in functional problems, because it would require a long list of specific notation and conditions. We will simply explain how the procedure is constructed; we refer to Ferraty *et al.* (2010) for a precise statement of its asymptotic validity.

Residual bootstraping uses a second bandwidth which is denoted by b; the corresponding kernel estimate is denoted by $\tilde{r}_{\text{ker}}$. The procedure consists in repeating the following steps N_B times:

(i) Compute $\hat{\epsilon}_i = Y_i - \tilde{r}_{\text{ker}}(X_i)$ for $i = 1, \ldots, n$.
(ii) Draw, for $i = 1, \ldots, n$, new residuals $\tilde{\epsilon}_i$ and then draw new responses $\tilde{Y}_i = \tilde{r}_{\text{ker}}(X_i) + \tilde{\epsilon}_i$ (this step will be explained in detail later at the end of this section).
(iii) Compute the kernel estimate as in Definition 4.1 but using the new sample $(X_i, \tilde{Y}_i), i = 1, \ldots, n$, (and denote by $\hat{r}_{\text{ker}}^b$ this new estimate).

Then we approximate the distribution of the theoretical error $(\hat{r}_{\text{ker}}(x) - r(x))$ using the empirical distribution (over the N_B replications) of the bootstrapped errors $(\hat{r}_{\text{ker}}^b(x) - \tilde{r}_{\text{ker}}(x))$. It is shown in Ferraty et al. (2010) that both distributions are asymptotically equivalent, subject to conditions controlling the respective asymptotic behaviors of h and b and subject to the way the bootstrapped residuals appearing in step (ii) are constructed. Without entering into the technical details of this proof, it is worth mentioning that step (i) of the procedure requires that the asymptotic behavior of the kernel estimate taken at each random element X_i is controlled and once again the uniformity result (and, more precisely, Corollary 4.2) plays a major role.

Let us mention that there are many ways to generate bootstrapped residuals $\tilde{\epsilon}_i$. The most popular ways for performing step (ii) are the following:

(ii*a*) The naive bootstrap, consisting in simply performing a random permutation of the original residuals $\hat{\epsilon}_i$; this approach is convenient for homoscedastic situations;
(ii*b*) The wild bootstrap, consisting in generating new variables U_i and defining $\tilde{\epsilon}_i = \hat{\epsilon}_i U_i$; this approach is convenient even in heteroscedastic situations as long as the U_i are chosen in such a way that $\tilde{\epsilon}_i$ and $\hat{\epsilon}_i$ have the same first moments (see Härdle and Marron (1991), for an example of a probability distribution available for the variables U_i).

In conclusion, let us mention that such a boostrapped approximation of the distribution error can be useful in various practical problems. Its use for confidence-interval construction has been developed in Ferraty et al. (2010). Its use for data-driven bandwidth selection is still an open problem, but the multivariate ideas discussed by Härdle and Bowman (1988) could certainly be adapted in some way to functional situations.

Other applications of uniform consistency results

It is not possible to list all the possible applications of the uniform consistency results stated above. The main reason for this is that such kinds of results (specifically, results like Corollary 4.2) are needed as long as asymptotic properties are expected for multistage estimation procedures. This can clearly be seen above in the sections on bandwidth selection using the leave-one-out estimate (4.45) and also the section on bootstrapping using the preliminary step of the estimation of residuals (step (i) of the procedure). Therefore an obvious field of application of this result is (or will be) the study of models involving more than one operator, such as the additive model that will be studied in Section 4.4 or the various semiparametric models that will be studied in Section 4.3. Since the study of semiparametric models/estimates for functional data is a very young field, future statistical research will certainly make extensive use of these uniform results.

4.2.5 Comments and brief bibliographical survey

The aim of this section is twofold. Firstly, we comment on the hypothesis introduced earlier in order to get asymptotic results. Note that, because the general structure of the results presented above is not specific to the nonparametric setting but is common to the next sections of this chapter, the comments on the general results themselves (and, more specifically, on rates of convergence) are referred to the concluding Section 4.6. Our second aim here is to provide a short bibliographical survey for nonparametric functional regression. At the end we pay specific attention to functional discrimination, which can be seen as a special case of the general functional regression framework.

Comments on the hypothesis

As we have said in the previous sections, we have not attempted to present the results under the most general hypotheses, but have rather preferred to construct general sets of possible hypotheses that could apply to more than one theorem. The bibliographical references mentioned earlier will help readers to obtain most of the technical details. However, even if some of our conditions may appear rather technical and complicated, it should be emphasized that they are sufficiently unrestrictive to open the way for many applications.

The most important conditions that allow us to deal with the functional setting are those linked with the small-ball probability function ϕ_x. It is easy to check that all the conditions are trivially satisfied if we restrict our results to the standard multivariate situation $E = \mathbb{R}^k$ and, furthermore, they allow us to extend many earlier papers in these fields to situations in which the explanatory variable does not necessarily admit a density with respect to the Lebesgue measure on $\mathbb{R}^k$. More generally, they are trivially satisfied as long as the functional process X is of fractal type (that is, when for some $a > 0$, $\phi_x(\epsilon) \sim C\epsilon^a$). Even if we know that we can always construct a topology associated with a suitable semi-metric d such that any process X is fractal, it is worth noting that our assumptions on ϕ_x are satisfied in much more general situations than fractal ones. All the details on these questions can be found in the general discussion provided in Chapter 13 of the monography by Ferraty and Vieu (2006a). In the same spirit and as discussed in Ferraty et al. (2009), the topological conditions imposed on the entropy function $\psi_S(\epsilon)$ are satisfied in a wide range of situations (including the multivariate and fractal ones, but also many others). This is also true of the conditions needed on the functions τ and ξ in Sections 4.4, 4.5, and 4.6 for specifying the exact leading terms of L_p errors (see Ferraty et al. (2007)).

Finally, the conditions imposed on the parameters of the estimates (namely, kernels and bandwidths) are not restrictive in the sense that they are similar to those widely used in multivariate nonparametric statistics.

Further literature on functional regression

Further literature on kernel functional regression

The kernel estimate $\hat{r}_{\text{ker}}$ has been studied quite a lot in the literature since its introduction by Ferraty and Vieu (2000, 2002). The study of its pointwise almost-sure consistency properties has been described in detail in Ferraty and Vieu (2006a), and these results have been extended to not-necessarily-independent pairs (X_i, Y_i) (with direct applications to functional time-series analysis) by Ferraty et al. (2002), Ferraty and Vieu (2004, 2008), and Benhenni et al. (2008). Chapter 5 of this book is especially devoted to this functional time-series setting. Regarding the extension of the L_p pointwise convergence of the kernel estimate under dependence conditions, the reader is referred to Delsol (2007), while asymptotic normality under dependence conditions has been obtained by Masry (2005) and Delsol (2008a). With regard to uniform consistency properties, the literature is restricted to Ferraty and Vieu (2004, 2008) and Ferraty et al. (2009). To conclude this short survey, let us note that a robust version of $\hat{r}_{\text{ker}}$ has been studied in Crambes et al. (2008). Let us also indicate that this estimate has been revisited in the special case when E is a finite-dimensional manifold by Pelletier (2006), and in the special discrimination problem (that is, when Y only takes a finite number of values) by Ferraty and Vieu (2003) and Abraham et al. (2006).

Further literature on local linear functional regression

The literature on local linear regression for functional variables is much less extensive. The first contribution to this topic was that of Barrientos-Marin (2007), and up to now only pointwise consistency results have been available; the result we presented earlier was taken from these. The reader may also look at the recent work by Baíllo and Grané (2008), Boj et al. (2008), and Barrientos-Marin et al. (2010) for similar results. See also Aneiros *et al.* (2008) for similar methods in a time-series framework.

Further literature on kNN regression

Regarding the asymptotic properties of the kNN functional estimate $\hat{r}_{kNN}$, the literature is in general restricted to the rate of pointwise almost-sure consistency, which we presented above and which was taken from Burba *et al.* (2009) (see also Burba *et al.* (2008) for a consistency result without rate). However, it is worth noting that kNN ideas are very popular in the classification community and a few papers have been devoted to the special case of the regression model (4.1) when Y takes only a finite number of values. This specific regression model has direct applications in functional data discrimination; references in this direction include Ferraty and Vieu (2003), Biau *et al.* (2005), and Cerou and Guyader (2006). The next section is especially devoted to such discrimination problems.

On functional discrimination

It is worth noting that all the results presented earlier in this chapter apply directly to situations in which the response variable only takes a finite number of values, let us say $Y \in \{1, \ldots, G\}$. This is typically the case when one has to deal with supervised clasification. In this situation, one has a sample of functional objects $X_1, \ldots, X_n$, each of them being connected to one of the specific groups $1, \ldots, G$, and one wishes to estimate, for each $g = 1, \ldots, G$, the function:

$$R_g(\cdot) = P(Y = g | X = \cdot).$$

Then, once each operator R_g has been estimated (let us say by some estimator $\hat{R}_g$), one can use this to assign a group to a new functional object x simply by taking the value $\hat{g}_x$ such that

$$\hat{g}_x = \max_{\{g=1,\ldots,G\}} \hat{R}_g(x).$$

Once we have noted that each operator R_g can be seen as a specific regression operator, such as those defined in (4.1), with response variable

$$Z_g = 1 \text{ if } Y = g$$
$$Z_g = 0 \text{ if } Y \neq g,$$

it is easy to see that all the literature devoted to functional regression (including all of what has been presented above in this chapter) can be directly used in functional discrimination. An extensive discussion on nonparametric functional discrimination can be found in Chapter 8 of Ferraty and Vieu (2006*a*).

4.3 Kernel methods in semiparametric functional regression

4.3.1 Presentation

Semiparametric ideas have been widely developed in standard multivariate statistics (see, for instance, Bickel *et al.* (1993), Horowitz (1998), and Härdle *et al.* (2004) for general monographs and Sperlich *et al.* (2006) for a selection of the most recent developments). The main goal of multivariate semiparametric statistics is to achieve a trade-off between models that are too flexible (as nonparametric ones can be when the dimension of the covariables is high) and too restrictive (as linear models, or more generally parametric models, can be). In other words, semiparametric modeling is a nice way to reduce dimensionality effects in multidimensional statistics. Therefore it is quite natural to expect a major development of semiparametric ideas in the functional setting which, roughly speaking, corresponds to some infinite-dimensional situation. The newness of this topic means that there are not many developments available in the literature, and as far as we know developments only concern two specific models: the single functional index model and the partial linear functional model. The aim of this section is to present the few theoretical advances that exist concerning these models. In Chapter 1 of this volume the reader can find a more general discussion on semiparametric functional models, including a precise definition of what is called a "semiparametric functional," as well as various other models whose investigation is still an open problem.

A major common feature of semiparametric estimation (see Definition 1.5 of Chapter 1) is that we involve one, or more than one, functional nonlinear operators which have to be estimated nonparametrically. This is why the general nonparametric functional estimation techniques presented above in Section 4.2 will play a major role in semiparametric statistics. Once again because of the youth of semiparametric functional statistics as a research area, all the advances in this field concern only kernel-type estimates (such as those defined in Definition 4.1). Thus all that follows in Section 4.3 concerns such kinds of estimates, while extensions to other types of estimates (such as kNN or local linear estimates) are still open and challenging problems.

4.3.2 Single functional index regression

Presentation of the model

The single functional index model consists in assuming that the functional variable X acts on the response Y only through its projection on some (unknown) fixed functional element θ_0. The formulation of this idea requires us to reinforce the conditions on the space E on which X takes its values. More precisely, to insure the existence of the projection it is necessary to assume that E is a Hilbert space endowed with an inner product $\langle \cdot, \cdot \rangle_E$ (and an associated norm $||\cdot||_E$). The following model was first introduced in Ferraty et al. (2003) where its identifiability was studied.

Definition 4.5 *The single functional index regression model is defined by assuming that the regression model defined in (4.1) can be rewritten as*

$$Y = r(X) + \epsilon = g(\langle X, \theta_0 \rangle_E) + \epsilon, \qquad (4.52)$$

where g is an unknown real function and θ_0 is an unknown parameter in E, and where $E(\epsilon|X) = 0$.

From now on we will assume the identifiability of this model. General conditions for insuring this identifiability are given in Ferraty et al. (2003). This is clearly a semiparametric model (see Definition 1.5 of Chapter 1) with a nonparametric component g and a parametric component $\theta \in E$. The statistical problem consists in estimating the operator g and the parameter θ_0 together from a sequence of n independent pairs (X_i, Y_i) each with the same distribution as (X, Y).

Kernel estimation

There are many ways to attack this problem in standard multivariate situations (that is, when $E = \mathbb{R}^p$), a selection of references includes Härdle et al. (1993), Hristache et al. (2001), and Delecroix et al. (2006). The literature for functional settings is much less extensive and, as far as we know, only the following two-stage estimation procedure has been studied.

First of all one considers, for each functional element $\theta \in E$, the following real-valued operator:

$$\forall x \in E, \; r_\theta(x) = E(Y|\langle X, \theta \rangle = x).$$

Note that, under model (4.52), we have:

$$r(\cdot) = g(\langle \cdot, \theta_0 \rangle_E) = r_{\theta_0}(\cdot).$$

It is natural to estimate r_θ using kernel smoothing ideas, and this leads to:

$$\forall x \in E, \; \hat{r}_\theta(x) = \sum_{i=1}^n \frac{Y_i K\left(\frac{|\langle x - X_i, \theta \rangle_E|}{h}\right)}{K\left(\frac{|\langle x - X_i, \theta \rangle_E|}{h}\right)}. \qquad (4.53)$$

It is worth noting that, for a fixed value of θ, the study of the estimate $\hat{r}_\theta$ is not that complicated since it can be viewed as a special case of the kernel functional estimate (see Definition 4.1) in which the semi-metric is taken to be

$$\forall (u, v) \in E^2, \ d_\theta(u, v) = |\langle u - v, \theta \rangle_E|. \tag{4.54}$$

In a second step, we look at the value of θ for which the estimate $\hat{r}_\theta$ is closest to the true unknown operator r_{θ_0} with respect to some measure of accuracy such as, for instance, the following mean integrated squared error:

$$MISE(\hat{r}_\theta) = \int_E E\left(\hat{r}_\theta(x) - r_{\theta_0}(x)\right)^2 dP_X(x). \tag{4.55}$$

Of course, because r_{θ_0} is unknown, this measure of error is uncomputable in practice, and a standard data-driven approximation of it consists in introducing the following cross-validation criterion:

$$CV(\theta) = \frac{1}{n} \sum_{j=1}^{n} \left(Y_j - \hat{r}_\theta^{-j}(X_j)\right)^2, \tag{4.56}$$

where

$$\forall y \in E, \ \hat{r}_\theta^{-j}(y) = \frac{\sum_{i \neq j}^{n} Y_i K\left(\frac{|\langle x - X_i, \theta \rangle_E|}{h}\right)}{\sum_{i \neq j}^{n} K\left(\frac{|\langle x - X_i, \theta \rangle_E|}{h}\right)}. \tag{4.57}$$

Now, the functional index θ is estimated by minimizing this criterion over some set of possible values Θ (which is allowed to depend on n), to be discussed later:

$$\hat{\theta} = \arg\min_{\theta \in \Theta} CV(\theta). \tag{4.58}$$

Finally, under model (4.52) the regression operator r is estimated by means of:

$$\forall x \in E, \ \hat{r}_{\text{SFIM}}(x) = \hat{r}_{\hat{\theta}}(x). \tag{4.59}$$

The section on asymptotics below will be devoted to the statement of theoretical results that prove the good asymptotic performance of this two-stage procedure; first let us state and comment on the general conditions needed to insure the validity of the procedure.

General assumptions

As discussed before, the first step of the procedure consists, for each θ, in estimating the real-valued operator r_θ. The estimator $\hat{r}_\theta$ defined by (4.53) can be seen as a special case of the kernel estimate $\hat{r}_{\text{ker}}$ with the particular semi-metric d_θ defined in (4.54). So it is not surprising that we need the same kinds of assumptions as those introduced in Section 4.2, subject of course to suitable adaptation to this new semi-metric d_θ.

General conditions on the model

The general modeling conditions on the distribution of (X, Y) are summarized, on the one hand (and in a similar way to what was done in (4.2)), by assuming that the unknown operators r_θ satisfy the following Lipschitz regularity assumption:

$$\exists \beta > 0, \exists C < +\infty, \forall \theta \in \Theta, \forall (x, x') \in E^2,$$

$$|r_\theta(x) - r_\theta(x')| \leq C d_\theta(x, x')^\beta, \tag{4.60}$$

and on the other hand by assuming the same moment conditions on the variable Y as in (4.7). It is now worth introducing some conditions on the small-ball probability function. These conditions will be much simpler than those used in the general case because the semi-metric d_θ acts only on the one-dimensional variable $\langle X, \theta \rangle$, and not really on the infinite-dimensional variable X as was the case in the general framework of Section 4.2. The conditions can be summarized by assuming that, when $\epsilon \to 0$, the following conditions are almost surely satisfied:

$$\forall \theta \in \Theta, \, P(d_\theta(X, X_1) \leq \epsilon | X) \sim \epsilon c_{X,\theta}, \text{ with } 0 < c_1 \leq c_{X,\theta} \leq c_2 < \infty. \tag{4.61}$$

It is easy to see that these conditions are not very restrictive since they are satisfied whenever the real random variables $\langle X, \theta \rangle$ have densities (having upper and lower bounds) with respect to the Lebesgue measure on $\mathbb{R}$.

Conditions needed for kernel estimation at a fixed value of θ

It now remains for us to state the conditions on the parameters of the estimate (that is, on the kernel function K and on the smoothing parameter h). The general assumptions (4.4), (4.5), and (4.6) have to be reinforced by assuming that the bandwidth is such that

$$\exists (d_1, d_2), \exists n_0, 0 < d_1 < d_2 < 1 \text{ and } \forall n > n_0, n^{-d_2} \leq h \leq n^{-d_1}, \tag{4.62}$$

and by assuming that the kernel function K is strictly decreasing and such that

$$\exists (k_1, k_2), 0 < k_1 \leq k_2 < \infty, \, k_1 1_{[0,1]}(t) \leq K(t) \leq k_2 1_{[0,1]}(t). \tag{4.63}$$

Conditions needed for the estimation of θ

To insure the good behavior of the cross-validation criterion CV as a data-driven approximation of the true theoretical error *MISE*, we must assume that the set Θ of possible values of θ is not too large. More precisely, the following cardinality condition is needed:

$$\exists a > 0, \exists C < \infty, \#\Theta = C n^a. \tag{4.64}$$

Finally, for simplicity of presentation, we assume that there exists some compact subset $S \subset E$ such that:

$$X \in S, \, a.s. \tag{4.65}$$

While this last condition looks rather restrictive, it can be easily passed over simply by introducing into the various criteria (that is, on *CV* and *MISE*) some weight function having support on S, exactly as we did before for bandwidth selection.

Some asymptotics

The main result of this section is the next theorem which states the L_2 rates of convergence of the two-stage estimate $\hat{r}_{\text{SFIM}}$. This result is taken from the recent paper by Ait-Saïdi *et al.* (2008). It will not be proved in detail here, but the main lines of argument will be described. The complete proof can be found in the above-mentioned paper.

Theorem 4.9 *Assume that the model* (4.52) *holds with* $\theta_0 \in \Theta$. *Under conditions* (4.7), (4.60), (4.61), (4.62), (4.63), (4.64), *and* (4.65), *we have:*

$$\int_E E\left(\hat{r}_{\text{SFIM}}(x) - r_{\theta_0}(x)\right)^2 dP_X(x) = O\left(h^{2\beta}\right) + O\left(\frac{1}{nh}\right), \text{ a.s.} \qquad (4.66)$$

Proof of Theorem 4.9 The complete proof can be found in Ait-Saïdi (2008). It is broken down into two main intermediary results.

(1) Firstly, we look for results concerning the kernel estimate $\hat{r}_\theta$, for each fixed functional element θ. One can apply the general uniform consistency results obtained earlier in Theorem 4.7 and Corollary 4.2, noting that condition (4.61) implies that the small-ball probability defined in (4.34) is now such that

$$\phi_S(\epsilon) \sim C\epsilon, \text{ as } \epsilon \to 0,$$

and that the entropy function ψ_S defined in (4.33) is such that

$$\psi_S(\epsilon) \sim -\log(\epsilon), \text{ as } \epsilon \to 0.$$

In summary, result (4.37) becomes:

$$\sup_{x \in S} |\hat{r}_\theta(x) - r_\theta(x)| = O\left(h^\beta\right) + O\left(\sqrt{\frac{\log n}{nh}}\right), \text{ a.s.} \qquad (4.67)$$

A similar route can be followed for the L_2 errors of the estimates $\hat{r}_\theta$. Since they are special cases of the estimate $\hat{r}_{\text{ker}}$, one can use the bounds obtained in (4.11), by taking $\phi_x(h) = h$, to show that:

$$\text{var}(\hat{r}_\theta(x)) = O\left(\frac{1}{nh}\right), \qquad (4.68)$$

the O being uniform on $x \in S$ because of the compactness of S. On the other hand, the first part of (4.67) insures, once again uniformly in $x \in S$, that:

$$E\hat{r}_\theta(x) - r_\theta(x) = O(h^\beta). \qquad (4.69)$$

Both results (4.68) and (4.69) can be summarized in the following:

$$\int_E E\,(\hat{r}_\theta(x) - r_\theta(x))^2 \, dP_X(x) = O(h^{2\beta}) + O\left(\frac{1}{nh}\right). \quad (4.70)$$

(2) Secondly, we look at the second step of the procedure linked with the cross-validated estimation $\hat{\theta}$ of θ_0. By proceeding as we did earlier for bandwidth selection, we can show by following standard arguments on fractional delta-sequence estimators (see Marron and Härdle (1986)) that the integrated error *MISE* is asymptotically equivalent (uniformly on $\theta \in \Theta$) to its empirical version:

$$A_1(\theta) = \frac{1}{n}\sum_{j=1}^n \left(r_{\theta_0}(X_j) - \hat{r}_\theta^{-j}(X_j)\right)^2.$$

Hence, for some quantity A_3 independent of $\theta \in \Theta$ we can write:

$$CV(\theta) = A_1(\theta) + \frac{2}{n}\sum_{j=1}^n \left(Y_j - r_{\theta_0}(X_j)\right)\left(r_{\theta_0}(X_j) - \hat{r}_\theta^{-j}(X_j)\right) + A_3, \quad (4.71)$$

and we can show that (once again uniformly in $\theta \in \Theta$) that

$$\frac{2}{n}\sum_{j=1}^n \left(Y_j - r_{\theta_0}(X_j)\right)\left(r_{\theta_0}(X_j) - \hat{r}_\theta^{-j}(X_j)\right) = o\,(MISE(\hat{r}_\theta)). \quad (4.72)$$

It is worth noting that, because of the structure of the terms of *MISE*, *CV*, and A_1, the uniform consistency result (4.67) will be used repeatedly both for proving the equivalence betwen *MISE* and A_1 and for checking (4.72). Finally one arrives at the following asymptotic optimality property:

$$\lim_{n\to\infty} \frac{MISE(\hat{r}_{\hat{\theta}})}{\inf_{\theta\in\Theta} MISE(\hat{r}_\theta)} = 1, \text{ a.s.} \quad (4.73)$$

Now, because $\theta_0 \in \Theta$, we get from (4.73) that

$$MISE(\hat{r}_{\hat{\theta}}) \leq MISE(\hat{r}_{\theta_0}), \quad (4.74)$$

and result (4.66) follows directly from (4.70). □

Comments

It is worth noting that the main point in the proof of Theorem 4.9 is the statement of the asymptotic optimality property (4.73) for the cross-validated functional index $\hat{\theta}$. This result may prove interesting for many other purposes than that of the statement of (4.66), and we note that it has been stated without needing to use the strong condition that $\theta_0 \in \Theta$. Indeed, this condition is only needed when writing (4.74) in order to close the gap between the optimality property (4.73) and the required result (4.66). Note that this last gap could also be closed without the assumption that $\theta_0 \in \Theta$, but the rate of convergence would be more complicated since it would include an additional term depending on the quantity $\Delta = \inf_{\theta \in \Theta} ||\theta - \theta_0||_E$. Clearly, consistency without the condition $\theta_0 \in \Theta$ would require (at least) that Δ tends to zero as n grows to infinity. These empirical ideas still need to be precisely formulated.

To highlight the interest of the optimality property (4.73), one may note that it can be used for stating asymptotic properties of the index estimate $\hat{\theta}$ itself. The following result is just an example of this point:

Corollary 4.3 *If the operator $\theta \to r_\theta$ is a one-to-one correspondance on Θ, and if the conditions of Theorem 4.9 are satisfied, then we have:*

$$||\hat{\theta} - \theta_0||_E \to 0, \text{ in probability.} \tag{4.75}$$

Proof of Corollary 4.3 To save space, this proof is just outlined here; a complete version can be found in Ait-Saïdi *et al.* (2008). First of all we note that the one-to-one correspondance between θ and r_θ insures that

$$\forall \theta \in \Theta, \forall \epsilon > 0, \exists \eta > 0, ||\theta - \theta_0||_E > \epsilon \Rightarrow I(\theta) > \eta, \tag{4.76}$$

where

$$I(\theta) = \int \left(r_\theta(x) - r_{\theta_0}(x)\right)^2 dP_X(x).$$

Now, by again using standard arguments on fractional delta-sequence estimates as described in Marron and Härdle (1986) and as we did earlier in the proof of Theorem 4.9 when checking that *MISE* was equivalent to its empirical version A_1, it can be shown that *MISE* is asymptotically equivalent (uniformly over $\theta \in \Theta$) to its stochastic version $I(\theta)$. Finally, this equivalence together with (4.66) and (4.76) will be enough to prove Corollary 4.3. □

By following the same route, and subject to extra technical conditions linked with the differentiability of the operator *MISE*, we could expect to obtain the rates of convergence given in Corollary 4.3, but this result still remains to be stated precisely. In fact, many open questions remain since, as far as we know, the literature on the single functional index model is restricted to three papers. The oldest one, Ferraty *et al.* (2003), studies the identifiability of the model (4.52) and states a pointwise version of (4.67). The most complete paper is that by Ait-Saïdi *et al.* (2008), from which all the results presented earlier are taken. The preliminary step (4.67) has been extended to dependent variables by Ait-Saïdi *et al.* (2005). Interesting open theoretical questions concern the use of alternative nonparametric functional estimates (such as, for instance, kNN or local linear estimates as described earlier in Definitions 4.2 and 4.3), or the use of alternative ways of estimating the functional parameter θ_0, or the question of simultaneously choosing a data-driven bandwidth $\hat{h}$ and a data-driven parameter $\hat{\theta}$, etc. From a practical point of view, the main open issue is the construction of the set of possible directions Θ.

4.3.3 Partial linear functional regression

This section is devoted to the presentation of recent advances in the partial linear modeling of functional variables. We wish to acknowledge the assistance of German Aneiros-Pérez, who is an international expert in this field, and who kindly agreed to review and correct a preliminary draft of this section.

Presentation of the model

Partial linear modeling ideas are developed in situations in which the explanatory variable X is composed of two parts, one acting nonparametrically on the response Y and the other one acting in a linear way. These ideas have been widely studied in multivariate settings, that is, when $X \in \mathbb{R}^{p_1} \times \mathbb{R}^{p_2}$. A selection of recent references includes Speckman (1988), Chen (1988), Härdle *et al.* (2000), Schick (1996), Aneiros-Pérez and Quintela del Rio (2001, 2002), Aneiros-Pérez (2002), Aneiros-Pérez *et al.* (2004), and Tong *et al.* (2008). The aim of this section is to present recent extensions of these ideas to situations in which X is functional.

Let us consider the standard regression model with functional covariate, as defined in (4.1), but we now assume that the functional covariate can be broken down into a functional component V and a multivariate one U. The partial linear functional model consists in assuming that the variable $X = (U, V)$ acts on the response Y in such a way that the action of the finite-dimensional component U is linear and the action of the infinite-dimensional component is nonparametric. In what follows we assume that $U = (U_1, \ldots, U_p)^t \in \mathbb{R}^p$ with $\mathbb{R}^p$ endowed with the Euclidean norm $\|\cdot\|$, and that V takes values in an abstract space F endowed with a semi-metric d_F.

Definition 4.6 *The single partial linear functional regression model is defined by assuming that the regression model defined in (4.1) can be rewritten as*

$$Y = r(X) + \epsilon = \gamma^t U + m(V) + \epsilon, \qquad (4.77)$$

where $\gamma = (\gamma_1, \ldots, \gamma_p)^t$ is an unknown vector in $\mathbb{R}^p$, where m is an unknown functional (not neceessarily linear) operator acting on the space F, and where $E(\epsilon|X) = 0$.

This is clearly a semiparametric model (see Definition 1.5 of Chapter 1), with a nonparametric component and a p-dimensional linear one. The statistical problem consists in estimating the operator m and the multivariate parameter $(\gamma_1, \ldots, \gamma_p)$ together, from a sequence of n independent variables $(X_i, Y_i) = (U_i, V_i, Y_i)$ each having the same distribution as (X, Y). We will also need to use the following vectors:

$$U_i = (U_{i,1}, \ldots, U_{i,p})^t, \ \gamma = (\gamma_1, \ldots, \gamma_p)^t, \ \mathbf{Y} = (Y_1, \ldots, Y_n)^t,$$

and the following $n \times p$ matrix:

$$\mathbf{U} = (\mathbf{U}_1, \ldots, \mathbf{U}_n)^t.$$

Kernel estimation

Exactly as before for the single functional index model, model (4.77) can be studied in two stages. The first stage consists in estimating the linear (finite-dimensional) component parameter γ by means of standard linear regression techniques (let us say that $\hat{\gamma}$ is such an estimate). Then a nonparametric functional technique can be used to estimate the nonlinear infinite-dimensional component m simply by performing kernel regression of the residual (one-dimensional) variable $Y - \hat{\gamma}^t U$ on the functional explanatory variable V.

The nonparametric functional methods presented above in Section 4.2 (and, more specifically, the kernel regression estimate proposed in Definition 4.1) will clearly be those of principal interest for dealing with the functional variable V. Let us first set out some general notation linked with kernel smoothing. Let us denote by $\mathbf{I_n}$ the $n \times n$ identity matrix, and by $\mathbf{W_{n,\text{ker}}}$ the $n \times n$ weighting matrix

$$\mathbf{W_{n,\text{ker}}} = \left(\frac{W_{n,\text{ker}}(V_i, V_j)}{\sum_{k=1}^n W_{n,\text{ker}}(V_i, V_k)} \right)_{i,j=1,\ldots,n}.$$

Recall that the weights $W_{n,\text{ker}}$ are given (see Definition 4.1) by

$$\forall (v, v') \in F \times F, \ W_{n,\text{ker}}(v, v') = K\left(\frac{d_F(v, v')}{h} \right).$$

We will make use of the random elements

$$\tilde{\mathbf{Y}} = \left(\mathbf{I_n} - \mathbf{W_{n,\text{ker}}} \right) \mathbf{Y} \text{ and } \tilde{\mathbf{U}} = \left(\mathbf{I_n} - \mathbf{W_{n,\text{ker}}} \right) \mathbf{U}.$$

The linear finite-dimensional component γ is estimated by means of

$$\hat{\gamma} = (\tilde{\mathbf{U}}^t \tilde{\mathbf{U}})^{-1} \tilde{\mathbf{U}}^t \tilde{\mathbf{Y}}. \tag{4.78}$$

Then the functional operator m is estimated by using kernel regression of the residual

$$E_i = \left(Y_i - \mathbf{U}_i^t \hat{\gamma} \right),$$

on the functional covariate V_i. More precisely, this idea yields the following kernel-type estimate

$$\hat{m}_{\text{ker}}(v) = \frac{\sum_{i=1}^n W_{n,\text{ker}}(v, V_i) \left(Y_i - \mathbf{U}_i^t \hat{\gamma} \right)}{\sum_{i=1}^n W_{n,\text{ker}}(v, V_i)}. \tag{4.79}$$

Finally, the estimate of the regression operator r under model (4.77) is defined by

$$\forall x = (\mathbf{u}, v) \in E = \mathbb{R}^p \times F, \ \hat{r}_{\text{PLFR}}(x) = \hat{\gamma}^t \mathbf{u} + \hat{m}_{\text{ker}}(v). \tag{4.80}$$

General assumptions

The first set of conditions that we need is composed of standard hypothesis to deal with the linear multivariate component γ. More precisely, one needs a link between the multivariate variable U and the functional variable V. It is assumed that

$$\eta = (\eta_1, \ldots, \eta_p)^t \text{ is independent of } \epsilon, \tag{4.81}$$

where, for any $j = 1, \ldots, p$, the variable η_j is defined through the regression model

$$U_j = g_j(V) + \eta_j = E(U_j|V) + \eta_j.$$

The following standard condition is required:

$$E\eta\eta^t \text{ is a positive definite matrix}. \tag{4.82}$$

The second set of conditions concerns the estimation of the functional component m and is, roughly speaking, composed of all the hypotheses which are necessary in order to make use of results on uniform consistency such as those in Theorem 4.7 and Corollary 4.2 stated earlier. More precisely, we need a nonparametric model for each functional component, that is, for $f \in \{m, g_1, \ldots, g_p\}$ we assume that:

$$\exists \beta > 0, \forall (v, v') \in S_F \times S_F, |f(v) - f(v')| \leq Cd_F(v, v')^\beta, \tag{4.83}$$

where S_F is a subset of F such that

$$P(V \in S_F) = 1. \tag{4.84}$$

Recall that d_F is the semi-metric on F, and denote by ϕ_{S_F} the corresponding small-ball probability function (such as that defined in (4.34)) and by ψ_{F_S} the corresponding entropy function (such as that defined in (4.33)). The following conditions need to be satisfied in order to make possible the use of results of the same kind as in Theorem 4.7:

$$\text{Conditions (4.34), (4.35), and (4.36) hold for } d_F, S_F, \phi_{S_F}, \text{ and } \psi_{S_F}. \tag{4.85}$$

Similarly, regarding the parameters of the estimate (that is, the bandwidth h and the kernel K) one requires that

$$\text{Conditions (4.5) and (4.6) hold for any } x \in S. \tag{4.86}$$

The conditional moment condition (4.7) has to be adapted by assuming existence and continuity of:

$$\forall j = 1, \ldots, p, \forall m \geq 2, E\left(|U_j|^m | V = \cdot\right) \text{ and } E\left(|Y - \gamma^t U|^m | V = \cdot\right). \tag{4.87}$$

Finally, because the construction of the linear estimate $\hat{\gamma}$ involves a preliminary nonparametric step (through the term $\tilde{Y}$), we need a last condition insuring that the error appearing in the nonparametric estimation does not alter the rate of convergence of the linear component. More precisely, one requires

$$\max\left\{nh^{4\beta};\ \frac{\log^4 n}{n\phi_{S_F}^2(h)};\ \frac{\psi_{S_F}^2\left(\frac{\log n}{n}\right)}{n\phi_{S_F}^2(h)};\ \frac{\log^2 n\psi_{S_F}\left(\frac{\log n}{n}\right)}{n\phi_{S_F}^2(h)}\right\} \to 0, \tag{4.88}$$

and

$$\exists \tau > 1/2,\ \frac{\phi_{S_F}(h)\log n^2}{n^{\tau-1}} \to \infty. \tag{4.89}$$

Some asymptotics

The following result is extracted from Aneiros-Pérez et al. (2006). We present it here in a slightly different form than in the above-mentioned paper, simply to avoid the introduction of additional technical conditions and to retain a basic common structure in all of this chapter. The next result concerns uniform almost-sure consistency (with rates) over some set $S = \mathcal{C} \times S_F \subset \mathbb{R}^p \times F$, $\mathcal{C}$ being a compact subset of $\mathbb{R}^p$.

Theorem 4.10 *Assume that model (4.77) holds. Under conditions (4.4), (4.81), (4.82), (4.83), (4.84), (4.85), (4.86), (4.87), (4.88), and (4.89), we have:*

$$\sup_{x \in \mathcal{C} \times S_F} |\hat{r}_{\text{PLFR}}(x) - r(x)| = O\left(\sqrt{\frac{\log\log n}{n}}\right) + O\left(h^\beta\right)$$

$$+ O\left(\sqrt{\frac{\psi_S\left(\frac{\log n}{n}\right)}{n\phi_S(h)}}\right),\ a.s. \tag{4.90}$$

Proof of Theorem 4.10 We just give the main steps of the proof, emphasizing aspects that are linked with the estimation of the functional operator m. A complete proof can be found in Aneiros-Pérez et al. (2006) under slightly different conditions.

(1) The first step of the proof consists in showing that the linear coefficient γ has the same asymptotic properties in the partial linear model (4.77) as it does when it is estimated in the simple linear multivariate model. More precisely, one can show by standard techniques that for any $j = 1, \ldots, p$ and for some finite real constants C_j we have

$$\limsup_{n \to \infty} \sqrt{\frac{n}{\log\log n}} |\hat{\gamma}_j - \gamma_j| = C_j. \tag{4.91}$$

We note that the proof of this result is rather long and technically difficult, but it is not surprising since it follows (globally) the same route as described, for instance, in Liang (2000) in the standard partial linear model when V is also finite dimensional.

The functional feature consists in using the uniform consistency result (4.42) to deal with the term $\tilde{Y}$ that appears in the construction of the estimate γ, and in using condition (4.86) to make sure that this does not alter the $\sqrt{n}$-consistency of the parametric estimate.

(2) The second step consists in introducing the following estimate:

$$\tilde{m}(v) = \frac{\sum_{i=1}^{n} W_{n,\ker}(v, V_i)\left(m(V_i) + \epsilon_i\right)}{\sum_{i=1}^{n} W_{n,\ker}(v, V_i)}.$$

This is a kernel-type estimator (see Definition 4.1) with the new response variable $m(V) + \epsilon$. We are now able to apply Theorem 4.7. More precisely, the result (4.37) becomes:

$$\sup_{v \in S_F} |\tilde{m}(v) - m(v)| = O\left(h^\beta\right) + O\left(\sqrt{\frac{\psi_{S_F}\left(\frac{\log n}{n}\right)}{n\phi_{S_F}(h)}}\right), \quad a.s. \quad (4.92)$$

On the other hand, we have the following inequality:

$$\sup_{v \in S_F} |\hat{m}_{\ker}(v) - m(v)| \leq \sup_{v \in S_F} |\tilde{m}(v) - m(v)|$$
$$+ \|\hat{\gamma} - \gamma\| \sup_{v \in S_F} \|W(v)\|, \quad (4.93)$$

where

$$W(v) = \frac{\sum_{i=1}^{n} W_{n,\ker}(v, V_i)\left(U_i^t\right)}{\sum_{i=1}^{n} W_{n,\ker}(v, V_i)}.$$

Finally, the proof of Theorem 4.10 follows directly from (4.91), (4.92), (4.93), and from the definition of $\hat{m}_{PLFR}$ (see (4.80)). □

Comments

Once again, the main point to be emphasized in the above proof is the important role of the uniform results stated in Section 4.2.4, which are not only used to deal with the estimation of the functional component m in the statement of (4.92) but are also used for the statement of asymptotic results on the estimation of γ (such as for instance (4.91)).

The literature on partial linear functional models is rather restricted. As far as we know there are only two papers in this field. The first one is by Aneiros-Pérez and Vieu (2006), and includes not only Theorem 4.10 but also the asymptotic distribution for the multivariate estimate γ^t. Note that in this paper, result (4.90) was obtained under slightly weaker conditions (for example, on the moments of the error terms ϵ and η) but was restricted to a rather specific compact set S_F. In Aneiros-Pérez and Vieu (2008) an extension of Theorem 4.10 to dependent samples is presented. This last result allows for time-series applications.

Many interesting open problems still remain. They include, for instance, extensions to other nonparametric estimates of the functional component m (including, for instance, kNN or local linear estimates as described before in Definitions 4.2 and 4.3), or the question of choosing a data-driven bandwidth $\hat{h}$. Alternative models in which the linear component is also functional, and many other results already known in the multivariate setting, also need suitable adaptation to functional variables...

4.4 USING KERNELS FOR ADDITIVE FUNCTIONAL REGRESSION

This section discusses some recent advances in the additive modeling of functional variables. Unlike the other parts of this chapter in which the results have been taken from the existing literature, this section contains a theorem (Theorem 4.11) that has not yet been published elsewhere.

4.4.1 Presentation of the model

The additive model has been developed in situations when the explanatory variable X is composed of various different parts, each of which has to be modeled in a nonparametric way. These ideas have been widely developed in the multivaiate setting, that is, when $X \in \mathbb{R}^p$. A selection of references here includes Stone (1985, 1986, 1994), Härdle and Hall (1993), Hastie and Tibshirani (1986, 1990), Schimek and Turlach (2000), Mammen and Park (2006), and Horowitz et al. (2006). The aim of this section is to present a recent extension of additive ideas to situations in which X is functional.

Let us consider the standard regression model with functional covariate, as defined in (4.1), but assume that the functional covariate can be decomposed into $X = (X^1, \ldots, X^p)$. Each component X^j takes values in some abstract space E^j endowed with a semi-metric d_j. So the space E in which the explanatory variable X takes its values is the product space $E = E^1 \times \cdots \times E^p$. Recall that E is endowed with a semi-metric d that can be (but is not necessarily) constructed from the d_j.

Definition 4.7 *The additive functional regression model is defined by assuming that the regression model defined in (4.1) can be rewritten as*

$$Y = r(X) + \epsilon = \sum_{j=1}^{p} r^j(X^j) + \epsilon, \qquad (4.94)$$

where each r^j is an unknown functional (*not necessarily linear*) operator acting on the marginal space E^j, and where $E(\epsilon|X) = 0$.

It is clear that the formulation (4.94) is not unique. It is therefore necessary to put some restrictions on the additive components to deal with identifiability of the model. This is achieved by introducing the jth stage residuals

$$\forall j = 1, \ldots, p, \ \epsilon^j = Y - \sum_{k=1}^{j} r^k(X^k),$$

and by assuming that

$$r^1(X^1) = E(Y|X^1) \text{ and } \forall j \geq 2, r^j(X^j) = E(\epsilon^{j-1}|X^j). \quad (4.95)$$

This is clearly a dimensionality-reduction model (see Definition 1.4 of Chapter 1). The statistical problem consists in estimating the operators r^j from a sequence of n independent variables $(X_i, Y_i) = (X_i^1, \ldots, X_i^p, Y_i)$. We will also use the notation, for $i = 1, \ldots, n$ and $j = 1, \ldots, p$:

$$\epsilon_i^j = Y_i - \sum_{k=1}^{j} r^k\left(X_i^k\right).$$

Kernel estimation

Exactly as before for the single functional index model or for the partial linear functional model, the model (4.94) can be studied in series of stages. The idea here is to use nonparametric functional kernel ideas to estimate each nonlinear operator r^j. Let us introduce some kernel functions $K^1, \ldots, K^p$ and bandwidths $h_1, \ldots, h_p$. As in Definition 4.1 we also introduce the following local weights:

$$\forall j = 1, \ldots, p, \ \forall (v, v') \in E^j \times E^j, \ W_{n,\text{ker}}^j(v, v') = K^j\left(\frac{d_j(v, v')}{h_j}\right).$$

In a first approximation, we estimate the first additive component by means of the following kernel estimate:

$$\forall x^1 \in E^1, \ \hat{r}_{\text{ker}}^1(x^1) = \frac{\sum_{i=1}^{n} W_{n,\text{ker}}^1\left(x^1, X_i^1\right) Y_i}{\sum_{i=1}^{n} W_{n,\text{ker}}^1\left(x^1, X_i^1\right)}. \quad (4.96)$$

Then, in an iterative way, we construct the other estimates by performing, for any $j = 2, \ldots, p$, the regression of the $(j-1)$th-order estimated residuals

$$\hat{\epsilon}_i^{j-1} = Y_i - \sum_{k=1}^{j-1} \hat{r}_{\text{ker}}^k\left(X_i^k\right), \ i = 1, \ldots, n,$$

on the next variable X_i^j, $i = 1, \ldots, n$. This leads, for $j \geq 2$, to the kernel estimates:

$$\hat{r}_{\text{ker}}^j(x^j) = \frac{\sum_{i=1}^n W_{n,\text{ker}}^j(x^j, X_i^j) \hat{\epsilon}_i^{j-1}}{\sum_{i=1}^n W_{n,\text{ker}}^j(x^j, X_i^j)}, \quad x^j \in E^j. \tag{4.97}$$

Finally, under model (4.4), the additive kernel-type estimate of the regression operator r is defined by

$$\forall x = (x^1, \ldots, x^p) \in E, \ \hat{r}_{\text{Add}}(x) = \sum_{j=1}^p \hat{r}_{\text{ker}}^j(x^j). \tag{4.98}$$

General assumptions

Because the additive estimate (4.98) is just a combination of p standard kernel regression estimates, the conditions required for the statement of asymptotic results are not really surprising. Indeed, these conditions are just those required to apply repeatedly the general uniform consistency results for kernel functional regressors as stated earlier. In what follows, we introduce the following subsets of the functional spaces

$$S^j \subset E^j, \ j = 1, \ldots, p.$$

The nonparametric modeling consists here in assuming, for each $j = 1, \ldots, p$, some Lipschitz-type condition on the operator r^j to be estimated:

$$\exists \beta^j > 0, \exists C_j < +\infty, \forall (x, x') \in E^j \times E^j, |r^j(x) - r^j(x')| \leq C_j d_j(x, x')^{\beta^j}. \tag{4.99}$$

We need general conditions on the small-ball probability functions, whose definition we recall:

$$\forall j = 1, \ldots, p, \forall x \in S^j, \ \phi_{j,x}(\epsilon) = P_{X^j}(d_j(x, \epsilon)).$$

We assume that there exist functions ϕ_{S^j} and constants a_1^j, a_2^j, and a_3^j such that for any $\epsilon > 0$ and for any $j = 1, \ldots, p$:

$$0 < a_1^j \phi_{S^j}(\epsilon) \leq \inf_{x \in S^j} \phi_{j,x}(\epsilon) \leq \sup_{x \in S^j} \phi_{j,x}(\epsilon) \leq a_2^j \phi_{S^j}(\epsilon) < \infty, \tag{4.100}$$

and

$$\phi'_{S^j}(\epsilon) \leq a_3^j. \tag{4.101}$$

We also need some conditions on the entropy ψ_{S^j} of each subset S^j (see Definition 4.4). Assume that for any $j = 1, \ldots, p$:

$$\exists n_0, \forall n > n_0, \ \frac{(\log n)^2}{n \phi_{S^j}(h_j)} < \psi_{S^j}\left(\frac{\log n}{n}\right) < \frac{n \phi_{S^j}(h_j)}{\log n}, \tag{4.102}$$

and

$$\exists \delta > 0, \sum_{i=1}^{n} e^{-\delta \psi_{S^j}\left(\frac{\log n}{n}\right)} < \infty. \qquad (4.103)$$

Each kernel K^j and each bandwidth h_j is assumed to satisfy the same general conditions as in Section 4.2. It is assumed that for any $j = 1, \ldots, p$:

$$K^j \geq 0, \int_{\mathbb{R}} K^j = 1, \ K^j \text{ is Lipschitz on } [0, 1),$$

$$\text{and support}(K^j) \subset [0, 1], \qquad (4.104)$$

$$\forall \epsilon \leq \epsilon_0, \forall x \in S^j, \ C_1 \phi_{S^j}(\epsilon) \leq E K^j \left(\frac{d^j(x, X)}{\epsilon}\right) \leq C_2 \phi_{S^j}(\epsilon), \qquad (4.105)$$

and

$$\lim_{n \to \infty} h_j = 0 \text{ and } \lim_{n \to \infty} \frac{n \phi_{S^j}(h_j)}{\log n} = \infty. \qquad (4.106)$$

Some asymptotic results

In Theorem 4.11 below we give the uniform rate of convergence of the additive estimate $\hat{r}_{\text{Add}}$. Because such a result has not been published before and because its proof is rather simple, we will provide a complete proof. The uniformity is stated over a subset $S = S^1 \times \cdots \times S^p \subset E$, such that

$$\forall j = 1, \ldots, p, \ P(X^j \in S^j) = 1. \qquad (4.107)$$

Theorem 4.11 *Assume that the model defined by (4.94), (4.95), and (4.7) holds. Assume that conditions (4.99), (4.100), (4.101), (4.102), (4.103), (4.104), (4.105), (4.106), and (4.107) hold. Then we have:*

$$\sup_{x \in \times S} |\hat{r}_{\text{Add}}(x) - r(x)| = O\left(\sum_{j=1}^{p} h_j^{\beta^j}\right) + O\left(\sum_{j=1}^{p} \sqrt{\frac{\psi_{S^j}\left(\frac{\log n}{n}\right)}{n \phi_{S^j}(h_j)}}\right), \ a.s. \quad (4.108)$$

Proof of Theorem 4.11 The proof is based on the decomposition below. For $x = (x^1, \ldots, x^p) \in E$, we have:

$$|r(x) - \hat{r}_{\text{Add}}(x)| \leq \sum_{j=1}^{p} \left|r^j(x^j) - \hat{r}_{\text{ker}}^j(x^j)\right|. \qquad (4.109)$$

(1) The first component of the sum in (4.109) can be treated directly, because the estimate $\hat{r}_{\text{ker}}^1$ is a special type of kernel estimate (see Definition 4.1) with response

Y and explanatory variable X^1. Therefore we have, by a direct application of Theorem 4.7:

$$\sup_{x^1 \in S^1} |r^1(x^1) - \hat{r}_{\ker}^1(x^1)| = O\left(h_1^{\beta^1}\right) + O\left(\sqrt{\frac{\psi_{S^1}\left(\frac{\log n}{n}\right)}{n \phi_{S^1}(h_1)}}\right), \quad a.s. \qquad (4.110)$$

(2) Let us now consider the case $j = 2$. We can split the estimate $\hat{r}_{\ker}^2$ into two parts:

$$\hat{r}_{\ker}^2 = \hat{r}_{\ker}^{2,1} + \hat{r}_{\ker}^{2,2},$$

with

$$\forall x^2 \in E^2, \; \hat{r}_{\ker}^{2,1}(x^2) = \frac{\sum_{i=1}^{n} W_{n,\ker}^2\left(x^2, X_i^2\right) \epsilon_i^1}{\sum_{i=1}^{n} W_{n,\ker}^2\left(x^2, X_i^2\right)},$$

and

$$\forall x^2 \in E^2, \; \hat{r}_{\ker}^{2,2}(x^2) = \frac{\sum_{i=1}^{n} W_{n,\ker}^2\left(x^2, X_i^2\right) \left(r^1\left(X_i^1\right) - \hat{r}_{\ker}^1\left(X_i^1\right)\right)}{\sum_{i=1}^{n} W_{n,\ker}^2\left(x^2, X_i^2\right)}.$$

The treatment of $\hat{r}_{\ker}^{2,1}$ is easy because it is a special type of kernel estimate (see Definition 4.1) with response ϵ^1 and explanatory variable X^2. Therefore we have, by a direct application of Theorem 4.7:

$$\sup_{x^2 \in S^2} |r^2(x^2) - \hat{r}_{\ker}^{2,1}(x^2)| = O\left(h_2^{\beta^2}\right) + O\left(\sqrt{\frac{\psi_{S^2}\left(\frac{\log n}{n}\right)}{n \phi_{S^2}(h_2)}}\right), \quad a.s. \qquad (4.111)$$

To deal with the other term involving the estimate $\hat{r}_{\ker}^{2,2}$ it suffices to use condition (4.107) together with Corollary 4.2 to see that

$$\forall x^2 \in E^2, \; \hat{r}_{\ker}^{2,2}(x^2) = O\left(h_1^{\beta^j}\right) + O\left(\sqrt{\frac{\psi_{S^1}\left(\frac{\log n}{n}\right)}{n \phi_{S^1}(h_1)}}\right), \quad a.s. \qquad (4.112)$$

Finally, we have almost surely:

$$\sup_{x^2 \in S^2} |r^2(x^2) - \hat{r}_{\ker}^2(x^2)| = O\left(h_2^{\beta^2}\right) + O\left(\sqrt{\frac{\psi_{S^2}\left(\frac{\log n}{n}\right)}{n \phi_{S^2}(h_2)}}\right) \quad a.s.$$

$$= O\left(h_1^{\beta^1}\right) + O\left(\sqrt{\frac{\psi_{S^1}\left(\frac{\log n}{n}\right)}{n \phi_{S^1}(h_1)}}\right) \quad a.s. \qquad (4.113)$$

(3) It is clear that the previous step can easily be iterated, and we arrive for any $j = 2, \ldots, p$ at:

$$\sup_{x^j \in S^j} \left| r^j(x^j) - \hat{r}_{\ker}^j(x^j) \right| = \sum_{k=1}^{j} \left(O\left(h_k^{\beta_k} \right) \right)$$

$$+ \sum_{k=1}^{j} \left(O\left(\sqrt{\frac{\psi_{S^k}\left(\frac{\log n}{n}\right)}{n \phi_{S^k}(h_k)}} \right) \right) \quad a.s. \quad (4.114)$$

Finally the result (4.108) follows directly from (4.109), (4.110), and (4.114). □

Comments

Once again, and as has already been pointed out for other multistage models such as the single functional index (see Section 4.3.2) or partial linear functional (see Section 4.3.3) models, the main point to be highlighted in the above proof is the important role of the uniform results stated in Section 4.2.4. Note that, even if one wishes to study pointwise consistency properties for each additive component, the statement of results like (4.112) needs to control for random terms such as $r^1\left(X_i^1\right) - \hat{r}_{\ker}^1\left(X_i^1\right)$, here Corollary 4.2 is a key tool.

To conclude this section, let us just note that the literature on additive functional modeling is not very developed. As far as we know, the first theoretical paper was that by Ferraty and Vieu (2009), in which complementary asymptotic results are given (in terms of squared prediction errors). See also Müller (2008) for recent developments (but in a slightly different context), and Aneiros-Pérez et al. (2006) for a real environmetrical curves dataset application. Of course, many open questions remain. The most important to be dealt with may concern data-driven bandwidth selection procedures, but it would be worth adapting many other points already well known in standard multivariate nonparametric statistics to the functional framework.

4.5 ON TESTING FUNCTIONAL REGRESSION MODELS

This section is devoted to the presentation of the few very recent results that exist for testing the procedures for structural regression. Here we wish to acknowledge Laurent Delsol, who is a pioneer in this field, and who kindly agreed to review and improve a preliminary draft of this section.

4.5.1 Introduction

Despite the fact that the literature on nonparametric functional regression estimation is relatively new, it can be seen from the previous sections of this chapter that we now have a wide range of models available in this area. They include purely nonparametric models (see Section 4.2), dimensionality-reduction models such as the single functional index model (Section 4.3.2), partial linear functional models (Section 4.3.3) and additive models (Section 4.4), and purely parametric models such as the linear functional regression model discussed in Chapter 2 of this book. A natural question is therefore to decide whether one of these models is more adapted than others to some given practical situation. This question has not yet been much studied in the literature, but a few recent advances exist on testing procedures for checking the validity of some model. The aim of this section is to describe the main approaches of the testing procedures proposed in Delsol (2008b, 2008c) and Delsol et al. (2010). A larger bibliographical discussion can be found in this last paper by Delsol et al. (2010).

We will not give details of the theoretical developments nor of the technical assumptions, because this would give rise to long and tedious notation. In Section 4.5.2, we will set out a general framework for structural regression test procedures and we will briefly describe various specific situations in which these procedures can be helpful. Then in Section 4.5.3 we will discuss how the various kernel-type estimates studied earlier in this chapter can be used to construct broad families of statistics tests. Some asymptotics will be briefly described in Section 4.5.4.

4.5.2 A general structural testing problem

Within the general framework of the functional regression problem (4.1), the question is to decide whether the true regression operator r belongs to some fixed specific family $\mathcal{F}_0$ of operators or not. In other words, we wish to test the null hypothesis

$$H_0 : \{\exists R \in \mathcal{F}_0, \ P(r(X) = R(X)) = 1\} \tag{4.115}$$

against some alternative that says that the operator r is *sufficiently far* from $\mathcal{F}_0$. For instance if $||\cdot||$ is some norm on the space of operators, we can measure the distance between r and the class $\mathcal{F}_0$ by looking at the distance between r and its closest approximation (let us say r_0) in $\mathcal{F}_0$:

$$\Delta(r, \mathcal{F}_0) = \inf_{r' \in \mathcal{F}_0} ||r - r'||.$$

For clarity, we will assume that there is an unique element r_0 such that

$$\Delta(r, \mathcal{F}_0) = ||r - r_0||, \tag{4.116}$$

but this condition can be weakened, as indicated in Delsol et al. (2010). This leads us, from some given real sequence η_n, to consider the following alternative hypothesis:

$$H_{1,n} : \{\Delta(r, \mathcal{F}_0) \geq \eta_n\}. \tag{4.117}$$

We will see later how a general test statistic can be constructed for this structural problem (see Section 4.5.3), for which an asymptotic normality distribution can be stated under H_0, and divergence can be observed under $H_{1,n}$ as soon as η_n is sufficiently large (see Section 4.5.4).

Before going into these details, let us first discuss a few specific examples of families $\mathcal{F}_0$ in order to show the high degree of generality of this approach.

- *Testing linearity*. One may wish to know whether the true operator is linear, and in this case one uses the family

$$\mathcal{F}_{0,\text{lin}} = \{r, \text{linear and continuous operators}\}.$$

- *Testing non-effect*. One may wish to know whether the functional variable X has an effect on the response Y, and in this case one uses the family

$$\mathcal{F}_{0,\text{ne}} = \{r, \text{constant operators}\}.$$

More generally in situations when $X = (X^1, X^2)$ one may wish to see whether some component, let us say X^2, has an effect on Y. In this case one uses the family

$$\mathcal{F}_{0,\text{ne2}} = \{r, \exists r^1, \forall x = (x^1, x^2), r(x) = r^1(x^1)\}.$$

- *Testing additivity*. In the situation in which the response X is multiple, that is, when $X = (X^1, \ldots, X^p)$, one may wish to decide whether the additive model studied in Section 4.4 is accurate or not. In this case one uses the family

$$\mathcal{F}_{0,\text{add}} = \left\{r, \exists r^1, \ldots, r^p, \forall x = (x^1, \ldots, x^p), r(x) = \sum_{j=1}^{p} r^j(x^j)\right\}.$$

- *Testing a specific semiparametric model*. One may wish to decide whether the effect of the variable X can be reduced to the effect of one single projection. In other words, one may wish to test the validity of the single functional index model studied in Section 4.3.2. For this one can use the family

$$\mathcal{F}_{0,\text{SFIM}} = \{r, \exists g, \exists \theta, \forall x, r(x) = g(\langle x, \theta \rangle)\}.$$

The same kind of question may also be posed when $X = (U, V)$ with $U \in \mathbb{R}^p$ to check the validity of the partial functional index model studied in Section 4.3.3. In this case one uses the family

$$\mathcal{F}_{0,\text{PLFR}} = \{r, \exists \gamma, \exists m, \forall x = (u, v), r(x) = m(v) + \gamma^t u\}.$$

- *Testing an unfunctional model.* Other types of questions may concern the functional features of the model. For instance, if X is a curve

$$X = \{X(t), t \in (a, b)\},$$

one may wish to check whether X acts on Y only through a few values $X(t_1), \ldots, X(t_p)$. In this case one uses the family

$$\mathcal{F}_{0,\text{Unf}} = \left\{ r, \exists g, \exists t_1, \ldots, t_p, \forall x, \, r(x) = g(x(t_1), \ldots, x(t_p)) \right\}.$$

These are just a few of the possible situations that could be modeled by hypotheses of the form (4.115), and one can easily imagine many other problems of this kind.

4.5.3 Construction of kernel-based test statistics

In the setting of standard finite-dimensional statistics, structural testing problems have been widely investigated in the literature (see, for instance, the monograph by Hart (1997)). One way of doing this is to build test statistics based on the difference of a purely nonparametric estimate and another estimate which is specific to the model $\mathcal{F}_0$ that one wishes to test. This idea has been popularized by Härdle and Mammen (1993) and has been used in many further papers to deal with a large variety of situations. Given the recent advances in estimation procedures for functional regression (see the previous sections of this chapter), it is natural to think of adapting Härdle–Mammen's ideas to functional data.

More precisely, we will use as a pilot estimate the kernel estimate $\hat{r}_{\text{ker}}$ (see Definition 4.1), and the first approach is to construct test statistics by looking at quantities such as

$$\int (\hat{r}_{\text{ker}}(x) - \hat{r}_0(x))^2 \, w(x) \, dP_X(x),$$

where $\hat{r}_0$ is an estimate of r under the specific model defined by the null hypothesis (4.115), and w is a known weight function. For technical reasons, and once again following the ideas in Härdle and Mammen (1993), it is more convenient to use the folllowing statistics:

$$T_{0,n} = \int \left(\sum_{i=1}^{n} (Y_i - \hat{r}_0(X_i)) W_{n,\,\text{ker}}(x, X_i) \right)^2 w(x) \, dP_X(x), \qquad (4.118)$$

where we recall that the local weights are defined from a kernel K and a bandwidth h by:

$$W_{n,\,\text{ker}}(x, X_i) = K\left(\frac{d(x, X_i)}{h} \right).$$

The principal advantages of this statistic are both to suppress the bias and to overcome the technical problems linked with the random denominator appearing in the kernel estimate $\hat{r}_{\text{ker}}$. In the functional setting, this statistic has been introduced by Delsol (2008b). We will briefly present its asymptotic behavior in the next section, but before that we would like to illustrate it through the few specific situations discussed above at the end of Section 4.5.2.

- *Testing linearity.* To test the family $\mathcal{F}_{0,\text{lin}}$, one can use any of the various existing linear functional regressors (see Chapter 2) as a linear estimate $\hat{r}_0$.
- *Testing non-effect.* To test the family $\mathcal{F}_{0,\text{ne}}$, one can use the naive constant estimator $\hat{r}_0(x) = \frac{1}{n}\sum_{i=1}^{n} Y_i$, $\forall x$. To test the family $\mathcal{F}_{0,\text{ne2}}$ one can use a functional estimate based only on the second covariate, such as for instance the kernel estimate $\hat{r}_0 = \hat{r}_{\text{ker}}^1$.
- *Testing additivity.* To test the family $\mathcal{F}_{0,\text{add}}$, one can use the additive estimate $\hat{r}_0 = \hat{r}_{\text{Add}}$ as defined in (4.98).
- *Testing a specific semiparametric model.* To test the family $\mathcal{F}_{0,\text{SFIM}}$ one can use the estimate $\hat{r}_0 = \hat{r}_{\text{SFIM}}$ as defined in (4.59), and to test $\mathcal{F}_{0,\text{PLFR}}$ one can use the estimate $\hat{r}_0 = \hat{r}_{\text{PLFR}}$ as defined in (4.80).
- *Testing an unfunctional model.* To test the family $\mathcal{F}_{0,\text{Unf}}$ the choice is rather extensive since as an estimate $\hat{r}_0$ one can use any of the wide range of well-known p-dimensional smoothers (kernel, splines, local polynomial, kNN, ...)

Of course all these applications will be possible only if, in each situation, the estimate $\hat{r}_0$ can be shown to satisfy the technical conditions required (see discussion below). Once again, these are just a few of the possible applications of the general methodology presented here. As long as our knowledge of functional regression estimation is growing, more applications for testing could be (and certainly will be) developed directly. Indeed, to apply this general methodology to any submodel $\mathcal{F}_0$ one needs to have at hand some estimate $\hat{r}_0$ for which the rates of convergence can be controlled under the null hypothesis H_0. More precisely, recalling that r_0 is defined by (4.116), one requires (for instance) conditions on $\hat{r}_0$ similar to the following:

$$\text{Under } H_0, \quad \delta(\hat{r}_0, r_0) = o_p(r_n), \qquad (4.119)$$

where δ is some measure of accuracy for estimation under the model H_0. Delsol et al. (2010) propose various specific choices of δ. These authors also extend the methodology to the situation in which the unicity condition (4.116) is not satisfied. They also relax (4.119) into a condition saying that, in some sense, $\hat{r}_0$ *is not too far from the family* $\mathcal{F}_0$.

4.5.4 Some theoretical advances

It is not our aim here to state precisely the various conditions needed to obtain asymptotic behavior of the statistic $T_{0,n}$. All the results presented below are stated under quadratic-type measures of accuracy, for the δ appearing in the null hypothesis (see (4.119)) as well as for the norm defining the alternative hypothesis (see (4.117)). All the details can be found in Delsol (2008b, 2008c) and Delsol et al. 2010. Our goal here is simply to highlight the main ideas.

First of all, by following the general approach as in multivariate settings, we may note that the statistic $T_{0,n}$ can be split into 6 terms in the following way:

$$T_{0,n} = T_1 + T_2 + T_3 + T_4 + T_5 + T_6,$$

with

$$T_1 = \int \sum_{i=1}^n (Y_i - r(X_i))^2 W_{n,\text{ker}}^2(x, X_i) w(x) \, dP_X(x),$$

$$T_2 = \int \sum_{i \neq j} (Y_i - r(X_i))(Y_j - r(X_j)) W_{n,\text{ker}}(x, X_i) W_{n,\text{ker}}(x, X_j) w(x) \, dP_X(x),$$

$$T_3 = \int \sum_{i=1}^n (r(X_i) - \hat{r}_0(X_i))^2 W_{n,\text{ker}}^2(x, X_i) w(x) \, dP_X(x),$$

$$T_4 = \int \sum_{i \neq j} (r(X_i) - \hat{r}_0(X_i))(r(X_j) - \hat{r}_0(X_j)) \times W_{n,\text{ker}}(x, X_i) W_{n,\text{ker}}(x, X_j) w(x) \, dP_X(x),$$

$$T_5 = 2 \int \sum_{i=1}^n (r(X_i) - \hat{r}_0(X_i))(Y_i - r(X_i)) W_{n,\text{ker}}^2(x, X_i) w(x) \, dP_X(x),$$

and

$$T_6 = 2 \int \sum_{i \neq j} (Y_i - r(X_i))(r(X_j) - \hat{r}_0(X_j)) \times W_{n,\text{ker}}(x, X_i) W_{n,\text{ker}}(x, X_j) w(x) \, dP_X(x).$$

It is worth noting that the terms T_1 and T_2 have the same behavior under both hypotheses H_0 and $H_{1,n}$. The bias of the statistic $T_{0,n}$ will be provided by the term T_1, while its variance will be provided by the term T_2. This leads us naturally to consider the following statistic:

$$\mathcal{T}_n = \frac{T_{0,n} - E T_1}{\sqrt{\text{var}(T_2)}}. \tag{4.120}$$

The first result that we state below concerns the asymptotic distribution of the centered and reduced statistic $\mathcal{T}_n$ under the null hypothesis. Roughly speaking (see Delsol (2008c) or Delsol et al. (2010)), if the rate of convergence of the estimate $\hat{r}_0$ is *sufficiently fast*, that is, if the sequence r_n defined in (4.119) is *small enough*, the terms T_j, $j = 3, \ldots, 6$ can be shown to be negligible. This results from the fact that each of these terms T_j, $j = 3, \ldots, 6$ involves quantities like $r(X) - \hat{r}_0(X)$, which are exactly equal under H_0 to $r_0(X) - \hat{r}_0(X)$. Once these terms are shown to be negligible, one can look at the precise behavior of both the previous terms T_1 and T_2 and establish, by means of some appropriate central limit theorem, results such as:

$$\text{Under } H_0, \quad \mathcal{T}_n \xrightarrow{\mathcal{L}} \mathcal{N}(0, 1). \tag{4.121}$$

The second result we will give concerns the asymptotic behaviour of $\mathcal{T}_n$ under the alternative hypothesis $H_{1,n}$. As mentioned before, the behavior of T_1 and T_2 is the same under $H_{1,n}$ as under H_0, and therefore the divergence of the test under the alternative hypothesis will depend on the remaining terms T_j, $j = 3, \ldots, 6$. Because these involve quantities like $r(X) - \hat{r}_0(X)$, they can be shown to be *as large as is* desired subject to conditions on $r - r_0$, that is, by assuming that the sequence η_n defining the null hypothesis (4.117) is *large enough* (see Delsol (2008c) or Delsol et al. (2010) for details). This leads to results of the form

$$\text{Under } H_{1,n}, \quad \mathcal{T}_n \xrightarrow{\mathcal{P}} \infty. \tag{4.122}$$

4.5.5 Comments

The statements of the results (4.121) and (4.122) mentioned above are technically difficult. While the main conditions needed are those discussed before to control the alternative hypothesis $H_{1,n}$ and also to control the quality of estimation under the null hypothesis H_0, various other conditions are now necessary. On the one hand, they include the same kinds of assumptions as those given in Section 4.2 in order to insure the good behavior of the pilot estimate $\hat{r}_{\text{ker}}$. On the other hand, they also include additional conditions on the small-ball probability of the pairs (X_i, X_j), because a U-statistics methodology is used for the treatment of the leading terms T_1 and T_2. Also, for technical reasons, it is necessary to construct the null estimate $\hat{r}_0$ using a second statistical sample independent of the original sample (X_i, Y_i), $i = 1, \ldots, n$. As discussed in Delsol (2008c) and Delsol et al. (2010), the high technical level of these conditions reflects the broad generality of the method without being too restrictive.

It is worth noting that exact asymptotic expansions of the leading terms ET_1 and $\text{var}(T_2)$ are also obtained in Delsol (2008c) and Delsol et al. (2010). It turns out that the rather complicated form of these expansions makes a direct use of the statistics $\mathcal{T}_n$ quite unrealistic in practice. While a naive point of view would lead us to conclude that these results are not very useful, it is worth noting that theoretical results like (4.121) and (4.122) are indispensable preliminary practical tools. This is already the case in standard multivariate situations, where such theoretical results are used for proving the asymptotic validity of bootstrapping procedures. We may conclude by saying that in our functional framework, the full power of the results (4.121) and (4.122) will be revealed only when the validity of resampling procedures can be checked. As far as we know, this is still an open question (see, however, Delsol (2008c) for preliminary empirical ideas on testing procedures by functional resampling that show the good behavior of bootstrapping ideas in the case of finite samples).

4.6 Comments

4.6.1 On rates of convergence in functional regression

Topological effects on pointwise results

The common feature of the various asymptotic results presented previously is the great importance of the topological structure that exists on the functional space E. We will now briefly discuss a few instances where these topological effects appear clearly.

The influence of the topology appears directly in all the pointwise rates of convergence, through the small-ball probability function

$$\phi_x(\cdot) = \{y \in E, d(x, y) \leq \cdot\}.$$

This is true for all kinds of estimates in nonparametric modeling (see Theorems 4.1, 4.2, and 4.3). It is also true for any alternative model such as the additive model (see Theorem 4.11), the single functional model (see Theorem 4.9), or the partial linear functional model (see Theorem 4.10). And it is also true independently of the mode of convergence, since it appears for pointwise consistency (see Theorem 4.1) as well as for L_2 consistency (see Theorem 4.4) and for asymptotic normality (see Theorem 4.5). This remark has a direct practical impact (see Section 4.6.2 below) since the choice of topology (that is, the choice of the semi-metric d) will influence the behavior of any non/semiparametric functional regression estimates directly. The reader may find more comments about the statistical effects of topological structure, as well as various standard examples of possible choices of semi-metric, in Ferraty and Vieu (2006a).

Let us now look more precisely at the rates of convergence obtained in semi-parametric modeling (see Section 4.3) or in additive modeling (see Section 4.4). Undoubtedly, the rates of convergence obtained for these models (see Theorems 4.9, 4.10, and 4.11) are much better than the rates obtained for nonparametric models (see Theorem 4.1). This is due to the fact that all these models are of dimensionality-reduction type (see Definition 1.4 of Chapter 1), and so the rates of convergence are controlled by the small-ball probability function of a variable lying in a lower-dimensional space (even if this new variable can also be infinite dimensional) than the original explanatory functional variable X.

We conclude these comments by mentioning some open questions. Indeed, while the standard method for stating rates of convergence in non- or semiparametric functional problems is through small-ball probability considerations, this is not the case in linear functional regression (see Chapter 2), where we usually control rates of convergence by means of conditions on the eigenvalues of the covariance operator of the functional variable X. While the probabilistic literature on links between small-ball probabilities and eigenvalues of covariance operators has been developed for various specific continuous-time Gaussian processes and specific topologies (see Nazarov and Nikitin (2004), Bronski (2003) and Gao et al. (2003) for a few recent references in this area), some additional work is still needed in order to allow for general statistical comparison between the results presented here in non/semiparametric functional statistics and those given in Chapter 2 in parametric functional statistics.

Topological effects on uniform results

If we consider uniform consistency results, topology is even more important. For instance, in Theorem 4.7, the topological structure acts directly on the rates of convergence through the entropy function ψ_S which measures the complexity of the set S on which uniformity is stated. In Ferraty et al. (2009) one can find various examples for which the entropy function can be (exactly or asymptotically) expressed, and for which the rate of convergence obtained in Theorem 4.7 is slower than the pointwise one. This confirms what the very form of functional regression estimates based on local weighting (see (4.3)) leads us to expect about the major importance of local features in the functional space E. This point will also be of great importance in practical situations (see Section 4.6.2 below).

On optimality of the rates of convergence

An interesting open question is whether the various rates of convergence presented earlier are optimal or not. More precisely, it would be nice to discover in the near future whether the results on optimal rates of convergence obtained by Stone (1982) in the multivariate setting can be extended in some sense to the functional setting. Even though we may remark (see below for more details) that, if we restrict all the results presented before to the special situation in which $E = \mathbb{R}^p$, we get back to Stones's rates, the general question of optimal rates is still completely open in the functional regression setting.

On density of the functional variable

It is worth noting that another advantage of small-ball probability considerations is that they avoid imposing conditions on the density of the functional variable X. This is a key point for our procedure because it is not clear what we could take, in a general abstract semi-space, as a reasonable measure of reference for defining the notion of density (see, however, the recent contribution by Delaigle and Hall (2010), which proposes a new concept of density for infinite-dimensional variables). On the one hand, this issue makes the proofs much more technical than they would be if we had assumed the existence of a density for X, but from another point of view it allows us to state all the results in a much more general setting (an even more general setting than the normal one in the standard multivariate literature).

Links with the standard multivariate literature

To fix our ideas even further, it may be helpful to look at how the previous results behave in the standard multivariate situation (that is, in the special case when $E = \mathbb{R}^p$ and d is Euclidean distance).

It is easy to see that, in this case, as long as X has a density with respect to the Lebesgue measure on $E = \mathbb{R}^p$, one can show that

$$\phi_x(\epsilon) \sim C_x \epsilon^p,$$

in such a way that all the pointwise results stated before match those in the earlier multivariate literature. This is true in nonparametric estimation (see Theorems 4.1, 4.2, 4.3, 4.4, and 4.5) as well as for additive models (see Theorem 4.11) and for semiparametric modeling (see Theorems 4.9 and 4.10). Furthermore, as discussed just above in Section 4.6.1, all the results presented in this chapter extend the standard multivariate literature to situations in which X has no density with respect to the Lebesgue measure.

We can make a similar observation for the uniform result stated in Theorem 4.7, since it is also very easy to see that in this situation the entropy function can be shown, for any compact set $S \subset \mathbb{R}^p$, to be

$$\psi_S(\epsilon) = \log\left(\frac{C_S}{\epsilon^p}\right).$$

Therefore the leading dispersion term in Theorem 4.7 can be rewritten as

$$\sqrt{\frac{\psi_S\left(\frac{\log n}{n}\right)}{n\phi_S(h)}} \sim C\sqrt{\frac{\log n}{nh^p}},$$

leading to the same rate of convergence in the pointwise and the uniform cases (compare Theorems 4.1 and 4.7), as is well known in standard multivariate nonparametric statistics.

4.6.2 On practical issues in functional regression

It is not our purpose here to provide a long discussion of practical issues. It is worth noting, however, that the nonparametric methodologies have been succesfully applied in many situations. The reader may find a free online package, including S/R+ routines, guidelines for users, and various real case studies in Ferraty and Vieu (2006b). These practical studies confirm the general theoretical comments previously given, at least on two points. The first point is the importance of the semi-metric; routines for various families of semi-metrics, as well as empirical ideas for chosing them, are also given in the package. The second point concerns the local features of functional data and the importance of data-driven location-adaptive selection procedures for the bandwidth. The various case studies presented in this package highlight the very good behavior of the local cross-validation procedure presented earlier in Section 4.2.4, as well as the kNN estimate presented there, because both of these approaches fully take into acount local features of the problem.

With regard to the other models (additive, partial linear functional, and single functional index model) the situation is not as developed and no similar package is yet available, but the reader may find many case studies in the various papers cited before on these topics (see Sections 4.3.2, 4.3.3, and 4.4). The same applies for the testing methodologies described in Section 4.5, for which there remain quite a lot of open problems before a completely automatic statistical package can be presented.

4.6.3 Using kernels in other functional problems

As in standard multivariate nonparametric estimation, kernel smoothing ideas are not only of importance in regression but are also important in any other problem that involves the nonparametric estimation of some functional object. This chapter has focused on regression problems, but it is worth noting that recent work has been devoted to the study of other kernel-type estimators involving functional variables. Kernel estimation of the conditional distribution function of a real variable Y given a functional one X (with a natural application to conditional functional quantile estimation) has been investigated previously in Ferraty *et al.* (2006) and revisited in Ezzahrioui and Ould-Saïd (2008a) (see also Ferraty and Vieu (2006, Chapter 6)). Kernel estimation of the conditional density function of a real variable Y given a functional one X (with a natural application to conditional functional mode estimation) has been investigated by Ferraty *et al.* (2006) and revisited in several papers, including Dabo-Niang and Laksaci (2007) and Ezzahrioui and Ould-Saïd (2008b) (see also Ferraty and Vieu (2006, Chapter 6)). Kernel estimation of the conditional hazard function of a real variable Y given a functional one X has been investigated by Quintela del Rio (2008) and Ferraty *et al.* (2008). Note that kernel ideas have also been investigated for estimating the density of functional variables by Dabo-Niang (2004) and Delaigle and Hall (2010), while unsupervised classification problems involving functional variables are considered by Dabo-Niang *et al.* (2006) (see also Ferraty and Vieu (2006, Chapter 9)).

To conclude this discussion, let us mention that all the techniques that use kernel estimation can be investigated by relaxing the independence conditions on the statistical sample into some kind of dependence structure, with the principal aim of mating the methodology directly usable in time-series analysis. For functional variables, this approach began with the contribution by Ferraty *et al.* (2002), and much recent work has been developed with this goal. The dependent extensions of all the work presented in this chapter are the topic of Chapter 5.

4.7 CONCLUSION

This chapter has presented, through selected asymptotic results, the main recent developments in kernel methods for regression analysis with a functional covariate, including nonparametric, semiparametric, and additive kernel estimation. It has also presented a survey of the literature in this field. Except for the additive modeling section, which contains a few new results, all the other results presented here are taken from the existing literature, some of them being slightly modified in order to maintain a common framework for all the various sections.

This chapter is presented in the context of the current infatuation with the development of statistical methods for functional data analysis; this is attested not only by the wide variety of other contributions in this book, but also by recent special issues of various top-level statistical journals (see, for instance, Davidian *et al.* (2004), González-Manteiga and Vieu (2007), Valderrama (2007), and Ferraty (2010), and by the recent books of Ramsay and Silverman (1997, 2002, 2005) and Ferraty and Vieu (2006a). The reader may also refer to the volume edited by Dabo Niang and Ferraty (2008), which is especially devoted to contributions given at the first international workshop on functional and operatorial statistics (IWFOS08) held in Toulouse in June 2008. We hope that this contribution will help readers to discover the main specificities of functional kernel regression and that it will contribute to motivating further advances in this challenging and active field of modern statistics.

Acknowledgements

We wish to acknowledge all those who have actively participated in the activities (through meetings, seminars, or just informal discussions) of the working group STAPH in Toulouse. This group intends to develop all the functional features of modern statistics, and its activities have obviously played an important role in our own research and have therefore had a great impact on this chapter. All the activities of this group are available online (see Staph (2009)).

Finally we wish to address specific thanks to German Aneiros-Pérez and Laurent Delsol. German Aneiros-Pérez is a top-level specialist on partial linear models and he has helped with great efficiency in the writing of Section 4.3.3 on this topic. Similarly, Laurent Delsol is a pioneer in the field of structral regression testing with functional variables and he has been of great help in producing Section 4.5 of this chapter.

References

ABRAHAM, C., BIAU, G., CADRE, B. (2006). On the kernel rule for function classification. *Ann. Inst. Statist. Math.*, **58**, 619–33.

AIT SAIDI A., FERRATY, F., KASSA, R. (2005). Single functional index model for time series. *Rom. J. Pure & Applied Math.*, **50**, 321–330.

AIT-SAÏDI, A., FERRATY, F., KASSA, R. and VIEU, P. (2008). Cross-validated estimations in the single-functional index model. *Statistics*, **42**, 475–494.

ANEIROS-PÉREZ, G. (2002). On bandwidth selection in partial linear regression models under dependence. *Statist. Probab. Lett.*, **57**, 393–401.

ANEIROS-PÉREZ, G., CAO, R., VILAR-FERNADEZ, J. (2008). Functional method for time series prediction: a nonparametric approach. In *Proceedings of IASC-2008* (Mizuta, M., Nakaro, J. eds), 91–100. Yokohama, Japan.

ANEIROS-PÉREZ, G., CARDOT, H., ESTÉVEZ-PÉREZ, G., VIEU, P. (2006). Maximum ozone concentration forecasting by functional nonparametric approaches. *Environmetrics*, **15**, 675–85.

ANEIROS-PÉREZ, G., GONZÀLEZ-MANTEIGA, W., VIEU, P. (2004). Estimation and testing in a partial linear regression model under long-memory dependence. *Bernoulli*, **10**, 49–78.

ANEIROS-PÉREZ, G., QUINTELA DEL RIO, A. (2001). Asymptotic properties in partial linear models under dependence. *Test*, **10**, 333–55.

ANEIROS-PÉREZ, G., QUINTELA DEL RIO, A. (2002). Plug-in bandwidth choice in partial linear models with autoregressive errors. *J. Statist. Plann. Inference*, **100**, 23–48.

ANEIROS-PÉREZ, G., VIEU, P. (2006). Semi-functional partial linear regression. *Statist. Probab. Lett.*, **76**, 1102–10.

ANEIROS-PÉREZ, G., VIEU, P. (2008). Nonparametric time series prediction: a semi-functional partial linear modeling. *J. Multivariate Anal.*, **99**, 834–57.

BAÍLLO, A., GRANÉ, A. (2008). Local linear regression for functional predictor and scalar response. In *Functional and Operatorial Statistics* (Dabo-Niang, S., Ferraty, F., eds), 47–52. Physica-Verlag, Heidelberg.

BARRIENTOS-MARIN, J. (2007). Some practical problems of recent nonparametric procedures: testing, estimation and application. PhD Thesis, Univ. of Alicante, Spain.

BARRIENTOS-MARIN, J., FERRATY, F., VIEU, P. (2010). Locally modelled regression and functional data. *J. Nonparametr. Statist.*, 22, 617–632.

BENHENNI, K., FERRATY, F., RACHDI, M., VIEU, P. (2007). Locally smoothing regression with functional data. *Computational Statistics*, 22, 353–70.

BENHENNI, K., HEDLI-GRICHE, S., RACHDI, M., VIEU, P. (2008). Consistency of the regression estimator with functional data under long memory conditions. *Statist. Probab. Lett*, 78, 1043–9.

BIAU, G., BUNEA, F., WEGKAMP, M.H. (2005). Functional classification in Hilbert spaces. *IEEE Trans. Inform. Theory*, 51, 2163–72.

BICKEL, P.J., KLAASSEN, C.A., RITOV, Y., WELLNER, J.A. (1993) *Efficient and Adaptive Estimation for Semiparametric Models.* Johns Hopkins University Press, Baltimore.

BOJ, E., DELICADO, P., FORTIANA, J. (2008). Local linear functional based on weighted-based regression. In *Functional and Operatorial Statistics* (Dabo-Niang, S., Ferraty, F., eds), 57–64. Physica-Verlag, Heidelberg.

BRONSKI, J. (2003). Small ball constants and tight eigenvalue asymptotics for fractional Brownian motions. *J. Theoret. Probab.*, 16, 87–100.

BURBA, F., FERRATY, F., VIEU, P. (2008). Convergence de l'estimateur des *k* plus proches voisins en régression pour variables fonctionnelles. *C.R.A.S., Série I*, 336, 339–42.

BURBA, F., FERRATY, F., VIEU, P. (2009). k-nearest neighbour method in functional nonparametric regression. *J. Nonparametr. Stat.*, 21 (4), 453–469.

CÉROU, F., GUYADER, A. (2006). Nearest neighbor classification in infinite dimension. *ESAIM: Probab. Stat.*, 10, 340–355.

CHEN, H. (1988). Convergence rates for parametric components in a partially linear model. *Ann. Statist.*, 16, 136–147.

COLLOMB, G. (1976). Estimation non-paramétrique de la régression. PhD Thesis, Univ. Paul Sabatier, Toulouse, France.

COLLOMB, G. (1980). Estimation de la régression par la méthode des k points les plus proches avec noyau: quelques propriétés de convergence ponctuelle. In *Nonparametric Asymptotic Statistics*, Lecture Notes in Mathematics, 821, 159–75. Springer, Berlin.

COLLOMB, G. (1981). Estimation non-paramétrique de la régression: revue bibliographique. *Internat. Statist. Rev.*, 49, 75–93.

COLLOMB, G. (1984). Propriétés de convergence presque complète du prédicteur la régression. *Z. Wahrscheinlichkeitstheorie verw. Gebiete*, 66, 441–60.

COLLOMB, G. (1985). Nonparametric regression: an up-to-date bibliography. *Statistics*, 16, 309–24.

CRAMBES, C., DELSOL, L., LAKSACI, A. (2008). Robust nonparametric estimation for functional data. *J. Nonparametr. Stat.*, 20, 573–98.

DABO-NIANG, S. (2004). Kernel density estimator in an infinite dimensional space with a rate of convergence in the case of diffusion process. *Appl. Math. Lett.*, 17, 381–6.

DABO-NIANG, S., FERRATY, F. (2008). *Functional and Operatorial Statstics.* Springer, New York.

DABO-NIANG, S., FERRATY, F., VIEU, P. (2006). Mode estimation for functional random variable and its application for curves classification. *Far East J. Theor. Stat.*, 18, 93–119.

DABO-NIANG, S., LAKSACI., A. (2007). Estimation non paramétrique du mode conditionnel pour variable explicative fonctionnelle. *C. R. Math. Acad. Sci. Paris.*, 344, 49–52.

DAVIDIAN, M., LIN, X., WANG, J.L. (2004). Introduction to the emerging issues in longitudinal and functional data analysis (with discussion). *Statistica Sinica*, **14**, 613–19.

DELAIGLE, A., HALL, P. (2010). Defining probability density for a distribution of random functions. *Annals of Statistics*, **38**, 1171–93.

DELECROIX, M., HRISTACHE, M., PATILEA, V. (2006). Semiparametric M-estimation in single-index regression. *J. Statist. Plann. Inference*, **136**, 730–69.

DELSOL, L. (2007). Régréssion non-paramétrique fonctionnelle: expressions asymptotiques des moments. *Revue de l'Inst. Statist. Univ Paris*, **LI**(3), 43–67.

DELSOL, L. (2008a). Advances on asymptotic normality in nonparametric functional time series analysis. *Statistics*, **43**(1), 13–33.

DELSOL, L. (2008b). Tests de structure en régression sur variable fonctionnelle. *C. R. Math. Acad. Sci. Paris*, **346**, 343–6.

DELSOL, L. (2008c). Régression sur variable fonctionnelle: estimation, tests et applications. PhD Thesis, Université Paul Sabatier, Toulouse, France.

DELSOL L., FERRATY F., and VIEU P. (2010). Structural test in regression on functional variables (submitted).

EZZAHRIOUI, M. and OULD-SAÏD, E. (2008a). Asymptotic normality of the kernel estimator of conditional quantiles in a normed space. *Far East J. Theor. Stat.*, **25**, 15–38.

EZZAHRIOUI, M. and OULD-SAÏD, E. (2008b). Asymptotic results of a nonparametric conditional quantile estimator for functional time series. *Comm. Statist. Theory Methods*, **37**, 2735–2759.

FERRATY, F., GOIA, A., VIEU, P. (2002). Functional nonparametric model for time series: a fractal approach for dimension reduction. *Test*, **11**, 317–44.

FERRATY, F., LAKSACI, A., VIEU, P. (2006). Estimating some characteristics of the conditional distribution in nonparametric functional models. *Statistical Inference for Stochastic Processes*, **9**, 47–76.

FERRATY, F., LAKSACI, A., VIEU, P., TADJ, A. (2009). Rate of uniform consistency for nonparametric estimates with functional variables. *J. Stat. Plan. Infer.*, **140**, 335–352.

FERRATY, F., VAN KEILEGOM, I. and VIEU, P. (2010). On the Validity of the Bootstrap in Non-Parametric functional regression. *Scand. J. Statist.*, **37**, 286–306.

FERRATY, F., MAS, A., VIEU, P. (2007). Nonparametric regression on functional data: inference and practical aspects. *Austral. New Zealand J. Statist.*, **49**(3), 267–86.

FERRATY, F., PEUCH, A., VIEU, P. (2003). Modèle à indice fonctionnel simple. *Comptes Rendus Académie Sciences Paris*, **336**, 1025–8.

FERRATY, F., RABHI, A., VIEU, P. (2008). Estimation non-paramétrique de la fonction de hasard avec variable explicative fonctionnelle. *Rom. J. Pure & Applied Math.*, **52**, 1–18.

FERRATY, F., VIEU, P. (2000). Dimension fractale et estimation de la régression dans des espaces vectoriels semi-normés. *Comptes Rendus Académie Sciences Paris*, **330**, 139–42.

FERRATY, F., VIEU, P. (2002). The functional nonparametric model and application to spectrometric data. *Computational Statistics*, **17**, 545–64.

FERRATY, F., VIEU, P. (2003). Curves discrimination: a nonparametric functional approach. *Comp. Statist. and Data Anal.*, **44**, 161–73.

FERRATY, F., VIEU, P. (2004). Nonparametric models for functional data, with application in regression, time-series prediction and curve discrimination. *J. Nonparametr. Stat.*, **16**, 111–125.

FERRATY, F., VIEU, P. (2006a). *Nonparametric Functional Data Analysis. Theory and Practice.* Springer, New York.

FERRATY, F., VIEU, P. (2006b). *NPFDA: R/S+ routines.* Free access online at http://www.math.univ-toulouse.fr/staph/npfda/.

FERRATY, F., VIEU, P. (2008) Erratum to: "Nonparametric models for functional data, with application in regression, time-series prediction and curve discrimination" (*J. Nonparametr. Stat.* (2004), **16** (1–2), 111–25). *J. Nonparametr. Stat.*, **20**, 187–9.

FERRATY, F., VIEU, P. (2009). On functional regression modelling. *Comp. Statist. Data Analysis*, **53**, 1400–13.

FERRATY, F. (2010). Special Issue: Statistical Methods and Problems in Infinite-dimensional Spaces (Ed.). *J. Mult. Anal.*, **101**, 305–490.

GAO, F., HANNIG, J., TORCASO, F. (2003). Integrated Brownian motions and exact L_2-small balls. *Ann. Probab.*, **31**, 1320–37.

GONZÁLEZ MANTEIGA, W., VIEU, P. (2007). Introduction to the special issue on statistics for functional data. *Comp. Statist. Data Analysis*, **51**, 4788–92.

GYÖRFI, L., KOHLER, M., KRZYZAK, A., WALK, H. (2002). *A distribution-free Theory of Nonparametric Regression.* Springer, New York.

HÄRDLE, W. (1990). *Applied Nonparametric Regression.* Oxford University Press, Oxford.

HÄRDLE, W., BOWMAN, A.W. (1988). Bootstrapping in nonparametric regression: local adaptive smoothing and confidence bands. *J. Amer. Statist. Assoc.*, **83**, 102–10.

HÄRDLE, W., HALL, P. (1993). On the backfitting algorithm for additive regression models. *Statist. Neerlandica*, **47**, 43–57.

HÄRDLE, W., HALL, P., ICHIMURA, H. (1993). Optimal smoothing in single index models. *Ann. Statist.*, **21**, 157–78.

HÄRDLE, W., LIANG, H., GAO, J. (2000). *Partially Linear Models.* Physica, Heidelberg.

HÄRDLE, W., MARRON, J.S. (1985). Optimal bandwidth selection in nonparametric regression function estimation. *Ann. Statist.*, **13**, 1465–81.

HÄRDLE, W., MARRON, J.S. (1991). Bootstrap simultaneous error bars in nonparametric regression. *Ann. Statist.*, **19**, 778–96.

HÄRDLE, W. and MAMMEN, E. (1993). Comparing nonparametric versus parametric regression fits. *Ann. Statist.*, **21**, 1926–1947.

HÄRDLE, W., MÜLLER, M., SPERLICH, S., WERWATZ, A. (2004). *Nonparametric and Semiparametric Models.* Springer, New York.

HART, J. (1997). Nonparametric smoothing and lack-of-fit tests. Springer Series in Statistics. Springer-Verlag, New York.

HASTIE, T., TIBSHIRANI, R. (1986). Generalized additive models (with discussion). *Statist. Sci.*, **1**, 297–318.

HASTIE, T., TIBSHIRANI, R. (1990). *Generalized Additive Models.* Chapman and Hall, London.

HOROWITZ, J. (1998). *Semiparametric Methods in Econometrics.* Lecture Notes in Statistics, **131**. Springer, New York.

HOROWITZ, J., KLEMELÄ, J., MAMMEN, E. (2006). Optimal estimation in additive regression models. *Bernoulli*, **12**, 271–98.

HRISTACHE, M., JUDITSKY, A., SPOKOINY, V. (2001). Direct estimation of the index coefficient in the single index model. *Ann. Statist.*, **29**, 595–623.

KOLMOGOROV, A., TIKHOMIROV, V. (1959). ϵ-entropy and ϵ-capacity. (in Russian). *Uspekhi Mat. nauk.*, **14**, 3–86.

LIANG, H. (2000). Asymptotic normality of parametric part in partly linear model with measurement errors in the nonparametric part. *J. Statist. Plann. Inference*, **86**, 51–62.

MAMMEN, E. (2000). Resampling methods in nonparametric regression. In *Smoothing and Regression. Approaches, Computation, and Application* (M. Schimek, ed.), 425–50. John Wiley & Sons, New York.

MAMMEN, E., PARK, B. (2006). A simple smooth backfitting method for additive models. *Ann. Statist.*, **34**, 2252–71.

MARRON, J.S., HÄRDLE, W. (1986). Random approximations to some measures of accuracy in nonparametric curve estimation. *J. Multivar. Anal.*, **20**, 91–113.

MASRY, E. (2005). Nonparametric regression estimation for dependent functional data: asymptotic normality. *Stochastic Process. Appl.*, **115**, 155–77.

MÜLLER, H.G. (2008). Functional additive modelling. In *Abstracts of ISNI'2008 Meeting, Vigo, Spain*. Available at http://www.isni2008.com.

NAZAROV, A., NIKITIN, Y. (2004). Exact L_2-small ball behavior of integrated Gaussian processes and spectral asymptotics of boundary value problems. *Probab. Theory Related Fields*, **129**, 469–94.

PELLETIER, B. (2006). Non-parametric regression estimation on closed Riemannian manifolds. *J. of Nonparametric Statistics*, **18**, 57–67.

QUINTELA DEL RIO, A. (2008). Hazard function given a functional variable: non-parametric estimation under strong mixing conditions. *J. Nonparametr. Stat.*, **20**, 413–30.

RACHDI, M., VIEU, P. (2007). Nonparametric regression for functional data: automatic smoothing parameter selection. *J. Statist. Plann. Inference*, **137**(9), 2784–801.

RAMSAY, J.O., SILVERMAN, B.W. (1997). *Functional Data Analysis.* Springer, New York.

RAMSAY, J.O., SILVERMAN, B.W. (2002). *Applied Functional Data Analysis: Methods and Case Studies.* Springer, New York.

RAMSAY, J.O., SILVERMAN, B.W. (2005). *Functional Data Analysis* (second edition). Springer, New York.

SARDA, P., VIEU, P. (2000). Kernel regression. In *Smoothing and Regression. Approaches, Computation, and Application* (M. Schimek, ed.), 43–70. John Wiley & Sons, New York.

SCHICK, A. (1996). Root-n consistent estimation in partly linear regression models. *Statist. Probab. Lett.*, **28**, 353–358.

SCHIMEK, M. (2000). *Smoothing and Regression. Approaches, Computation, and Application.* John Wiley & Sons, New York.

SCHIMEK, M., TURLACH, B. (2000). Additive and generalized additive models. In *Smoothing and Regression. Approaches, Computation, and Application* (M. Schimek, ed.), 277–328. John Wiley & Sons, New York.

STAPH. (2009). Groupe de travail en statistique fonctionnelle et opératorielle, Toulouse, France. Activities online at http://www.math.univ-toulouse.fr/staph/.

STONE, C. (1982). Optimal global rates of convergence for nonparametric estimates. *Ann. Statist.*, **10**, 1040–53.

STONE, C. (1985). Additive regression and other nonparametric models. *Ann. Statist.*, **13**, 689–705.

STONE, C. (1986). The dimensionality reduction principle for generalized additive models. *Ann. Statist.*, **14**, 590–606.

STONE, C. (1994). The use of polynomial splines and their tensor products in multivariate function estimation. *Ann. Statist.*, **22**, 118–84.

SPERLICH, S., HÄRDLE, W., AYDINH, G. (2006). *The Art of Semiparametrics*. Physica, Heidelberg.

SPECKMAN, P. (1988). Kernel smoothing in partial linear models. *J. Roy. Statist. Soc. Ser. B*, **50**, 413–436.

TONG, X.W., CUI, H.J., YU, P. (2008). Consistency and normality of Huber-Dutter estimators for partial linear model. *Sci. China Ser. A*, **51**, 1831–42.

VALDERRAMA, M. (2007). Introduction to the special issue on modelling functional data in practice. *Comput. Statist.*, **22**, 331–4.

VIEU, P. (1991). Nonparametric regression: optimal local bandwidth choice. *J. R. Statist. Soc.*, **53**, 453–64.

CHAPTER 5

NONPARAMETRIC METHODS FOR α-MIXING FUNCTIONAL RANDOM VARIABLES

LAURENT DELSOL

5.1 INTRODUCTION

THE study of statistical techniques for functional data is currently an area of great interest, both because of its great variety of potential applications and the theoretical challenges that still have to be overcome (see the monographs by Ramsay and Silverman (1997, 2002, 2005), Bosq (2000), and Ferraty and Vieu (2006)). Chapter 1 of this book has provided an introduction to parametric and nonparametric functional regression models (in addition, see Ferraty and Vieu (2006) for a more general definition of functional parametric and nonparametric models). Parametric models have been widely considered in the independent case (see Chapter 2 for an overview of this topic) and in the case of linear functional processes (see Chapter 3 for recent advances and more references). In order to avoid structural assumptions, the study of functional nonparametric models has received a lot of attention. In such models the usual estimators are based on functional kernel smoothing methods (see Chapter 4 for an introduction to these tools in the independent case). These methods correspond to an extension of classical multivariate kernel smoothing to the case where the variable takes values in an infinite-dimensional space $\mathcal{E}$. In such a case, the notion of distance used to compute the kernel weights is of crucial importance because all metrics are not equivalent. Taking this fact into account, Ferraty and Vieu (2000) have proposed functional kernel smoothing methods in which the functional variables are compared using a semi-metric. Their ideas have led to many generalizations of standard multivariate kernel smoothing methods to the functional case (see, for instance, Chapter 4 of this book and Section 5.4 in particular).

Some problems require taking into account the dependence structure that may exist within the dataset. Many modelizations of this dependence structure have been considered in the literature. Strong mixing, also known in the literature as weak dependence, seems to be a very popular approach. The study of multivariate kernel methods for such dependence conditions has been widely considered (see, for instance, Collomb (1984), Györfi *et al.* (1989), Yoshihara (1994), and Bosq (1996, 1998)). The α-mixing condition introduced by Rosenblatt (1956) is one of the most general weak-dependence modelizations (see Section 5.3).

The aim of this chapter is to explain how functional kernel methods can be used to study α-mixing datasets. We link the existing literature devoted to the multivariate α-mixing case with recent advances in functional kernel methods with independent variables (see Chapter 4 for an overview of regression issues and Ferraty and Vieu (2006) for a more general discussion). In this chapter we have deliberately chosen to take time to present the motivation for this research, some useful tools, methods, results, and interesting prospects for the future. We hope that this will be useful for readers interested in discovering and understanding the main ideas of functional kernel smoothing methods with α-mixing variables. Moreover, many references are given to allow readers more familiar with these notions to find additional results and detailed proofs. Later, Chapter 14 of this volume focuses on Berstein and maximal inequalities for α-mixing sequences of Banach-valued random variables. The content of chapter 14 provides an interesting complement to the tools presented in this chapter.

The remainder of this chapter is set out as follows. In Section 5.2 we discuss the way in which prediction problems involving dependent functional datasets may arise from the study of time series. This example motivates the methods developed in the remainder of this chapter and presents an interesting potential application. Then Section 5.3 focuses on strong mixing conditions. We first recall the notion of α-mixing coefficients and α-mixing variables introduced by Rosenblatt (1956). Then we present some conditions for a Markov chain to be α-mixing. Finally, we recall some useful tools that provide covariance inequalities, exponential inequalities, and central limit theorems (CLTs) for α-mixing sequences. These results are used in Section 5.4.2 to study asymptotic properties of functional kernel estimators. Section 5.4 focuses on the use of kernel smoothing methods with α-mixing datasets. Various functional kernel estimators corresponding to different prediction methods are presented in Section 5.4.1. We then state some results dealing with their asymptotic properties: almost sure convergence, asymptotic normality, and $\mathbb{L}^p$ errors. Finally, in Section 5.5 we present some interesting prospects for further research.

5.2 A FUNCTIONAL APPROACH TO TIME-SERIES STUDIES

In this section we discuss the way in which time series can be interpreted as a set of dependent functional variables. This provides an interesting example of problems that may lead us to consider and study datasets of dependent functional variables.

5.2.1 The standard discretized model

In time-series studies one typically observes the evolution of a quantity (for instance, ozone concentration, the price of a product, electricity consumption, or temperature) over time. Using the observations collected upto a given time, it is often interesting to try to predict the future values of the variable under consideration. Predicted values can be useful in taking decisions concerning the near future (for instance, to prevent pollution surges, to choose to buy or sell a product, or to adapt electricity production to demand).

Mathematically, over a given time τ, the evolution of the quantity of interest can be represented by a real random stationary process $Z_t, t \in [0; \tau]$. Our aim is to try to predict the future values of Z_t for $t > \tau$ from the values observed up to time τ. For example, in order to study the El Niño phenomenon, monthly sea temperature measurements have been made for 58 years (see Fig. 5.1) from June 1950 to May 2008, and we would like to try to predict the temperatures for the 59th year. This dataset is available on the website http://www.cpc.ncep.noaa.gov/data/indices/. In practice, the process Z_t is only observed on a discretized time grid $0 \leq t_0 < t_1 < \cdots < t_i \leq \tau$. For many years statisticians have considered the following discretized model:

$$Z_{t_{i+1}} = r\left(Z_{t_i}, Z_{t_{i-1}}, \ldots, Z_{t_{i-p}}\right) + \epsilon_i, \qquad (5.1)$$

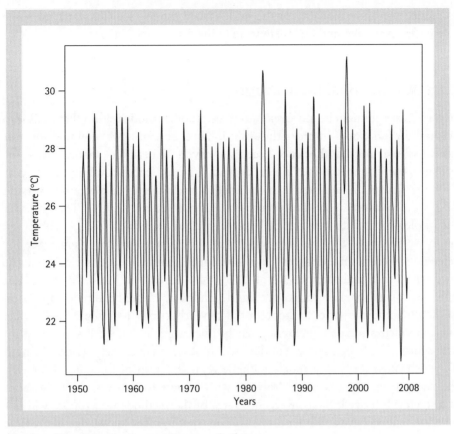

Fig. 5.1 Monthly sea temperature evolution.

where p is a parameter linked to the number of past values we want to take into account, r is an unknown function defined on $\mathbb{R}^p$, and ϵ_i is a centered residual not correlated with the variables $Z_{t_i}, \ldots, Z_{t_{i-p}}$. The existence of a function r independent of i is not guaranteed and requires some additional assumptions on the process Z_t. This is true, for instance, when the process is strictly stationary and square integrable. However, the strict stationarity of the process is not a necessary assumption and the existence of r can be obtained whenever Z_t is square integrable and

$$m(z_1, \ldots, z_p) = \mathbb{E}\left[Z_{t_{i+1}} | Z_{t_i} = z_1, \ldots, Z_{t_{i-p}} = z_p\right]$$

does not depend on i. In the nonparametric case, i.e. when r cannot be indexed by a finite number of real parameters, the prediction results are affected by the curse of dimensionality when the discretization becomes thinner or when we want to take more past values into account (i.e. as p grows). Indeed, Stone (1982) has shown that the optimal convergence rates of nonparametric estimators decrease quickly when the number of regressors grows. Various alternative methods have been proposed to reduce the dimensionality of the multivariate model (see Chapter 1 for some references, examples, and their extension to the functional context).

5.2.2 A functional modelization

In some cases, it might be more relevant to use another modelization that takes into account the functional nature of the evolution of the quantity we want to study over time. Assume now that the process seems to have a "T-periodic" shape. This is, for instance, the case in the El Niño study, with a period corresponding to one year (T = 12). The main idea here is to consider Z_t no longer in its discretized version $(Z_{t_0}, \ldots, Z_{t_i})$ but as a continuous process Z_t observed on $[0, TN]$. Then we split the whole process into a dataset of N successive curves $X_i(t) = Z_{(i-1)T+t}$ observed on the segment $[0, T]$. This idea was initially introduced by Bosq (1991) (see the references therein and Chapter 3 for additional comments). It is clear that the successive functional variables $(\mathcal{X}_1, \ldots, \mathcal{X}_N)$ are dependent. We will explain later how we propose to take into account this dependence. We apply this decomposition to the El Niño time series and obtain the 58 yearly curves represented in Fig. 5.2. Such a modelization of a time series may be interesting in that it takes into account the regularity and the structure of the process observed over each period of length T, which often leads to a more synthetic representation of the data. Moreover, the resulting method is more "robust" with respect to the way the time series is discretized. Indeed, the convergence rates of the methods we propose will not deteriorate if we have a thinner discretization (that is, more knowledge of the time series). Furthermore, these methods can be used with nonequidistant discretization grids or even in the case where the nature and number of discretization points differs from one period to another.

Finally, in some cases where the time series is measured with errors, smoothing the original discretized values to obtain a curve can reduce the effect of these errors on the prediction results. In this chapter, we assume that we have at our disposal a dataset of functional variables. Of course, in practice, a pre-treatment of the data is necessary to obtain functional observations from discretized values. A more detailed discussion on this specific topic is given, for instance, in Ramsay and Silverman (2005), Ferraty and Vieu (2006), Crambes (2007), or Delsol (2009). These pre-treatment methods are also discussed and the end of Chapter 1 and in Chapter 9, focusing on registration issues.

Using this functional modelization, we are interested in regression models of the following form:

$$G(\mathcal{X}_i) = r(\mathcal{X}_{i-1}, \ldots, \mathcal{X}_{i-p}) + \epsilon_i, \qquad (5.2)$$

where G is a known operator and r is an unknown operator defined on $\mathcal{E}^p$. The issue of the existence of an operator r independent of i is similar to that discussed in the discretized case. This model is a specific case of the functional regression

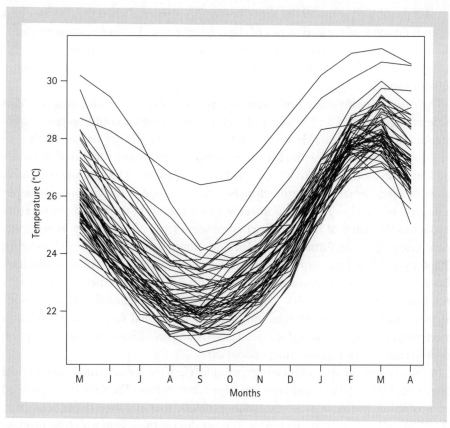

Fig. 5.2 Yearly sea temperature curves.

model presented in Chapter 1 (model (1.1)) with response variable $Y_i = G(\mathcal{X}_i)$ and explanatory variable $X_i = (\mathcal{X}_{i-1}, \ldots, \mathcal{X}_{i-p})$. More generally, it may also be interesting to consider other characteristics of the conditional law of Y_i given X_i (see the examples presented in Section 5.4.1 below).

In this chapter, we focus more specifically on the case in which G is a known real-valued operator and $p = 1$. The choice of the operator G depends on the objective we have. For instance, to predict the value at a given time-point t_0, one may take G as the operator G_{t_0} that associates to a curve $\mathcal{X} \in \mathcal{E}$ its value at t_0:

$$G_{t_0}(\mathcal{X}) = \mathcal{X}(t_0).$$

In other situations, it may be better to choose G in a different way. Here is a non-exhaustive list of interesting examples:

- $G(\mathcal{X}) = \sup_{t \in [0,T]} \mathcal{X}(t)$ (prediction of extreme values);
- $G(\mathcal{X}) = \int_0^T \mathcal{X}(t)\,dt$ (prediction of cumulative effects);
- $G(\mathcal{X}) = 1_{\{\mathcal{X} \in \mathcal{C}\}}$, where $\mathcal{C}$ is a given family of elements of $\mathcal{E}$ (discrimination).

It is also possible to consider the above prediction problems when the operator also depends on some of the derivatives of the original curve $\mathcal{X}$. Finally, all methodologies and theoretical results presented in this chapter can be trivially adapted to the case where $p > 1$ because the vector $(\mathcal{X}_i, \ldots, \mathcal{X}_{i-p})$ is still a functional explanatory random variable.

The case of multivariate prediction problems is also interesting. It seems that some of the methods and results presented in this chapter could be fairly easily extended to this context. However, we have chosen to restrict our discussion to the univariate case because the literature devoted to α-mixing functional variables and kernel methods seems to be restricted to such models. The case where G takes values in a functional space is therefore beyond the scope of this chapter. However, in the case where $G = Id$ and r is linear, we refer readers to the abundant literature devoted to the study of linear Hilbertian processes (see, for instance, Bosq (1989, 2000, 2007), Mas and Pumo (2007), Turbillon (2007), and the references therein). Chapter 3 of this volume provides an overview of these methods. A few results have also been proposed for nonlinear autoregressive processes (see Besse et al. (2000) and Antoniadis et al. (2006)). The extension of the functional kernel methods discussed in Section 5.4.1 to take into account functional responses provides an interesting prospect for the future (see Dabo-Niang and Rhomari (2009) for a first result in this direction in the independent case).

In conclusion, in this motivating example we have transformed the original time series into a dataset composed of dependent curves. Then we have modeled the prediction problem as a regression problem that involves dependent pairs $(\mathcal{X}_i, G(\mathcal{X}_{i+1})) \in \mathcal{E} \times \mathbb{R}$. In the next section we discuss the way in which we propose to take into account the dependence that exists between these variables.

5.3 Strong mixing assumptions

In the remainder of this chapter we will assume that we have at our disposal a dataset corresponding to the observation of n dependent random pairs $(X_i, Y_i) \in \mathcal{E} \times \mathbb{R}$. We now discuss the kinds of dependence we will consider, and provide some useful tools to deal with such datasets.

5.3.1 The α-mixing condition

As discussed in the previous section, it may be interesting and necessary in some situations to propose methods that take into account the dependence that may exist within a dataset. There are many ways to model this dependence (see Section 5.3.4). In this chapter we consider one of the most general cases of weak dependence, known in the literature as α-mixing. This notion of dependence was first introduced by Rosenblatt (1956). We recall the definition of α-mixing coefficients and α-mixing variables:

Definition 5.1 *Let* $(Z_i)_{i \in \mathbb{Z}}$ *be a sequence of random variables.*

- *The α-mixing coefficients associated with the sequence* $(Z_i)_{i \in \mathbb{Z}}$ *are defined by:*

$$\alpha(j) = \sup_{k \in \mathbb{Z}} \sup_{A \in \mathcal{G}_k,\, B \in \mathcal{F}_{k+j}} |\mathbb{P}(A \cap B) - \mathbb{P}(A)\mathbb{P}(B)|, \; j \in \mathbb{N}^*$$

where $\mathcal{G}_k = \sigma(\{Z_i, i \leq k\})$ *and* $\mathcal{F}_\ell = \sigma(\{Z_i, i \geq \ell\})$.

- *A sequence of random variables* $(Z_i)_{i \in \mathbb{Z}}$ *is said to be α-mixing if the corresponding α-mixing coefficients fulfill the condition* $\alpha(n) \xrightarrow{n \to +\infty} 0$.

When the sequence is only defined for some set of indices $T \subset \mathbb{Z}$, the sequence is said to be α-mixing if its completed version

$$\tilde{Z}_i = \begin{cases} Z_i, & i \in T \\ 0, & i \notin T \end{cases}$$

is α-mixing.

It follows directly from the above definition that the notion of an α-mixing sequence does not depend on the space in which the variables take values. The α-mixing condition is closely related to the property that for all k, the dependence between the variables Z_k and Z_{k+j} vanishes when j tends to infinity. It is clear that the α-mixing condition is trivially fulfilled in the independent case or in the m-dependent case ($m \in \mathbb{N}$).

By definition, the sequence of α-mixing coefficients is nonincreasing and nonnegative. Therefore it converges to a limit value $\bar{\alpha}$. When the sequence is α-mixing this limit value is zero. The way in which the sequence of α-mixing coefficients decreases to zero is sometimes assumed to be arithmetic or geometric. We follow the standard definitions and say the α-mixing coefficients are arithmetic of order a (with $a > 0$) if they have the following form:

$$\exists C > 0, \; \forall j \in \mathbb{N}^*, \; \alpha(j) \leq C j^{-a},$$

and geometric when

$$\exists \rho, \; 0 \leq \rho < 1, \; \exists C > 0, \; \forall j \in \mathbb{N}^*, \; \alpha(j) \leq C \rho^j.$$

Because the ratio ρ^j / j^{-a} tends to zero for any fixed value of a, it is clear that geometric α-mixing coefficients are arithmetic of order a for any a. For the sake of simplicity, in this chapter we focus on the specific situation of arithmetic α-mixing coefficients and provide references for more general situations. However, it is worth noting that our results also hold for geometric α-mixing coefficients (without any restriction on a). Moreover, many examples of dependent processes are arithemtically α-mixing (see the references given in Section 5.3.2).

Before proceding further and giving some useful tools for the study of α-mixing sequences of variables, it would seem interesting to present some examples or counter examples of processes for which the above α-mixing condition may hold.

5.3.2 α-mixing processes

The year of citations is not consistently given. It is sometimes given with brackets when citations are made between brackets dealing with sufficient conditions for a process to be α-mixing. The first results were obtained for Markov chains Rosenblatt (1972); Davydov (1973), linear processes Withers (1981), and autoregressive processes Athreya and Pantula (1986); Mokkadem (1990); Lee (1995). Our aim is not to propose an exhaustive overview of the sufficient conditions that insure a sequence is strong-mixing. However, for those readers not familiar with strong-mixing conditions, it is worth presenting briefly a result explaining what the mixing assumption means for Markov chains.

Proposition 5.1 (Bradley 2005) *Suppose $Z := (Z_i)_{i \in \mathbb{Z}}$ is a strictly stationary Markov chain.*

1. *If Z is a countable-state Markov chain, then the following two statements are equivalent:*
   - *Z is irreducible and aperiodic;*
   - *$\alpha(j) \to 0$ as $j \to +\infty$.*

2. *If Z is ergodic and aperiodic, then the following property holds:*

$$\text{if } \alpha(j) < \frac{1}{4} \text{ for some } j \geq 1, \text{ then } \alpha(j) \to 0 \text{ as } j \to 0.$$

Remark: When Z is a strictly stationary, finite-state, irreducible, and aperiodic Markov chain the corresponding α-mixing coefficient sequence decreases at least exponentially fast.

Even though not directly within the scope of this chapter, the introductory examples of α-mixing and non-α-mixing real-valued linear processes discussed in Bosq (1998) might be useful references in order to become more familiar with α-mixing conditions. In order to consider further real-valued dependent processes (such as the counterexample presented in Bosq (1998)), some generalizations of α-mixing coefficients have been proposed (see, for instance, Dedecker and Prieur (2005) and the references therein).

A more detailed discussion of α-mixing conditions can be found, for instance, in the reviews by Bradley (2005, 2007), Doukhan (1994), Bosq (1998), Rio (2000), and Dedecker and Doukhan (2003) and the references they contain. Recently, Allam and Mourid (2002) focus on the case of Banach autoregressive processes, while Veretennikov (2006) discusses the special case of diffusion processes.

5.3.3 Some useful tools

Functional kernel smoothing estimators are constructed from weighted averages (see Chapter 4 and Section 4.4.1 of this volume for more details). If the sequence of variables $(\mathcal{X}_i, Y_i)$ is arithmetically α-mixing of order a, then the sequence of kernel weights $W_{n,\ker}(x, \mathcal{X}_i)$ (using the same notation as in Chapter 4) is arithmetically α-mixing with the same order whenever the kernel is measurable (see Ferraty and Vieu (2006, Section 10.3)). The asymptotic properties of the functional kernel smoothing estimators discussed in this chapter can be obtained from results concerning real-valued α-mixing sequences. Consequently, we restrict our overview to the real-valued case. However, various results have been proved for α-mixing sequences of functional variables (see, for instance, Dehling (1983), Politis and Romano (1994), Dehling and Sharipov (2005), and Chapter 14 for recent advances). Such tools should be useful in extending our results to more general cases in which the quantity to be predicted is functional.

From the early 1970s, many interesting tools have been introduced to deal with real-valued sequences of α-mixing variables. There is an extensive literature devoted, for example, to exponential inequalities, the strong law of large numbers, the law of the iterated logarithm, covariance inequalities, central limit theorems, and Bahadur representations. In this chapter we simply present three results concerning covariance inequalities, exponential inequalities, and central limit theorems, which will be useful in obtaining asymptotic properties for kernel estimators with α-mixing datasets. Additional references are given in the monographs by Doukhan (1994), Bosq (1998), Rio (2000), and Yoshihara (2004).

Proposition 5.2 (see Rio (1993), Bosq (1998), or Rio (2000)) *Let X and Y be two real-valued random variables.*

- *Let $Q_X(u) = \inf\{t, \ P(|X| > t) \leq u\}$ be the quantile function of $|X|$. Then, if $Q_X Q_Y$ is integrable over $]0;1[$ we have*

$$|\text{Cov}(X, Y)| \leq 2 \int_0^{2\alpha} Q_X Q_Y \, du,$$

where $\alpha = \sup_{A \in \sigma(X), B \in \sigma(Y)} |P(A \cap B) - P(A) P(B)|$.
- *If $X \in \mathbb{L}^r(P)$ and $Y \in \mathbb{L}^q(P)$, where $q > 1$, $r > 1$, and $\frac{1}{q} + \frac{1}{r} = 1 - \frac{1}{p}$, then*

$$|\text{Cov}(X, Y)| \leq 2p(2\alpha)^{\frac{1}{p}} \|X\|_q \|Y\|_r.$$

- *If in particular $X, Y \in \mathbb{L}^\infty(P)$, then*

$$|\text{Cov}(X, Y)| \leq 4\alpha \|X\|_\infty \|Y\|_\infty.$$

These inequalities are very useful for studying the variance of sums of α-mixing variables. They are also useful for controlling additional covariance terms and for making them negligible with respect to the variance terms (see the proof of Theorem 5.1 for more details). A version of this inequality for α-mixing arrays is discussed in Liebscher (2001).

Proposition 5.3 (see Rio (2000)) *Let $(Z_t)_{t\in\mathbb{N}}$ be a sequence of identically distributed zero-mean real random variables. Denote by $\alpha(n)$ the corresponding α-mixing coefficients and*

$$s_n^2 = \sum_{i=1}^n \sum_{j=1}^n |\text{Cov}(Z_i, Z_j)|.$$

Assume now that there exist $c \geq 1$ and $a > 1$ such that $\alpha(j) \leq cj^{-a}$ for all $j > 0$. Then, for all $\lambda > 0$ and $r \geq 1$, the following inequalities hold:

1. *If there exists $p > 2$ such that $P(|Z_1| > t) \leq t^{-p}$, then one has*

$$P(|S_n| \geq 4\lambda) \leq 4\left(1 + \frac{\lambda^2}{rs_n^2}\right)^{-\frac{r}{2}} + 4Cnr^{-1}\left(\frac{\lambda}{r}\right)^{-\frac{(a+1)p}{a+p}},$$

where $C = 2p(2p-1)^{-1}(2^a c)^{\frac{p-1}{a+p}}$.

2. *If $\|X_i\|_\infty \leq 1$, $i \in \mathbb{N}$, then one has*

$$P(|S_n| \geq 4\lambda) \leq 4\left(1 + \frac{\lambda^2}{rs_n^2}\right)^{-\frac{r}{2}} + 2cnr^{-1}\left(\frac{2r}{\lambda}\right)^{a+1}.$$

Such inequalities may be seen as an equivalent of Fuk–Nagaev inequalities for the case of α-mixing sequences. They are very useful in proving the almost-sure convergence of sums of α-mixing variables. We will use them to prove the almost-sure convergence of kernel estimators (see the proof of Theorem 5.1 for more details). A general version of such exponential inequalities in the case of non-arithmetic α-mixing coefficients is discussed in Rio (2000). Another formulation of this general result, adapted specifically for nonparametric estimation, is also given in Ferraty and Vieu (2006). Other exponential inequalities have been proposed in order to extend the Hoeffding and Bernstein inequalities (see, for instance, Bosq (1998)) to the case of α-mixing variables. These may be used to show the almost-sure convergence of functional kernel estimators under other sets of assumptions. Finally Rhomari (2002) and Chapter 14 of this volume focus on Bernstein inequalities for random variables taking values in an infinite-dimensional space.

We now recall a useful asymptotic normality result for arrays of α-mixing random variables. For this, we need to introduce the notion of α-mixing coefficients for triangular arrays $\{Z_{n,i}, 1 \leq i \leq n, n \in \mathbb{N}\}$:

$$a_j := \max_{n,k:1\leq k\leq n-j} \sup_{A\in {}^n\mathcal{F}_{j+k}^n, B\in {}^n\mathcal{F}_1^k} |P(A\cap B) - P(A)P(B)|,$$

where ${}^n\mathcal{F}_k^m := \sigma(Z_{n,k}, \ldots, Z_{n,m})$, $n \in \mathbb{N}$, $1 \leq k \leq m \leq n$. We also have to introduce some notation:

$$R_n := \max_{1 \leq i \leq n} \mathbb{E}\left[Z_{i,n}^2\right], \quad r_n := \max_{j,k: 1 \leq j,k \leq n,\, j \neq k} \left(\mathbb{E}\left[|Z_{n,j} Z_{n,k}|\right] + \mathbb{E}\left[|Z_{n,j}|\right] \mathbb{E}\left[|Z_{n,k}|\right]\right)$$

and the following conditions:

$$R_n = O\left(n^{-1}\right) \text{ and } \exists p > 2,\ \mathbb{E}\left[|Z_{n,i}|^p\right]^{\frac{1}{p}} < +\infty,\ \sum_{k=1}^{\infty} k^{\frac{2}{p-2}} \alpha_k < +\infty, \quad (5.3)$$

$$\exists (m_n)_{n \in \mathbb{N}} \subset \mathbb{N}^*,\ m_n \to +\infty,\ n m_n r_n = o(1),$$

$$\text{and } \left(\sum_{m_n+1}^{\infty} j^{\frac{2}{p-2}} \alpha_j\right)^{\frac{p-2}{p}} \sum_{i=1}^{n} \left(\mathbb{E}\left[|Z_{n,i}|^p\right]\right)^{\frac{2}{p}} = o(1), \quad (5.4)$$

$$\exists (\tau_n)_{n \in \mathbb{N}} \subset \mathbb{N}^*,\ m_n \to +\infty,\ \tau_n = o\left(\sqrt{n}\right),\ \forall \epsilon_0 > 0,\ \frac{n}{\tau_n} \alpha_{[\epsilon_0 \tau_n]} \xrightarrow{n \to +\infty} 0,$$

$$\text{and } \sum_{i=1}^{n} \mathbb{E}\left[Z_{n,i}^2 \mathbf{1}_{\{|Z_{i,n}| > \tau_n^{-1}\}}\right] = o(1) \quad (5.5)$$

Proposition 5.4 (Theorem 2.1 in Liebscher (2001)) *Let $\{Z_{n,i}, 1 \leq i \leq n, n \in \mathbb{N}\}$ be a triangular array of random variables with $\mathbb{E}[Z_{n,i}] = 0$. Assume assumptions (5.3)–(5.5) hold and*

$$\lim_{n \to +\infty} \sum_{i=1}^{n} \mathbb{E}\left[Z_{n,i}^2\right] = \sigma^2.$$

Then one has $\sum_{i=1}^{n} Z_{i,n} \xrightarrow{\mathcal{L}} \mathcal{N}(0, \sigma^2)$.

A complete proof of this result is given in Liebscher (2001), where other sets of assumptions are also considered. Many other results dealing with central limit theorems for sums of α-mixing variables can be found in the literature (see, for instance, Bosq (1998) and Rio (1995, 2000)). Such results are useful in proving the asymptotic normality of functional kernel estimators (see the proof of Theorem 5.2).

5.3.4 Other dependence modelizations

In this chapter we discuss the way in which asymptotic properties of functional kernel estimators may be extended from the independent to the α-mixing case. Even if it is not actually within the scope of this chapter, it seems worth including a brief discussion on other possible ways to model the dependence that may exist within a dataset.

There are various other modelizations of weak dependence based, for instance, on the notion of β-mixing (Volonskii and Rozanov 1959), ϕ-mixing (Ibragimov 1962), or ρ-mixing variables (Kolmogorov and Rozanov 1960). The reviews by Doukhan (1994), Bosq (1998), and Bradley (2005) provide an overview and comparison of the various weak-dependence modelizations. In particular, it appears in these comparisons that the notion of α-mixing variables is one of the most general weak-dependence assumptions. Indeed, it is less restrictive than β-, ϕ-, or ρ-mixing conditions. The existence of many useful tools (for example, those presented in Section 5.3.3) for studying α-mixing variables is actually another argument for investigating the case of α-mixing datasets. Recently, some generalizations of these notions have been introduced by Dedecker and Prieur (2005). Moreover, Doukhan and Louhichi (1999) have introduced a new notion of weak dependence that may hold for more general processes. See also Dedecker *et al.* (2007) for a more detailed discussion and more examples. Because of the newness of these conditions, we have at our disposal only a small number of tools and results for such weak-dependence processes. It is possible to imagine the extension of our results to the dependence assumptions proposed by Doukhan and Louhichi (1999) or Dedecker and Prieur (2005), since some of the results we use have recently been extended to such dependence conditions (see, for instance, Coulon-Prieur and Doukhan (2000), Dedecker and Prieur (2005), Doukhan and Neumann (2007), and Dedecker *et al.* (2007)). Moreover, the asymptotic normality result presented in Section 5.4.2 has been extended to the non-stationnary case (Aspirot *et al.* 2008). Finally, it would also seem interesting to consider strong dependence conditions. Strong dependence phenomena are usually modeled by long memory type processes (see, for instance, Mandelbrot and Van Ness (1968), Beran (1994), and Palma (2007)). The results one can obtain for such datasets are usually of a different nature to those proved in the independent case (see, for instance, Estevez and Vieu (2003), Benhenni *et al.* (2008), and Hedli-Griche (2008) for a few seminal results concerning the regression on functional variables).

5.4 Kernel smoothing with functional data

Many results have been obtained for kernel methods with dependent multivariate variables (see, for instance, Collomb (1984), Györfi *et al.* (1989), Yoshihara (1994), and Bosq (1996, 1998)). We now discuss the ways in which kernel smoothing methods may be useful in different situations involving α-mixing functional datasets. We aim to extend recent advances in functional kernel methods to the independent case (see Ferraty and Vieu (2006, Chapter 4) and the references therein).

Essentially, we make assumptions on the α-mixing coefficients and the dependence that exists within the dataset in order to get similar results to those stated in Chapter 4.

5.4.1 Some functional kernel estimators

Because this chapter concerns the use of kernel methods for α-mixing functional variables, we have deliberately chosen to restrict our bibliographical review to this topic. Ferraty and Vieu (2006) provide many references for kernel estimation with both multivariate and functional α-mixing datasets.

We are interested here in nonparametric models (in the sense of Chapter 1 or of Ferraty and Vieu (2006)) in which the distribution of the real variable of interest Y depends on a functional (that is, taking values in a semi-metric space $(\mathcal{E}, d)$) variable $\mathcal{X}$. In this section, we discuss various problems for which functional kernel methods have been used with α-mixing variables $(\mathcal{X}_i, Y_i)$ identically distributed as $(\mathcal{X}, Y)$. We use the same notation as in Chapter 4 and denote kernel weights by $W_{n,\ker}(x, \mathcal{X}) = K\left(\frac{d(\mathcal{X}, x)}{h}\right)$, take the sequence h of smoothing parameters, and the small-ball probabilities $\phi_x(t) = \mathbb{P}(d(\mathcal{X}, x) \leq t)$. Moreover we introduce a sequence g of strictly positive numbers and a primitive H of a standard non-negative symmetric kernel K_0 whose integral over $\mathbb{R}$ equals one:

$$\forall u \in \mathbb{R}, \ H(u) = \int_{-\infty}^{u} K_0(t) \, dt.$$

In order to avoid overlap with the content of Chapter 4, we have deliberately chosen to present a brief overview of nonparametric problems (based on the different characteristics of the conditional distribution of Y given $\mathcal{X}$) and the corresponding functional kernel estimators (which extend classical mutivariate kernel estimators). Interested readers will find a more detailed discussion on the main ideas that led to the consideration of such estimators, and the way the kernel, the semi-metric, and the smoothing parameter may be chosen in practice, in Chapter 4 or in Ferraty and Vieu (2006).

The conditional mean:

$$\forall x \in \mathcal{E}, \ \hat{r}_{\ker}(x) = \begin{cases} \frac{\sum_{i=1}^{n} W_{n,\ker}(x, \mathcal{X}_i) Y_i}{\sum_{i=1}^{n} W_{n,\ker}(x, \mathcal{X}_i)} & \text{if } \sum_{i=1}^{n} W_{n,\ker}(x, \mathcal{X}_i) > 0, \\ 0 & \text{if } \sum_{i=1}^{n} W_{n,\ker}(x, \mathcal{X}_i) = 0. \end{cases} \qquad (5.6)$$

The conditional cumulative distribution function:

$$\hat{F}_Y^{\mathcal{X}}(x, y) = \begin{cases} \frac{\sum_{i=1}^{n} W_{n,\ker}(x, \mathcal{X}_i) H\left(\frac{Y_i - y}{g}\right)}{\sum_{i=1}^{n} W_{n,\ker}(x, \mathcal{X}_i)} & \text{if } \sum_{i=1}^{n} W_{n,\ker}(x, \mathcal{X}_i) > 0, \\ 0 & \text{if } \sum_{i=1}^{n} W_{n,\ker}(x, \mathcal{X}_i) = 0. \end{cases} \qquad (5.7)$$

Conditional quantiles:

$$\hat{t}_\gamma(x) = \inf\left\{y, \hat{F}_Y^{\mathcal{X}}(x, y) \geq \gamma\right\}. \tag{5.8}$$

The conditional density function:

$$\hat{f}_Y^{\mathcal{X}}(x, y) = \begin{cases} \frac{\sum_{i=1}^n W_{n,\ker}(x,\mathcal{X}_i)\frac{1}{g}K_0\left(\frac{Y_i-y}{g}\right)}{\sum_{i=1}^n W_{n,\ker}(x,\mathcal{X}_i)} & \text{if } \sum_{i=1}^n W_{n,\ker}(x, \mathcal{X}_i) > 0, \\ 0 & \text{if } \sum_{i=1}^n W_{n,\ker}(x, \mathcal{X}_i) = 0. \end{cases} \tag{5.9}$$

The conditional mode:

$$\hat{\theta}(x) \in \arg\max_{y \in \Theta} \hat{f}_Y^{\mathcal{X}}(x, y). \tag{5.10}$$

The conditional hazard function:

$$\hat{H}_Y^{\mathcal{X}}(x, y) = \frac{\hat{f}_Y^{\mathcal{X}}(x, y)}{1 - \hat{F}_Y^{\mathcal{X}}(x, y)} \quad \text{defined if } \hat{F}_Y^{\mathcal{X}}(x, y) < 1. \tag{5.11}$$

Conditional M- and Z-estimators:

$\forall x \in \mathcal{E}$, the conditional Z-estimator $\hat{Z}(x)$ and M-estimator $\hat{M}(x)$ fulfill:

$$\hat{C}\left(\hat{Z}(x), x\right) = o_p(1) \text{ and } \hat{C}\left(\hat{M}(x), x\right) = \max_{t \in \mathbb{R}} \hat{C}(t, x) + o_p(1), \tag{5.12}$$

where

$$\hat{C}(t, x) = \begin{cases} \frac{\sum_{i=1}^n W_{n,\ker}(x,\mathcal{X}_i)\Phi(x,Y,t)}{\sum_{i=1}^n W_{n,\ker}(x,\mathcal{X}_i)} & \text{if } \sum_{i=1}^n W_{n,\ker}(x, \mathcal{X}_i) > 0, \\ 0 & \text{if } \sum_{i=1}^n W_{n,\ker}(x, \mathcal{X}_i) = 0. \end{cases}$$

Discrimination into G classes:

$$\forall x \in \mathcal{E}, \hat{C}_x = \arg\max_{g \in \{1,\ldots,G\}} \hat{p}_g(x), \tag{5.13}$$

where

$$\forall x \in \mathcal{E}, \hat{p}_g(x) = \begin{cases} \frac{\sum_{i=1}^n W_{n,\ker}(x,\mathcal{X}_i)\mathbf{1}_{\{C(\mathcal{X}_i)=g\}}}{\sum_{i=1}^n W_{n,\ker}(x,\mathcal{X}_i)} & \text{if } \sum_{i=1}^n W_{n,\ker}(x, \mathcal{X}_i) > 0, \\ 0 & \text{if } \sum_{i=1}^n W_{n,\ker}(x, \mathcal{X}_i) = 0. \end{cases}$$

Before ending this section, it would seem important to make some comments. Firstly, the estimators (5.8), (5.10), (5.12), and (5.13) are not defined on the set

$$A_n^c(x) := \left\{\sum_{i=1}^n W_{n,\ker}(x, \mathcal{X}_i) = 0\right\}$$

whose probability tends to 0. Secondly, one needs to make some assumptions to ensure the uniqueness of the quantity one wants to predict. These assumptions are very similar to those made when $\mathcal{X}$ is a multivariate variable. It is worth noting that the estimators $\hat{p}_g$ are specific cases of conditional mean estimators. Consequently, the results stated here for conditional mean estimators are also directly linked to discrimination issues (see Ferraty and Vieu (2006, Chapter 8 and Section 11.6) for more details and references). The study of the conditional hazard function is closely related to reliability issues. Finally, it is well known that the conditional mean prediction method is not robust in the presence of outliers. The conditional mode and conditional quantiles are robust alternatives. More generally, one may also consider conditional robust M- or Z-estimators obtained for specific values of Φ (for instance, the Huber function). Because of space restrictions we have chosen to pay more attention to the specificities of functional kernel smoothing with α-mixing functional data. Comments on the prediction problems listed above and on the construction of the corresponding functional kernel estimators would not differ from the independent case. Readers are referred to Ferraty and Vieu (2006, Chapter 4) and the references given in Section 5.4.2 for more details on these topics.

The adaptation to the dependent context of the local linear smoothing and kNN-estimation ideas presented in Chapter 4 has not yet been carried out. Moreover, some of the functional kernel estimators considered in the independent case (see Chapter 4 or, more generally, Ferraty and Vieu (2006)) have not been considered in an α-mixing context. Their study would be a relevant topic for the future.

5.4.2 Asymptotic properties

In this section we show how some of the results presented in Chapter 4 may be extended to the α-mixing case. The main problem is that we need to control various additional terms arising from the dependence that exists within the dataset. The tools presented in Section 5.3.3 are useful in this context (for example, to control the covariance terms). As we will see, the proofs differ from those presented in Chapter 4 mainly in the way we break up the sum of the covariance terms and in the nature of the asymptotic tools available. For the sake of simplicity, we have chosen to present in detail only a few results concerning the functional kernel estimator of the regression operator introduced in (5.6). However, we provide many references in which the interested reader can find similar results for the other kernel estimators presented in Section 5.4.1. Although we focus mainly on three asymptotic properties: almost-complete convergence, asymptotic normality, and $\mathbb{L}^p$ convergence, we also provide references in which other asymptotic properties are considered. The nature of the results presented in this section is similar to that of the result presented for the independent case in Chapter 4. However, we need to make additional assumptions here (mainly concerning the α-mixing coefficients) on the dependence that exists within the dataset, in order to control the way it influences the asymptotic results. For the sake of simplicity, we present the details for the case of arithmetic α-mixing coefficients and provide additional references in which our results are extended to more general dependence modelizations. In the remainder of this section, x stands for a fixed element of $\mathcal{E}$.

Almost-sure convergence

The aim of this section is to provide a result on pointwise almost-sure convergence for the kernel estimator $\hat{r}_{\text{ker}}$ (see (5.6)), in the case of an arithmetic α-mixing dataset. Consider the (nonparametric) functional regression model (already introduced and considered in Chapters 1 and 4)

$$Y = r(\mathcal{X}) + \epsilon, \quad \text{with } \mathbb{E}[\epsilon \mid \mathcal{X}] = 0 \text{ a.s.}, \tag{5.14}$$

and write $E[r(\mathcal{X}) - r(x)|d(\mathcal{X}, x) = s]$ as in Chapter 4. To state our result, we propose making the following assumptions on the model (5.14):

$$\exists \beta > 0, \exists C > 0, \exists \delta > 0, \forall 0 \leq s \leq \delta, |\xi(s)| \leq Cs^\beta, \tag{5.15}$$

$$\exists p > 2, t \mapsto \sigma_p(t) := \mathbb{E}\left[|Y|^p \mid \mathcal{X} = t\right] \text{ is continuous at } x \text{ (w.r. to } d), \tag{5.16}$$

$$\forall \epsilon > 0, \phi_x(\epsilon) > 0. \tag{5.17}$$

Assumption (5.15), introduced in Ferraty et al. (2007), is a little more general that the classical assumption that r is a Hölderian operator. Assumptions (5.16) and (5.17) are standard (see, for instance, Ferraty et al. (2002), Masry (2005), or Ferraty and Vieu (2006)). Assumption (5.16) can be replaced by the more general condition that there exists $\eta > 0$ such that

$$\mathbb{E}\left[|Y|^p \mid d(\mathcal{X}, x)\right] 1_{\{d(\mathcal{X}, x) \leq \eta\}}$$

is almost-surely bounded.

We now introduce some assumptions concerning the construction of the kernel estimator:

$$K \text{ is measurable and } \exists 0 < C_1 \leq C_2 < +\infty, C_1 1_{[0;1]} \leq K \leq C_2 1_{[0;1]}, \tag{5.18}$$

$$h \subset \mathbb{R}_*^+, h \overset{n \to +\infty}{\to} 0, \text{ and } \exists \eta_1 > \frac{4}{p}, n^{1-\eta_1} \phi_x(h) \overset{n \to +\infty}{\to} +\infty. \tag{5.19}$$

The assumption made on the nature of the kernel function K is very standard (see, for instance, Masry (2005), Ferraty and Vieu (2006), or Delsol (2008, 2009)). However, an alternative that avoids the condition that K is positive on its support but assumes more regularity is proposed in Ferraty et al. (2002) and Ferraty and Vieu (2006). See also Chapter 4 for a slightly different condition. Note that in the independent case one only needs to assume that $n\phi_x(h) \overset{n \to +\infty}{\to} +\infty$. The dependence within the dataset requires us to use other exponential inequalities, which lead to stronger assumptions on the small-ball probabilities.

Finally, we introduce the notation:
$$\psi_{i,j}(h) := P\left(d(\mathcal{X}_i, x) \leq h, d(\mathcal{X}_j, x) \leq h\right),$$
and we propose some assumptions concerning the nature of the dependence:

$$\exists \gamma > 0, \exists C > 0, \forall i \neq j, |\mathbb{E}\left[Y_i Y_j \,|\, d(\mathcal{X}_i, x), d(\mathcal{X}_j, x)\right]| 1_{B_d^2(x,y)}(\mathcal{X}_i, \mathcal{X}_j) \leq C, \quad (5.20)$$

$$\exists 0 < \eta_2 \leq 1, \; 0 < \psi_x(h) := \max_{i \neq j} \psi_{i,j}(h) = O\left(\phi_x^{1+\eta_2}(h)\right), \quad (5.21)$$

$$\exists a > \max\left(\frac{(1+\eta_2)p - 2}{\eta_2(p-2)}, \frac{p(4-\eta_1)}{\eta_1 p - 4}\right), \; \exists c, \; \forall j \in \mathbb{N}^*, \; \alpha(j) \leq cj^{-a}. \quad (5.22)$$

Assumption (5.22) is directly related to the exponential inequalities stated in Proposition 5.3. Assumptions (5.20)–(5.22) are used along with Proposition (5.2) to make the additional sum of covariance terms negligible with respect to the sum of variance terms. Obviously, these assumptions vanish if the variables are independent.

Theorem 5.1 *Under assumptions (5.15)–(5.22), one obtains:*

$$\hat{r}_{\text{ker}}(x) - r(x) = O\left(h^\beta\right) + O_{a.s.}\left(\sqrt{\frac{\log n}{n\phi_x(h)}}\right).$$

Sketch of the proof We use arguments very similar to those given in Ferraty et al. (2002) and Ferraty and Vieu (2006). See also Delsol (2009) for the sharper bound of the sum of covariances we use here.

We introduce the notation:

$$\hat{r}_{\text{ker},1}(x) = \frac{\sum_{i=1}^n W_{n,\text{ker}}(x, \mathcal{X}_i)}{n\mathbb{E}\left[W_{n,\text{ker}}(x, \mathcal{X}_i)\right]} \text{ and } \hat{r}_{\text{ker},0}(x) = \frac{\sum_{i=1}^n Y_i W_{n,\text{ker}}(x, \mathcal{X}_i)}{n\mathbb{E}\left[W_{n,\text{ker}}(x, \mathcal{X}_i)\right]}. \quad (5.23)$$

Then we propose using the decomposition:

$$\hat{r}_{\text{ker}}(x) - r(x)$$
$$= \left[\frac{1}{\hat{r}_{\text{ker},1}(x)}\left(\left(\hat{r}_{\text{ker},0}(x) - \mathbb{E}\left[\hat{r}_{\text{ker},0}(x)\right]\right) - \left(r(x) - \mathbb{E}\left[\hat{r}_{\text{ker},0}(x)\right]\right)\right)\right.$$
$$\left. - \frac{r(x)}{\hat{r}_{\text{ker},1}(x)}\left(\hat{r}_{\text{ker},1}(x) - 1\right)\right] 1_{A_n(x)} - r(x) 1_{A_n^c(x)}, \quad (5.24)$$

where $A_n(x) = \left\{\sum_{i=1}^n W_{n,\text{ker}}(x, \mathcal{X}_i) > 0\right\}$.

It is important to note that because K is measurable and the pairs $(\mathcal{X}_i, Y_i)_{1 \leq i \leq n}$ are arithmetic α-mixing variables of order a, the sequences of real variables $\Gamma_{i,n}^1 := W_{n,\text{ker}}(\mathcal{X}_i, x)$ and $\Gamma_{i,n}^2 := Y_i W_{n,\text{ker}}(\mathcal{X}_i, x)$ have arithmetic α-mixing coefficients of the same order (see Ferraty and Vieu (2006, Chapter 10)). The main step in this proof is to use the exponential inequality given in Proposition 5.3 to show that:

Lemma 5.1 Under assumptions (5.15)–(5.22), $\hat{r}_{\text{ker},1}(x) - 1 = O_{a.s.}\left(\sqrt{\frac{\log n}{n\phi_x(h)}}\right)$, and
$\hat{r}_{\text{ker},0}(x) - \mathbb{E}\left[\hat{r}_{\text{ker},0}(x)\right] = O_{a.s.}\left(\sqrt{\frac{\log n}{n\phi_x(h)}}\right)$.

The main difficulty is to show that

$$s_{\ell,n}^2 := \sum_{1 \leq i,\, j \leq n} \left|\text{Cov}\left(\Gamma_{i,n}^\ell, \Gamma_{j,n}^\ell\right)\right| = O\left(n\phi_x(h)\right), \quad \ell = 1, 2. \tag{5.25}$$

To do this, we use the following decomposition:

$$s_{\ell,n}^2 = \sum_{1 \leq |i-j| \leq u_n} \left|\text{Cov}\left(\Gamma_{i,n}^\ell, \Gamma_{j,n}^\ell\right)\right| + \sum_{u_n < |i-j| \leq n} \left|\text{Cov}\left(\Gamma_{i,n}^\ell, \Gamma_{j,n}^\ell\right)\right|$$
$$+ n\text{Var}\left(\Gamma_{1,n}^\ell\right), \tag{5.26}$$

where u_n is a sequence of positive numbers. Then, from assumptions (5.15)–(5.22) and Proposition 5.2 one obtains the following bounds:

$$\text{Var}\left(\Gamma_{1,n}^\ell\right) = O\left(\phi_x(h)\right), \tag{5.27}$$

$$\sum_{1 \leq |i-j| \leq u_n} \left|\text{Cov}\left(\Gamma_{i,n}^\ell, \Gamma_{j,n}^\ell\right)\right| = O\left(nu_n \phi_x^{1+\eta_2}(h)\right), \tag{5.28}$$

$$\forall \epsilon > 0, \sum_{u_n < |i-j| \leq n} \left|\text{Cov}\left(\Gamma_{i,n}^\ell, \Gamma_{j,n}^\ell\right)\right| = O\left(nF_x^{\frac{2}{p}}(h) u_n^{-a\frac{p-2}{p}+1+\epsilon}\right). \tag{5.29}$$

Then one takes

$$u_n = (\phi_x(h))^{\frac{2-p(1+\eta_2)}{a(p-2)-\epsilon p}},$$

and (5.25) comes directly from assumption (5.19) and statements (5.26)–(5.29) for small enough ϵ. Finally, we use Proposition 5.3 for (large enough) fixed n with $r = \log^2(n)$ and $\lambda = C_0\sqrt{n\phi_x(h)\log n}$ and find that for large enough C_0 (not depending on n), one gets:

$$P\left(\left|\sum_{i=1}^n \Gamma_{i,n}^\ell - \mathbb{E}\left[\Gamma_{i,n}^\ell\right]\right| > 4C_0\sqrt{n\phi_x(h)\log n}\right) = O\left(n^{-1-\epsilon_0}\right), \tag{5.30}$$

for some $\epsilon_0 > 0$. This ends the proof of Lemma 5.1.

The bound (5.30) can also be used to show that for any $C > 0$, for large enough n there exists $\epsilon > 0$ such that:

$$P\left(|r(x)|1_{A_n^c(x)} > C\sqrt{\frac{\log n}{n\phi_x(h)}}\right)$$

$$\leq P\left(\left|\sum_{i=1}^n (\Gamma_{i,n}^1 - \mathbb{E}\left[\Gamma_{i,n}^1\right])\right| \geq C_1\sqrt{n\phi_x(h)\log n}\sqrt{\frac{n\phi_x(h)}{\log n}}\right)$$

$$= O\left(n^{-1-\epsilon}\right).$$

The following lemma is enough to complete the proof.

Lemma 5.2 *Under assumptions (5.15)–(5.22), one gets:*

$$r(x) - \mathbb{E}\left[\hat{r}_{\text{ker},0}(x)\right] = O\left(h^\beta\right).$$

This is a direct consequence of (5.15) and (5.18) (see Ferraty and Vieu (2006)).

Additional comments
The pointwise almost-sure convergence result presented here can be used as a preliminary step to obtaining uniform almost-sure convergence over a compact set S. The arguments used to obtain such an extension are similar to those discussed in Chapter 4 (see also Ferraty *et al.* (2009)) for the independent case. Any interested reader may also find additional results and comments for the α-mixing case in Ferraty *et al.* (2002), Ferraty and Vieu (2004), or Ferraty and Vieu (2006). The results stated in Chapter 14 would seem to provide the interesting possibility of extending those presented here to the case of a functional response variable.

Almost-sure convergence results have also been given for other functional kernel estimators. Almost-sure convergence properties of functional kernel estimators of the conditional cumulative distribution function and the conditional quantiles have been considered by Ezzahrioui and Ould Saïd (2008). Other papers focus on almost-sure convergence of functional kernel estimators of the conditional density and the conditional mode (see, for instance, Ferraty *et al.* (2005b) and Ferraty *et al.* (2005a)). Similar results have also been obtained for the functional kernel estimator of the hazard function by Ferraty *et al.* (2008). Finally, the almost-sure convergence of general functional kernel conditional Z-estimators has been considered by Attouch *et al.* (2007) and Attouch (2009).

Asymptotic normality
In this section we present a result dealing with the asymptotic normality of the kernel estimator $\hat{r}_{\text{ker}}(x)$ (see (5.6)). This asymptotic property was first considered by Masry (2005) in the arithmetic α-mixing case. However, the expression of asymptotic bias and variance terms is not explicitly given. Recently, Ferraty *et al.* (2007) give the explicit expression of these terms in the case of an independent dataset. Their result is extended to the case of general α-mixing datasets in Delsol (2009); the construction of pointwise asymptotic confidence bands for r is also discussed here.

To prove the asymptotic normality of the kernel estimator, we introduce the notation:

$$\forall u > 0, \forall s \in [0;1], \tau_u(s) = \frac{\phi_x(us)}{\phi_x(u)},$$

$$\mathbb{E}\left[\epsilon^2 | \mathcal{X}\right] = \sigma_\epsilon^2(\mathcal{X}),$$

and propose the following assumptions on model (5.14):

$$\forall \eta > 0, \; \phi_x(\eta) > 0 \text{ and } \phi_x(0) = 0, \tag{5.31}$$

$$\xi(0) = 0 \text{ and } \xi'(0) \text{ exists}, \tag{5.32}$$

$$\sigma_\epsilon^2 \text{ is continuous at } x \text{ and } \sigma_\epsilon^2 := \sigma_\epsilon^2(x) > 0, \tag{5.33}$$

$$\forall s \in [0;1], \; \lim_{n \to +\infty} \tau_h(s) = \tau_0(s) \text{ with } \tau_0(s) \neq 1_{[0;1]}(s). \tag{5.34}$$

Let us make some comments on these assumptions. Assumptions regarding the law of the functional variable $\mathcal{X}$ principally concern small-ball probabilities $\phi_x(\cdot)$ and do not require the existence of a density with respect to a given measure. Assumption (5.34) holds for a large selection of functional variables, for instance for Gaussian processes (see the discussion in Ferraty et al. (2007)). If assumption (5.33) is very standard, assumption (5.16) is an alternative to classical Hölderian conditions. This condition was introduced by Ferraty et al. (2007) and allows us to obtain, instead of convergence rates, the explicit expression of the dominant term for bias.

We now make some additional assumptions on the nature of the kernel function:

$$K \text{ has for support } [0;1], \text{ is } C^1 \text{ and nonincreasing on }]0;1[, K(1) > 0. \tag{5.35}$$

Assumption (5.35) is quite common (see, for instance, Masry (2005), Ferraty et al. (2007), or Delsol (2009)) and is satisfied by many standard kernel functions restricted to the segment $[0;1]$.

Of course we also need some assumptions to deal with conditional distributions and dependence:

$$\exists p > 2, \; \exists \gamma_1 > 0, \; \exists M > 0, \; \mathbb{E}\left[|Y|^p | d(\mathcal{X}, x)\right] 1_{B_d(x,\gamma)}(\mathcal{X}) \leq M, \tag{5.36}$$

$$\exists \gamma_2 > 0, \; \exists C > 0, \; \forall i, j \in \mathbb{Z}, \; \forall \ell_1, \ell_2 \in \{0,1\},$$

$$\mathbb{E}\left[|Y_i|^{\ell_1} |Y_j|^{\ell_2} | d(\mathcal{X}_i, x), d(\mathcal{X}_j, x)\right] 1_{B_d(x,\gamma)}(\mathcal{X}_i, \mathcal{X}_j) \leq C, \tag{5.37}$$

$$\exists a > \max\left(\frac{2}{\eta_1} + 1, \frac{(1+\eta_2)p - 2}{\eta_2(p-2)}\right), \; \exists c > 0, \; \forall j \in \mathbb{N}, \; \alpha(j) \leq cj^{-a}, \tag{5.38}$$

where η_1 and η_2 come from conditions (5.19) and (5.21). If r is bounded on a neighbourhood of x, assumptions (5.37) and (5.38) hold whenever the same properties hold for the residuals, which may result from the fact that the residuals are either bounded or independent of the $\mathcal{X}_i$'s and in $\mathbb{L}^p$. As in Section 5.1, assumptions (5.36)–(5.38) are useful for controlling the additional sum of covariances and for making it negligible with respect to the sum of variances. The condition $a > 2/\eta_1 + 1$ is used to get the asymptotic normality of the kernel estimator, and can be replaced by the condition:

$$\exists (v_n)_n \in \mathbb{N}^{\mathbb{N}}, \ v_n = o\left(\sqrt{n\phi_x(h)}\right) \text{ and } (n/\phi_x(h))^{\frac{1}{2}} v_n^{-a} \to 0$$

(see Masry (2005) and Delsol (2009) for more details). Finally, the result presented below remains true under more general α-mixing conditions as those discussed in Delsol (2009).

Now, in order to give explicit expressions for the asymptotic dominant bias and variance terms, we introduce, as in Ferraty et al. (2007), the following constants:

$$M_0 = K(1) - \int_0^1 (sK(s))' \tau_0(s) \, ds,$$

$$M_1 = K(1) - \int_0^1 K'(s) \tau_0(s) \, ds,$$

$$M_2 = K^2(1) - \int_0^1 \left(K^2\right)'(s) \tau_0(s) \, ds.$$

Their expression depends on the nature of the kernel function and the limit function τ_0. Moreover, because of the assumptions made on K and τ_0, the constants M_1 and M_2 are positive. It also seems important to note that they have more explicit expressions when the small balls are fractal or when the kernel is the indicator function (see Delsol (2008) for more details).

Theorem 5.2 *Let* $\hat{F}(t) = \frac{1}{n}\sum_{i=1}^n 1_{[d(X_i,x),+\infty[}(t)$. *Under assumptions* (5.19), (5.21), *and* (5.31)–(5.38), *we have*

$$\frac{M_1}{\sqrt{M_2 \sigma_\epsilon^2}} \sqrt{n\hat{F}(h)} \left(\hat{r}_{\ker}(x) - \mathbb{E}\left[\hat{r}_{\ker,0}(x)\right]\right) \to \mathcal{N}(0,1),$$

where $\hat{r}_{\ker,0}$ *is defined in* (5.23).
If, in addition, $h\sqrt{n\phi_x(h)} = O(1)$, *then we have*

$$\frac{M_1}{\sqrt{M_2 \sigma_\epsilon^2}} \sqrt{n\hat{F}(h)} \left(\hat{r}_{\ker}(x) - r(x) - B_n\right) \to \mathcal{N}(0,1),$$

where $B_n = h\xi'(0)\frac{M_0}{M_1}$.

This result opens up interesting prospects for constructing asymptotic pointwise confidence bands for the regression operator r. A procedure based on the estimation of the constants is discussed in Delsol (2009), while Aspirot et al. (2008) propose using bootstrap methods adapted to dependent datasets (see, for example, Politis and Romano (1994)). Furthermore, resampling methods for independent functional data are discussed in Chapter 7. Another potential application of this asymptotic normality result is for providing the expressions of asymptotic dominant terms in $\mathbb{L}^p$ errors (see the following section). A result on consistency under weaker assumptions is also discussed in Masry (2005).

Sketch of the proof The complete proof of Theorem 5.2 is given under slightly different assumptions in Delsol (2009). Some similar arguments are also given in Masry (2005) and Ferraty et al. (2007).

We start from the decomposition:

$$\sqrt{nF(h)}\left(\hat{r}_{\text{ker}}(x) - \mathbb{E}\left[\hat{r}_{\text{ker},0}(x)\right]\right)$$

$$= \sqrt{nF(h)}\left(\frac{\left(\hat{r}_{\text{ker},0}(x) - \mathbb{E}\left[\hat{r}_{\text{ker},0}(x)\right]\right) - \left(\hat{r}_{\text{ker},1}(x) - 1\right)\mathbb{E}\left[\hat{r}_{\text{ker},0}(x)\right]}{\hat{r}_{\text{ker},1}(x)}\right)1_{A_n(x)}$$

$$- \sqrt{nF(h)}\mathbb{E}\left[\hat{r}_{\text{ker},0}(x)\right]1_{A_n^c(x)}, \qquad (5.39)$$

and use the notation $\Gamma_{i,n}^\ell$ introduced in the proof of Theorem 5.1 to define:

$$U_{i,n} := \sqrt{n\phi_x(h)}\frac{\left(\Gamma_{i,n}^2 - \mathbb{E}[\Gamma_{i,n}^2]\right) - \left(\Gamma_{i,n}^1 - \mathbb{E}[\Gamma_{i,n}^1]\right)\mathbb{E}[\Gamma_{i,n}^2]}{n\mathbb{E}[\Gamma_{i,n}^1]}.$$

Note that, for a fixed value of n, the variables $U_{i,n}$ are identically distributed and that $\{U_{i,n}, 1 \leq i \leq n, n \in \mathbb{N}\}$ is an α-mixing array because the sequence $(X_i, Y_i)_{i \in \mathbb{N}}$ is α-mixing. The proof is based on the following facts, whose proof is given in Delsol (2009) and Ferraty et al. (2007):

$$\mathbb{E}\left[\hat{r}_{\text{ker},0}(x)\right] - r(x) = h\phi'(0)\frac{M_0}{M_1} + o(h), \qquad (5.40)$$

$$\mathbb{E}\left[U_{1,n}^2\right] = \frac{M_2\sigma_\epsilon^2}{nM_1^2} + o\left(\frac{1}{n}\right), \qquad (5.41)$$

$$\hat{r}_{\text{ker},1}(x) \rightsquigarrow 1, \quad \frac{\hat{F}(h)}{F(h)} \rightsquigarrow 1, \quad \text{where } \rightsquigarrow \text{ stands for convergence in law.} \qquad (5.42)$$

From assumption (5.38), there exists $\eta_0 > 0$ such that

$$\eta_2\left(a - \eta_0 - \frac{p}{p-2}\right) - 1 > 0.$$

First use (5.41) and Proposition 5.4 with

$$m_n = \left[(F(h))^{-\frac{\nu}{2} - \frac{p-2}{2(a(p-2)-\eta_0(p-2)-p)}}\right] \text{ and } \tau_n = \sqrt{n\phi_x(h)}(\log(n))^{-\frac{a}{a+1}}$$

to show (see Delsol (2009) for technical details) that

$$\sqrt{n\phi_x(h)}\left(\hat{r}_{\text{ker},0}(x) - \mathbb{E}\left[\hat{r}_{\text{ker},0}(x)\right]\right)$$

$$- \left(\hat{r}_{\text{ker},1}(x) - 1\right)\mathbb{E}\left[\hat{r}_{\text{ker},0}(x)\right] \rightsquigarrow \mathcal{N}\left(0, \frac{M_2\sigma_\epsilon^2}{M_1^2}\right).$$

Then from (5.42) we obtain that

$$P(A_n(x)) \leq P\left(|\hat{r}_{\text{ker},1}(x) - 1| = 1\right) \to 0,$$

which implies in particular that

$$1_{A_n^c(x)} \rightharpoonup 1 \text{ and } \sqrt{nF(h)} \mathbb{E}\left[\hat{r}_{\text{ker},0}(x)\right] 1_{A_n(x)} \rightharpoonup 0.$$

Finally, the proof is concluded using statements (5.40) and (5.41) together with Slutsky's Theorem.

Additional comments
Similar results have been obtained for the same kernel estimator under other dependence assumptions. The case of non-stationary sequences of functional variables is considered in Aspirot et al. (2008). In the case of long memory functional processes, some results concerning consistency and asymptotic normality have been obtained by Estevez and Vieu (2003), Benhenni et al. (2008), and Hedli-Griche (2008).

Asymptotic normality has also been proved for other functional kernel estimators. The asymptotic normality of the functional kernel estimators of the conditional cumulative distribution function and conditional quantiles is discussed in Ezzahrioui (2007) and Ezzahrioui and Ould Saïd (2008).

The asymptotic normality of the functional kernel estimator of the conditional hazard function is given in Ezzahrioui (2007). Finally, Attouch et al. (2008) and Attouch (2009) give a result on the asymptotic normality of conditional functional kernel Z-estimators.

$\mathbb{L}^q$ convergence

In this section, we focus on results dealing with the $\mathbb{L}^q$ convergence of functional kernel estimators. As in the two previous sections, we focus on a result concerning the functional kernel estimator $\hat{r}_{\text{ker}}$ of the regression operator, and propose some additional references in which similar results are discussed for other functional kernel estimators. First results dealing with the $\mathbb{L}^q$ convergence rate of $\hat{r}_{\text{ker}}$ were given in the independent case by Dabo-Niang and Rhomari (2003). More recently, Ferraty et al. (2007) give, as a by-product of the study of asymptotic normality, the explicit expression of the dominant bias and variance terms, which leads them to make explicit the expression of the dominant $\mathbb{L}^2$ error terms. In the α-mixing case, it seems that the first result was provided by Delsol (2007), who gave the expression for the dominant terms of $\mathbb{L}^q$ errors for various values of q. The result presented below is one of those set out in Delsol (2007), where the specific case of independent variables is also discussed. The main idea consists in showing that, for some positive value q, the sequence of variables

$$Z_{n,q} := \left| \frac{M_1}{\sqrt{M_2 \sigma_\epsilon^2}} \sqrt{nF(h)} \left(\hat{r}_{\text{ker}}(x) - \mathbb{E}\left[\hat{r}_{\text{ker},0}(x)\right] \right) \right|^q$$

is uniformly integrable, and then using the asymptotic normality given by Theorem 5.2 to get the convergence of the moment of order q.

To prove the uniform integrability result, we introduce the following notation:

$\forall k \geq 2$,

$$\Theta_k(s) := \max\left(\max_{1 \leq i_1 < \cdots < i_k \leq n} P\left(d\left(X_{i_j}, x\right) \leq s, 1 \leq j \leq k\right), F^k(s)\right)$$

and propose the following assumptions:

$$\exists t, \ 2 \leq t < p, \ \forall k \leq t, \ \exists v_k > 0, \ \Theta_k(s) = O\left(F(s)^{1+v_k}\right), \tag{5.43}$$

$$\exists u, \ 2 \leq u \leq p, \ \exists \eta > 0, \ \exists C > 0,$$

$$\max_{i,n} \mathbb{E}\left[|Y_i|^u \mid d\left(\mathcal{X}_j, x\right), 1 \leq j \leq n\right] 1_{B_d(x,\eta)}(\mathcal{X}_i) \leq C, \tag{5.44}$$

$$a > \max_{2 \leq k \leq t} (k-1) \frac{(1+\mu_k) p - t}{\mu_k (p-t)}, \tag{5.45}$$

where $\mu_1 = 1$ and $\forall k > 1$, $\mu_k = \min(v_k, \mu_{k-1})$.

Assumptions (5.43) and (5.45) are extensions of assumptions (5.21) and (5.22). They are used to control the additional terms that arise from the dependence of the variables that appear in the mean of $W_{n,q}$. Moreover, when one assumes that r is bounded on a neighbourhood of x, assumption (5.44) is satisfied whenever the same assumption holds for the ϵ's instead of the Y's. In this case, assumption (5.44) is trivially fulfilled when the residuals $(\epsilon_i)_{i\in\mathbb{Z}}$ are bounded or when they are independent of the curves $(\mathcal{X}_i)_{i\in\mathbb{Z}}$ and have a finite moment of order u.

We are now in a position to give the expressions for the dominant terms in centered moments for $\mathbb{L}^q$ errors.

Theorem 5.3 *Let W be a standard Gaussian random variable and write $\ell = 2\left[\frac{\min(t,u)}{2}\right]$, $B = \phi'(0) \frac{M_0}{M_1}$, and $V = \sqrt{M_2 \sigma_\epsilon^2 / M_1^2}$. Under assumptions (5.19), (5.21), (5.31)–(5.38), and (5.43)–(5.45), one gets:*

$$\forall 0 \leq q < \ell, \ \mathbb{E}\left[|\hat{r}_{\ker}(x) - \mathbb{E}[\hat{r}_{\ker}(x)]|^q\right] = \frac{V}{(nF(h))^{\frac{q}{2}}} \left(\mathbb{E}[|W|^q] + o(1)\right).$$

If, in addition, $\sqrt{n\phi_x(h)}h = O(1)$, then one also has that for all $0 \leq q < \ell$:

$$\mathbb{E}\left[|\hat{r}_{\ker}(x) - r(x)|^q\right] = \mathbb{E}\left[\left|hB + W\frac{V}{\sqrt{nF(h)}}\right|^q\right] + o\left(\frac{1}{(nF(h))^{\frac{q}{2}}}\right).$$

The same results also hold without the absolute values when q is an integer (see Delsol (2007)). Moreover, when q is an integer it is possible to give the explicit expressions of $\mathbb{L}^q$ errors. That is the aim of the next corollary. In the remainder of this chapter we denote by G (respectively, g) the cumulative distribution function (respectively, density) of the standard Gaussian law. When q is odd, we need to introduce the following sequences of polynomials:

$$P_{2m+1}(u) = \sum_{l=0}^{m} a_{m,l} u^{2l+1} \text{ and } Q_{2m+1}(u) = \sum_{l=0}^{m} b_{m,l} u^{2l},$$

where

$$a_{m,l} = \frac{(2m+1)!}{(2l+1)! 2^{(m-l)}(m-l)!},$$

$$b_{m,l} = \sum_{j=m-l+1}^{m} \left[C_{2m+1}^{2j+1} \frac{2^j j!}{2^{j+l-m}(j+l-m)!} - C_{2m+1}^{2j} \frac{(2j)! 2^{j+l-m}(j+l-m)!}{2^j j! (2(j+l-m))!} \right]$$

$$+ C_{2m+1}^{2(m-l)+1} 2^{m-l}(m-l)!,$$

and the function $\psi_m(u) = (2G(u) - 1) P_{2m+1}(u) + 2g(u) Q_{2m+1}(u)$ (with the convention that a sum over an empty set of indices equals zero).

Corollary 5.1 *Under the assumptions of Theorem 5.3, if ϕ_x it is possible to show that:*

$\forall m \in \mathbb{N}, \ 0 \leq 2m < \ell,$

$$\mathbb{E}\left[|\hat{r}_{\ker}(x) - r(x)|^{2m}\right] = \sum_{k=0}^{m} \frac{V^{2k} B^{2(m-k)}(2m)!}{(2(m-k))! k! 2^k} \frac{h^{2(m-k)}}{(n\phi_x(h))^k} + o\left(\frac{1}{(n\phi_x(h))^m}\right),$$

$\forall m \in \mathbb{N}, \ 0 \leq 2m+1 < \ell,$

$$\mathbb{E}\left[|\hat{r}_{\ker}(x) - r(x)|^{2m+1}\right] = \frac{1}{(n\phi_x(h))^{m+\frac{1}{2}}} \left(V^{2m+1} \psi_m\left(\frac{Bh\sqrt{n\phi_x(h)}h}{V}\right) = O(1)\right),$$

The previous results open up interesting perspectives concerning the issue of choosing the best bandwidth, in the sense of $\mathbb{L}^p$ error. These results also seem innovative in the multivariate context where, for instance, the study of the explicit expression of the $\mathbb{L}^1$ error seems to be restricted to Wand (1990) for the specific case of a real fixed design. The results generalize previous work of Dabo-Niang and Rhomari (2003) and Ferraty *et al.* (2007) to the case of dependent variables. A generalization of the results given above to the case where Y takes values in a Banach space is stated, for the independent case, in Dabo-Niang and Rhomari (2009). The use of CLTs and covariance inequalities for functional α-mixing variables seems necessary in order to extend this result to the α-mixing case.

Sketch of the proof of Theorem 5.3 A detailed proof is given in Delsol (2007). The main idea here is to start by showing that, for all $0 \leq q < \ell$, the sequence of variables $Z_{n,q}$ is uniformly integrable. This follows from the fact that $\mathbb{E}\left[Z_{n,\ell} 1_{A_n(x)}\right]$ is bounded (see details in Delsol (2007)). Moreover, one also gets, as a by-product of the proof, that

$$\sqrt{nF(h)}^{\ell} \mathbb{E}\left[|\hat{r}_{\text{ker},1}(x) - 1|^{\ell}\right] = O(1).$$

This allows us to deduce the following bound:

$$\left|\frac{M_1}{\sqrt{M_2 \sigma_\epsilon^2}} \sqrt{nF(h)} \mathbb{E}\left[\hat{r}_{\text{ker},0}(x)\right]\right|^q P(A_n(x))$$

$$= O\left(\sqrt{nF(h)}^q P\left(|\hat{r}_{\text{ker},1}(x) - 1| \geq 1\right)\right)$$

$$= O\left(\sqrt{nF(h)}^q \mathbb{E}\left[|\hat{r}_{\text{ker},1}(x) - 1|^q\right]\right)$$

$$= O(1).$$

Then we perform the following decomposition:

$$\mathbb{E}\left[\hat{r}_{\text{ker}}(x)\right] = \mathbb{E}\left[\hat{r}_{\text{ker},0}(x)\right] + \mathbb{E}\left[\hat{r}_{\text{ker},0}(x)\left(1 - \hat{r}_{\text{ker},1}(x)\right)\right]$$

$$+ \mathbb{E}\left[\hat{r}_{\text{ker}}(x)\left(1 - \hat{r}_{\text{ker},1}(x)\right)^2\right]$$

$$= \mathbb{E}\left[\hat{r}_{\text{ker},0}(x)\right] + \text{Cov}\left(\hat{r}_{\text{ker},0}(x), \hat{r}_{\text{ker},1}(x)\right)$$

$$+ \mathbb{E}\left[\mathbb{E}\left[\hat{r}_{\text{ker}}(x) \,|\, d\left(\mathcal{X}_i, x\right), 1 \leq i \leq n\right]\left(1 - \hat{r}_{\text{ker},1}(x)\right)^2\right],$$

and use assumption (5.44) to show that $\mathbb{E}\left[\hat{r}_{\text{ker}}(x) \,|\, d\left(\mathcal{X}_i, x\right), 1 \leq i \leq n\right]$ is almost-surely bounded. Finally, from various by-products of the proof of Theorem 5.1, we obtain

$$\mathbb{E}\left[\hat{r}_{\text{ker}}(x)\right] = \mathbb{E}\left[\hat{r}_{\text{ker},0}(x)\right] + O\left(\frac{1}{n\phi_x(h)}\right). \tag{5.46}$$

It follows directly from (5.46), together with the uniform integrability of $Z_{n,q}$, that the sequence of variables

$$W_{n,q} := \left|V^{-1}\sqrt{nF(h)}\left(\hat{r}_{\text{ker}}(x) - \mathbb{E}\left[\hat{r}_{\text{ker}}(x)\right]\right)\right|^q$$

is uniformly integrable. Moreover, it follows directly from Theorem 5.2 and (5.46) that the variables $W_{n,q}$ converge in distribution to $|W|^q$, where W is a standard Gaussian variable. These properties are enough to insure the convergence of the mean of the variables $W_{n,q}$ to the mean of the variable $|W|^q$, which leads directly to the first part of Corollary 5.1.

The proof of the second part uses a by-product of the proof of Theorem 5.2, statement (5.40), which allows us to prove that the random sequence

$$\left|V^{-1}\sqrt{nF(h)}\left(\hat{r}_{\text{ker}}(x) - r(x)\right)\right|^q$$

is uniformly integrable. This uniform integrability, together with the following lemma, is used to complete the proof.

Lemma 5.3 *Let $f : \mathbb{R} \to \mathbb{R}$ be a continuous function; we introduce the notation*

$$\mathcal{F}_{C,M,f} = \left\{ f_{\zeta,M}(x) = 1_{|x| \leq M} f(x + \zeta), |\zeta| \leq C \right\}.$$

If Z_n converges in law to Z, whose law does not contain atoms, then for any fixed M

$$\sup_{|\zeta| \leq C} \left| \mathbb{E}\left[f_{\zeta,M}(Z_n) \right] - \mathbb{E}\left[f_{\zeta,M}(Z) \right] \right| \to 0.$$

Sketch of the proof of Corollary 5.1 More details are given in Delsol (2007). The proof of the first statement (where q is even) follows directly from Theorem 5.3 and the values of the moments of the standard Gaussian law.

To prove the second statement, we introduce the notation

$$U_n = \sqrt{\frac{M_1^2 n F(h)}{M_2 \sigma_\epsilon^2}} Bh,$$

and we use the following decomposition:

$$\mathbb{E}\left[|U_n + W|^{2m+1} \right] = \sum_{k=0}^{2m+1} C_{2m+1}^k (U_n)^{2m+1-k} I_k,$$

where

$$\forall k \in \mathbb{N}, \; I_k = -\int_{-\infty}^{-U_n} \frac{w^k e^{-\frac{w^2}{2}}}{\sqrt{2\pi}} dw + \int_{-U_n}^{+\infty} \frac{w^k e^{-\frac{w^2}{2}}}{\sqrt{2\pi}} dw.$$

The proof follows directly from the following properties:

$$I_{2k+1} = 2g(U_n) \sum_{j=0}^{k} (U_n)^{2(k-j)} \frac{2^k k!}{2^{k-j} (k-j)!},$$

$$I_{2k} = -2g(U_n) \sum_{j=0}^{k-1} (U_n)^{2(k-j)-1} \frac{(2k)! 2^{k-j} (k-j)!}{2^k k! (2(k-j))!} + [2G(U_n) - 1] \frac{(2k)!}{2^k k!},$$

with the convention that a sum over an empty set of parameters is null (see the detailed proof in Delsol (2007)).

Additional comments

A few results exist concerning the $\mathbb{L}^p$ convergence of other functional kernel estimators. The $\mathbb{L}^p$-convergence rates of the functional kernel estimator of the conditional mode are given in Dabo-Niang and Laksaci (2008). A result of the same type as Theorem 5.3 in the case of conditional functional kernel Z-estimators is stated in Crambes et al. (2008).

5.5 CONCLUSION AND PROSPECTS

For many years, kernel methods have been standard tools for studying nonparametric models involving multivariate random variables. When the variables lie in an infinite-dimensional space, the nature of the underlying topology has a more important effect. In order to take into account this observation, it is worth following the idea of Ferraty and Vieu (2000) and considering the topology as a statistical parameter. From this basic idea, many kernel methods have been adapted to take into account variables with values in a semi-metric space. Moreover, motivated by potential applications to time-series prediction, some results have been proved in the context of α-mixing datasets. In this chapter we have presented some useful tools for studying the asymptotic properties of functional kernel methods with α-mixing variables. In Section 5.4 we have discussed a wide range of nonparametric models in which kernel methods have already been investigated, and we have stated some asymptotic properties of these estimators. In addition to focusing on existing kernel methods for functional α-mixing variables, this chapter has also aimed to bring out some interesting prospects for future research and to provide some tools that may be useful in dealing with them.

Some of the methods that already exist in the independent case have not yet been considered in the α-mixing case. This is, for example, true of some of the estimators discussed in Chapter 4 (the functional kNN estimator, the local linear functional kernel estimator, etc.) and of the kernel density estimator considered by Dabo-Niang (2002, 2004) and Dabo-Niang et al. (2006, 2007). Moreover, the bibliographical references presented in Section 5.4 highlight the fact that a few of the asymptotic properties of some functional kernel estimators still have to be extended to the case of α-mixing functional random variables. The assumptions and proofs discussed in this chapter are starting points for the attainment of this goal. Furthermore, it should be worth studying how our results can be extended to other weak dependence conditions such as those introduced by Doukhan and Louhichi (1999) or Dedecker and Prieur (2005). Some results have recently been obtained for the functional kernel estimator of the conditional mean under non-stationarity (Aspirot et al. 2008). Such dependence conditions should also be considered for more general estimators. Finally, it would be interesting but more challenging to investigate the case of strong dependence and extend the early results of Estevez and Vieu (2003), Benhenni et al. (2008), and Hedli-Griche (2008).

The choice of the semi-metric is a crucial issue in functional kernel methods. The topology under consideration has a direct influence on the regularity assumption made on r and on the convergence rates of the functional kernel estimators. There are no theoretical arguments for deciding if one particular topology is adapted to a given dataset. However, from experimental observations, it seems to be worth considering semi-metrics based on derivatives when the curves being observed are very smooth, and projection semi-metrics when they are rough. The most common way to choose the semi-metric in practice is to propose a family of semi-metrics (from the knowledge we have on the data) and then choose the one within this family that is the most adapted using cross validation (see Ferraty and Vieu (2006)). The use of new semi-metrics, or new procedures for constructing a semi-metric that is more adapted to specific datasets, is a crucial challenge for the future. Such improvements should be useful for increasing the performance of the functional kernel estimators presented above.

The use of kernel methods to study time series of images is actually an attractive prospect. From a theoretical point of view, the methods and results presented in this chapter can be used directly to study α-mixing sets of images. However, the problem of finding an appropriate notion of distance between images is not easy and has not received enough attention up to now. Fourier's decomposition appears to be a standard tool in image analysis (see, for instance, Bigot et al. (2009)). It would be interesting to investigate whether the semi-metrics constructed from a comparison of the first components of these decompositions led to relevant results. Other notions of distance could also be constructed from a comparison of the position of specific points (called labels). In some situations, it may also be interesting to use semi-metrics that are invariant under various transformations (rotation, scaling, translation, ...). There is much work to be done to understand how to construct relevant notions of distance between images and to study the efficiency of functional kernel methods. However, this would lead to very interesting prospects in terms of potential applications.

Many of the results and methods presented in this chapter deal with the estimation of some characteristic of the conditional law of a real variable of interest, given a functional explanatory variable. In many situations, it is more interesting to consider problems in which the variable of interest also belongs to an infinite-dimensional space. For instance, this would be a useful approach for predicting the entire trajectory of a process over a time period, instead of only giving pointwise predictions. Some results have been proved in this direction for $\mathbb{L}^p$ errors and almost-sure convergence in Dabo-Niang and Rhomari (2009), for the independent case. The extension of these results to α-mixing conditions is an interesting prospect that should be investigated in the next few years. Moreover, such results would be useful in considering prediction issues for nonparametric autoregressive functional processes, and would make a link between the methodologies presented in Chapter 3 and the use of functional kernel ideas with dependent variables. To achieve this, we need to use tools adapted to the study of α-mixing sequences of functional random variables, such as those discussed in Chapter 14.

To conclude this chapter, I would like to say a few words on structural testing procedures constructed from functional kernel methods. Functional kernel methods are appropriate for making an explanatory analysis of functional datasets because they do not require structural assumptions. At the same time, various other specific estimators have been proposed in models in which a structural assumption has been made. For instance, in the case of the regression model presented in Chapter 1, some estimation methods have been designed for the case where r is linear, is a single index operator,... (see Chapters 2, 3, and 4). Before trying to estimate the regression operator, it seems relevant to ask if the explanatory variable has an effect on the response variable and if this effect has a specific structure. Extending the approach proposed in Härdle and Mammen (1993), Delsol et al. (2010) have suggested using functional kernel methods to construct general structural testing procedures in regression on functional variables with an independent dataset. The extension of their approach to the dependent case is an interesting prospect for the future. Moreover, it would be worth considering structural tests dealing with other characteristics of the conditional law of Y given $\mathcal{X}$, or considering the case where Y is a functional variable.

Acknowledgements

I would like to thank Frédéric Ferraty and Yves Romain for their kind invitation to write this chapter, and three anonymous referees for their constructive comments and suggestions. More generally, I am grateful to the members of the working group STAPH at Toulouse for many interesting discussions and advice, with special thanks to Christophe Crambes, Frédéric Ferraty, Ali Laksaci, and Philippe Vieu who have contributed directly to the research I have done on α-mixing functional kernel smoothing.

References

ALLAM, A., MOURID, T. (2002). Geometric absolute regularity of Banach space-valued autoregressive processes. *Statist. Probab. Lett.*, **60**(3), 241–52.

ANTONIADIS, A., PAPAROTDITIS, E., SAPATINAS, T. (2006). A functional wavelet-kernel approach for time series prediction. *J. R. Stat. Soc. Ser. B Stat. Methodol.*, **68**(5), 837–57.

ASPIROT, L., BERTIN, K., PERERA, G. (2008). Asymptotic normality of the Nadaraya-Watson estimator for non-stationary functional data and applications to telecommunications. *J. Nonparametric Statistics*, **21**(5), 535–51.

ATHREYA, K.B., PANTULA, S.G. (1986). Mixing properties of Harris chains and autoregressive processes. *J. Appl. Probab.*, **23**(4), 880–92.

ATTOUCH, M. (2009). Estimation robuste de la fonction de regression pour variables fonctionnelles. PhD Thesis, Univ. du Littoral Côte d'Opale.

ATTOUCH, M., LAKSACI, A., OULD-SAÏD, E. (2007). Strong uniform convergence rate of robust estimator of the regression function for functional and dependent processes. Technical Report L.M.P.A.

ATTOUCH, M., LAKSACI, A., OULD-SAÏD, E. (2008). Asymptotic normality of a robust estimator of the regression function for functional time series. Preprint, LMPA No. 378, Janvier 2008. Univ. du Littoral Côte d'Opale. Submitted.

ATTOUCH, M., LAKSACI, A., SAID, E.O. Asymptotic normality of a robust estimator of the regression function for functional time series data. Journal of the Korean Statistical Society (in Press).

BENHENNI, K., HEDLI-GRICHE, S., RACHDI, M., VIEU, P. (2008). Consistency of the regression estimator with functional data under long memory conditions. *Statist. Probab. Lett.*, 78(8), 1043–9.

BERAN, J. (1994). *Statistics for Long-memory Processes*. Monographs on Statistics and Applied Probability, 61. Chapman and Hall, New York.

BESSE, P., CARDOT, H., STEPHENSON, D. (2000). Autoregressive forecasting of some functional climatic variations. *Scandinavian Journal of Statistics*, 27, 673–87.

BIGOT, J., GAMBOA, F., VIMOND, M. (2009). Estimation of translation, rotation and scaling between noisy images using the Fourier Mellin transform. *SIAM Journal on Imaging Sciences*, 2(2), 614–45.

BOSQ, D. (1989). Propriétés des opérateurs de covariance empiriques d'un processus stationnaire hilbertien. [Properties of empirical covariance operators for a Hilbertian stationary process] *C. R. Acad. Sci. Paris Sér. I Math.*, 309(14), 873–5.

BOSQ, D. (1991). Modelization, nonparametric estimation and prediction for continuous time processes. In *Nonparametric Functional Estimation and Related Topics* (G. Roussas, ed.), 509–29. NATO Adv. Sci. Inst. Ser. C Math. Phys. Sci., 335. Kluwer, Dordrecht.

BOSQ, D. (1993). Bernstein-type large deviations inequalities for partial sums of strong mixing processes. *Statistics*, 24(1), 59–70.

BOSQ, D. (1996). *Nonparametric Statistics for Stochastic Processes. Estimation and Prediction*. Lecture Notes in Statistics, 110. Springer, New York.

BOSQ, D. (1998). *Nonparametric Statistics for Stochastic Processes. Estimation and Prediction* (second edition). Lecture Notes in Statistics, 110. Springer, New York.

BOSQ, D. (2000). *Linear Processes in Function Spaces: Theory and Applications*. Lecture Notes in Statistics, 149. Springer, New York.

BOSQ, D. (2007). General linear processes in Hilbert spaces and prediction. *J. Statist. Plann. Inference*, 137(3), 879–94.

BRADLEY, R.C. (2005). Basic properties of strong mixing conditions. A survey and some open questions. Update of, and a supplement to, the 1986 original. *Probab. Surv.*, 2, 107–44 (electronic).

BRADLEY, R.C. (2007). *Introduction to Strong Mixing Conditions*. Vols 1, 2, and 3. Kendrick Press, Heber City, UT.

COLLOMB, G. (1981). Estimation non-paramétrique de la régression: revue bibliographique. (In French with English summary). *Internat. Statist. Rev.*, 49(1), 75–93.

COLLOMB, G. (1984). Propriétés de convergence presque complète du prédicteur à noyau. [Almost complete convergence properties of kernel predictors] *Z. Wahrsch. Verw. Gebiete*, 66(3), 441–60.

COULON-PRIEUR, C., DOUKHAN, P. (2000). A triangular central limit theorem under a new weak dependence condition. *Statist. Probab. Lett.*, 47(1), 61–8.

CRAMBES, C. (2007). Régression fonctionnelle sur composantes principales pour variable explicative bruitée. *Comptes Rendus de l'Académie des Sciences*, **345**(9), 519–22.

CRAMBES, C., DELSOL, L., LAKSACI, A. (2008). Robust nonparametric estimation for functional data. *J. Nonparametr. Stat.*, **20**(7), 573–98.

DABO-NIANG, S., RHOMARI, N. (2009). Kernel regression estimation in a Banach space. *Journal of Statistical Planning and Inference*, **139**, 1421–34.

DABO-NIANG, S. (2002). Sur l'estimation de la densité dans un espace de dimension infinie : application aux diffusions. (in french) [On Density estimation in an infinite-dimensional space : application to diffusion processes] PhD Paris VI.

DABO-NIANG, S. and RHOMARI, N. (2003). Estimation non paramétrique de la régression avec variable explicative dans un espace métrique. (French. English, French summary) [Kernel regression estimation when the regressor takes values in metric space] *C. R. Math. Acad. Sci.* Paris **336**(1), 75–80.

DABO-NIANG, S. (2004). Density estimation by orthogonal series in an infinite dimensional space : application to process diffusion type I. The International Conference on Recent Trends and Directions in *J. Nonparametric Statist.*, **16**, 171–186.

Dabo-Niang, S., Ferraty, F. and Vieu, P. (2006). Mode estimation for functional random variable and its application for curves classication. *Far East J. Theor. Stat.* **18**(1), 93–119.

DABO-NIANG, S., FERRATY, F. and VIEU, P. (2007). On the using of modal curves for radar wave curves classication. *Comp. Statist. & Data Anal.* **51**(10), 4878–4890.

DABO-NIANG, S and LAKSACI, A. (2010). Note on conditional mode estimation for functional dependent data. *Statistica* In press.

DAVYDOV, J.A. (1973). Mixing conditions for Markov chains. (In Russian) *Teor. Verojatnost. i Primenen*, **18**, 321–38.

DEDECKER, J., DOUKHAN, P. (2003). A new covariance inequality and applications. *Stochastic Process. Appl.*, **106**(1), 63–80.

DEDECKER, J., DOUKHAN, P., LANG, G., LEON, R., JOSE R., LOUHICHI, S., PRIEUR, C. (2007). *Weak Dependence: with Examples and Applications*. Lecture Notes in Statistics, **190**. Springer, New York.

DEDECKER, J., PRIEUR, C. (2005). New dependence coefficients. Examples and applications to statistics. *Probab. Theory Related Fields*, **132**(2), 203–36.

DEHLING, H. (1983). Limit theorems for sums of weakly dependent Banach space valued random variables. *Z. Wahrsch. Verw. Gebiete*, **63**(3), 393–432.

DEHLING, H., SHARIPOV, O.S. (2005). Estimation of mean and covariance operator for Banach space valued autoregressive processes with dependent innovations. *Stat. Inference Stoch. Process.*, **8**(2), 137–49.

DELSOL, L. (2007). Régression non-paramétrique fonctionnelle: expressions asymptotiques des moments. *Annales de l'I.S.U.P.*, **51**(3), 43–67.

DELSOL, L. (2008). Régression sur variable fonctionnelle: estimation, tests de structure et applications. PhD Thesis, University of Toulouse.

DELSOL, L. (2009). Advances on asymptotic normality in nonparametric functional time series analysis. *Statistics*, **43**, 13–33.

DELSOL, L., FERRATY, F., VIEU, P. (2010). Structural test in regression on functional variables (submitted).

DOUKHAN, P. (1994). *Mixing. Properties and Examples.* Lecture Notes in Statistics, **85**. Springer, New York.

DOUKHAN, P., LOUHICHI, S. (1999). A new weak dependence condition and applications to moment inequalities. *Stochastic Process. Appl.*, 84(2), 313–42.

DOUKHAN, P., NEUMANN, M.H. (2007). Probability and moment inequalities for sums of weakly dependent random variables, with applications. *Stochastic Process. Appl.*, 117(7), 878–903.

ESTEVEZ, G., VIEU, P. (2003). Nonparametric estimation under long memory dependence. (English summary.) *J. Nonparametr. Stat.*, 15(4)–(5), 535–51.

EZZAHRIOUI, M. (2007). Prévision dans les modèles conditionnels en dimension infinie. PhD Thesis, Univ. du Littoral Côte d'Opale.

EZZAHRIOUI, M., OULD-SAÏD, E. (2008). Asymptotic results of a nonparametric conditional quantile estimator for functional time series. *Comm. Statist. Theory Methods*, 37(16)–(17), 2735–59.

FERRATY F., GOIA A., VIEU P. (2002). Functional nonparametric model for time series: a fractal approach for dimension reduction. *Test*, 11(2), 317–44.

FERRATY, F., LAKSACI, A., VIEU, P. (2005a). Functional time series prediction via conditional mode estimation. *C. R. Math. Acad. Sci. Paris*, 340(5), 389–92.

FERRATY, F., LAKSACI, A., VIEU, P., TADJ, A. (2009). Rate of uniform consistency for nonparametric estimates with functional data. *J. Stat. Plan. Infer.*, 140(2), 335–52.

FERRATY, F., MAS, A., VIEU, P. (2007). Advances on nonparametric regression for functional data. *ANZ Journal of Statistics*, 49, 267–86.

FERRATY, F., RABHI, A., VIEU, P. (2005b). Conditional quantiles for dependent functional data with application to the climatic El Niño phenomenon. *Sankhyā*, 67(2), 378–98.

FERRATY, F., RABHI, A., VIEU, P. (2008). Estimation non-paramétrique de la fonction de hasard avec variable explicative fonctionnelle. [Nonparametric estimation of the hazard function with explanatory functional variable] *Rev. Roumaine Math. Pures Appl.*, 53(1), 1–18.

FERRATY, F., VIEU, P. (2000). Dimension fractale et estimation de la régression dans des espaces vectoriels semi-normés *Compte Rendus de l'Académie des Sciences* Paris, 330, 403–6.

FERRATY, F. et VIEU, P. (2004). Nonparametric models for functional data, with application in regression, time-series prediction and curve discrimination. The International Conference on Recent Trends and Directions in Nonparametric Statistics. *J. Nonparametr. Stat.* 16(1-2), 111–125.

FERRATY, F., VIEU, P. (2006). *Nonparametric Functional Data Analysis.* Springer, New York.

GYÖRFI, L., HÄRDLE, W., SARDA, P., VIEU, P. (1989). *Nonparametric Curve Estimation from Time Series.* Lecture Notes in Statistics, 60. Springer, Berlin.

HÄRDLE, W., MAMMEN, E. (1993). Comparing nonparametric versus parametric regression fits. *Annals of Statistics*, 21(4), 1926–47.

HEDLI-GRICHE, S. (2008). Estimation de l'opérateur de régression pour des données fonctionnelles et des erreurs corrélées. PhD Thesis, UPMF, Grenoble.

IBRAGIMOV, I.A. (1962). Some limit theorems for stationary processes. *Theor. Prob. Appl.*, 7, 349–82.

KOLMOGOROV, A.N., ROZANOV, Y.A. (1960). On strong mixing conditions for stationary Gaussian processes. *Theor. Probab. Appl.*, 5, 204–8.

LEE, S. (1995). A note on the strong mixing property for a random coefficient autoregressive process. *J. Korean Statist. Soc.*, 24(1), 243–8.

LIEBSCHER, E. (2001). Central limit theorems for α-mixing triangular arrays with applications to nonparametric statistics. *Math. Methods Statist.* , 10(2), 194–214.

MANDELBROT, B.B., VAN NESS, J.W. (1968). Fractional Brownian motions, fractional noises and applications. *SIAM Review*, 10(4), 422–37.

MAS, A., PUMO, B. (2007). The ARHD model. *J. Statist. Plann. Inference*, 137(2), 538–53.

MASRY, E. (2005). Nonparametric regression estimation for dependent functional data: asymptotic normality *Stochastic Process. Appl.*, 115(1), 155–77.

MOKKADEM, A. (1990). Propriétés de mélange des processus autorégressifs polynomiaux. [Mixing properties of polynomial autoregressive processes] *Ann. Inst. H. Poincaré Probab. Statist.*, 26(2), 219–60.

PALMA, W. (2007). *Long-memory Time Series. Theory and Methods*. John Wiley & Sons, Hoboken, NJ.

POLITIS, D.N., ROMANO, J.P. (1994). Limit theorems for weakly dependent Hilbert space valued random variables with application to the stationary bootstrap. *Statist. Sinica*, 4(2), 461–76.

RAMSAY, J., SILVERMAN, B. (1997). *Functional Data Analysis*. Springer, New York.

RAMSAY, J., SILVERMAN, B. (2002). *Applied Functional Data Analysis: Methods and Case Studies*. Spinger, New York.

RAMSAY, J., SILVERMAN, B. (2005). *Functional Data Analysis* (second edition). Spinger, New York.

RHOMARI, N. (2002). Approximation et inégalités exponentielles pour les sommes de vecteurs aléatoires dépendants. (French) [Approximation and exponential inequalities for sums of dependent random vectors] *C. R. Math. Acad. Sci.* Paris 334(2), 149–154.

RIO, E. (1993). Covariance inequalities for strongly mixing processes. *Ann. Inst. H. Poincaré Probab. Statist.*, 29(4), 587–97.

RIO, E. (1995). About the Lindeberg method for strongly mixing sequences. *ESAIM Probab. Statist.*, 1, 35–61 (electronic).

RIO, E. (2000). *Théorie Asymptotique des Processus Aléatoires Faiblement Dépendants.* [Asymptotic theory of weakly dependent random processes.] Mathematics & Applications, 31, Springer, Berlin.

ROSENBLATT, M. (1956). A central limit theorem and a strong mixing condition. *Proc. Nat. Acad. Sci. U. S. A.*, 42, 43–7.

ROSENBLATT, M. (1972). Uniform ergodicity and strong mixing. *Z. Wahrscheinlichkeitstheorie und Verw. Gebiete*, 24, 79–84.

STONE, C.J. (1982). Optimal global rates of convergence for nonparametric regression. *Ann. Statist.*, 10(4), 1040–53.

TURBILLON, C. (2007). Estimation et prévision des processus moyenne mobile fonctionnels. PhD Thesis, Univ. Paris 6.

VERETENNIKOV, A.Y. (2006). On lower bounds for mixing coefficients of Markov diffusions. In *From Stochastic Calculus to Mathematical Finance* (Y. Kabarov, R. Liptser, J. Stoyanov, eds), 623–33. Springer, Berlin.

VOLONSKII, V.A., ROZANOV, Y.A. (1959). Some limit theorems for random functions: I. *Theory Probab. Appl.*, 4, 178–97.

WAND, M.P. (1990). On exact $\mathbb{L}^1$ rates in nonparametric kernel regression. *Scand. J. Statist*, 17 N°3, 251–6.

WITHERS, C.S. (1981). Conditions for linear processes to be strong-mixing. *Z. Wahrsch. Verw. Gebiete*, 57(4), 477–80.

YOSHIHARA, K. (1994). *Weakly Dependent Stochastic Sequences and their Applications. Vol. IV. Curve Estimation Based on Weakly Dependent Data.* Sanseido, Chiyoda-ku.

YOSHIHARA, K. (2004). *Weakly Dependent Stochastic Sequences and their Applications. Vol. XIV. Recent Topics on Weak and Strong Limit Theorems.* Sanseido, Chiyoda-ku.

CHAPTER 6

FUNCTIONAL COEFFICIENT MODELS FOR ECONOMIC AND FINANCIAL DATA

ZONGWU CAI

6.1 Introduction

OVER the past two decades, the literature on methodological and theoretical research into functional coefficient models has been rapidly growing. In particular, due to their great flexibility and interpretability, functional coefficient models have been extensively applied in economics and finance in order to capture the dynamic changes and evolution in economic and financial phenomena. Given the limitations of space, it is impossible to survey here all the important recent developments and applications of functional coefficient models. Therefore, I choose to focus on the following areas. In Section 6.2, I review recent developments in the nonparametric estimation and testing of functional coefficient models. Section 6.3 is devoted to two major real-life applications in economics and finance. For more details on the methodology, theory, and applications of functional coefficient models to statistics, economics, finance and other fields, the reader is referred to the review papers by Fan and Zhang (2008), Cai and Hong (2009), and Cai *et al.* (2009*c*).

6.2 FUNCTIONAL COEFFICIENT MODELS

6.2.1 The models

A general nonparametric regression model (see (1.1) in Chapter 1) can be written as

$$Y_t = g(X_t, Z_t) + \varepsilon_t, \quad 1 \le t \le T, \tag{6.1}$$

where Y_t is the dependent variable, both $X_t \in \mathbb{R}^p$ and $Z_t \in \mathbb{R}^q$ are regressors, and the regression function $g(x, z)$ is a $\mathbb{R}^{p+q}$-dimensional surface. A functional (varying) coefficient model (see (1.20) in Chapter 1) has the following particular form:

$$g(X_t, Z_t) = \sum_{j=0}^{p} a_j(Z_t) X_{jt} = \mathbf{a}(Z_t)^\top \mathbf{X}_t, \tag{6.2}$$

which is linear in X_t and nonlinear in Z_t; the nonparametric coefficient functions are in $\mathbb{R}^q$ rather than in $\mathbb{R}^{p+q}$. Here, $X_{0t} = 1$, $\mathbf{a}(Z_t) = (a_0(Z_t), \ldots, a_p(Z_t))^\top$, $\mathbf{X}_t = (1, X_{1t}, \ldots, X_{pt})^\top$, and $\mathbf{A}^\top$ denotes the transpose of a matrix or vector $\mathbf{A}$. As in (1.21) in Chapter 1, one can assume that

$$E[\varepsilon_t | X_t, Z_t] = 0. \tag{6.3}$$

Then both variables X_t and Z_t are exogenous. However, (6.3) might not be true for many applications in economics and finance. In such cases, some of the components of X_t are called endogenous variables; see (6.13) and (6.14) later for details of this situation. We note that the functional coefficient model given in (6.2) and (6.3) was proposed in Cleveland et al. (1991) and studied extensively by Hastie and Tibshirani (1993). For more about the statistical properties of functional coefficient models, the reader is referred to the statistical survey paper by Fan and Zhang (2008).

It is clear from (6.2) and (6.3) that the identification condition can be derived as follows:

$$E\left[\mathbf{X}_t \mathbf{X}_t^\top | Z_t\right] \mathbf{a}(Z_t) = E[Y_t \mathbf{X}_t | Z_t]$$

and

$$\mathbf{a}(z) = \mathbf{\Omega}(z)^{-1} E[Y_t \mathbf{X}_t | Z_t = z],$$

provided that $\mathbf{\Omega}(z) = E\left[\mathbf{X}_t \mathbf{X}_t^\top | Z_t = z\right]$ is nonsingular for all z. Therefore, a sufficient and necessary condition to identify $\mathbf{a}(z)$ is $\mathbf{\Omega}(z) > 0$ for all z, which is an identification condition. When (6.3) does not hold, one needs instrumental variables (IV) to estimate and identify functionals in (6.2); see Cai et al. (2006) for details. In what follows, it is assumed that the model is identified.

As elaborated by Cai et al. (2006), functional coefficient models are appropriate and flexible enough for many applications, in particular when the additive separability of covariates is unsuitable for the problem at hand. For simplicity of notation, we assume here that $p = 1$ and $q = 1$. Indeed, by assuming that $g(x, z)$ has a higher-order partial derivative with respect to x and by applying a Taylor expansion to $g(x, z)$, one obtains

$$g(x, z) = \sum_{j=1}^{\infty} \frac{\partial^j g(0, z)}{\partial x^j} \frac{x^j}{j!} \approx \sum_{j=0}^{d} a_j(z) x_j \qquad (6.4)$$

for some d (large), where $a_j(z) = (j!)^{-1} \partial^j g(0, z)/\partial x^j$ and $x_j = x^j$. Equation (6.4) implies that a functional coefficient model given by (6.2) might be a good approximation to the general nonparametric model defined in (6.1).

More importantly, the functional coefficient model defined by (6.2) and (6.3) has the ability to capture heteroscedasticity. To understand this better, it is easy to see that

$$\mathrm{Var}(Y_t|Z_t) = \mathbf{a}(Z_t)^\top \mathrm{Var}(\mathbf{X}_t|Z_t)\mathbf{a}(Z_t) + \sigma_\varepsilon^2(Z_t),$$

where $\sigma_\varepsilon^2(Z_t) = \mathrm{Var}(\varepsilon_t|Z_t)$. Therefore, the first term in the above expression behaves as an ARCH-type model (Autoregressive Conditional Heteroscedasticity-type). Furthermore, the functional coefficient approach allows appreciable flexibility in the structure of the models being fitted without suffering from the "curse of dimensionality," since the nonparametric estimation is conducted in $\mathbb{R}^q$ instead of $\mathbb{R}^{p+q}$. Finally, functional coefficient models can be used as a tool to study covariate-adjusted regression for situations in which both predictors and response in a regression model are not directly observable, but are contaminated by a multiplicative factor that is determined by the value of an unknown function of an observable covariate (confounding variable); see Şentürk and Müller (2005) and Cai and Xu (2008) for more details.

The model given in (6.2) and (6.3) covers many familiar models popularly used in the literature. For example, if Z_t is time, it becomes the following trending time-varying coefficient model:

$$Y_t = \sum_{j=0}^{p} a_j(t) X_{jt} + \varepsilon_t, \quad 1 \leq t \leq T, \qquad (6.5)$$

which is able to capture dynamic changes and evolution over time. The trending time-varying time-series model given in (6.5) has attracted significant attention over the past two decades due to its many applications in economics and finance; some examples follow.

The market model in finance is an example that relates the return of an individual stock to the return of a market index or another individual stock, and the coefficient is usually called a beta coefficient in the capital asset pricing model (CAPM); see the papers by Cochrane (1996) and Cai (2007) and the book by Tsay (2005) for more details on the theory and for real examples. However, some recent studies show that the beta coefficients might vary over time; see Cai (2007) and the references therein.[1]

The term structure of interest rates is another example in which the evolution over time of the relationship between interest rates with different maturities is revealed; see Tsay (2005) for details. Our last example is the relationship between prototype electricity demand and other variables such as income or production, the real price of electricity, and temperature. Indeed, Chang and Martinez-Chombo (2003) have found that this relationship may change over time; their finding is based on the empirical study of the demand equation using monthly Mexican electricity data for residential, commercial, and industrial sectors. Although the literature is already vast and continues to grow swiftly, as pointed out by Phillips (2001), research in this area is only just beginning.

Also, notice that if X_{jt} in (6.5) is a lagged variable (say, $X_{jt} = Y_{t-j}$) and $\{a_j(t)\}$ satisfy some conditions, model (6.5) becomes the well-known locally-stationary time-series model proposed by Dahlhaus (1997), which has a great ability to capture nonstationarity and nonlinearity; see Dahlhaus (1997) and Dahlhaus and Subba Rao (2006) for details on the theory of locally-stationary time-series models and their applications in finance.

If $\{a_j(t)\}$ are piecewise-constant functions of time that is

$$a_j(t) = \sum_{l=1}^{m} a_{jl} \mathbb{I}(T_{l-1} \leq t < T_l), \tag{6.6}$$

where $1 = T_0 < T_1 < \cdots < T_m < T$, $\mathbb{I}(A)$ is the indicator function of the event A, and $\{T_j\}$ are the structural change points which might be known or unknown, the model defined in (6.6) includes the class of multiple structural-change models that are popular in economics and finance and can be used to characterize parameter instability. Parameter instability in economic and financial models is a common phenomenon. This is particularly true for time-series data covering an extended period, as it is more likely that the underlying data-generating mechanism will be disturbed over a longer time horizon by various factors such as policy shifts, macroeconomic announcements, global or regional financial crises, the announcement of unusually large unemployment figures by a government, a dramatic interest rate cut by the Federal Reserve, and so on. For example, for the empirical problem discussed in the paper by Bai (1997), the finding is that the response pattern of interest rates to changes in discount rates varies over time. The timing of the variation is consistent with the timing of changes to the Fed's operating procedures. It is well

---

[1] There have been many recent developments on time-varying beta coefficient CAPMs; see Bansal et al. (1993), Bansal and Viswanathan (1993), Jagannathan and Wang (1996), Ghysels (1998), Reyes (1999), Cui et al. (2002), Akdeniz et al. (2003), Wang (2002, 2003), and You and Jiang (2007).

known that failure to take into account parameter changes, when they are present, may lead to incorrect policy implications and predictions. On the other hand, a proper treatment of parameter changes can be useful in uncovering the underlying factors that fostered the changes, in identifying misspecification, and in analyzing the effect of a policy change. There is a vast literature on structural change models; see the papers by Bai (1997), Bai and Perron (1998), and Bai and Perron (2003) and the references therein. Finally, if X_{jt} is a lagged variable ($X_{jt} = Y_{t-j}$), the model becomes the piecewise autoregressive model (see Davis et al. 2006 for recent advances), which has the ability to depict nonstationarity and can approximate well the locally-stationary time-series models of Dahlhaus (1997); see Davis et al. (2006) for a more detailed discussion.

Now let the $\{a_j(Z_t)\}$ in (6.2) have the following particular parametric form:

$$a_j(Z_t) = \sum_{k=1}^{m} a_{jk} \mathbb{I}(Z_t \in \Omega_k),$$

where $\{\Omega_k\}$ form a (non-overlapping) partition of the whole domain of Z_t. This model is called a threshold model; it is a special case of a nonlinear model. The theoretical properties and practical implementation of threshold models have been dealt with by Tong (1990). Threshold regression theory has gained a lot of momentum recently in a variety of economic and financial studies; see Hansen (2000), Caner and Hansen (2001), Akdeniz et al. (2003), and Caner and Hansen (2004).

Note that since structural change models and threshold models are parametric models, and my focus is on nonparametric models, I will not review structural change models, locally-stationary time-series models, threshold models, and piecewise stationary processes here. The reader is referred to the above-mentioned literature on these models.

Finally, it is worth pointing out that the functional coefficient model defined in (6.2) and (6.3) can be used to analyze functional data, as in Ramsay and Silverman (1997). For example, Ramsay and Silverman (1997), Wei et al. (2006), Wei and He (2006), and Şentürk and Müller (2008) have extended the model given in (6.2) and (6.3) to the following form:

$$Y_i(t_{ij}) = \mathbf{a}(t_{ij})^\top \mathbf{X}_i(t_{ij}) + \varepsilon_i(t_{ij}), \quad 1 \leq j \leq n_i, \quad 1 \leq i \leq n, \tag{6.7}$$

where $Y_{ij} = Y_i(t_{ij})$, $\mathbf{X}_{ij} = \mathbf{X}_i(t_{ij})$, and $\varepsilon_i(t_{ij})$ is a zero-mean process with covariance function $\delta(t, s) = \text{Cov}(\varepsilon_i(t), \varepsilon_i(s))$, and they use this model for longitudinal growth studies. Model (6.7) might have potential applications in economics and finance; see Cai et al. (2009a) for a study of trending panel data. Cardot and Sarda (2008) consider a generalization of the functional coefficient regression model which takes the form

$$Y = \int \mathbf{a}(Z, t)^\top \mathbf{X}(t) \, dt + \varepsilon, \tag{6.8}$$

where Z and ε are real random variables such that $E(\varepsilon|Z) = 0$, $E(X\varepsilon|Z) = 0$, and $\text{Var}(\varepsilon|Z) = \sigma^2$, and they use this model for ozone pollution forecasting. Indeed, model (6.8) can be modified and generalized to become a functional volatility process, which provides a new tool for modeling volatility trajectories in financial markets; see Müller et al. (2007) for a more detailed discussion.

6.2.2 Nonparametric modeling procedures

There are three major approaches to estimating the $\{a_j(\cdot)\}$ in model (6.2) if they are assumed to be continuous. The first is kernel local polynomial smoothing; see Cai et al. (2000a) and Cai et al. (2000b). The second uses polynomial splines; see Huang and Shen (2004). The last uses smoothing splines; see Hastie and Tibshirani (1993). Local linear fitting has several very useful properties, such as high statistical efficiency in an asymptotic minimax sense, an adaptive design, and automatic edge correction (see Fan and Gijbels (1996)). Therefore, in what follows, I will outline only the kernel local polynomial smoothing method; for the other methods, the reader is referred to the above-mentioned papers.

Nonparametric estimation of functional coefficients

For simplicity, in what follows, I assume that $q = 1$. I estimate the functions $\{a_j(\cdot)\}$ in (6.2) using the local linear regression method from observations $\{(X_t, Y_t, Z_t)\}_{t=1}^T$. It is assumed throughout this chapter that $a_j(\cdot)$ has a continuous second derivative. Notice that one may approximate $a_j(Z_t)$ locally at any grid point $z \in \mathbb{R}$ by a linear function $a_j(Z_t) \approx a_j + b_j(Z_t - z)$. The local linear estimator is defined as $\hat{a}_j(z) = \hat{a}_j$, where $\{(\hat{a}_j, \hat{b}_j)\}$ minimize the sum of locally weighted least squares:

$$\sum_{t=1}^{T} \left[Y_t - \sum_{j=0}^{p} \{a_j + b_j(Z_t - z)\} X_{jt} \right]^2 K_h(Z_t - z), \quad (6.9)$$

where $K_h(x) = h^{-1} K(x/h)$, $K(\cdot)$ is a kernel function on $\mathbb{R}$, and $h > 0$ is a bandwidth which controls the degree of smoothing in the estimation and satisfies $h \to 0$ and $hT \to \infty$ as $T \to \infty$.

Notice that the local linear estimator can be viewed as the least squares estimator of the following working linear (parametric) model:

$$K_h^{1/2}(Z_t - z) Y_t = K_h^{1/2}(Z_t - z) \sum_{j=0}^{p} \{a_j + b_j(Z_t - z)\} X_{jt} + u_t.$$

Therefore the estimator $\hat{a}_j(z)$ is a linear estimator of $a_j(z)$ (a linear combination of $Y_1, \ldots, Y_T$) and computational implementation can be easily carried out using any standard statistical software.

Remark 6.1 The restriction to the locally-weighted least squares method suggests that normality is at least being considered as a baseline. However, when abnormality is clearly present, a local quasi-likelihood approach can be used; see Cai (2003). If there are any outliers, one can use a robust local linear fitting scheme; see Cai and Ould-Said (2003). If some of the X_t are endogenous variables, the various instrumental-variable-type estimates of linear and nonlinear simultaneous equations and transformation models can be easily applied here with some modifications; see Cai et al. (2006) and Das (2005). Although such methods (appropriately modified) can be applied to the current setting, their asymptotic properties are not obvious (see Section 6.3 later).

When Z_t is random, Cai et al. (2000b) show that, under some regularity conditions, $\hat{a}(z)$ is asymptotically normally distributed; that is

$$\sqrt{hT}\left[\hat{a}(z) - a(z) - \frac{h^2}{2}\mu_2 a''(z) + o_p(h^2)\right] \to N(0, \nu_0 \Sigma(z)), \qquad (6.10)$$

where $\mu_2 = \int u^2 K(u)\, du$, $\nu_0 = \int K(u)^2\, du$, and $\Sigma(z) = \Omega(z)^{-1}\Omega_1(z)\Omega(z)^{-1}/f_z(z)$. Here, $f_z(z)$ is the marginal density of Z_t and

$$\Omega_1(z) = E\left[\sigma^2(X_t, Z_t)\, X_t X_t^\top | Z_t = z\right],$$

where $\sigma^2(X_t, Z_t) = \text{Var}(\varepsilon_t | X_t, Z_t)$. When Z_t is time (indeed, Z_t is normalized as $Z_t = s_t = t/T$; see Cai (2007) for details), Cai (2007) has shown that, under some regularity conditions, for any $s \in [0, 1]$,

$$\sqrt{hT}\left[\hat{a}(s) - a(s) - \frac{h^2}{2}\mu_2 a''(s) + o_p(h^2)\right] \to N(0, \nu_0 \Sigma_s), \qquad (6.11)$$

where $\Sigma_s = \Omega^{-1}\Omega_1(s)\Omega^{-1}$, $\Omega = E\left[X_t X_t^\top\right]$, and $\Omega_1(s) = \sum_{k=-\infty}^{\infty} R_k(s)$ with $R_k(t) = \text{Cov}(\sigma(X_{i+k}, t)u_{i+k}X_{i+k}, \sigma(X_i, t)u_i X_i)$, $\varepsilon_t = \sigma(X_t, t)u_t$, and $\{u_t\}$ is stationary and $\text{Var}(u_t) = 1$.

From (6.10) and (6.11), it is easy to derive the asymptotic mean square error (AMSE), which is the asymptotic variance plus the square of the asymptotic bias, and to derive the optimal bandwidth by minimizing the AMSE. Clearly, the optimal bandwidth is $h_{\text{opt}} = c T^{-1/5}$ for some unknown positive constant c which can be estimated in a data-driven fashion, described below.

Bandwidth selection

It is well known that the bandwidth plays an essential role in the trade-off between reducing bias and variance. By following a similar idea to those in Cai and Tiwari (2000), Cai (2002), and Cai (2007), here I adapt a simple and quick method for selecting the bandwidth for the above estimation procedures, as follows. For the given observed values $\{Y_t\}_{t=1}^T$, the fitted values $\{\hat{Y}_t\}_{t=1}^T$ can be expressed as $\hat{Y} = H_h \tilde{Y}$, where $Y = (Y_1, \ldots, Y_T)^\top$ and H_h is called the $T \times T$ smoother (or hat) matrix associated with the smoothing parameter h. Motivated by the ideas in Cai and Tiwari (2000) and Cai (2002), I use the following nonparametric version of AIC (Akaike's Information Criterion) to select the optimal bandwidth h_{opt} by minimizing

$$\text{AIC}(h) = \log(\hat{\sigma}^2) + \frac{2(T_h + 1)}{(T - T_h - 2)},$$

where $\hat{\sigma}^2 = \sum_{t=1}^{T}(\hat{Y}_t - Y_t)^2/T$ and T_h is the trace of the hat matrix H_h. This selection criterion counteracts the over/under-fitting tendency of generalized cross-validation and the classical AIC; see Cai and Tiwari (2000) and Cai (2002) for more details. Alternatively, one might use an existing method from the time-series literature, although it may require more computing; see Fan and Gijbels (1996), Cai et al. (2000b), and Cai (2007). This bandwidth-selection criterion will be used in Section 6.3 for real examples.

6.2.3 Misspecification testing

An important econometric question when fitting model (6.2) or (6.5) is the need to test for the following scenarios

(1) whether all the coefficient functions actually vary (that is, if a linear model would be adequate);
(2) more generally, if a parametric model fits the given data such as testing for structural breaks as in Bai (1997) and Bai and Perron (1998, 2003), or testing a threshold model as in Hansen (2000), Caner and Hansen (2001), Akdeniz et al. (2003), and Caner and Hansen (2004), or a specific parametric form as in Ferson and Harvey (1998, 1999), Cai et al. (2000b), and Cai et al. (2000a);
(3) if there is no $a_0(\cdot)$ at all;
(4) whether there are some economic variables that are not statistically significant.

This amounts to testing whether some or all of the coefficient functions are constant or zero or take a certain parametric form. This testing problem can be formulated as

$$H_0 : a_j(z) = a_j^*(z, \gamma), \tag{6.12}$$

where $a_j^*(z, \gamma)$ is a given family of functions indexed by an unknown parameter vector γ. Some tests similar to (6.12) have been considered in the econometrics and finance literature; see, for example, Ghysels (1998) who uses the supreme Lagrange multiplier (LM) test proposed by Andrews (1993) for testing the structural break, and Akdeniz et al. (2003) who apply the heteroskedasticity consistent LM test for a threshold, as in Hansen (1996).

For the purposes of easy implementation, I adapt a misspecification test based on comparing the residual sum of squares (RSS) from both parametric (H_0) and nonparametric (H_a) fittings, described as follows. Let $\hat{\gamma}$ be a consistent estimator of γ (say MLE or LSE). The RSS under the null hypothesis is $\text{RSS}_0 = T^{-1} \sum_{t=1}^{T} e_{t,0}^2$, where $e_{t,0} = Y_t - \sum_{j=0}^{p} a_j^*(Z_t, \hat{\gamma}) X_{tj}$, and the RSS under H_a is $\text{RSS}_1 = T^{-1} \sum_{t=1}^{T} e_{t,1}^2$, where $e_{t,1} = Y_t - \sum_{j=0}^{p} \hat{a}_j(Z_t) X_{tj}$. The test statistic is defined as

$$J_T = \frac{\text{RSS}_0 - \text{RSS}_1}{\text{RSS}_1} = \frac{\text{RSS}_0}{\text{RSS}_1} - 1,$$

which can be regarded as a generalized F-test statistic (see Cai and Tiwari (2000), Cai (2002)) and a generalized likelihood ratio test statistic (see Fan et al. (2001)). The null hypothesis (6.12) is rejected for a large value of J_T. For simplicity, the p-value is computed by using the following nonparametric wild bootstrap approach that can accommodate heteroscedasticity in the model. Notice that this kind of test has been used in the statistics and econometrics literature by several authors; see, for example, Cai et al. (2000b), Cai and Tiwari (2000), and Cai (2007) for various applications in economics and finance and Cai et al. (2000a) and Fan et al. (2001) for applications in other areas.

The steps for the wild bootstrap sampling scheme are described as follows:

1. Generate the residuals $\{e_t^b\}_{t=1}^{T}$ from the centered nonparametric residuals $\{e_t^0\}_{t=1}^{T}$, where $e_t^0 = e_{t,1} - \bar{e}_{t,1}$ with $\bar{e}_{t,1} = T^{-1} \sum_{t=1}^{T} e_{t,1}$.
2. Define the bootstrap sample $Y_t^b = \sum_{j=0}^{p} a_j^*(Z_t, \hat{\gamma}) X_{tj} + e_t^b$. In practice, one can define $e_t^b = e_t^0 \cdot \eta_t$, where $\{\eta_t\}$ is a sequence of i.i.d. random variables with mean zero and unit variance.
3. Calculate the bootstrap test statistic J_T^* based on the bootstrap sampling sample $\{(Y_t^b, X_t, Z_t)\}_{t=1}^{T}$. Notice that, for simplicity, the same bandwidth might be used in calculating both J_T^* and J_T.
4. Compute the p-value of the test based on the relative frequency of the event $\{J_T^* \geq J_T\}$ in the replications of the bootstrap sampling.

Remark 6.2 *In the first step, the reason why one bootstraps the centralized residuals from the nonparametric fit instead of the parametric fit is that the nonparametric estimate of residuals is always consistent, no matter whether the null or the alternative hypothesis is correct. Therefore, the method should provide a consistent estimator of the null distribution even when the null hypothesis does not hold. The consistency issue addressed in Cai et al. (2000b) can be applied to the setting here (see Cai et al. (2000b) for details). This testing procedure will be used in Example 6.2 in Section 6.3 for a real application.*

6.3 APPLICATIONS IN ECONOMICS AND FINANCE

There are many applications of functional coefficient models in economics and finance; I only present two real examples in the next two subsections, due to space limitations. For further empirical examples in economics and finance, the reader is referred to the above-mentioned papers and the additional papers by Li et al. (2002), Hong and Lee (2003), Fan and Zhang (2003), Fan et al. (2003), Cai et al. (2009b), and Cai and Wang (2009) and the references therein.

6.3.1 Functional coefficient instrumental-variable models

Functional coefficient models are appropriate for many applications in economics. For example, we give a labor economics problem. A large body of work has established that while positive, marginal returns on education vary with the level of schooling (see Schultz (1997)), if work experience is also an attribute valued by employers, then the marginal returns on education should vary with experience. In fact, Card (2001) has suggested that if a wage model assumes the additive separability of education and experience, the returns on education will be understated at higher levels of education because the marginal return on education plausibly increases with work experience. This setting is therefore a natural one for a functional coefficient model.

Under a functional coefficient representation, the nonparametric structural model no longer exhibits the ill-posed problem of Newey and Powell (2003). Cai et al. (2006) have shown that, under standard regularity conditions, the model is identified and the estimators are well defined with known asymptotic distribution. They also show that under this representation the estimators obtain faster convergence rates than those of analogous structural models that do not admit a functional coefficient representation.

Das (2005) has considered a nonparametric IV model with discrete endogenous variables,

$$Y_i = g(X_i, Z_{1i}) + \varepsilon_i, \tag{6.13}$$

where X_i is a discrete endogenous variable and Z_{1i} is an exogenous variable. Here $E[\varepsilon_i | X_i, Z_{1i}] \neq 0$. Without loss of generality, I assume that $X_i = 0$ or 1. Then, $g(x, z_1)$ in (6.13) can be rewritten as

$$g(x, z_1) = g(0, z_1)\mathbb{I}(x = 0) + g(1, z_1)\mathbb{I}(x = 1) = a_0(z_1) + a_1(z_1)x,$$

where $a_0(z_1) = g(0, z_1)$ and $a_1(z_1) = g(1, z_1) - g(0, z_1)$. Therefore, $g(x, z_1)$ is linear in the endogenous variables but nonlinear in the exogenous variable Z_1. Clearly, this model belongs to the class of functional coefficient models. Therefore, Cai et al. (2006) have studied the following functional coefficient IV model:

$$Y_i = \sum_{j=1}^{d} a_j(Z_{i1}) X_{ij} + u_i = \mathbf{a}(Z_{i1})^\top \mathbf{X}_i + u_i, \quad E[u_i|Z_i] = 0, \tag{6.14}$$

where Y_i is an observable scalar random variable, $\{a_j(\cdot)\}$ are the unknown structural functions of interest, $X_{i0} \equiv 1$, $\mathbf{X}_i = (X_{i0}, X_{i1}, \ldots, X_{id})^\top$ is a $(d+1)$-dimensional vector consisting of d endogenous regressors, $\mathbf{a}(Z_{i1}) = (a_0(Z_{i1}), \ldots, a_d(Z_{i1}))^\top$, and Z_i is a $(d_1 + d_2)$-dimensional vector consisting of a d_1-dimensional vector Z_{i1} of exogenous variables and a d_2-dimensional vector Z_{i2} of instrumental variables.

Model (6.14) includes the following nonparametric IV model with binary endogenous variable D_i as a special case:

$$Y_i = a_0(Z_{i1}) + a_1(Z_{i1})D_i + \varepsilon_i,$$

which, as noted above, is analyzed in Das (2005). Further, if $a_j(\cdot)$ is a threshold function such as

$$a_j(\mathbf{z}) = a_{j1}\mathbb{I}(\mathbf{z} \leq \mathbf{r}_j) + a_{j2}\mathbb{I}(\mathbf{z} > \mathbf{r}_j)$$

for some $\mathbf{r}_j$, then model (6.14) may describe a threshold IV regression model. Indeed, Caner and Hansen (2004) have considered a threshold model related to this one with endogenous covariates. In this way, the class of models in (6.14) includes some interesting special cases that commonly arise in empirical research in economics.

To estimate $\{a_j(\mathbf{z}_1)\}$ in (6.14) nonparametrically, I propose using a two-stage nonparametric method as in Cai et al. (2006), described as follows. In the first stage I obtain $\hat{\pi}_j(Z_i)$, the fitted value for

$$\pi_j(Z_i) = E[X_{ij}|Z_i](1 \leq j \leq d; 1 \leq i \leq n).$$

To this end, I apply the local linear fitting technique and the jackknife (leave-one-out) idea, as follows. Assuming that $\{\pi_j(\cdot)\}$ has a continuous second derivative, when Z_k falls in a neighborhood of Z_i, a Taylor expansion approximates $\pi_j(Z_k)$ by

$$\pi_j(Z_k) \approx \pi_j(Z_i) + (Z_k - Z_i)^\top \pi'_j(Z_i) = \alpha_{ij} + (Z_k - Z_i)^\top \boldsymbol{\beta}_{ij}.$$

The jackknife idea is to use all the observations except the ith observation in estimating $\pi_j(Z_i)$. Then the least squares estimator with a local weight (i.e. locally-weighted least squares) is given by

$$\sum_{k \neq i}^{n} \{X_{kj} - \alpha_{ij} - (Z_k - Z_i)^\top \boldsymbol{\beta}_{ij}\}^2 K_{h_1}(Z_k - Z_i).$$

Minimizing the above locally-weighted least square with respect to a_{ij} and β_{ij} gives the local linear estimate of $\pi_j(Z_i)$ by $\hat{\pi}_{j,-i}(Z_i) = \hat{a}_{ij}$. Now, I derive the local linear estimator of $\{a_j(\cdot)\}$. The local linear estimators $\hat{b}_j$ and $\hat{c}_j$ are defined as the minimizers of the sum of weighted least squares

$$\sum_{i=1}^{n}\left[Y_i - \sum_{j=0}^{d}\{b_j + (Z_{i1}-z_1)^\top c_j\}\, \hat{\pi}_{j,-i}(Z_i)\right]^2 L_{h_2}(Z_{i1}-z_1),$$

and $\hat{a}_j(z_1) = \hat{b}_j$. Cai et al. (2006) have shown that, under some regularity conditions, $\hat{a}_j(z_1)$ is asymptotically normally distributed. Cai et al. (2006) have also suggested an ad hoc bandwidth selection procedure for selecting two bandwidths in a data-driven fashion; see Cai et al. (2006) for details. In the first step, the bandwidth is chosen to be as small as possible and in the second step, one can use the data-driven method mentioned in Section 6.2.2 to choose the optimal bandwidth. This bandwidth selection criterion will be used in Example 6.1 below.

Example 6.1 I investigate the empirical relation between wages and education, using a random sample of young Australian female workers from the 1985 wave of the Australian Longitudinal Survey. The endogeneity of education in a wage model due to unobservable heterogeneity in schooling choices is well known in the literature; see, for example, the review in Card (2001). I consider the following functional coefficient specification:

$$Y = \delta_0(Z_{12})^\top Z_{11} + g_0(Z_{12}) + g_1(Z_{12})X + \varepsilon$$

and

$$E(X \mid Z_{11}, Z_{12}, Z_2) = \pi(Z_{11}, Z_{12}, Z_2),$$

where Y is the natural logarithm of the hourly wage, Z_{11} includes indicators for marital status, government employment, union status, and whether Australian-born, Z_{12} is a measure of work experience measured in years, X is the measure of (endogenous) education ("Schooling"), Z_2 is an instrumental variable, and $g_0(\cdot)$, $g_1(\cdot)$, and $\pi(\cdot)$ are unknown functions. The object of interest is $g_1(\cdot)$, the functional coefficient of education, which depends on the level of experience.

The main results from the estimation of this model are summarized in Fig. 6.1 which plots the estimators of the functional coefficient $g_1(\cdot)$ correcting for endogeneity (the smooth solid line), and without correcting for endogeneity (the dashed line). First, notice that the profile without correcting for endogeneity is almost constant. In the profile correcting for endogeneity, one finds that the range of $\widehat{g}_1(\cdot)$ is positive and nonlinear for all values of experience in the sample. This implies that, holding experience fixed at any level, the marginal wage returns on schooling (given by the functional coefficient) are strictly positive. In addition, I provide the 95 percent pointwise confidence interval (dotted lines) for the profile, which shows

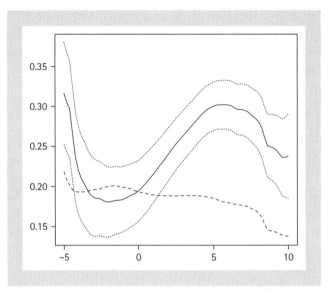

Fig. 6.1 Functional coefficient estimates. The figure corresponds to the functional coefficient $g_1(\cdot)$, plotting the two-stage local linear estimate (solid line) with pointwise 95% confidence intervals (dotted lines), and the ordinary nonparametric estimate (dashed line).

clearly that the pointwise confidence interval does not contain a constant function. This implies that $\hat{g}_1(\cdot)$ is indeed not a constant function. This result shows that the functional coefficient model captures the known nonlinear effect of education on wages, as discussed in Card (2001). However, the confidence intervals indicate that the results correcting for endogeneity are statistically significant only for experience between 0 and 15. Indeed, Cai et al. (2006) have given details on how to construct a pointwise confidence interval for functional coefficient IV models.

Finally, notice also that the derivative of $\hat{g}_1(\cdot)$ changes over its range, being negative at both low and high levels of experience but positive in the middle range of experience. This suggests that while the marginal returns on education are positive, these returns are themselves declining in experience for both low-experienced and high-experienced workers.

6.3.2 Functional coefficient beta models

Although there is a vast amount of empirical evidence on time variation in betas and risk premia, there is no theoretical guidance on how betas and risk premia vary with time, or variables that represent conditioning information. Many recent studies focus on modeling the variation in betas using continuous approximation and the theoretical framework of the conditional CAPM (Capital Asset Pricing Model); see Cochrane (1996), Jaganathan and Wang (1996, 2002), Wang (2002, 2003), and Ang and Liu (2004), and the references therein. Recently, Ghysels (1998) has discussed the problem in detail and stressed the impact of the misspecification of beta risk dynamics on inference and estimation; he argues that betas change over time very slowly and linear factor models like the conditional CAPM may have a tendency to overstate the time variation. Furthermore, he shows that for several well-known time-varying beta models, a serious misspecification produces time variation in betas that is highly volatile and leads to large pricing errors.

To combine the above-mentioned varying-coefficient beta models under a unified framework, I consider a general nonparametric econometric model

$$r_{i,t} = \beta_{i,0}(Z_{i,t}) + \beta_{i,1}(Z_{i,t})^\top r_{m,t} + \varepsilon_{i,t}, \quad 1 \leq i \leq N \text{ and } 1 \leq t \leq T, \quad (6.15)$$

where $r_{i,t}$ is the ith excess return on any asset or portfolio, $\beta_{i,1}(\cdot)$ is a $p \times 1$ vector of the varying-coefficient betas, the prime denotes the transpose of a matrix or vector, and $r_{m,t}$ represents a $p \times 1$ vector of the excess returns on the market portfolios or indices. Here, $Z_{i,t}$ is either a set of instruments or time, both $\beta_{i,0}(\cdot)$ and $\beta_{i,1}(\cdot)$ are unknown functions, and $\varepsilon_{i,t}$ is the error term satisfying $E[\varepsilon_{i,t} | r_{m,t}, Z_{i,t}] = 0$. It is common in the finance literature to assume that the market return $r_{m,t}$ and the state variable $Z_{i,t}$ are uncorrelated with the error term $\varepsilon_{i,t}$; see Akdeniz et al. (2003). Here, I allow for the possibility that the error terms $\{\varepsilon_{i,t}\}$ might be autocorrelated among both i and t and that the conditional variance $\sigma_{i,t}^2 = \text{Var}[\varepsilon_{i,t} | r_{m,t}, Z_{i,t}]$ might not be constant. This time-varying conditional heteroscedasticity can commonly be seen in many financial applications; see Reyes (1999) and Cho and Engle (2000) and the references therein. Furthermore, I assume that the series $\{(u_{i,t}, r_{m,t})\}$ is strictly stationary α-mixing, where $u_{i,t} = \varepsilon_{i,t}/\sigma_{i,t}$. Clearly, model (6.15) includes the above-mentioned models as a special case, and also some other models in the finance literature.

Some specific examples of model (6.15) have been studied by several authors in the literature; see, for example, Ferson and Harvey (1998, 1999) in which the betas are a linear function of instruments (an index model). To measure the risk of an individual stock against the market of US stocks, Cui et al. (2002) have considered the following structural change model:

$$r_t = \beta_0(t, T_0) + \beta_1 r_{m,t} + \varepsilon_t,$$

where r_t is the daily return of Microsoft stock, $r_{m,t}$ is Standard & Poor's 100 Index, as an approximation to this market, and $\beta_0(t, T_0)$ is a structure-change function with unknown change point T_0. You and Jiang (2007) have extended the above model to a semi-parametric setting

$$r_t = \beta_{0,1} \mathbb{I}(t \leq T_0) + \beta_{0,2} \mathbb{I}(t > T_0) + \beta_1(t) X_t + \varepsilon_t$$

with $T_0 = 64$, where $\beta_1(\cdot)$ is an unknown smooth function. Recently, Ferson and Harvey (1998, 1999) and Harvey (1989) have studied various parametric models by assuming the betas to be linear combinations of world market-wide information variables and/or attributes for security, whereas Akdeniz et al. (2003) have investigated the threshold CAPM with economic variable(s)

$$r_t = \beta_0 + \beta_{11}\,\mathbb{I}(\xi_t \leq \lambda) + \beta_{12}\,\mathbb{I}(\xi_t > \lambda)\,r_{m,t} + \varepsilon_t,$$

where λ is an unknown threshold parameter and ξ_t is an economic variable such as the one month real t-bill rate, the dividend yield of the CRSP value weighted NYSE stock index, the de-trended stock price level, a measure of the slope of the term structure, or a quality-related yield spread in the corporate bond market.

If $Z_{i,t}$ is simply time t, $\beta_{i,j}(t)$ depends on time t. To estimate $\beta_{i,j}(t)$ nonparametrically, as argued by Robinson (1989), it is necessary to assume that $\beta_{i,j}(t)$ depends on the sample size T in order to provide the asymptotic justification for any nonparametric smoothing estimators. Following this convention, I assume that $\beta_{i,j}(t) = \beta_{i,j}(Z_t)$, $0 \leq j \leq 1$, where $Z_t = t/T$ and $\beta_{i,j}(\cdot)$ is an unknown function. The intuitive explanation for this "intensity" assumption is that we need an increasingly intense sampling of data points to derive a consistent estimation; see Robinson (1989) and Cai (2007) for further discussion. Similarly, I might assume that $\sigma_{i,t} = \sigma_i(r_{m,t}, Z_t)$ for some unknown function $\sigma_i(\cdot, \cdot)$. To estimate the beta functions $\beta_{i,j}(z_0)$ nonparametrically at any given grid point z_0, one can apply the formulation in (6.9) to this setting with a minor modification (the details are omitted due to their similarity with the above formulation). The bandwidth selection criterion described in Section 6.2.2 can be applied here, in particular in our implementation in Example 6.2 below.

Example 6.2 I will apply the time-varying beta model proposed in (6.15) and its modeling procedures to analyze the common stock price (P_{1t}) of Microsoft stock (MSFT) during the year 2000 using the daily closing prices. In order to measure its risk relative to the market of US blue-chip stocks, I take Standard & Poor's 100 index (P_{2t}) as an approximation to this market. For the first 10-month period with 206 observations, Cui et al. (2002) have modeled the stock price gains Y_t (the price on the tth day divided by the price on day 1) and X_t (the change in the market index from day 1 to the tth day) using the following threshold model:

$$Y_t = \beta_0(t, T_0) + \beta\,X_t + \varepsilon_t,$$

where $\beta_0(t, T_0)$ is a threshold function with unknown change point T_0 with the estimated value $\hat{T}_0 = 64$. Recently, You and Jiang (2007) have extended the above model to a semiparametric setting:

$$Y_t = \beta_{0,1}\,\mathbb{I}(t \leq 64) + \beta_{0,2}\mathbb{I}(t > 64) + \beta_1(t)\,X_t + \varepsilon_t,$$

and they have used a penalized spline method to estimate the unknown slope function $\beta_1(t)$. Following the convention in the finance literature, here I consider the simple daily stock returns $r_t = P_{1t}/P_{1,t-1} - 1$ for the MSFT price and $r_{m,t} = P_{2t}/P_{2,t-1} - 1$ for the S&P 100 Index. It should be noted that the returns of the S&P 100 Index may not be as non-stationary as the Microsoft stock returns.

To establish the empirical relationship between the MSFT returns and the S&P 100 Index, in a similar way to Tsay (2005), who has considered the linear relationship between the 1-year Treasury constant maturity rate and the 3-year Treasury constant maturity rate, I first fit the following simple beta model:

$$r_t = a_0 + a_1 r_{m,t} + \varepsilon_t.$$

Notice that in finance and security analysis, a_1 measures the risk of an individual stock or portfolio as its (standardized) beta coefficient in CAPM against a market index or portfolio. If a_1 is greater than 1, the change in this stock price is expected to be more than that in the market index and thus the stock is regarded as a risky stock. As a result, the least squares estimates of a_0 and a_1 are $-0.0027(0.0018)$ and $1.3612(0.1243)$, respectively, which are plotted (dashed line) in Fig. 6.2. By comparing these results with those from Cui et al. (2002) and You and Jiang (2007), I suspect that the coefficients a_0 and a_1 might change over time. To provide more empirical evidence, I examine the covariance between r_t and $r_{m,t}$, and I find that the covariance does change over time; this is not presented here due to space limitations. Therefore, I have sufficient reasons to fit the following time-varying beta model:

$$r_t = \beta_0(t) + \beta_1(t) r_{m,t} + \varepsilon_t.$$

The local linear estimators $\hat{\beta}_0(\cdot)$ and $\hat{\beta}_1(\cdot)$ are computed, and the estimated curves $\hat{\beta}_0(\cdot)$ (panel (a)) and $\hat{\beta}_1(\cdot)$ (panel (b)) are depicted in Fig. 6.2.

It is evident from Fig. 6.2 that both the trend $\beta_0(\cdot)$ in (a) and the slope $\beta_1(\cdot)$ in (b) do change over time; the slope $\beta_1(\cdot)$ is almost above 1 except during the period of trading days from 141 to 171. (The corresponding calendar days are from 26 July 2000 to 7 September 2000). For the trend function $\beta_0(\cdot)$ that reflects the dynamic change in MSFT itself, although it goes up and down during this period, the overall trend increases slightly for the first three quarters. But the trend decreases dramatically for the last quarter. In contrast, the beta function $\beta_1(\cdot)$ remains constant (around 1.17) during the first 111 trading days of the year (13 June 2000), it decreases afterwards until the 161st trading day (23 August 2000), and finally it increases (to the end of the year). Therefore, we conclude that MFST is a stock that was more volatile than the US blue-chip market as a whole.

Finally, to support the above-mentioned conclusions statistically, I consider the null testing hypothesis $H_0 : \beta_0(\cdot) = a_0$ and $\beta_1(\cdot) = a_1$. The testing procedure described in Section 6.2.3 was used with bootstrap sampling 1000 times. The

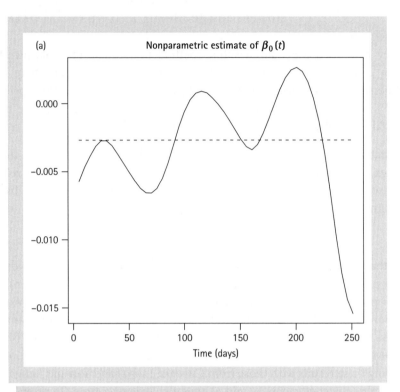
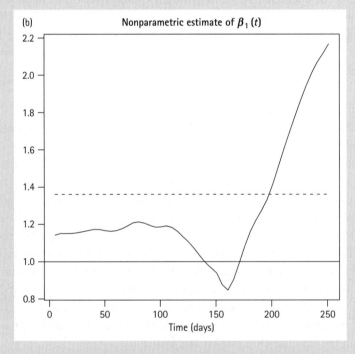

Fig. 6.2 Results for Example 2. Panel (a) The local linear estimator (solid line) of the trend function $\beta_0(\cdot)$ and the least square estimate of α_0 (dashed line). Panel (b) The local linear estimator (solid line) of the beta coefficient $\beta_1(\cdot)$ and the least square estimate of α_1 (dashed line).

resulting p-value was less than 0.001. Therefore, this test result further supports the finding that both the trend function $\beta_0(\cdot)$ and the beta function $\beta_1(\cdot)$ do change over time.

6.4 Concluding remarks

In this chapter, I have presented a selective overview of recent developments in functional coefficient models, with particular application to economics and finance. Indeed, numerous papers have appeared that address various types of functional coefficient models over the past two decades. The references given in this chapter are not exhaustive, due to space limitations. In addition to their applications to economics and finance, functional coefficient models have also been used in other areas of statistics such as time series, longitudinal data analysis, and survival analysis (see Fan and Zhang (2008)). Finally, Cai *et al.* (2009c) have surveyed some of the recent developments in nonparametric econometric models, including some applications of functional coefficient models to economics, while Cai and Hong (2009) have given a review of recent developments in nonparametric estimation and testing of financial econometric models, including functional coefficient diffusion models which are frequently used to describe the dynamics of an underlying process, including stock and bond prices and various interest rates.

Acknowledgements

The author thanks Frédéric Ferraty and two referees for their constructive comments and suggestions. This research was supported, in part, by the National Nature Science Foundation of China grants #70871003 and #70971113, and funds provided by the Cheung Kong Scholarship from the Chinese Ministry of Education, the Minjiang Scholarship from Fujian Province, China, and Xiamen University.

References

Akdeniz, L., Altay-Salih, A., Caner, M. (2003). Time-varying betas help in asset pricing: the threshold CAPM. *Studies in Nonlinear Dynamics and Econometrics*, **6**, 1–16.

Andrews, D.W.K. (1993). Tests for parameter instability and structural change with unknown change point. *Econometrica*, **61**, 821–56.

ANG, A., LIU, J. (2004). How to discount cashflows with time-varying expected return. *The Journal of Finance*, **59**, 2745–83.
BAI, J. (1997). Estimation of a change point in multiple regression models. *Review of Economics and Statistics*, **79**, 551–63.
BAI, J., PERRON, P. (1998). Estimating and testing linear models with multiple structural changes. *Econometrica*, **66**, 47–78.
BAI, J., PERRON, P. (2003). Computation and analysis of multiple structural change models. *Journal of Applied Econometrics*, **18**, 1–22.
BANSAL, R., HSIEH, D.A., VISWANATHAN, S. (1993). A new approach to international arbitrage pricing. *The Journal of Finance*, **48**, 1719–47.
BANSAL, R., VISWANATHAN, S. (1993). No arbitrage and arbitrage pricing: a new approach. *The Journal of Finance*, **47**, 1231–62.
CAI, Z. (2002). A two-stage approach to additive time series models. *Statistica Neerlandica*, **56**, 415–33.
CAI, Z. (2003). Local quasi-likelihood approach to varying-coefficient discrete-valued time series models. *Journal of Nonparametric Statistics*, **15**, 693–711.
CAI, Z. (2007). Trending time varying coefficient time series models with serially correlated errors. *Journal of Econometrics*, **136**, 163–88.
CAI, Z., DAS, M., XIONG, H., WU, X. (2006). Functional coefficient instrumental variables models. *Journal of Econometrics*, **133**, 207–41
CAI, Z., FAN, J., LI, R. (2000*a*). Efficient estimation and inferences for varying-coefficient models. *Journal of American Statistical Association*, **95**, 888–902.
CAI, Z., FAN, J., YAO, Q. (2000*b*). Functional-coefficient regression models for nonlinear time series. *Journal of American Statistical Association*, **95**, 941–56.
CAI, Z., HONG, Y. (2009). Some recent developments in nonparametric finance. *Advances in Econometrics*, **25**, 379–432.
CAI, Z., HSIAO, C., ZHU, Y. (2009*a*). Trending panel models. Working paper, Department of Mathematics and Statistics, University of North Carolina at Charlotte.
CAI, Z., KUAN, C.-M., SUN, L. (2009*b*). Nonparametric pricing kernel models. Working paper, Department of Mathematics and Statistics, University of North Carolina at Charlotte.
CAI, Z., GU, J., LI, Q. (2009*c*). Recent developments in nonparametric econometrics. *Advances in Econometrics*, **25**, 495–549.
CAI, Z., OULD-SAID, E. (2003). Local robust regression estimation for time series. *Statistics and Probability Letters*, **65**, 433–49.
CAI, Z., TIWARI, R.C. (2000). Application of a local linear autoregressive model to BOD time series. *Environmetrics*, **11**, 341–50.
CAI, Z., WANG, Y. (2009). Instability of predictability of asset returns. Working paper, Department of Mathematics and Statistics, University of North Carolina at Charlotte.
CAI, Z., XU, X. (2008). Nonparametric quantile estimations for dynamic smooth coefficient models. *Journal of the American Statistical Association*, **103**, 1596–608.
CANER, M., HANSEN, B.E. (2001). Threshold autoregressions with a near unit root. *Econometrica*, **69**, 1555–97.
CANER, M., HANSEN, B.E. (2004). Instrumental variable estimation of a threshold model. *Econometric Theory*, **20**, 813–43.
CARD, D. (2001). Estimating the return to schooling: progress on some persistent econometric problems. *Econometrica*, **69**, 1127–60.

CARDOT, H., SARDA, P. (2008). Varying-coefficient functional linear regression models. *Communications in Statistics: Theory and Methods*, 37, 3186–203.

CHANG, Y., MARTINEZ-CHOMBO, E. (2003). Electricity demand analysis using cointegration and error-correction models with time varying parameters: the Mexican case. Working paper, Department of Economics, Rice University.

CHO, Y.-H., ENGLE, R.F. (2000). Time-varying betas and asymmetric effects of news: empirical analysis of blue chip stocks. Working paper, Department of Finance, New York University.

CLEVELAND, W.S., GROSSE, E., SHYU, W.M. (1991). Local regression models. In *Statistical Models in S*, (Chambers, J.M. and Hastie, T.J, eds), 309–76. Wadsworth & Brooks, Pacific Grove.

COCHRANE, J.H. (1996). A cross-sectional test of an investment-based asset pricing model. *Journal of Political Economy*, 104, 572–621.

CUI, H., HE, X., ZHU, L. (2002). On regression estimators with de-noised variables. *Statistica Sinica*, 12, 1191–205.

DAHLHAUS, R. (1997). Fitting time series models to nonstationary processes. *The Annals of Statistics*, 25, 1–37.

DAHLHAUS, R., SUBBA RAO, S. (2006). Statistical inference for locally stationary ARCH models. *The Annals of Statistics*, 34, 1075–114.

DAS, M. (2005). Instrumental variables estimators for nonparametric models with discrete endogenous regressors. *Journal of Econometrics*, 124, 335–61.

DAVIS, R.A., LEE, T.C.M., RODRIGUEZ-YAM, G.A. (2006). Structural break estimation for nonstationary time series models. *Journal of American Statistical Association*, 101, 223–38.

FAN, J., GIJBELS, I. (1996). *Local Polynomial Modelling and Its Applications*. Chapman and Hall, London

FAN, J., JIANG, J., ZHANG, C., ZHOU, Z. (2003). Time-dependent diffusion models for term structure dynamics and the stock price volatility. *Statistica Sinica*, 13, 965–92.

FAN, J., ZHANG, C. (2003). A re-examination of diffusion estimators with applications to financial model validation. *Journal of the American Statistical Association*, 98, 118–34.

FAN, J., ZHANG, W. (2008). Statistical methods with varying coefficient models. *Statistics and its Interface*, 1, 179–95.

FAN, J., ZHANG, C., ZHANG, J. (2001). Generalized likelihood ratio statistics and Wilks phenomenon. *The Annals of Statistics*, 29, 153–93.

FERSON, W.E., HARVEY, C.R. (1998). Fundamental determinants of national equity market returns: a perspective on conditional asset pricing. *Journal of Banking and Finance*, 21, 1625–65.

FERSON, W.E., HARVEY, C.R. (1999). Conditional variables and the cross section of stock return. *The Journal of Finance*, 54, 1325–60.

GHYSELS, E. (1998). On stable factor structures in the pricing of risk: do time varying betas help or hurt. *Journal of Finance*, 53, 549–74.

HANSEN, B.E. (1996). Inference when a nuisance parameter is not identified under the null hypothesis. *Econometrica*, 64, 413–30.

HANSEN, B.E. (2000). Sample splitting and threshold estimation. *Econometrica*, 68, 575–605.

HARVEY, C.R. (1989). Time-varying conditional covariances in tests of asset pricing models. *Journal of Financial Economics*, 24, 289–317.

HASTIE, T.J., TIBSHIRANI, R.J. (1993). Varying coefficient models (with discussion). *Journal of the Royal Statistical Society, Series B*, 44, 646–85.

Hong, Y., Lee, T.-H. (2003). Inference and forecast of exchange rates via generalized spectrum and nonlinear time series models. *Review of Economics and Statistics*, 85, 1048–62.

Huang, J.Z., Shen, H. (2004). Functional coefficient regression models for nonlinear time series: a polynomial spline approach. *Scandinavian Journal of Statistics*, 31, 515–34.

Jagannathan, R., Wang, Z. (1996). The conditional CAPM and the cross-section of expected returns. *Journal of Finance*, 51, 3–53.

Jagannathan, R., Wang, Z. (2002). Empirical evaluation of asset pricing models: a comparison of the SDF and beta methods. *The Journal of Finance*, 57, 2337–67.

Li, Q., Huang, C., Li, D., Fu, T. (2002). Semiparametric smooth coefficient models. *Journal of Business and Economic Statistics*, 20, 412–22.

Müller, H.-G., Sen, R., Stadtmüller, U. (2007). Functional data analysis for volatility. Working paper, Department of Statistics, University of California at Davis.

Newey, W.K., Powell, J.L. (2003). Nonparametric instrumental variables estimation. *Econometrica*, 71, 1565–78.

Phillips, P.C.B. (2001). Trending time series and macroeconomic activity: some present and future challenges. *Journal of Econometrics*, 100, 21–7.

Ramsay, J.O., Silverman, B.W. (1997). *Functional Data Analysis*. Springer, New York.

Reyes, M.G. (1999). Size time-varying beta and conditional heteroscedasticity in UK stock return. *Review of Financial Economics*, 8, 1–10.

Robinson, P.M. (1989). Nonparametric estimation of time-varying parameters. In *Statistical Analysis and Forecasting of Economic Structural Change* (Hackl, P. and Westland, A.H., eds), 253–64. Springer, Berlin.

Schultz, T.P. (1997). *Human Capital, Schooling and Health*. IUSSP, XXIII, General Population Conference, Yale University.

Şentürk, D., Müller, H.G. (2005). Covariate adjusted correlation analysis via varying coefficient models. *Scandinavian Journal of Statistics*, 32, 365–83.

Şentürk, D., Müller, H.G. (2008). Generalized varying-coefficient models for longitudinal data. *Biometrika*, 95, 653–66.

Tong, H. (1990). *Nonlinear Time Series: A Dynamical System Approach*. Oxford University Press, Oxford.

Tsay, R.S. (2005). *Analysis of Financial Time Series*. Wiley, New York.

Wang, K.Q. (2002). Nonparametric tests of conditional mean-variance efficiency of a benchmark portfolio. *Journal of Empirical Finance*, 9, 133–69.

Wang, K.Q. (2003). Asset pricing with conditioning information: a new test. *Journal of Finance*, 58, 161–96.

Wei, Y., He, X. (2006). Conditional growth charts (with discussion). *The Annals of Statistics*, 34, 2069–97.

Wei, Y., Pere, A., Koenker, R., He, X. (2006). Quantile regression methods for reference growth charts. *Statistics in Medicine*, 25, 1369–82.

You, J., Jiang, J. (2007). Inferences for varying-coefficient partially linear models with serially correlated errors. In *Advances in Statistical Modeling and Inference: Essays in Honor of Kjell A. Doksum* (Nair, V., ed.), 175–95. Series in Biostatistics, 3. World Scientific, Singapore.

PART II

BENCHMARK METHODS FOR FDA

CHAPTER 7

RESAMPLING METHODS FOR FUNCTIONAL DATA

TIMOTHY MCMURRY

DIMITRIS POLITIS

IN this chapter we discuss the current state of methodological and practical developments for resampling inference techniques in situations where either the data and/or the parameters being estimated take values in a space of functions.

7.1 INTRODUCTION

Denote by $\mathcal{X}_1, \ldots, \mathcal{X}_n$ a set of observations that take values in an abstract space D; for simplicity of presentation, our data will be assumed to be independent and identically distributed (i.i.d.) from probability distribution P. The goal of the analysis is, typically, inference for a parameter $\theta(P)$ that could also be denoted simply as θ when there is no ambiguity as to the underlying measure.

Depending on the application, the parameter θ may be finite dimensional or function valued; our primary focus will be on the latter situation. If θ is finite dimensional, traditional tools for establishing bootstrap consistency as well as central limit theorems often apply. In the situation where θ is functional, resampling often becomes one of the only practical tools for inference.

Our goal in this chapter is to highlight the current state of resampling methods in the context of functional data and functional parameters. We use *resampling* as a generic term which encompasses inference techniques based on recalculating the estimator of $\theta(P)$ either on subsets of the data (*subsampling*) or on samples chosen with replacement from the original data set (*bootstrap*).

The problem of resampling-based inference for functional data and parameters has been explored from several perspectives, including nonparametric smoothing, bootstrap, subsampling, and empirical processes. Extensive theoretical results for functional data have been achieved in the context of subsampling and in the empirical process literature; the bootstrap is less developed in this respect, and numerical properties and finite sample performance have received only modest attention.

The remainder of this survey is laid out as follows: Section 7.2 contains basic background and notation which will be used throughout the chapter; Section 7.3 discusses results from nonparametric smoothing; Section 7.4 summarizes the major results in subsampling; Section 7.5 discusses what is known about the bootstrap; Section 7.6 presents a few recent real-data applications of bootstrapping with functional data; and Section 7.7 offers possible directions for further research and exploration.

7.2 Background and notation

We will consider a sequence of i.i.d. random elements $\mathcal{X}_1, \ldots, \mathcal{X}_n$. The elements are assumed to be D-valued, where D is, unless otherwise noted, a metric space with probability measure P. If D is strictly finite-dimensional, we will emphasize this fact by denoting the data by $X_1, \ldots, X_n$ (i.e. using Roman fonts).

Bootstrap data will be indicated by asterisks, for example $\mathcal{X}_1^*$. Similarly, P^* and $\mathbb{E}^*$ will refer to probability and expectation based on bootstrap randomness, conditional on the original random elements $\mathcal{X}_1, \ldots, \mathcal{X}_n$.

Our primary goal is to estimate the sampling distribution of $\hat{\theta}_n$, a statistic estimating the parameter of interest $\theta(P)$. In our discussion of the functional data bootstrap in Section 7.5, the estimators considered will be limited to those of the form $\hat{\theta}_n = f(\bar{\mathcal{X}})$, where $\bar{\mathcal{X}}$ denotes the average of $\mathcal{X}_1, \ldots, \mathcal{X}_n$, or $\hat{\theta}_n = \overline{f(\mathcal{X})}$ for a suitably smooth function f. Bootstrap consistency results in nonparametric smoothing have been developed in many specific applications; density estimation and regression are discussed in Section 7.3. By contrast, subsampling has been shown to apply in quite broad settings which we discuss in full generality in Section 7.4.

In each case the primary tool is that of weak convergence. To start with, the distribution of the root $\tau_n(\hat{\theta}_n - \theta(P))$ is shown (or assumed) to have a well defined limit as $n \to \infty$. Here, τ_n is the rate of convergence of $\hat{\theta}_n - \theta(P)$; in regular situations it is of order $n^{1/2}$ but this is not always the case. Then, this asymptotic distribution is approximated by its resampled or subsampled analog which is basically an empirical distribution of *pseudo-statistics* that are appropriately centered and scaled; when this approximation is valid asymptotically, we say that the bootstrap mechanism works.

In general spaces, a sequence of probability measures P_n converges weakly to a measure P if $\int f dP_n \to \int f dP$ for all bounded continuous functions f; on the real line this is equivalent to the usual notion of weak convergence by the Portmanteau theorem. In order to simplify notation, we define

$$Pf := \int f dP.$$

Weak convergence can be metrized by looking at the rate of the above convergence across a slightly restricted class of functions. Let BL_1 denote the class of Lipschitz continuous functions which are bounded by 1 and with Lipschitz constant bounded by 1. The bounded Lipschitz metric on a collection of measures is then given by

$$\rho_{BL}(P_1, P_2) = \sup_{f \in BL_1} \left| \int f dP_1 - \int f dP_2 \right|.$$

Convergence in a bounded Lipschitz metric is equivalent to weak convergence.

The preceding discussion sweeps under the rug many issues of measurability. In many situations, our random elements and estimators in function space cannot be assumed to be integrable. We will require all estimators to be asymptotically measurable, and any expectations of nonmeasurable elements should be regarded as outer expectations in the sense discussed in van der Vaart and Wellner (1996).

7.3 NONPARAMETRIC SMOOTHING

In this section, we give a brief discussion of bootstrap results as applied in nonparametric smoothing. The data in this setting have traditionally been finite dimensional, but the parameter of interest is functional, which ties these problems to related problems involving functional data. Very recent work, discussed at the end of this section, has begun to extend these results to functional data.

The goal of nonparametric smoothing is typically to use the data to estimate an unknown function, such as a probability density or conditional expectation (regression function), which is only assumed to possess some degree of smoothness. Unfortunately, the flexibility allowed by most estimators makes it difficult to ascertain whether observed features are physically meaningful or due to randomness in the data and the estimation process. These difficulties make effective inference for the entire estimated function vitally important, and the functional nature of the problem makes the bootstrap a natural approach.

In the remainder of this section we highlight some major results in density estimation and nonparametric regression.

7.3.1 Confidence bands in density estimation

Let $X_1, \ldots, X_n$ be i.i.d. real-valued observations from a density $f(x)$ which is unknown but assumed to be smooth in some sense. The classical kernel estimate of $f(x)$ proposed by Rosenblatt (1956) is given by

$$\hat{f}(x) = \frac{1}{nh} \sum_{i=1}^{n} K\left(\frac{X_i - x}{h}\right), \tag{7.1}$$

where $K(\cdot)$ is typically a symmetric probability density such as the normal distribution.

Letting the bandwidth h be proportional to $n^{-1/5}$, the estimator (7.1) has a mean square error of order $n^{-4/5}$ provided that $f(x)$ is twice differentiable; see, for example, Wand and Jones (1995). However, finite samples often produce estimates with irregular shapes and spurious modes, which necessitates the use of effective inference techniques even for exploratory analysis.

Hall (1993) proposes a resampling algorithm to produce confidence bands for f, and develops the Edgeworth expansions necessary to calculate their convergence rates. His approach is to use the bootstrap to estimate the distribution of the suprema of the approximately pivotal quantity

$$R(x) = \frac{\hat{f}(x) - \mathbb{E}\hat{f}(x)}{[\mathbb{E}\hat{f}(x)]^{1/2}}.$$

$R(x)$ is approximately pivotal, where a *pivot* is a statistic whose distribution does not depend on any unknown quantities. Working with pivots often improves bootstrap performance; see Efron and Tibshirani (1993).

The bootstrap sample is constructed by drawing $X_1^*, \ldots, X_n^*$ i.i.d. from the empirical distribution of $X_1, \ldots, X_n$; $\hat{f}^*(x)$ is constructed in the same way as $\hat{f}(x)$, just using the resampled data. It is easy to see that $\mathbb{E}^* \hat{f}^*(x) = \hat{f}(x)$, which leads to the bootstrap pivot

$$R^*(x) = \frac{\hat{f}^*(x) - \hat{f}(x)}{\hat{f}(x)^{1/2}}.$$

A $(1 - a)100\%$ confidence band for $\mathbb{E}\hat{f}(x)$ on the interval $[a, b]$ can be produced by choosing the lower limit L^* and upper limit U^* such that

$$P^*\left[L^* \leq \inf_{x \in [a,b]} R^*(x) \leq \sup_{x \in [a,b]} R^*(x) \leq U^*\right] = 1 - a.$$

The bootstrap approximation to the distribution of the pivotal statistic would then imply that

$$P\left[L^* \leq \inf_{x \in [a,b]} R(x) \leq \sup_{x \in [a,b]} R(x) \leq U^*\right] \simeq 1 - a.$$

The above yields an approximate $(1 - a)100\%$ confidence band for $\mathbb{E}\hat{f}(\cdot)$; this is generally different from a band for $f(\cdot)$ since $\hat{f}(\cdot)$ is a biased estimate. To obtain a confidence band for $f(\cdot)$, either an explicit bias correction must be employed, or the bandwidth h must be chosen in a suboptimal way, i.e. of smaller order than $n^{-1/5}$. In the latter case, the confidence band is consequently being based on an under-smoothed estimator whose bias is negligible (as compared to its standard deviation).

The issues in this setting are similar to those in nonparametric regression, and the theory has been more developed in that setting. For these reasons, we defer discussion until the next section.

7.3.2 Confidence bands in nonparametric regression and autoregression

In nonparametric regression, data $(X_1, Y_1), \ldots, (X_n, Y_n)$ are observed and the goal is to estimate the conditional mean

$$r(x) = \mathbb{E}(Y|X = x).$$

There are several well-investigated estimators of $r(\cdot)$, the most common being the Nadaraya–Watson estimator and the local linear estimator. Each roughly takes the form of a weighted average

$$\hat{r}(x) = \sum_{i=1}^{n} w(X_i) Y_i,$$

where the weights sum to 1, reflect the distance from X_i to x, the degree of smoothing chosen by the practitioner, and aspects of the local spacing of the X_i's. A thorough exposition is given in many texts on kernel smoothing; see, for example, Wand and Jones (1995).

Several authors have investigated the construction of bootstrap confidence intervals and bands for these regression estimates. Algorithms differ greatly in their details and assumptions. One approach is to resample the pairs of data; this is discussed in Efron and Tibshirani (1986, 1993) and Mammen (1992). Claeskens and van Keilegom (2003) use the data pairs to estimate a bivariate density and then use the smoothed bootstrap of Silverman and Young (1987), which resamples from the estimated joint density of X and Y.

Other implementations perform the resampling on the residuals. In some cases, for example Härdle and Bowman (1988), the residuals themselves are bootstrapped. In others, such as Härdle and Marron (1991) and Neumann and Polzehl (1998), the wild bootstrap of Wu (1986) is used; the wild bootstrap is typically used with heteroskedastic errors and involves producing a simulated residual e_i^* from only the single observed residual e_i. A final possible approach is to use the local bootstrap of Shi (1991). For concreteness, we outline the approach used in the residual and wild bootstraps.

In this class of methods, an initial pilot estimator is used to smooth the data and estimate the residuals, giving

$$e_i = Y_i - \hat{r}_1(x).$$

Next, the residuals are bootstrapped, using either the residual or wild bootstrap, and added to a possibly different estimate of the regression function $r(\cdot)$, producing bootstrap *pseudo*-data

$$Y_i^* = e_i^* + \hat{r}_2(x).$$

The pseudo-data are then smoothed again to produce a *pseudo*-statistic, i.e. a bootstrap regression curve. Repeating this procedure a number of times results in a collection of bootstrap curves that can be used to produce a confidence band for the true underlying curve. Neumann and Polzehl (1998) have investigated confidence bands for regression, and these ideas have been extended to include derivatives of the regression function by Claeskens and van Keilegom (2003). Auestad and Tjøstheim (1991) suggest using the bootstrap for confidence bands in time series *auto*regression; the theoretical underpinnings of this approach have been established by Neumann and Kreiss (1998) who have shown the formal equivalence of autoregression to regression in large samples.

The need for multiple estimators $\hat{r}_1(x), \hat{r}_2(x), \ldots$ is one of the more undesirable aspects of the theory and is due to the fact that nonparametric regression estimators are inherently biased. Since the estimates are not centered at the correct value, confidence intervals need to be explicitly re-centered or constructed in a way that mimics the bias of the original estimator, which can be achieved through multiple levels of smoothing. Härdle and Bowman (1988) used explicit bias correction; Hall (1992) advocates under-smoothing, which has guided most subsequent investigations. Interestingly, however, McMurry and Politis (2008) have showed that in some settings the need for bias correction is obviated through the use of infinite-order, flat-top kernels.

7.3.3 Bootstrap in nonparametric functional regression

Recently, Ferraty et al. (2008) have established the consistency of the bootstrap for confidence intervals in nonparametric functional regression. In this setting the observed data are assumed to be i.i.d. pairs of the form $(\mathcal{X}_1, Y_1), \ldots, (\mathcal{X}_n, Y_n)$, where the $\mathcal{X}_i$ take values in a function space with semi-metric $d(\cdot, \cdot)$; the Y_i are univariate. They are related by

$$Y_i = r(\mathcal{X}_i) + \epsilon_i,$$

where $r(\cdot)$ is the unknown regression functional and ϵ_i is a real-valued, mean zero error. The functional Nadaraya–Watson estimator is then given by

$$\hat{r}_h(\mathcal{X}) = \frac{\sum_{i=1}^n Y_i K\left(d(\mathcal{X}_i, \mathcal{X})/h\right)}{\sum_{i=1}^n K\left(d(\mathcal{X}_i, \mathcal{X})/h\right)}, \quad (7.2)$$

where K is a non-negative kernel, and h is the bandwidth. The asymptotic properties of (7.2) are discussed in Ferraty and Vieu (2006) and more recent developments are presented in Chapter 4 of this volume.

Ferraty et al. (2008) prove that both the residual bootstrap and wild bootstrap are consistent, and they use their results to construct asymptotically valid pointwise confidence intervals. Their results are strikingly similar to those achieved in traditional nonparametric regression; once the estimator (7.2) is defined, the procedures remain substantially unchanged. Multiple levels of smoothing are used for bias correction, and the usefulness of the approach is demonstrated through a simulation study and a real data example.

7.4 SUBSAMPLING

The theory of subsampling for high-dimensional and even functional data has been developed in Politis et al. (1999). More recently, a simplified and unified way of looking at functional data and parameters has been presented in a K-sample setting in Politis and Romano (2008, 2010). In both settings the theory is fully developed for functional stationary time-series data satisfying strong mixing conditions. While such generality is desirable in many applications, for simplicity of exposition we focus on the situation of an i.i.d. sequence that can be generalized to triangular array data whose rows are i.i.d.

Throughout, let $\mathcal{X}_1, \ldots, \mathcal{X}_n$ be an i.i.d. sequence with distribution P, and assume the goal of the analysis is inference for the parameter $\theta(P)$, which takes values in a metric space Θ. To this end, we study the probabilistic distribution of a *root*, which is simply a Θ-valued function of the data $\mathcal{X}_1, \ldots, \mathcal{X}_n$, the parameter

$\theta(P)$, and the sample size n, with the property that if the distribution of the root were known, the desired inference could be performed.

An easy example of a root can be found in the classical procedure for estimating the mean, μ, of real-valued i.i.d. data $X_1, \ldots, X_n$, in which case the quantity $R_n := n^{1/2}(\bar{X} - \mu)$ has the desired property under quite general conditions. In the preceding example, $n^{1/2}$ is the normalizing factor which gives R_n a nondegenerate limiting distribution; while $n^{1/2}$ is the most typical rate, in order to accommodate different estimators, we denote a general normalizing rate by τ_n.

We let $R_n(\mathcal{X}_1, \ldots, \mathcal{X}_n, \theta(P))$ denote the root and $J_n(P)$ denote its probability law, where the root is regarded as a random element of Θ. We will show that the probability law $J_n(P)$ can be estimated through use of the appropriate subsampling procedure.

The primary idea behind subsampling is that the distribution of the root can be approximated by an appropriately scaled version of the empirical distribution of the root calculated on subsets of the data of size b, where $b < n$. For i.i.d. data we typically examine all $\binom{n}{b}$ subsamples, or if this is too large, a random sample thereof.

We begin with the primary result which establishes conditions under which the empirical distribution of random elements from a triangular array converges to their marginal distribution. Throughout, we assume random elements, denoted here by $\mathcal{Y}$, take values in a metric space D, equipped with metric $d(\cdot, \cdot)$. As discussed previously, the random elements are not necessarily assumed to be either separable or measurable.

Theorem 7.1 (Politis et al. 1999) *Let $\mathcal{Y}_1, \ldots, \mathcal{Y}_n$ be independent D-valued observations from distribution P. Let $\hat{P}_n$ denote the empirical measure of the $\mathcal{Y}_i$, $1 \leq i \leq n$. In addition, assume P is tight; that is, for every $\epsilon > 0$ there exists a compact set K so that $P(K) > 1 - \epsilon$. Then $\rho_{BL}(\hat{P}_n, P) \to_P 0$, which implies weak convergence.*

The above result can be used to establish the validity of subsampling for a wide range of applications, of which we now explore several.

7.4.1 Parameter estimation

As before, we denote the distribution of the root by $J_n(P)$. Let $L_{b,n}$ be the empirical distribution of $R_b(\mathcal{X}_{i_1}, \ldots, \mathcal{X}_{i_b}; \hat{\theta}_n)$, which puts mass $1/\binom{n}{b}$ at each of its $\binom{n}{b}$ possible values. Finally denote the empirical distribution of the subsampled values of $R_b(\mathcal{X}_{i_1}, \ldots, \mathcal{X}_{i_b}; \theta(P))$ by $H_{b,n}$. Consistency of the subsampling distribution can be proved by showing first that $H_{b,n}$ converges weakly to $J_n(P)$, and then extending the result to the typical situation where $\theta(P)$ is unknown.

Lemma 7.1 (Politis et al. 1999) *Let $\mathcal{X}_{n,1}, \ldots, \mathcal{X}_{n,n}$, for $n \in \mathbb{N}$, be a triangular array of row-wise independent random observations. Assume $J_n(P)$ converges weakly to a* *limit law $J(P)$ which concentrates on a separable subset of S, $b/n \to 0$, and $b \to \infty$ as $n \to \infty$. Then $\rho_{BL}(H_{n,b}, J_n(P)) \to_P 0$.*

Lemma 7.1 is a consequence of a generalization of Theorem 7.1. The theorem is stated for independent $\mathcal{Y}$'s but it can be extended to the setting where the $\mathcal{Y}$'s are the $\binom{n}{b}$ distinct subsampled values of the root rather than the original data.

With this generalization in hand, let the ith subsample consist of observations $i_1, \ldots, i_b$. We let $\mathcal{Y}_i = R_b(\mathcal{X}_{i_1}, \ldots, \mathcal{X}_{i_b}; \theta(P))$, and then apply the theorem. In this case $\hat{P}_n = H_{n,b}$, and $P_n = J_n$, and the immediate result is $\rho_{BL}(H_{n,b}, J_n(P)) \to_P 0$, as desired.

Of course, the primary concern is inference for $\theta(P)$ which, except in the case of hypothesis testing, is unknown. The following theorem shows that under the additional mild assumption $\tau_b/\tau_n \to 0$, Lemma 7.1 remains true when $\theta(P)$ is replaced by $\hat{\theta}_n$.

Theorem 7.2 (Politis et al. 1999) *Under the assumptions of Theorem 7.1, and the additional assumption $\tau_b/\tau_n \to 0$ as $n \to \infty$, $\rho_{BL}(L_{n,b}, J_n(P)) \to_P 0$.*

It is interesting to see how easily Theorem 7.2 follows from Lemma 7.1. By the triangle inequality, it suffices to show $\rho_{BL}(L_{n,b}, H_{n,b}) \to_P 0$. We note that $L_{n,b}$ is the empirical distribution of the values

$$\tau_b[\hat{\theta}_{b,i,n} - \hat{\theta}_n] = \tau_b[\hat{\theta}_{b,i,n} - \theta(P)] + \tau_b[\theta(P) - \hat{\theta}_n],$$

which is exactly the distribution of $H_{n,b}$ shifted by

$$\tau_b[\theta(P) - \hat{\theta}_n] = (\tau_b/\tau_n)\left(\tau_n[\theta(P) - \hat{\theta}_n]\right).$$

Since $\tau_n[\theta(P) - \hat{\theta}_n]$ has a nondegenerate limiting distribution, and $\tau_b/\tau_n \to 0$, it follows immediately that this shift is asymptotically negligible.

7.4.2 K-sample subsampling and a unified approach to functional parameters

The results of the previous section can be extended to the situation where, rather than estimating a single parameter, the goal is to estimate a parameter which depends on several independent samples; this is the standard set-up in a classical one-way analysis of variance (ANOVA), but the generality of Theorem 7.1 makes it possible to explore more general situations as well.

In the present subsection, we assume the data consist of K independent samples with $K \geq 1$; the kth sample will be denoted $\mathcal{X}^{(k)} = \left(\mathcal{X}_1^{(k)}, \ldots, \mathcal{X}_{n_k}^{(k)}\right)$. The random variables $\mathcal{X}_j^{(k)}$ are assumed to take values in a metric space D, which may be a functional space or a simpler space like $\mathbb{R}^d$.

It will be assumed throughout that $\mathcal{X}_1^{(k)}, \ldots, \mathcal{X}_{n_k}^{(k)}$ is an i.i.d. sample from distribution $P^{(k)}$ and that $\mathcal{X}^{(k)}$ and $\mathcal{X}^{(k')}$ are independent for $k \neq k'$; it is possible to relax the assumption of independence within samples. Taken together, the probability law for the entire experiment is $P = (P^{(1)}, \ldots, P^{(k)})$, and the goal will be inference for some parameter $\theta(P)$ which takes values in a normed linear space Θ with norm denoted by $||\cdot||$.

Since the aim of the analysis is typically to quantify the distance between $\theta(P)$ and its estimator $\hat{\theta}_n$, rather than working with them directly, we consider a real-valued function of the root, $g[\tau_n(\hat{\theta}_n - \theta(P))]$, where $g: \Theta \to \mathbb{R}$ is a continuous function; this provides a unified and simplified approach which remains valid even when $K = 1$. Let $J_n(x, P)$ denote the probability law of $g[\tau_n(\hat{\theta}_n - \theta(P))]$. The function g will typically be a norm on Θ or a projection of Θ onto $\mathbb{R}$. Allowing g to remain unspecified allows the present work to be immediately applicable to a wide range of interesting situations.

As in the previous section, the main idea is to reconstruct the statistic of interest on subsamples of the original data. We let $\boldsymbol{n} = (n_1, \ldots, n_K)$ denote sizes for each of the K samples and $\boldsymbol{b} = (b_1, \ldots, b_K)$ denote the subsamples' sizes, which are chosen so that $b_k/n_k \to 0$ for all k.

Since the K samples are assumed to be independent of each other, we calculate all combinations of subsamples on each of the samples. For simplicity of notation, let q_k denote the number of possible subsamples from the kth sample, and let $i_k = 1, \ldots, q_k$ index these samples, and $q = \prod_{k=1}^{K} q_k$. With this notation, we can denote the empirical subsampling approximation to J_n by the empirical distribution of the subsampled values, given by

$$L_{n,b}(x) = \frac{1}{q} \sum_{i_1=1}^{q_1} \sum_{i_2=1}^{q_2} \cdots \sum_{i_K=1}^{q_K} 1\{g[\tau_b(\hat{\theta}_{i,b} - \hat{\theta}_n)] \leq x\}.$$

Under the modest conditions described in the following theorem, $L_{n,b}$ provides a consistent estimate for the true law J_n.

Theorem 7.3 (Politis and Romano 2008) *Assume there exists a nondegenerate limiting law $J(P)$ such that $J_n(P)$ converges weakly to $J(P)$ as $\min_k n_k \to 0$; as $\min_k n_k \to 0$, $\tau_b ||\hat{\theta}_n - \theta(P)|| = o_P(1)$; $b_k/n_k \to 0$, $\tau_b/\tau_n \to 0$, and $b_k \to \infty$ as $\min_k n_k \to \infty$; and that g is uniformly continuous. Then*

(i) $L_{n,b}(\cdot) \to_P J(x, P)$ *for all points of continuity of* $J(\cdot, P)$.

(ii) *If* $J(\cdot, P)$ *is continuous at* $J^{-1}(1 - \alpha, P)$, *then the event*

$$g[\tau_n(\hat{\theta}_n - \theta(P))] \leq L_{n,b}^{-1}(1 - \alpha)$$

has asymptotic probability equal to $1 - \alpha$.

Here, $J^{-1}(s, P) \equiv \inf\{x : J(x, P) \geq s\}$, with a similar definition for $L_{n,b}^{-1}(s)$.

7.4.3 Subsampling the empirical process

Let $\mathcal{X}_1, \ldots, \mathcal{X}_n$ be i.i.d. D-valued observations with marginal distribution P. Let $\hat{P}_n$ denote the empirical measure. The empirical process indexed by a class of functions $f \in \mathcal{F}$ is given by

$$Z_n(f) = n^{1/2}[\hat{P}_n f - Pf], \qquad (7.3)$$

where $f: D \to \mathbb{R}$. Empirical processes have received considerable attention in statistics and probability literature; an excellent overview can be found in van der Vaart and Wellner (1996). Very often, interest focuses on uniform properties of the process given by equation (7.3) across $\mathcal{F}$. Necessary and sufficient conditions for uniform consistency of the bootstrap have been studied extensively for i.i.d. functional and finite-dimensional data; these will be discussed in Section 7.5.3. Comparable results for real-valued time series are found in Bühlmann (1994), Naik-Nimbalkar and Rajarshi (1994), and Radulović (2002).

With regards to subsampling, Politis et al. (1999) have developed convergence criteria in a general time-series framework which includes both finite-dimensional and functional data. Here we discuss the i.i.d. version of their results. In the following, let $L^\infty(\mathcal{F})$ denote the metric space of real-valued bounded functions on $\mathcal{F}$ equipped with the sup norm, $||\cdot||_\infty$. The following theorem is a direct consequence of Theorem 7.1 using $R_n = n^{1/2}[\hat{P}_n(\cdot) - P_1(\cdot)]$ and $D = L^\infty(\mathcal{F})$.

Theorem 7.4 (Politis et al. 1999) *Let $\mathcal{X}_1, \ldots, \mathcal{X}_n$ be i.i.d. D-valued observations. Assume the law of Z_n converges weakly to a limiting process Z that concentrates on a separable subset of $L^\infty(\mathcal{F})$. Finally, assume $b/n \to 0$ as $n \to \infty$. Then $\rho_{BL}(J_n(P), L_{n,b}) \to_P 0$.*

Theorem 7.4 is not as general as the one presented in Section 7.5.3 but it requires substantially fewer conditions; this is typical since, in contrast with the bootstrap, subsampling works under almost minimal assumptions.

7.5 BOOTSTRAP

Bootstrap consistency results for functional data can be split into two major categories depending on whether or not the statistic of interest is also functional. Politis and Romano (1994a) examine the bootstrap for the mean of Hilbert-space-valued time series; Section 7.5.1 follows their work. Cuevas et al. (2006) have conducted a substantial simulation experiment which is described in Section 7.5.2. Theorems 2.4 and 3.1 of Giné and Zinn (1990) establish consistency for many real-valued statistics and, as a corollary, for the bootstrap for the mean of Banach-space-valued random variables; these results are presented in Section 7.5.3.

7.5.1 Bootstrap for functional statistics

The results in Politis and Romano (1994a) were motivated by the study of estimators which are smooth functionals of the empirical process for time-series data. The empirical process takes values in an appropriate Hilbert space, which leads to a general theory for a class of Hilbert-space-valued random variables. The results apply very broadly and are stated in terms of triangular arrays of random elements satisfying strong mixing conditions.

The dependence condition somewhat complicates matters because the bootstrap must be modified so that the resampled data has an asymptotically similar dependence structure to the original data; this is accomplished through the use of the stationary bootstrap introduced in Politis and Romano (1994b). The idea is that the resampled data contain random length strings of consecutive observations from the original data, where the average length of the strings grows as n becomes large.

For simplicity, as in preceding sections, we focus on the i.i.d. setting; notably, the stationary bootstrap reduces to the usual Efron's (1979) i.i.d. bootstrap simply by letting the average length of the above-mentioned strings equal 1. In the following analysis, let H be a Hilbert space. The core of the theory rests on the following theorem, which establishes consistency of the bootstrap for the mean of the functional data. In the present form the theorem is weaker than Corollary 7.2 (see below), but as shown in Politis and Romano (1994a), it also holds for data satisfying dependence conditions.

Theorem 7.5 (Politis and Romano 1994a) *Let $\mathcal{X}_1, \ldots, \mathcal{X}_n$ be an i.i.d. sequence of essentially bounded H-valued random variables with mean θ. Let $\bar{\mathcal{X}}_n = n^{-1} \sum_{i=1}^{n} \mathcal{X}_i$ and $\mathcal{Z}_n = \sqrt{n}(\bar{\mathcal{X}}_n - \theta)$. Denote the law of $\mathcal{Z}_n$ by $L(\mathcal{Z}_n)$. Let $\mathcal{Z}_n^* = n^{1/2} (\mathcal{X}_n^* - \bar{\mathcal{X}}_n)$, and denote its law by $L(\mathcal{Z}_n^* | \mathcal{X}_1, \ldots, \mathcal{X}_n)$. Then $\rho(L(\mathcal{Z}_n), L(\mathcal{Z}_n^* | \mathcal{X}_1, \ldots, \mathcal{X}_n)) \to_P 0$, where ρ is any metric metrizing weak convergence on H.*

If $Y_1, \ldots, Y_n$ are i.i.d. real-valued random variables with distribution F, then $\mathcal{X}_i(t) = 1_{[Y_i \leq t]} - F(t)$ can be regarded as a functional observation in the Hilbert space $L^2(v)$, where v is a finite measure on $\mathbb{R}$. If we combine this reasoning with the previous theorem and the notion of Fréchet differentiability, we obtain an immediate corollary extending the consistency to smooth functionals of the empirical distribution.

In the following, let v denote a finite measure on the real line, F and G denote distributions in some class $\mathcal{F} \subset L^2(v)$, and $T(\cdot)$ denote a functional on $\mathcal{F}$ which is Fréchet differentiable in the sense that for fixed F and as $||G - F|| \to 0$, $T(G) - T(F) = \langle \psi, G - F \rangle + o(||G - F||)$, for some $\psi \in L^2(v)$; here $\langle \cdot, \cdot \rangle$ denotes the inner product on H.

Corollary 7.1 (Politis and Romano 1994a) *Let L_n denote the true distribution of $n^{1/2}(T(\hat{f}_n) - T(F))$ and let $\hat{L}_n$ denote the bootstrap distribution of*

$n^{1/2}\left(T\left(\hat{f}_n^*\right) - T(\hat{f}_n)\right)$, then $\rho_1(\hat{L}_n, L_n) \to 0$, where ρ_1 is any metric metrizing weak convergence on the real line. Moreover, L_n converges weakly to a Gaussian distribution with mean 0 and variance σ_ψ^2, where

$$\sigma_\psi^2 = \text{Var}\left(\langle X_1, \psi \rangle\right).$$

In the context of functional data, the parameters of interest are often also functional, so it is desirable to study the case where T takes on values in a function space. Corollary 7.1 can be extended to this setting using an appropriate differentiability condition and a "functional delta method." We briefly outline the approach; technical details can be found in Cuevas and Fraiman (2004), and Cuevas et al. (2006) provides additional insights.

Let T be an operator defined on $\mathcal{P}$, a suitable space of probability measures, with values in a Banach space F. Assume that T is differentiable at $P \in \mathcal{P}$ in the sense that

$$T(Q) = T(P) + T'_P(Q - P) + o(\rho_{BL}(P, Q)), \tag{7.4}$$

where T'_P is a linear operator on an appropriate space of finite signed measures. In order to simplify notation, let $S_n = n^{1/2}(T(P_n) - T(P))$ and $S_n^* = n^{1/2}\left(T\left(P_n^*\right) - T(P_n)\right)$, and let $L(S_n)$ and $L\left(S_n^*|X_1, \ldots, X_n\right)$ respectively denote their laws. In order to establish bootstrap consistency we need to show

$$\rho_{BL}\left(L\left(S_n^*|X_1, \ldots, X_n\right), L(S_n)\right) \to 0 \tag{7.5}$$

either in probability or almost surely. The convergence (7.5) can be established via the linearization which results from the differentiability condition (7.4).

$$S_n = n^{1/2} T'_P(P_n - P) + n^{1/2} o\left(\rho_{BL}(P_n, P)\right)$$
$$= L_n + n^{1/2} o\left(\rho_{BL}(P_n, P)\right),$$

where L_n denotes the linearization of S_n. Let L_n^* denote the corresponding linearization of S_n^*. The linearizations L_n and L_n^* tend to the same Gaussian limit by the Central Limit Theorem and by the bootstrap consistency results given in Theorem 7.5 (above) or Corollary 7.2 (below).

The remainder terms $n^{1/2} o\left(\rho_{BL}(P_n, P)\right)$ and $n^{1/2} o\left(\rho_{BL}\left(P_n^*, P_n\right)\right)$ can be handled provided we are able to establish boundedness in probability, uniform on $\mathcal{P}$, of the sequences $n^{1/2} \rho_{BL}(P_n, P)$.

7.5.2 Simulation studies

Cuevas et al. (2006) have performed an extensive simulation study to assess the performance of the functional data bootstrap. To our knowledge this is the only large-scale investigation to date. We will briefly summarize their experiments and findings. They have investigated bootstraps for several functional measures of center and one measure of spread for data in $C[0, 1]$. The data were generated by the following two processes:

1. A Wiener process on $[0, 1]$ given by $\mathcal{X}(t) = 10t(1 - t) + B(t)$, where $B(t)$ is a standard Brownian motion.
2. A Gaussian process $\mathcal{X}(t)$ with mean $10t(1 - t)$ and covariance function $\text{Cov}(\mathcal{X}(s), \mathcal{X}(t)) = \exp(-|s - t|/0.3)$.

They constructed bootstrap confidence intervals based on the following estimators:

- the sample mean $\bar{\mathcal{X}}$
- the sample variance
- the FM median
- the α-trimmed FM mean with $\alpha = 0.25$
- the "kernel mode" estimate for the underlying density using a Gaussian kernel.

The FM trimmed mean and median were based on Fraiman and Muniz's (2001) notion of data depth which is defined as follows. For each $t \in [0, 1]$, let $F_{n,t}$ be the empirical distribution of the sample $x_1(t), \ldots, x_n(t)$. Let $Z_i(t) = 1 - |1/2 - F_{n,t}(x_i(t))|$ denote the univariate depth of the datum $x_i(t)$. These values are aggregated across time by

$$I_i = \int_0^1 Z_i(t) dt. \tag{7.6}$$

Observations x_i can then be ranked according to the corresponding value of I_i, with the "deepest" data having the largest values of I. Using this measure, the α-trimmed mean becomes the average of the $100(1 - \alpha)\%$ deepest functions in the sample, and the FM median is the deepest function in the sample.

The "kernel mode" estimator, an exploratory data analysis tool, is given by

$$\text{argmax}_{x \in E} \frac{1}{nh} \sum_{i=1}^{n} K\left(\frac{||x_i - x||}{h}\right),$$

which is approximated for computational efficiency by

$$\text{argmax}_{x \in \{x_1, \ldots, x_n\}} \frac{1}{nh} \sum_{i=1}^{n} K\left(\frac{||x_i - x||}{h}\right).$$

Cuevas et al. (2006) have also tested three different bootstraps, the standard bootstrap, a smoothed bootstrap, and, for the mean, a parametric bootstrap. Confidence intervals were constructed using both the L^2 and L^∞ norms. Sample sizes were $n = 25$ and $n = 100$. Each combination of estimator, norm, bootstrap procedure, and sample size was run 500 times.

Their simulations show great promise. To summarize their conclusions briefly:

1. The standard bootstrap proved unreliable for the sample median and kernel mode; this appears to be because the resampling produces a very small number of curves. These results were substantially improved by using a smoothed bootstrap.
2. With the exception of the failure of the standard bootstrap for the median and kernel mode, all other estimators performed quite well, but the parametric bootstrap did not provide any substantial advantage.
3. Performance improved substantially from $n = 25$ to $n = 100$. For almost all estimators at the larger sample size, observed coverage was close to expected coverage; interval width also decreased as expected.
4. Both the L^2 and L^∞ norms performed satisfactorily.

7.5.3 Bootstrapping the empirical process

The bootstrap for empirical processes has been extensively investigated by several authors including Giné and Zinn (1990) and Dudley (1990), and nicely summarized in Section 3.6 of van der Vaart and Wellner (1996). The results are extremely general and apply equally well to functional and finite-dimensional data. The results in this setting are substantially more complex than results in the preceding sections, so we focus on the highlights of the theory and refer readers to the preceding references for additional details.

Let $\mathcal{X}_1, \ldots, \mathcal{X}_n$ be a D-valued i.i.d. sample with probability distribution P, where D is again a metric space with metric $d(\cdot, \cdot)$. For consistency with notation elsewhere, we let $\hat{P}_n$ denote the empirical distribution. The bootstrap sample $\mathcal{X}_1^*, \ldots, \mathcal{X}_n^*$ is an i.i.d. sample from $\hat{P}_n$, and its empirical distribution will be denoted $\hat{P}_n^*$. The process is typically viewed as a random functional on a class of functions $\mathcal{F} \subset \{f : D \to \mathbb{R}\}$, for which it is assumed that $\sup_{f \in \mathcal{F}} |f(x) - Pf| < \infty$.

As in Section 7.4.3, we define the empirical process as follows:

$$Z_n(\cdot) = n^{1/2}[\hat{P}_n(\cdot) - P(\cdot)] \tag{7.7}$$

and its bootstrap version

$$Z_n^*(\cdot) = n^{1/2}\left[\hat{P}_n^*(\cdot) - \hat{P}_n(\cdot)\right] = n^{1/2}\sum_{i=1}^{n}(M_{ni} - 1)\delta_{\mathcal{X}_i}(\cdot), \tag{7.8}$$

where M_{ni} is the number of times $\mathcal{X}_i$ appeared in the resample. For any particular $f \in \mathcal{F}$ the consistency of the bootstrap is not particularly surprising because $\hat{P}_n f - Pf = (1/n)\sum_i f(\mathcal{X}_i) - \mathbb{E}(f(X_i))$, an average of real-valued random variables, and a situation in which the bootstrap is typically consistent.

The major success of empirical process bootstrap has been to establish conditions on $\mathcal{F}$ under which a sort of uniform convergence occurs that implies the bootstrap validity for all $f \in \mathcal{F}$ simultaneously. Roughly speaking, these conditions require that $\mathcal{F}$ not be too "big." We first start with a definition that describes a class of functions for which a uniform central limit theorem holds. We then discuss how this is sufficient for bootstrap consistency in probability, and along with an additional moment assumption implies almost sure bootstrap consistency.

Definition 7.1 *Let $\mathcal{F}$ be a collection of square integrable functions $f : D \to \mathbb{R}$ for which $\sup_{f \in \mathcal{F}} |f(x) - Pf| < \infty$ and $Z_n = n^{1/2}(\hat{P}_n - P) \Rightarrow G$ in $L^\infty(\mathcal{F})$, where G is a tight Borel-measurable element in $L^\infty(\mathcal{F})$. Then $\mathcal{F}$ is called a Donsker class.*

Let $\mathcal{F}$ be a Donsker class. The multivariate central limit theorem shows that for any functions $f_1, \ldots, f_k \in \mathcal{F}$, $(Z_n f_1, \ldots, Z_n f_k) \Rightarrow N(0, \Sigma_k)$, where $[\Sigma_k]_{i,j} = P(f_i - Pf_i)(f_j - Pf_j)$; this can be used to show that the limiting process G is by necessity Gaussian.

The Donsker assumption is almost enough to ensure uniform bootstrap consistency; we discuss these results first and then return to conditions that imply a class is Donsker.

Definition 7.2 *An* envelope function *for a class of functions $\mathcal{F}$ is any function $F(x)$ such that $|f(x)| \le F(x)$ for all $x \in D$ and $f \in \mathcal{F}$.*

The two preceding conditions imply convergence to a mean zero Gaussian process G in outer probability.

Theorem 7.6 (van der Vaart and Wellner 1996) *Let $\mathcal{F}$ be a class of measurable functions with finite envelope function. Then $\mathcal{F}$ is Donsker if and only if $\rho_{BL}(Z_n^*, G) \to_P 0$ and Z_n^* is asymptotically measurable, where G is the same limiting Gaussian process as in Definition 7.1.*

With an additional moment assumption, the preceding result can be strengthened to outer almost sure convergence.

Theorem 7.7 (van der Vaart and Wellner 1996) *Let $\mathcal{F}$ be a class of measurable functions with finite envelope function. Then $\mathcal{F}$ is Donsker and $P\,||f - Pf||_\mathcal{F}^2 < \infty$ if and only if $\rho_{BL}(Z_n^*, G) \to 0$ almost surely (a.s.) and Z_n^* is asymptotically measurable, where G is the same limiting Gaussian process as in Definition 7.1.*

Theorem 7.6 is sufficient for bootstrap consistency. An interesting consequence of Theorem 7.7 is the following corollary which implies bootstrap consistency for functional data and is discussed in Remark 2.5 in Giné and Zinn (1990).

Corollary 7.2 (Giné and Zinn 1990) *If $\mathcal{X}_1, \ldots, \mathcal{X}_n$ are i.i.d. B-valued random variables, where B is a separable Banach space, then $\mathbb{E}\left(||\mathcal{X}_i||^2\right) < \infty$ and a central limit theorem holds for the $\mathcal{X}_i$ if and only if $n^{-1/2} \sum_{i=1}^n (\mathcal{X}_i^* - \bar{\mathcal{X}}) \Rightarrow G_X$ a.s.*

The preceding results require that $\mathcal{F}$ is a Donsker class. There are numerous sufficient conditions for a class to be Donsker, and a set of necessary and sufficient conditions in Giné and Zinn (1990). Here we present one simple sufficient condition from van der Vaart and Wellner (1996).

All of the criteria rest on ways to quantify the size of a class of functions $\mathcal{F}$. Given two functions l and u we denote by the bracket $[l, u]$ the set of functions f such that $l \leq f \leq u$. An ϵ-bracket is a bracket $[l, u]$ where $||u - l|| < \epsilon$. The bracketing number $N_{[]}(\epsilon, \mathcal{F}, || \cdot ||)$ is the minimum number of ϵ-brackets needed to cover $\mathcal{F}$. A sufficient condition for $\mathcal{F}$ to be Donsker is

$$\int_0^\infty \sqrt{\log N_{[]}(\epsilon, \mathcal{F}, L_2(P))} \, d\epsilon < \infty.$$

There are other less restrictive criteria. Interested readers should consult the two references mentioned previously.

7.6 REAL-DATA APPLICATIONS

In this section we briefly discuss three recent applications of the bootstrap in functional data settings.

7.6.1 fMRI data

Locascio et al. (1997) use a residual resampling approach to detect areas of brain activation in fMRI brain imaging experiments. In their research, areas of brain activity are measured over time, while the subject is exposed to various experimental conditions, with the goal of estimating the areas of the brain which are activated during certain tasks. Traditional analytical tools are substantially inadequate in this situation because the problem is in essence that of performing a hypothesis test on every pixel with data that exhibit both spatial and temporal dependence. In general terms, their approach is to use an ARMA (autoregressive moving average) model to estimate the effects of the experimental condition at each pixel. Once the pixel-by-pixel data has been modeled, a classical technique, such as a p-value from a t-test, is used to test the significance of the activation level at each pixel. A test based on permutation of the residuals across time and experimental conditions, but not space, is then used to make the necessary adjustments for multiple testing.

7.6.2 Sulfur dioxide concentrations

Fernández de Castro et al. (2005) use a functional bootstrap approach to make predictions 30 minutes in advance of sulfur dioxide concentrations near a coal-fired power plant. The power plant had regulatory obligations to eliminate periods of high sulfur dioxide concentration in its vicinity, so its management was hoping to have warning of upcoming spikes so they could take preventative measures, such as switching types of coal.

The data were in the form of a time series $\mathcal{X}_1, \mathcal{X}_2, \ldots$, and were grouped in pairs $(\mathcal{X}_i, \mathcal{X}_{i+1})$, where the first coordinate function is the sulfur dioxide over the current half-hour period and the second coordinate function is the concentration over the subsequent half hour, which is to be predicted. Two functional approaches were used to model $\mathcal{X}_{i+1}$ as a function of $\mathcal{X}_i$, a Nadaraya–Watson type functional kernel estimator, and an autoregressive Hilbertian model; see Bosq (2000). The bootstrap was used to construct prediction bands for each estimator.

The bootstrap procedure for the kernel estimator works as follows. Denote the current time period curve by $\mathcal{Y}_0$, with the goal of forecasting $\mathcal{Y}_1$, the next half-hour period. Pairs of data $(\mathcal{X}_i, \mathcal{X}_{i+1})$ in the historical database are then resampled with probabilities depending on a distance from $\mathcal{X}_i$ to $\mathcal{Y}_0$; this procedure is similar to the local bootstrap of Paparoditis and Politis (2002). The second coordinates of the bootstrapped pairs then give an estimate of the range of possible outcomes through the Fraiman and Muniz notion of data depth, given in equation (7.6).

The bootstrap procedure for the autoregressive Hilbertian model follows a more traditional residual bootstrap. One-step-forward residuals are calculated for all forecasts using pairs of data in the database. The residuals are then decomposed in terms of their principal components, and the pseudo-data is constructed by separately bootstrapping each of the principal component coordinates.

7.6.3 Language classification

Cuesta-Albertos et al. (2007) use a functional bootstrap hypothesis test to cluster languages according to rhythmic features. It has been conjectured in the linguistics literature that languages can be classified as *stress-timed*, *syllable-timed*, or *mora-timed*, and the goal of the analysis was to perform this classification empirically. The approach taken by Cuesta-Albertos et al. was to look at *sonority*, a local index of regularity of the speech signal which takes values close to 1 for sonorant portions of the signal and 0 for obstruent portions.

The sonority of a speech sample can be viewed as a function in $L^2([0, T])$, where T is the length of the sample. Each individual sample sonority $S_{l,i}$, where l indexes language and i indexes sample within a language, was projected onto a (fixed) Brownian motion, through the L^2 inner product

$$\langle S_{l,i}, B \rangle = \int_0^T S_{l,i}(t) B(t) dt,$$

which reduces the curve to a univariate datum. A Kolmogorov–Smirnov type test is then used to compare the distribution of samples from two languages to see if they appear to group in the hypothesized way. Since the preceding test depends on the random Brownian path, its stability is increased by repeating the test for several independent Brownian motions.

The bootstrap is used to estimate the p-value for the preceding tests. The samples from the two languages l and l' are pooled and the bootstrap is used to draw independent samples from the pooled data set, thus mimicking a sampling environment that satisfies the null hypothesis.

Finally the authors employ a Monte Carlo simulation to empirically assess the ability of their procedure to distinguish between different sonority patterns. They construct Markov chains which mimic the features which are expected to be observed in sonority patterns in the different types of languages, and they use the modified Kolmogorov–Smirnov test to try to distinguish the different chains; the results are encouraging.

7.7 Discussion

As functional data proliferate, we expect bootstrap and subsampling to become very popular approaches to inference since they are straightforward to implement and have broad applicability. Nonetheless, the field remains young and offers rich avenues for further development; we mention some areas of particular interest.

In finite-dimensional problems, approaches such as studentization have been shown to produce bootstrap confidence intervals with higher-order accuracy, i.e. coverage that is very close to the nominal. While it seems likely that analogous properties hold for functional data, the question of convergence rates for the bootstrap has not been addressed in any setting, and there has been little discussion as to whether higher-order accuracy might be feasible. Furthermore, it is not trivial to see how the studentization can be optimally implemented with functional statistics.

Which method is preferable, bootstrap or subsampling? Of course, subsampling has in its favor its very wide applicability; by contrast, the bootstrap has thus far only been shown to be valid for the mean of functional data, the empirical process, and smooth functionals thereof, as outlined in Section 7.5, and regression, as discussed in Section 7.3. Results establishing the validity of the bootstrap in other settings would be highly desirable.

Focusing on the sample mean under regularity, both methods (bootstrap and subsampling) are consistent for a wide range of functional data but no work has been done to assess their performance relative to each other. In addition, very few simulation studies have explored the finite sample performance of these techniques.

In the coming years we expect these and other related problems to offer fruitful areas for theoretical development and practical implementation of resampling methods for functional data and/or statistics.

References

Auestad, B., Tjøstheim, D. (1991). Functional identification in nonlinear time series. In *Nonparametric Functional Estimation and Related Topics* (G. Roussas, ed.), 493–507. Kluwer, Dordrecht.

Bosq, D. (2000). *Linear Processes in Function Spaces: Theory and Applications.* Springer, Berlin.

Bühlmann, P. (1994). Blockwise bootstrapped empirical process for stationary sequences. *Annals of Statistics,* 22, 995–1012.

Claeskens, G., van Keilegom, I. (2003). Bootstrap confidence bands for regression curves and their derivatives. *Annals of Statistics,* 31, 1852–84.

Cuesta-Albertos, J., Fraiman, R., Galves, A., Garcia, J., Svarc, M. (2007). Classifying speech sonority functional data using a projected Kolmogorov-Smirnov approach. *Journal of Applied Statistics,* 34, 749–61.

Cuevas, A., Febrero, M., Fraiman, R. (2006). On the use of the bootstrap for estimating functions with functional data. *Computational Statistics & Data Analysis,* 51, 1063–74.

Cuevas, A., Fraiman, R. (2004). On the bootstrap methodology for functional data. In *COMPSTAT 2004 – Proceedings in Computational Statistics,* (J. Antoch, ed.), 127–35. Physica, Heidelberg.

Dudley, R.M. (1990). Nonlinear functionals of empirical measures and the bootstrap. In *Probability in Banach Spaces 7,* (E. Eberlein, J. Kuelbs, M. Marcus, eds), 63–82. Birkhäuser, Basel.

Efron, B. (1979). Bootstrap methods: another look at the jackknife. *Annals of Statistics,* 7, 1–26.

Efron, B., Tibshirani, R. (1986). Bootstrap methods for standard errors, confidence intervals, and other measures of statistical accuracy (with discussion). *Statistical Science,* 1, 54–96.

Efron, B., Tibshirani, R. (1993). *An Introduction to the Bootstrap.* Chapman and Hall, New York.

Fernández de Castro, B., Guillas, S., González Manteiga, W. (2005). Functional samples and bootstrap for predicting sulfur dioxide levels. *Technometrics,* 47, 212–22.

Ferraty, F., van Keilegom, I., and Vieu, P. (2008). On the validity of the bootstrap in nonparametric functional regression. IAP Statistics Network, Technical Report 08027.

Ferraty, F., Vieu, P. (2006). *Nonparametric Functional Data Analysis: Theory and Practice.* Springer, New York.

Fraiman, R., Muniz, G. (2001). Trimmed means for functional data. *Test,* 10, 419–40.

Giné, E., Zinn, J. (1990). Bootstrapping general empirical measures. *Annals of Probability,* 18, 851–69.

Hall, P. (1992). Effect of bias estimation on coverage accuracy of bootstrap confidence intervals for a probability density. *Annals of Statistics,* 20, 675–94.

HALL, P. (1993). On Edgeworth expansion and bootstrap confidence bands in nonparametric curve estimation. *Journal of the Royal Statistical Society, Series B*, 55, 291–304.

HÄRDLE, W., BOWMAN, A. (1988). Bootstrapping in nonparametric regression: local adaptive smoothing and confidence bands. *Journal of the American Statistical Association*, 83, 102–10.

Härdle, W. and Marron, J.S. (1991). Bootstrap simultaneous error bars for nonparametric regression. *Annals of Statistics*, 19, 778–96.

LOCASCIO, J., JENNINGS, P., MOORE, C., CORKIN, S. (1997). Time series analysis in the time domain and resampling methods for studies of functional magnetic resonance brain imaging. *Human Brain Mapping*, 5, 168–93.

McMURRY, T., POLITIS, D.N. (2008). Bootstrap confidence intervals in nonparametric regression with built-in bias correction. *Statistics and Probability Letters*, 78, 2463–9.

MAMMEN, E. (1992). *When Does Bootstrap Work? Asymptotic Results and Simulations.* Springer Lecture Notes in Statistics, 77. Springer, New York.

NAIK-NIMBALKAR, U.V., RAJARSHI, M.B. (1994). Validity of blockwise bootstrap for empirical processes with stationary observations. *Annals of Statistics*, 22, 980–94.

NEUMANN, M., KREISS, J.-P. (1998). Regression-type inference in nonparametric autoregression. *Annals of Statistics*, 26, 1570–613.

NEUMANN, M., POLZEHL, J. (1998). Simultaneous bootstrap confidence bands in nonparametric regression. *Journal of Nonparametric Statistics*, 9, 307–33.

PAPARODITIS, E., POLITIS, D.N. (2002). The local bootstrap for Markov processes. *Journal of Statistical Planning and Inference*, 108, 301–28.

POLITIS, D.N., ROMANO, J.P. (1994a). Limit theorems for weakly dependent Hilbert space valued random variables with application to the stationary bootstrap. *Statistica Sinica*, 4, 461–76.

POLITIS, D.N., ROMANO, J.P. (1994b). The stationary bootstrap. *Journal of the American Statistical Association*, 89, 1303–13.

POLITIS, D.N., ROMANO, J.P. (2008). K-sample subsampling. In *Functional and Operatorial Statistics* (S. Dabo-Niang, F. Ferraty, eds), 247–53. Physica, Heidelberg.

POLITIS, D.N., ROMANO, J.P. (2010). K-sample subsampling in general spaces: the case of independent time series. *Journal of Multivariate Analysis*, 101, 316–26.

POLITIS, D.N., ROMANO, J.P., WOLF, M. (1999). *Subsampling.* Springer, New York.

RADULOVIĆ, D. (2002). On the bootstrap and empirical processes for dependent sequences. In *Empirical Process Techniques for Dependent Data*, (H. Dehling, T. Mikosch, M. Sörensen, eds), 345–64 Birkhäuser, Basel.

ROSENBLATT, M. (1956). Remarks on some nonparametric estimates of a density function. *Annals of Mathematical Statistics*, 27, 832–7.

SHI, S. (1991). Local bootstrap. *Annals of the Institute of Statistical Mathematics*, 43, 667–76.

SILVERMAN, B.W. and YOUNG, G.A. (1987). The bootstrap: to smooth or not to smooth? *Biometrika*, 74, 469–79.

VAN DER VAART, A.W., WELLNER, J.A. (1996). *Weak Convergence and Empirical Processes.* Springer, New York.

WAND, M., JONES, M. (1995). *Kernel Smoothing.* Chapman and Hall/CRC Press, London.

WU, C.F.J. (1986). Jackknife, bootstrap and other resampling methods in regression analysis. *Annals of Statistics*, 14, 1262–350.

CHAPTER 8

PRINCIPAL COMPONENT ANALYSIS FOR FUNCTIONAL DATA

METHODOLOGY, THEORY, AND DISCUSSION

PETER HALL

8.1 INTRODUCTION

PRINCIPAL component analysis (PCA) has been variously described as a technique for dimension reduction, or a method for locating interesting and important features in multivariate data, or an approach to finding patterns in data. Of course, these descriptions are not contradictory. It could be argued that they do little more than repeat the idea that PCA is an approach to determining the dimensions, or rather the projections of dimensions, which convey most information or most mass, where mass is measured in the sense of variability or variance.

Indeed, PCA can be characterized as the operation of choosing, and ordering, the projections of high-dimensional space so that, at each successive step, the projections account for the greatest amount of variability in the given dataset. The fact that PCA involves projections, i.e. linear combinations, of dimensions characterizes it as a linear technique.

In this sense PCA has much in common with, and in fact can be regarded as a subset of, several classes of now-classical approaches to dimension reduction, such as projection pursuit (see, for example, Friedman and Tukey 1974). These techniques use the notion of "interestingness," rather than variability or variance, as the quantity to be maximized when ordering linear combinations of dimensions. Typically in methods such as projection pursuit the measure of interestingness is chosen to be antithetic to Gaussian-ness, but if we wished to emphasize interestingness from a Gaussian perspective then projection pursuit would lead us quickly to PCA.

The discussion above is rather standard in the context of high- but finite-dimensional data, not necessarily just for functional data, where we must contend with an infinite number of dimensions. The main change that has to be made to encompass the infinite-dimensional case is to recognize that there are arguably no natural dimensions, or components or variables, to start with. It is fair to think of PCA as a tool for variable choice for data in the form of functions, and in this context PCA has connections to a very large contemporary literature on variable selection for very high-dimensional data. However, in drawing this linkage we should observe that the nature of the variables chosen, as well as the order in which we rank them, is determined by PCA.

It is worth mentioning that there are, in many settings, important differences between infinite-dimensional functional data and the very high, but finite, dimensional data encountered today in data mining problems, particularly in genomics. In the latter cases the ingredient of sparsity is widespread; it implies that many of the vector components are of little or no value, being essentially noise. Moreover it is often the case that the majority of components do not have significant correlation with the relatively small number of components that are informative. Usually the opposite position obtains in the case of functional data, where a natural degree of correlation arises through the relationships of points on a smooth curve. Nevertheless, functional data analysis (FDA) has been successfully applied to solve problems involving small samples of very high-dimensional genomic data (see, for example, Leng and Müller 2006).

Early research on principal component analysis for functional data was undertaken by Besse and Ramsay (1986), Ramsay and Dalzell (1991), Rice and Silverman (1991), and Silverman (1995, 1996), among others. More recent contributions to techniques in the area include those by Brumback and Rice (1998), Cardot (2000), Cardot et al. (2000, 2003a), Girard (2000), James et al. (2000), Boente and Fraiman (2000), He et al. (2003), Yamanishi and Tanaka (2005), Cai and Hall (2006), Yao and Lee (2006), Cardot (2007), Hall and Horowitz (2007), Izem and Marron (2007), Ocaña et al. (2007), Reiss and Ogden (2007), Gervini (2008), Huang et al. (2008) and Mas (2008). Related contributions include those on general functional linear models or covariate analysis (see, for example, Capra and Müller 1997; Faraway 1997; Cardot *et al.* 1999, 2003*a*; Ocaña *et al.* 1999; Ferraty and Vieu 2000; Cuevas *et al.* 2002; Fan and Zhang 2000; Cardot and Sarda 2005, 2006; and Chapter 2 of this volume); on hypothesis testing for linear models (e.g. Cardot *et al.* 2003*b*); on functional quasi-likelihood (e.g. Chiou *et al.* 2003); on longitudinal data analysis (e.g. Staniswalis and Lee 1998; Yao *et al.* 2003*a*; Hall *et al.* 2006; Yao 2007); on curve alignment (e.g. Wang and Gasser 1997, 1999); on time-series analysis for functional data (e.g. Besse and Cardot 1996; Aguilera *et al.* 1999*a*,*b*; Ferraty *et al.* 2002*b*). The case of dependent functional data is discussed in Chapter 5 of this volume.

Particular applications involving functional data, where principal component analysis plays an important role, include those of Grambsch *et al.* (1995), Yu and Lambert (1999), Kneip and Utikal (2001), Lee *et al.* (2002), Pfeiffer *et al.* (2002), Yao *et al.* (2003*b*), and Escabias *et al.* (2005). The work of Dauxois *et al.* (1982), Bosq (1989), Besse (1992), Huang *et al.* (2002), Mas (2002), and Hall and Hosseini-Nasab (2006, 2009), for example, addresses empirical basis function approximations and approximations of covariance operators. Preda and Saporta (2005*a*,*b*) introduce a partial least squares analog of principal components in the context of functional linear regression. Zhang (2006) discusses connections between projection pursuit and principal component analysis.

Chapter 9 of the present monograph discusses the role played by curve registration as a prelude to PCA. Chapter 10 addresses the importance of PCA to classification and clustering methods for functional data. Chapter 11 treats the effect that sparsity, and methods for dealing with sparsity, have on PCA. Chapter 16 discusses the role of PCA in an operator-theoretic view of principal component analysis. Chapter 6 of Ramsay and Silverman (2005) surveys functional-data methods based on principal component analysis.

In the remainder of this chapter we shall introduce principal component analysis, first (in Section 8.2.1) in the relatively well-known context of multivariate analysis and then (in Section 8.2.2) for functional data. In the latter treatment we shall address functional linear regression, which in subsequent work we shall use to motivate discussion of a number of aspects of PCA for FDA. In Section 8.3 we shall illustrate the features of PCA for FDA by considering (in Section 8.3.1) its uses, principally for reducing and ordering dimension, and addressing different views of the best way of tackling prediction problems (Section 8.3.2). In that analysis particular attention will be paid to adaptive methods for prediction, and in that setting Section 8.3.3 will treat weighted least squares methods. Finally, Section 8.4 will discuss the connections which exist between the elusive concept of "density" for functional data, and PCA. In particular, we shall show that principal component functions are linked in a rather simple way to the amount of probability mass contained in a small ball around a given, fixed function, and show

how this property can be used to define a simple, easily estimable surrogate for density.

As can be seen from the summary above, the description that we shall give of PCA for FDA is definitely a personal one. Although we avoid technical details, our approach is strongly influenced by theoretical issues and by the different levels of accuracy of different estimator types.

8.2 Defining principal components and their estimators

8.2.1 Finite-dimensional case

Most of us are familiar with PCA in the context of finite-dimensional data. There, if X is a p-vector with zero mean then PCA is based on the particular sequence of orthonormal vectors $\psi_1, \ldots, \psi_p$ that have the property that, for each k in the range $1 \leq k \leq p$, and given $\psi_1, \ldots, \psi_{k-1}$, the distance

$$d(k) = E \left\| X - \sum_{j=1}^{k} \psi_j \psi_j^T X \right\|^2 \tag{8.1}$$

is minimized over all possible choices of ψ_k subject to $\|\psi_k\| = 1$. (In (8.1) we define $\|x\|$ to be the usual Euclidean norm of a p-vector x.) This definition of $\psi_1, \ldots, \psi_p$ implies that the random variables $\psi_j^T X$, which are proportional to the "principal component scores," are uncorrelated. Their respective variances, $\theta_j = \text{var}\left(\psi_j^T X\right)$, form a nonincreasing, non-negative sequence, i.e.

$$\theta_1 \geq \theta_2 \geq \cdots \geq \theta_p \geq 0. \tag{8.2}$$

In this notation the principal component scores, sometimes referred to as the principal components, are usually defined as $\theta_j^{-1/2} \psi_j^T X$ for $1 \leq j \leq p$, and in particular they are standardized for variance.

The $p \times p$ covariance matrix, $K = \text{var}(X)$, can be represented very simply in terms of the vectors ψ_j and variances θ_j. Indeed, $\psi_1, \ldots, \psi_p$ comprise an orthonormal basis for the space of p-vectors, and

$$K = \sum_{j=1}^{p} \theta_j \psi_j \psi_j^T. \tag{8.3}$$

Hence, for each j,

$$K\psi_j = \theta_j \psi_j \quad \text{for} \quad 1 \le j \le p. \tag{8.4}$$

In particular,

$$\text{for } 1 \le j \le p, \psi_j \text{ is an eigenvector of } K \text{ with eigenvalue } \theta_j \ge 0. \tag{8.5}$$

Results (8.3) and (8.5) can be interpreted as consequences of a finite-dimensional form of *Mercer's Theorem*, which states that if K is a symmetric and non-negative kernel, meaning that K is a symmetric function of two variables and that

$$\sum_{j=1}^{p} \sum_{k=1}^{p} K(j,k) c(j) c(k) \ge 0$$

for any sequence of real numbers $c(j)$, then there exists an orthonormal basis $\psi_1, \ldots, \psi_p$ of eigenvectors of the transformation K defined by

$$(K\psi)(k) = \sum_{j=1}^{p} K(j,k) \psi(j),$$

with respective eigenvalues $\theta_j \ge 0$, such that the eigenvalue sequence is nonnegative and (8.3) holds. (We write $K(j,k)$ for the (j,k)th component of the matrix K, and $\psi(j)$ for the jth component of the vector ψ.)

The *Karhunen–Loève* expansion of the random vector X is simply its representation in terms of the principal component basis ψ_j:

$$X(\ell) = \sum_{j=1}^{p} \psi_j(\ell) \sum_{k=1}^{p} \psi_j(k) X(k). \tag{8.6}$$

Equivalently, (8.6) is the unique representation of X as a linear combination of the principal component scores.

Equation (8.3) suggests estimators of the ψ_j's and θ_j's, as follows. Suppose we have independent p-vectors $X_1, \ldots, X_n$, each distributed as X, and let us define $\mathbf{X} = n^{-1} \sum_i X_i$ and take

$$\widehat{K} = \frac{1}{n} \sum_{i=1}^{n} (X_i - \mathbf{X})(X_i - \mathbf{X})^{\mathrm{T}}$$

to be our estimator of K. (Thus, $\widehat{K}$ is a $p \times p$ matrix.) Mercer's Theorem implies that, analogously to (8.3) and (8.5), we may write

$$\widehat{K} = \sum_{j=1}^{p} \hat{\theta}_j \hat{\psi}_j \hat{\psi}_j^{\mathrm{T}} \tag{8.7}$$

where

$$\text{for } 1 \leq j \leq p, \hat{\psi}_j \text{ is an eigenvector of } \widehat{K} \text{ with eigenvalue } \hat{\theta}_j \geq 0. \tag{8.8}$$

We can of course order the eigenvalues $\hat{\theta}_j$, obtaining, if we rearrange the indices appropriately,

$$\hat{\theta}_1 \geq \hat{\theta}_2 \geq \cdots \geq \hat{\theta}_p \geq 0. \tag{8.9}$$

In view of (8.2) and (8.9) we should interpret $\hat{\theta}_j$ as our estimator of θ_j for $1 \leq j \leq p$. Then, comparing (8.3) and (8.7), we see that it is appropriate to interpret $\hat{\psi}_j$ as our estimator of ψ_j. These are the estimators usually employed in PCA.

The resulting estimator of $\theta_j^{-1/2} \psi_j^T X$, a principal component score, when X is replaced by the data value X_i and the data are centred at $\mathbf{X}$, is $\hat{\theta}_j^{-1/2} \hat{\psi}_j^T (X_i - \mathbf{X})$. Since the eigenvalue and eigenvector estimators $\hat{\theta}_j$ and $\hat{\psi}_j$ depend on all the data $X_1, \ldots, X_n$, the estimated scores are not exactly uncorrelated, although they are up to terms of order n^{-1}.

8.2.2 Infinite-dimensional case

In principle the case where X is a random function rather than a random vector does not differ from the finite-dimensional one discussed in Section 8.2.1. Of course, a random function generally needs an infinite basis for representation, and so the series in (8.3), (8.6) and (8.7), and the sequences in (8.2), (8.4), (8.5), (8.8) and (8.9), are now required to be infinite. This means that convergence issues sometimes need to be checked, although for the most part they follow directly from a square-integrability assumption, given at (8.10) below. Modulo this issue, and an ambiguity in the definitions of $\hat{\psi}_j$ for $j \geq n+1$ (this point will be discussed later in the present section), the case of random functional data is very similar to that of data in the form of random vectors.

Specifically, let X be a random function supported on a compact interval $\mathcal{I}$ and satisfying

$$\int_{\mathcal{I}} E(X^2) < \infty. \tag{8.10}$$

The covariance function of X admits a spectral decomposition analogous to that at (8.3):

$$K(s,t) \equiv \text{cov}\{X(s), X(t)\} = \sum_{j=1}^{\infty} \theta_j \psi_j(s) \psi_j(t), \tag{8.11}$$

where, in consequence of (8.10), the expansion converges in L_2 on $\mathcal{I}^2$. Analogously to (8.2) we index the summands in (8.11) so that

$$\theta_1 \geq \theta_2 \geq \cdots \geq 0 \tag{8.12}$$

are the eigenvalues, with respect orthonormal eigenvectors (in fact, eigenfunctions) ψ_j of the linear operator with kernel K.

As in Section 8.2.1, and in particular analogously to (8.5), the sequence of functions $\psi_1, \psi_2, \ldots$ comprises a complete orthonomal basis for the space of square-integrable functions on $\mathcal{I}$. The function ψ_k can be defined inductively as the vector that minimizes, subject to $\|\psi_k\| = 1$, the distance criterion

$$d(k) = E \left\| X - \sum_{j=1}^{k} \psi_j \int_{\mathcal{I}} \psi_j X \right\|^2 \tag{8.13}$$

(compare (8.1)), given that $\psi_1, \ldots, \psi_{k-1}$ have already been determined. In the functional case we interpret $\|x\|$ as the square root of $\int_{\mathcal{I}} x(t)^2 \, dt$.

By analogy with (8.6), a Karhunen–Loève expansion of the random function X is its representation in the basis ψ_j, the main changes being that the sum over j in (8.6) is now an infinite series, and the sum over k, which represented an inner product, is replaced by an integral:

$$X(s) = \sum_{j=1}^{\infty} \psi_j(s) \int_{\mathcal{I}} \psi_j(t) X(t) \, dt. \tag{8.14}$$

(In this paragraph it is convenient to assume that $E(X) = 0$.) Again in view of (8.10), the infinite series in (8.14) converges in L_2. The expansion (8.14) represents X in terms of the principal component scores, which in the present infinite-dimensional setting are $\theta_j^{-1/2} \int_{\mathcal{I}} \psi_j X$ for $j \geq 1$.

The method for estimating θ_j and ψ_j is identical to that in Section 8.2.1: we construct an empirical version of (8.11) and, referring directly to that analogy, we determine estimators $\hat{\theta}_j$ and $\hat{\psi}_j$. In particular, our estimator of $K(s, t)$, computed from independent data $X_1, \ldots, X_n$ identically distributed as the random function X, is

$$\widehat{K}(s, t) = \frac{1}{n} \sum_{i=1}^{n} \{X_i(s) - \overline{X}(s)\} \{X_i(t) - \overline{X}(t)\}, \tag{8.15}$$

where again $\overline{X} = n^{-1} \sum_i X_i$. Positive definiteness and symmetry of the linear operator with kernel $\widehat{K}$, and the version of Mercer's Theorem in the functional-data setting, imply the existence of eigenvalues

$$\hat{\theta}_1 \geq \hat{\theta}_2 \geq \cdots \geq 0 \tag{8.16}$$

(compare (8.12)) and orthonormal eigenvectors (i.e. eigenfunctions) $\hat{\psi}_j$, such that

$$\widehat{K}(s, t) = \sum_{j=1}^{\infty} \hat{\theta}_j \hat{\psi}_j(s) \hat{\psi}_j(t), \tag{8.17}$$

where, just as in (8.11), the expansion converges in L_2 on $\mathcal{I}^2$. Property (8.16) amounts to a decision to order the indices of the estimators $\hat\theta_j$ so that these estimators are listed in decreasing order. This relabelling carries over concomitantly to the associated eigenfunctions $\hat\psi_j$.

Although (8.16) might be interpreted as suggesting that the sequence $\hat\theta_j$ could contain infinitely many nonzero values, in fact $\hat\theta_j = 0$ for all $j \geq n+1$. This is a consequence of the fact that $\hat\psi_1, \ldots, \hat\psi_n$ span the space of functions $X_1, \ldots, X_n$, and so if a function ψ is orthogonal to each of $\hat\psi_1, \ldots, \hat\psi_n$ then it is also orthogonal to $X_1, \ldots, X_n$, and therefore, in view of the definition of $\widehat K$ at (8.15),

$$\int_\mathcal{I} \widehat K(s,t)\,\psi(t)\,dt = 0. \tag{8.18}$$

In particular, (8.18) holds if we take $\psi = \hat\psi_k$ for any $k \geq n+1$. Therefore (8.17) and (8.18) imply that, for $k \geq n+1$ and for each $s \in \mathcal{I}$,

$$0 = \int_\mathcal{I} \left\{ \sum_{j=1}^\infty \hat\theta_j\,\hat\psi_j(s)\,\hat\psi_j(t) \right\} \hat\psi_k(t)\,dt = \sum_{j=1}^\infty \hat\theta_j\,\hat\psi_j(s)\,\delta_{jk} = \hat\theta_k\,\hat\psi_k(s), \tag{8.19}$$

where δ_{jk} denotes the Kronecker delta. Since $\hat\psi_k$ satisfies $\int_\mathcal{I} \hat\psi_k^2 = 1$ then it cannot vanish identically, and therefore (8.19) implies that $\hat\theta_k = 0$ for each $k \geq 1$. Consequently, (8.16) can be written more informatively as

$$\hat\theta_1 \geq \cdots \geq \hat\theta_n \geq \hat\theta_{n+1} = \hat\theta_{n+2} = \cdots = 0. \tag{8.20}$$

This property has implications for the choice of the functions $\hat\psi_j$ when $j \geq n+1$. In particular, (8.17) and (8.20) imply that these functions $\hat\psi_j$ are not determined in any way by $\widehat K$, except of course that they should be orthogonal to $\hat\psi_1, \ldots, \hat\psi_n$, and should be such that the infinite orthonormal sequence $\hat\psi_1, \hat\psi_2, \ldots$ is complete in the class of square-integrable functions. The functions $\hat\psi_j$, for $j \geq n+1$, can be chosen arbitrarily subject to these properties. However, if $\hat\theta_n \neq 0$ then $\hat\psi_1, \ldots, \hat\psi_n$ are completely determined, except for their signs.

8.3 APPLICATIONS, AND DIFFERING VIEWPOINTS, OF PCA

8.3.1 Dimension reduction

Arguably the main purposes to which principal component analysis is put are dimension reduction and dimension ordering. The definition of the functions $\hat\psi_1, \hat\psi_2, \ldots$ as successive minimizers of the low-dimensional "approximation" at

(8.13) provides some justification for these applications. The definition characterizes ψ_j as the vector such that $\int_\mathcal{I} \psi_j X$ explains the greatest possible amount of variation of X, beyond the level of variation explained using previous ψ_k's; see Section 8.2.1. Therefore the sequence $\psi_1, \psi_2, \ldots$ can be interpreted as being arranged in order of the amount of variability that successive components capture, and from that viewpoint the set of elements in the sequence, and their order there, are canonical.

There is a vast variety of examples of the application of this idea. It has its roots in high-, but finite-, dimensional problems discussed more than half a century ago. In particular, Kendall (1957) recommended the use of principal component analysis to select and order variables in multivariate analysis. Note, however, that the nature of "variable selection" here is quite different from that in much current usage, for example in high-dimensional data problems. Even in applications to multivariate analysis, let alone to functional data analysis, the variables in question are not vector components; instead they are linear forms in the data vectors (in multivariate analysis) or in the data functions (in the case of functional data).

Prediction problems comprise a popular area for applying this variable selection approach, for example prediction in functional linear regression. In this example, on which most of the discussion in the remainder of Section 8.3 will be based, it is assumed that we observe data pairs (X_i, Y_i) for $1 \leq i \leq n$, and that they are generated by the model

$$Y_i = \alpha + \int_\mathcal{I} \beta X_i + \text{error}, \tag{8.21}$$

where X_i denotes a random function, Y_i and the error term are scalars, and the "parameters" of interest are α, the scalar intercept, and β, the slope function. See Chapter 2 of this volume for an overview of the linear functional regression model.

If we express X_i in terms of its principal component basis, obtaining

$$X_i = \sum_{j=1}^{\infty} \psi_j \int_\mathcal{I} X_i \psi_j;$$

and if we write $\beta = \sum_{j \geq 1} \beta_j \psi_j$ in the same manner, where $\beta_j = \int_\mathcal{I} \beta \psi_j$; then we can depict the model (8.21) equivalently as

$$Y_i = \alpha + \sum_{j=1}^{\infty} \beta_j \int_\mathcal{I} X_i \psi_j + \text{error}. \tag{8.22}$$

The error term in (8.22) is assumed to have zero mean, conditional on the explanatory variable X_i, so that the model is one of regression:

$$\mu(X_i) \equiv E(Y_i \mid X_i) = \alpha + \sum_{j=1}^{\infty} \beta_j \int_\mathcal{I} X_i \psi_j. \tag{8.23}$$

The latter formula suggests a simple approximation, which in turn motivates methods for estimation:

$$E(Y_i \mid X_i) \approx a + \sum_{j=1}^{p} \beta_j \int_{\mathcal{I}} X_i \hat{\psi}_j, \tag{8.24}$$

where the frequency cut-off, p, is a smoothing parameter and we might choose the unknowns a and $\beta_1, \ldots, \beta_p$ by least squares. (The eigenfunction estimator $\hat{\psi}_j$, in (8.24) and below, is defined as suggested in Section 8.2.2.) In particular, these quantities might be selected to minimize

$$S(a, \beta_1, \ldots, \beta_p) = \sum_{i=1}^{n} \left(Y_i - a - \sum_{j=1}^{p} \beta_j \int_{\mathcal{I}} X_i \hat{\psi}_j \right)^2,$$

or equivalently, $\beta_1, \ldots, \beta_p$ might be estimated by choosing $(\beta_1, \ldots, \beta_p) = (\hat{\beta}_1, \ldots, \hat{\beta}_p)$ to minimize

$$S(\beta_1, \ldots, \beta_p) = \sum_{i=1}^{n} \left\{ Y_i - \bar{X} - \sum_{j=1}^{p} \beta_j \int_{\mathcal{I}} (X_i - \bar{X}) \hat{\psi}_j \right\}^2 \tag{8.25}$$

and then defining $\hat{a} = \bar{X} + \sum_{j \leq p} \hat{\beta}_j \int_{\mathcal{I}} (X_i - \bar{X}) \hat{\psi}_j$. The resulting estimator of the function β is $\hat{\beta} = \sum_{j \leq p} \hat{\beta}_j \hat{\psi}_j$.

The least squares estimators $\hat{\beta}_j$ defined by minimizing the sum of squares at (8.25) are given equivalently by

$$\sum_{j=1}^{p} \hat{\beta}_j \hat{\sigma}_{jk} = \hat{c}_k, \quad 1 \leq k \leq p,$$

where

$$\hat{\sigma}_{jk} = \int_{\mathcal{I}} \int_{\mathcal{I}} \widehat{K}(s, t) \hat{\psi}_j(s) \hat{\psi}_k(t) \, ds \, dt = \hat{\theta}_j,$$

$$\hat{c}_j = \frac{1}{n} \sum_{i=1}^{n} (Y_i - \bar{Y}) \int_{\mathcal{I}} (X_i - \bar{X}) \hat{\psi}_j \tag{8.26}$$

and $\bar{Y} = n^{-1} \sum_i Y_i$. Therefore

$$\hat{\beta}_j = \hat{\theta}_j^{-1} \hat{c}_j. \tag{8.27}$$

Our estimator of the regression mean, $\mu(x) = E(Y \mid X = s)$ (see (8.23)), is

$$\hat{\mu}(x) = \bar{Y} + \sum_{j=1}^{p} \hat{\beta}_j \int_{\mathcal{I}} (x - \bar{X}) \hat{\psi}_j = \bar{Y} + \sum_{j=1}^{p} \hat{\theta}_j^{-1} \hat{c}_j \int_{\mathcal{I}} (x - \bar{X}) \hat{\psi}_j. \tag{8.28}$$

The smoothing parameter p can be chosen by cross-validation, as follows. Let $\mathcal{Z} = \{(X_1, Y_1), \ldots, (X_n, Y_n)\}$ denote the full dataset, define $\hat{\beta}_j$ as at (8.27), and compute the version $\hat{\mu}_{-i}$ of $\hat{\mu}$, at (8.28), from the $(n-1)$-sample $\mathcal{Z} \setminus \{(X_i, Y_i)\}$. Then calculate the cross-validatory criterion

$$\mathrm{CV}(p) = \frac{1}{n} \sum_{i=1}^{n} \{Y_i - \hat{\mu}_{-i}(X_i \mid p)\}^2, \qquad (8.29)$$

and, finally, take p to minimize $\mathrm{CV}(p)$. This approach can be interpreted, in approximate terms, as minimizing the conditional mean summed squared error,

$$\mathrm{MSSE}(p) = \frac{1}{n} \sum_{i=1}^{n} E\left[\{\mu(X_i) - \hat{\mu}(X_i \mid p)\}^2 \mid \mathcal{X}\right]. \qquad (8.30)$$

An alternative, computationally more straightforward approach is to use leave-one-out methods only to compute $\hat{c}_j$ at (8.26), and $\bar{Y}$, but leave $\hat{\psi}_j$ and $\hat{\theta}_j$ unchanged. In particular, if

$$\bar{Y}_{-i} = n^{-1} \sum_{k: k \neq i} Y_k$$

is the leave-one-out version of $\bar{Y}$ but with divisor n, rather than $n-1$, and if we define

$$\hat{c}_{j,-i} = \frac{1}{n} \sum_{k: k \neq i} (Y_k - \bar{Y}_{-i}) \int_{\mathcal{I}} (X_i - \mathbf{X}) \hat{\psi}_j$$

and redefine

$$\hat{\mu}_{-i}(x) = \bar{Y}_{-i} + \sum_{j=1}^{p} \hat{\theta}_j^{-1} \hat{c}_{j,-i} \int_{\mathcal{I}} (x - \mathbf{X}) \hat{\psi}_j,$$

then we can continue to take p to minimize $\mathrm{CV}(p)$, at (8.29), and interpret it as an empirical attempt at minimizing $\mathrm{MSSE}(p)$, at (8.30). This cross-validation algorithm motivates others of the same type.

8.3.2 Interpreting PCA-based choice of dimension

In problems where interest focuses on the function X alone, there is perhaps not a great deal to challenge when interpreting PCA as a device for choosing, or ordering, variables. Indeed, the estimators $\hat{\psi}_1, \hat{\psi}_2, \ldots$ of the functions $\psi_1, \psi_2, \ldots$ are commonly considered to be ways of representing successively more complex, and higher-order, influences on the shapes of realizations of X. Perhaps the main drawback of this viewpoint is the rather abstract nature of the basis functions, defined as they are in terms of the refined mathematical distance $d(k)$ at (8.13).

Alternatives, which arguably have easier interpretation in terms of realizations, have been suggested by, for example, Cox (1968), Mosteller and Tukey (1977) and Hall et al. (2007).

However, in cases where there are further variables, for example a response Y as in the regression model at (8.21), a principal-component-based ordering of variables can be criticized because it does not take into account the relationship between the X's and the other information in the dataset. For example, the ordering of components in the definition of the estimator $\hat{\mu}(x)$, at (8.28), of the regression mean $\mu(x) = E(Y \mid X = s)$, takes into account only the explanatory functions X_i, not the responses Y_i. It is perhaps also reasonable to consider accommodating the regressand, x, when considering how we might order the components when the task is to predict a future Y, rather than to make inferences about the distribution of X. One might ask too whether we should choose a different orthonormal sequence, alternative to $\hat{\psi}_1, \hat{\psi}_2, \ldots$, taking into account both the response variables and the regressand. The approach suggested in Section 8.3.1 does none of those things.

Cox (1968, p. 272) has argued that x should be taken into account, and in particular has suggested that:

A difficulty [with the approach discussed in Section 8.3.1] seems to be that there is no logical reason why the dependent variable should not be closely tied to the least important principal component.

To express in mathematical terms some of the implications of Cox's remarks, suppose that we are in the fortunate position of actually knowing the principal component basis $\psi_1, \psi_2, \ldots$ In this case our regression estimator would have the form

$$\tilde{\mu}(x) = \bar{Y} + \sum_{j=1}^{p} \tilde{\beta}_j \int_{\mathcal{I}} (x - \bar{X}) \psi_j \tag{8.31}$$

(compare (8.28)), where $(\beta_1, \ldots, \beta_p) = (\tilde{\beta}_1, \ldots, \tilde{\beta}_p)$ minimizes

$$\sum_{i=1}^{n} \left\{ Y_i - \bar{X} - \sum_{j=1}^{p} \beta_j \int_{\mathcal{I}} (X_i - \bar{X}) \psi_j \right\}^2$$

(compare (8.25). If x is virtually orthogonal to $\psi_1, \ldots, \psi_p$, but depends substantially on $\psi_{p+1}, \psi_{p+2}, \ldots$, then, by including the first p components in the definition of $\tilde{\mu}(x)$ at (8.31), we may be doing little more than adding noise to the estimator; but by excluding higher-order terms we are ensuring relatively poor performance of the estimator $\hat{\mu}(x)$. The cross-validation algorithm in Section 8.3.1 does little to help, for example because it addresses the choice of p for an average value of x, not specifically for an x that might be virtually orthogonal to $\psi_1, \ldots, \psi_p$. However, ψ_j's that are orthogonal to x do not actually make a contribution to ψ_j, over and above their role in estimating the intercept term, a, in the model at (8.22). This should ameliorate our concern about the order of the ψ_j's.

More to the point, we defined our estimator $\tilde{\mu}(x)$ (and also its more noisy version $\hat{\mu}(x)$) in such a manner that, if it is to include components corresponding to ψ_j for $j \geq p+1$ (specifically, the components on which our contrary x depends), then it must also include the terms corresponding to $\psi_1, \ldots, \psi_p$, which are not needed to the same extent. Cox (1968) argues, in part, that this is not necessarily helpful; principal components ψ_j for $j \geq p+1$, which might be less important than $\psi_1, \ldots, \psi_p$ in terms of representing a typical X, could be more important for representing $E(Y \mid X = x)$ for our particular realization x of X.

Mosteller and Tukey (1977, p. 397) observe, however, that this view is pessimistic since it is relatively unlikely that a realization of X will be so different from a typical X. They feel that it is unlikely that nature conspires to make the problem as challenging as perhaps was suggested by Cox:

A malicious person who knew our xs and our plan for them could always invent a y to make our choices look horrible. But we don't believe nature works that way — more nearly that nature is, as Einstein put it (in German), "tricky, but not downright mean." And so we offer a technique that frequently helps ...

Cox (1968) had suggested an alternative approach than principal component ordering, in which "simple combinations, not the principal components, can be used as regressor variables." Mosteller and Tukey (1977) lend support to this idea, without citing Cox's (1968) paper, by arguing that a statistician could "choose new linear combinations of the few largest principal components so as to make them more interpretable." Nevertheless, Mosteller and Tukey express a degree of support for ordering components in the eigenvalue order of principal component analysis, at least to the extent that "separating big components from little components can be effective."

These issues, and others, are discussed in more detail by Cook (2007), by the discussants Christensen (2007), B. Li (2007) and L. Li and Nachtsheim (2007) of Cook (2007), and by Hall and Yang (2010). Methods based on the eigenvalue ordering of principal components have long had supporters, apparently starting with Kendall (1957) and including, in the 1960s and 1970s, Spurrell (1963) and Hocking (1976).

Support for variable selection based on eigenvalue ordering can be provided in terms of a theoretical, but non-asymptotic, minimax property given in Theorem 8.1 below. This asserts that if we have only rather basic prior knowledge of the slope function β then, in a linear prediction property based on functional data, we cannot do much better than found our estimator on the basis $\hat{\psi}_1, \hat{\psi}_2, \ldots$, arranged in that order. Specifically, assume that $1 \leq p \leq n$ and take $\chi_1, \ldots, \chi_p$ to be a subsequence of $\hat{\psi}_1, \ldots, \hat{\psi}_n$. (The functions $\chi_{p+1}, \chi_{p+2}, \ldots$ can be defined arbitrarily, subject to $\chi_1, \chi_2, \ldots$ forming a complete orthonormal basis.) In particular, $\chi_j = \hat{\psi}_{r_j}$, say, for $1 \leq j \leq p$, where $r_1, \ldots, r_p$ are distinct members of the set $\{1, \ldots, n\}$. Assume that the error term in (8.22) has zero mean conditional on X, and finite variance, and define (conditional) mean squared error by:

$$\text{MSE} = E\left[\{\hat{\mu}(x) - \mu(x)\}^2 \mid \mathcal{X}\right].$$

The following theorem is discussed and proved by Hall and Yang (2010).

Theorem 8.1 *If the only information we have about β is that its norm, $\|\beta\| = (\int_{\mathcal{I}} \beta^2)^{1/2}$, equals a given constant $C > 0$ (or alternatively, that $\|\beta\| \leq C$), then the choice of $r_1, \ldots, r_p$ that produces, for each C and p, the least mean squared error when this quantity assumes the largest value that it is permitted for the class of slopes β satisfying $\|\beta\| = C$ (or, respectively, $\|\beta\| \leq C$), is $r_j = j$ for each j. This is identical to the choice that orders the eigenvectors $\hat{\psi}_j$ canonically by insisting on a decreasing ranking of eigenvalues, as at (8.9).*

If we have additional information about β, beyond that admitted in Theorem 8.1, then we may be able to reduce mean squared error by reordering the basis. For example, suppose the problem is one of linear regression for functional data; that the interval $\mathcal{I}$ on which the random functions X and X_i are defined is the interval $[-1, 1]$; and that we know that β is an even function on $\mathcal{I}$. Then we can ignore any basis functions ψ_j that are odd functions, since their contribution to accurate prediction will be zero. Equivalently, when we rank the functions ψ_j, as a prelude to defining the estimator $\tilde{\mu}(x)$ at (8.31), we should rank last all ψ_j's that are odd functions.

However, this toy example is one where the additional information comes from outside the dataset, and for that reason, among others, it is not particularly realistic. In practice it can be particularly difficult to obtain performance increases in cases where we must rely solely on the data, specifically the pairs $(X_i, Y_i), \ldots, (X_n, Y_n)$, to provide all the information we may use to construct a predictor which departs from one based on variable selection through eigenvalue ordering.

To expand on this point we briefly consider alternatives to the predictor $\hat{\mu}$, defined at (8.28). They include a weighted predictor, where we keep all components but downweight those that correspond to relatively small eigenvalue estimates, and an explicitly ranked predictor, where we order components to reflect an empirical measure of their pairwise correlation with Y. Specifically, as alternatives to $\hat{\mu}(x)$ at (8.28) we can consider:

$$\hat{\mu}_{\text{weight}}(x) = \bar{Y} + \sum_{j=1}^{p} \hat{\beta}_j \, \hat{\theta}_j^r \int_{\mathcal{I}} (x - \bar{X}) \hat{\psi}_j, \tag{8.32}$$

$$\hat{\mu}_{\text{rank}}(x) = \bar{Y} + \sum_{j=1}^{p} \hat{\beta}_{k_j} \int_{\mathcal{I}} (x - \bar{X}) \hat{\psi}_{k_j}. \tag{8.33}$$

To calculate $\hat{\mu}_{\text{rank}}(x)$ we compute, for each j, the set of pairs

$$\left(\int_{\mathcal{I}} X_i(t)\hat{\psi}_j(t)\,dt,\ Y_i\right)$$

for $1 \leq i \leq n$; next we derive the correlation coefficient, $\hat{\rho}_j$, for these n pairs; and then we take $\hat{k}_1, \ldots, \hat{k}_n$ to be the permutation of $1, \ldots, n$ which ensures that $|\hat{\rho}_{\hat{k}_1}| \geq \cdots \geq |\hat{\rho}_{\hat{k}_n}|$. The respective values of $\hat{\beta}$ for the predictors $\hat{\mu}_{\text{weight}}$ and $\hat{\mu}_{\text{rank}}$ are $\hat{\beta} = \sum_{j\geq 1} \hat{\beta}_j \hat{\theta}_j^r \hat{\psi}_j$ and $\hat{\beta} = \sum_{1\leq j\leq p} \hat{\beta}_{\hat{k}_j} \hat{\psi}_{\hat{k}_j}$, and the respective tuning parameters are r, a general positive number, and p, a general positive integer.

In numerical practice, however, the predictors at (8.32) and (8.33) seldom perform better than the simpler $\hat{\mu}$ at (8.28). This is due to the additional noise introduced by including the eigenvalue estimators $\hat{\theta}_j$, and the correlation estimators $\hat{\rho}_j$, when constructing $\hat{\mu}_{\text{weight}}$ and $\hat{\mu}_{\text{rank}}$. This difficulty is particularly apparent for small to moderate sample sizes, even when n is as large as 200. Admittedly the estimator $\hat{\mu}_{\text{weight}}$ reweights components in only a rudimentary way, and a method that used two or more smoothing parameters to determine the weights, rather than the single parameter r, might be attractive. However, using a more sophisticated approach generally only adds further noise to the predictor, and so reduces performance to an even greater extent. Further details are given by Hall and Yang (2010).

8.3.3 Weighted least squares in functional linear regression

Although the methods suggested at (8.32) and (8.33) generally do not provide an improvement over more conventional predictors, in particular $\hat{\mu}$ at (8.28), the use of weighted rather than ordinary least squares when computing the estimators $\hat{\beta}_j$, for example when constructing the criterion $S(\beta_1, \ldots, \beta_p)$ at (8.25), can prove advantageous.

It is common in parametric regression (see, for example, Carroll and Ruppert 1988) to model variance as a function of the assumed parametric form of $E(Y|X)$. Likewise, for functional data we could model $\sigma(X)^2 = \text{var}(\epsilon \mid X)$ as

$$\sigma(X)^2 = g\left(\alpha + \sum_{j=1}^{p} \beta_j S_j\right),$$

where $\alpha, \beta_1, \ldots, \beta_p$ are as in the model at (8.22), $S_j = \theta_j^{-1/2} \int_{\mathcal{I}} (X - EX)\psi_j$ is the jth principal component score (see Section 8.2.2), and g is a univariate function. We recommend modelling g parametrically, in order not to introduce too much extra noise into the problem, and to this end the model

$$g(u) = |u|^c, \tag{8.34}$$

where $c > 0$, is attractive. In parametric problems it is more common to take $g(u) = c_1 |u|^{c_2}$, where c_1 and c_2 are positive constants (see Carroll and Ruppert 1988), but numerical experiments suggest that this approach also produces too high a level of stochastic variability in the context of functional linear regression.

We estimate c, in (8.34), by first computing unweighted least squares "pilot estimators" $\hat{\beta}_j$, as in Section 8.3.1, and then calculating the residuals

$$\hat{\epsilon}_i = Y_i - \bar{Y} - \sum_{j=1}^{p} \hat{\beta}_j \int_{\mathcal{I}} (X_i - \bar{X}) \hat{\psi}_j \, .$$

Next we fit the model

$$\sigma(X)^2 = \left| \alpha + \sum_{j=1}^{p} \beta_j S_j \right|^c \tag{8.35}$$

in another least squares step, choosing $c = \hat{c}$ to minimize

$$\sum_{i=1}^{n} \left\{ \hat{\epsilon}_i^2 - \left| \bar{Y} + \sum_{j=1}^{p} \hat{\beta}_j \int_{\mathcal{I}} (X_i - \bar{X}) \hat{\psi}_j \right|^c \right\}^2 .$$

Our estimator of $\text{var}(Y - \alpha - \int_{\mathcal{I}} \beta X \mid X = x)$ is, when $x = X_i$,

$$\hat{w}(X_i)^{-1} = \left| \bar{Y} + \sum_{j=1}^{p} \hat{\beta}_j \int_{\mathcal{I}} (X_i - \bar{X}) \hat{\psi}_j \right|^{\hat{c}} .$$

Using these weights we modify the least squares problem originally given by the objective function at (8.25), obtaining the new objective function:

$$\sum_{i=1}^{n} \left\{ Y_i - \bar{Y}_w - \sum_{j=1}^{p} \beta_j \int_{\mathcal{I}} (X_i - \bar{X}_w) \hat{\psi}_j \right\}^2 \hat{w}(X_i), \tag{8.36}$$

where

$$\bar{X}_w = \left\{ \sum_{i=1}^{n} \hat{w}(X_i) \right\}^{-1} \sum_{i=1}^{n} \hat{w}(X_i) X_i ,$$

$$\bar{Y}_w = \left\{ \sum_{i=1}^{n} \hat{w}(X_i) \right\}^{-1} \sum_{i=1}^{n} \hat{w}(X_i) Y_i ;$$

and we choose $\check{\beta}_1, \ldots, \check{\beta}_p$ to minimize the objective function at (8.36). Our new estimator of $\mu(x)$ is thus:

$$\tilde{\mu}(x) = \bar{Y}_w + \sum_{j=1}^{t} \check{\beta}_j \int_{\mathcal{I}} (x - \mathbf{X}_w) \hat{\psi}_j .$$

To reduce notational complexity in the discussion above we have used the same notation p (i.e. the same frequency cut-off) when calculating the pilot estimators $\hat{\beta}_j$, and when computing the final estimators $\check{\beta}_j$. However, in practice one would select a particular value of p for use with the pilot estimators, for example employing cross-validation; then keep that p fixed at its empirical value when computing the objective function at (8.36); and then choose $\check{\beta}_1, \ldots, \check{\beta}_p$ to minimize that function. We could use cross-validation to select the latter p, bearing in mind that the different, earlier value of p was based on an empirical calculation. Then, strictly speaking, data pairs would have to be successively omitted from the calculation of the first p when using a leave-one-out approach to selecting the second p.

Further details of weighted least squares in the context of functional data analysis may be obtained from Delaigle et al. (2009).

8.4 ROLE OF PRINCIPAL COMPONENTS IN THE ASSESSMENT OF DENSITY FOR FUNCTIONAL DATA

8.4.1 The notion of density for functional data

The notion of probability density for a random function is being used increasingly in functional data analysis. For example, nonparametric or structure-free methods for estimating functionals, using data in the form of functions, often involve discussion of the concept of density because they tend to be based on estimators of Nadaraya–Watson type, which require division by an estimator of a small-ball probability. See, for example, Ferraty et al. (2002a,b, 2007a,b), Ferraty and Vieu (2002, 2003, 2004, 2006a,b), Niang (2002), and Chapter 4 of this volume for an overview of kernel estimators in nonparametric regression. The concept of density also underpins discussion of the mode of the distribution of a random function, addressed for example by Gasser et al. (1998), Hall and Heckman (2002) and Dabo-Niang et al. (2004a,b, 2006).

However, a density function generally does not exist for functional data. We shall elaborate on this point in Section 8.4.2, but for the present we give formulae which will lead us to that conclusion, and we point to the close relationship between the notion of "density" for the distribution of a random function, and principal components.

Let X and x be, respectively, a random and a fixed function defined on $\mathcal{I}$, and let their principal component scores be $S_j = \theta_j^{-1/2} \int_\mathcal{I} X \psi_j$ and $s_j = \theta_j^{-1/2} \int_\mathcal{I} x \psi_j$. For simplicity we assume that $E(X) = 0$, although if this condition does not hold then the mean of X can be incorporated into s_j by replacing X and x by $X - E(X)$ and $x - E(X)$, respectively. Let $\theta_1, \theta_2, \ldots$ denote the eigenvalue sequence in the spectral decomposition of the covariance function, K, at (8.11), and assume that the sequence is, as usual, decreasing; see (8.2).

The assumptions above are all conventional; we used them frequently earlier in this chapter. Now, however, for the first time we impose detailed conditions on the distributions of the principal component scores, which we assume to be absolutely continuous. We write f_j for the probability density function of S_j. Recall that, by definition of the scores, the S_j's are uncorrelated and have mean zero and variance 1; in this work we also assume that they are independent.

Put $W_j = S_j - s_j$ and note that

$$\|X - x\|^2 = \int_\mathcal{I} (X - x)^2 = \sum_{j=1}^\infty \theta_j W_j^2.$$

Define the small-ball probability

$$p(x \mid h) = P(\|X - x\| \leq h) = P\left(\sum_{j=1}^\infty \theta_j W_j^2 \leq h^2\right),$$

where $h > 0$ will be taken small. Under regularity conditions, which are given by Delaigle and Hall (2010) and include the assumption that the scores are independent, it can be proved that

$$p(x \mid h) = \frac{(h \pi^{1/2})^r}{\Gamma(\frac{1}{2}r + 1)} \left\{\prod_{j=1}^r \theta_j^{-1/2} f_j(s_j)\right\} \exp\{o(r)\}, \tag{8.37}$$

where $r = r(h)$ denotes the "effective dimension" for a given value of scale, h, and is either a value of r that is sufficiently close to h^2, or, if no such r exists, the unique r such that $\theta_{r+1} < h^2 < \theta_r$. In particular, r is completely determined by the eigenvalue sequence and by h; it does not depend on the score densities f_j or on the ball centre x.

The approximation (8.37) connects the notion of "density," which is represented by the small-ball probability on the left-hand side of (8.37), with that of principal components. Indeed, we can write (8.37) equivalently as

$$\log p(x \mid h) = C(r, \theta) + r \, \ell(x \mid r) + o(r), \tag{8.38}$$

where

$$\ell(x \mid r) = \frac{1}{r} \sum_{j=1}^{r} \log f_j(s_j) \qquad (8.39)$$

and the constant $C(r, \theta)$ is defined by

$$\exp\{C(h, \theta)\} = \left(h\,\pi^{1/2}\right)^r \, \Gamma\left(\tfrac{1}{2}r + 1\right)^{-1} \prod_{j=1}^{r} \theta_j^{-1/2}. \qquad (8.40)$$

(Here $\theta = (\theta_1, \theta_2, \ldots)$ and Γ denotes the conventional gamma function.) Note that $C(r, \theta)$ depends only on the eigenvalue sequence and the value of h; it does not depend on the distributions of the principal component scores or on x.

The small-ball probability approximation at (8.38) implies that the "log-density" $\ell(x \mid r)$ captures the first-order effect that x has on $p(x \mid h)$. Note that, up to terms that are negligibly small, $\ell(x \mid r)$ is a monotone increasing function of $p(x \mid h)$. Moreover, we can arbitrarily permute the densities $f_1, \ldots, f_r$ without influencing anything other than the $o(r)$ term on the right-hand side of (8.38); that is, without having other than a negligible impact. Additionally, (8.38) makes it clear that the densities of the principal component scores are important only through their aggregate, defined by $\ell(x \mid r)$, and are of relatively little individual relevance.

8.4.2 A surrogate for density in the case of functional data

These properties of $\ell(x \mid r)$, defined at (8.39), suggest that this function might be viewed as a surrogate for density. Although $\ell(x \mid r)$ cannot, in general, be employed to compare densities for different random function distributions (because the constant defined in (8.40) represents at least some of the necessary differences in scale), it can be used as the basis for comparing density at different points x for the same random function distribution.

The distribution of a random function generally does not have a probability density in the usual sense, so the notion that $\ell(x \mid r)$ can, in an asymptotic sense, play the role of one is attractive. To appreciate why a density generally does not exist in the traditional sense, suppose there were to exist a function H of the small-ball radius h, and a function f (the density), with the property that

$$f(x) = \lim_{h \downarrow 0} H(h)^{-1}\, P(\|X - x\| \leq h)$$

is well defined. Then, taking logarithms of both sides, we would have:

$$\log P(\|X - x\| \leq h) = C_1(h) + \log f(x) + o(1), \qquad (8.41)$$

where $C_1 = \log H(h)$ does not depend on x, and $f(x)$ does not depend on h. However, except in degenerate cases where the function f is not informative, the results (8.38) and (8.41) cannot both be true.

The fact that the concept of "density" can be addressed particularly well by the log-density at (8.39) suggests a simple definition of the mode of a distribution of random functions: the modal function is $x_{\text{mode}} = \sum_j \theta_j^{1/2} m_j \psi_j$, where m_j denotes the mode of the distribution with density f_j. In a finite sample, x_{mode} can be estimated by

$$\hat{x}_{\text{mode}} = \sum_{j=1}^{p} \hat{\theta}_j^{1/2} \hat{m}_j \hat{\psi}_j ,$$

where $\hat{m}_j$ is the mode of an estimator $\hat{f}_j$ of f_j (see the next section for a definition of $\hat{f}_j$) and p is again a frequency cut-off.

8.4.3 Estimating the log-density

An attractive feature of the log-density is that it is easily estimated. Indeed, an estimator, $\hat{f}_j$, of the probability density function f_j of $S_j = \theta_j^{-1/2} \int_{\mathcal{I}} (X - EX) \psi_j$ can be computed using standard kernel methods:

$$\hat{f}_j(u) = \frac{1}{nh} \sum_{i=1}^{n} Q\left[\frac{1}{h}\left\{\frac{1}{\hat{\theta}_j^{1/2}} \int_{\mathcal{I}} (X_i - \bar{X}) \hat{\psi}_j - u\right\}\right], \qquad (8.42)$$

where h denotes a bandwidth and Q is a kernel function. The value of h can be chosen very effectively using standard methods for random data. Provided there is no tie for the eigenvalue θ_j,

$$\hat{\theta}_j \text{ and } \hat{\psi}_j \text{ are root-}n \text{ consistent for } \theta_j \text{ and } \psi_j, \text{ respectively.} \qquad (8.43)$$

(The eigenfunctions ψ_j and $\hat{\psi}_j$ are defined only up to a sign change, and it can be assumed without loss of generality that the sign of $\hat{\psi}$ is chosen so that $\int_{\mathcal{I}} \psi \hat{\psi} > 0$.)

If we take $u = \hat{s}_j = \hat{\theta}_j^{-1/2} \int_{\mathcal{I}} (x - \bar{X}) \hat{\psi}_j$ in (8.42) then the value of $\bar{X}$ cancels from the numerator inside the kernel in the definition of $\hat{f}_j(\hat{s}_j)$, and we obtain the following formula:

$$\hat{f}_j(\hat{s}_j) = \frac{1}{nh} \sum_{i=1}^{n} Q\left\{\frac{\int_{\mathcal{I}} (X_i - x) \hat{\psi}_j}{h \hat{\theta}_j^{1/2}}\right\}.$$

Using (8.43) it can be shown that $\hat{f}_j(\hat{s}_j)$ is first-order equivalent to its "ideal" counterpart,

$$\tilde{f}_j(s_j) = \frac{1}{nh} \sum_{i=1}^{n} Q\left\{\frac{\int_{\mathcal{I}} (X_i - x) \psi_j}{h \theta_j^{1/2}}\right\},$$

which we would use if we knew the true values of θ_j and ψ_j.

The first-order properties of $\bar{f}_j(s_j)$ as an estimator of $f_j(s_j)$, and hence also those of $\hat{f}_j(\hat{s}_j)$ as an estimator of the same quantity, can be worked out using standard arguments. In particular, it can be shown that $\bar{f}_j(s_j)$ has variance and bias asymptotic to $q\, f_j(s_j)/(nh)$ and $\frac{1}{2} q_2\, f''_j(s_j)\, h^2$, respectively, where $q = \int Q^2$ and $q_2 = \int u^2\, Q(u)\, du$. An estimator of the log-density $\ell(x \mid r)$, in (8.39), is given by

$$\hat{\ell}(\hat{x} \mid r) = \frac{1}{r} \sum_{j=1}^{r} \log \hat{f}_j(\hat{s}_j).$$

Here $\hat{x}$ denotes $\sum_j \hat{\theta}_j^{1/2}\, \hat{s}_j\, \hat{\psi}_j$.

References

AGUILERA, A.M., OCAÑA, F.A., VALDERRAMA, M.J. (1999a). Forecasting time series by functional PCA. Discussion of several weighted approaches. *Comput. Statist.*, **14**, 443–67.

AGUILERA, A.M., OCAÑA, F.A., VALDERRAMA, M.J. (1999b). Forecasting with unequally spaced data by a functional principal component approach. *Test*, **8**, 233–53.

BESSE, P. (1992). PCA stability and choice of dimensionality. *Statist. Probab. Lett.*, **13**, 405–10.

BESSE, P.C., CARDOT, H. (1996). Spline approximation of the forecast of a first-order autoregressive functional process. *Canad. J. Statist.*, **24**, 467–87.

BESSE, P., RAMSAY, J.O. (1986). Principal components analysis of sampled functions. *Psychometrika*, **51**, 285–311.

BOENTE, G., FRAIMAN, R. (2000). Kernel-based functional principal components. *Statist. Probab. Lett.*, **48**, 335–45.

BOSQ, D. (1989). Propriétés des opérateurs de covariance empiriques d'un processus stationnaire hilbertien. *C. R. Acad. Sci. Paris Sér. I*, **309**, 873–5.

BRUMBACK, B.A., RICE, J.A. (1998). Smoothing spline models for the analysis of nested and crossed samples of curves. *J. Amer. Statist. Assoc.*, **93**, 961–76.

CAI, T.T., HALL, P. (2006). Prediction in functional linear regression. *Ann. Statist.*, **34**, 2159–79.

CAPRA, B., MÜLLER, H.-G. (1997). An accelerated-time model for response curves. *J. Amer. Statist. Assoc.*, **92**, 72–83.

CARDOT, H. (2000). Nonparametric estimation of smoothed principal components analysis of sampled noisy functions. *J. Nonparam. Statist.*, **12**, 503–38.

CARDOT, H. (2007). Conditional functional principal components analysis. *Scand. J. Statist.*, **34**, 317–35.

CARDOT, H., FERRATY, F., SARDA, P. (1999). Functional linear model. *Statist. Probab. Lett.*, **45**, 11–22.

CARDOT, H., FERRATY, F., SARDA, P. (2000). Étude asymptotique d'un estimateur spline hybride pour le modèle linéaire fonctionnel. *C. R. Acad. Sci. Paris Sér. I*, **330**, 501–4.

CARDOT, H., FERRATY, F., SARDA, P. (2003a). Spline estimators for the functional linear model. *Statist. Sinica*, **13**, 571–91.

CARDOT, H., FERRATY, F., MAS, A., SARDA, P. (2003b). Testing hypotheses in the functional linear model. *Scand. J. Statist.*, **30**, 241–55.
CARDOT, H., SARDA, P. (2005). Estimation in generalized linear models for functional data via penalized likelihood. *J. Multivariate Anal.*, **92**, 24–41.
CARDOT, H., SARDA, P. (2006). Linear regression models for functional data. In *The Art of Semiparametrics* (W. Härdle, ed.), 49–66. Physica/Springer, Heidelberg.
CARROLL, R.J. and RUPPERT, D. (1988). *Transformation and Weighting in Regression*. Chapman and Hall, New York.
CHIOU, J.M., MÜLLER, H.-G., WANG, J.-L. (2003). Functional quasi-likelihood regression models with smooth random effects. *J. R. Statist. Soc. Ser. B*, **65**, 405–23.
CHRISTENSEN, R. (2007). Comment: Fisher Lecture: Dimension reduction in regression. *Statistical Science*, **22**, 27–31.
COOK, R.D. (2007). Fisher Lecture: Dimension reduction in regression. *Statistical Science*, **22**, 1–26.
COX, D.R. (1968). Notes on some aspects of regression analysis. *J. R. Statist. Soc. Ser. A*, **131**, 265–79.
CRAMBES, C. (2007). Régression fonctionnelle sur composantes principales pour variable explicative bruitée. *C. R. Math. Acad. Sci. Paris*, **345**, 519–22.
CUEVAS, A., FEBRERO, M., FRAIMAN, R. (2002). Linear functional regression: the case of fixed design and functional response. *Canad. J. Statist.*, **30**, 285–300.
DABO-NIANG, S., FERRATY, F., VIEU, P. (2004a). Estimation du mode dans un espace vectoriel semi-normé. *C. R. Math. Acad. Sci.*, **339**, 659–62.
DABO-NIANG, S., FERRATY, F., VIEU, P. (2004b). Nonparametric unsupervised classification of satellite wave altimeter forms. In *COMPSTAT 2004—Proceedings in Computational Statistics*, 879–86. Physica, Heidelberg.
DABO-NIANG, S., FERRATY, F., VIEU, P. (2006). Mode estimation for functional random variable and its application for curves classification. *Far East J. Theor. Stat.*, **18**, 93–119.
DAUXOIS, J., POUSSE, A., ROMAIN, Y. (1982). Asymptotic theory for the principal component analysis of a vector random function: some applications to statistical inference. *J. Multivariate Anal.*, **12**, 136–54.
DELAIGLE, A., HALL, P., APANASOVICH, T.V. (2009). Adaptive methods for prediction in the functional data linear model. Manuscript.
DELAIGLE, A., HALL, P., APANASOVICH, T.V. (2009). Weighted least squares methods for prediction in the functional data linear model. *Electronic J. statist.* **3**, 865–885.
DELAIGLE, A., HALL, P. (2010). Defining probability density for a distribution of random functions. *Annals of Statistics*, **38**, 1171–93.
ESCABIAS, M., AGUILERA, A.M., VALDERRAMA, M.J. (2005). Modeling environmental data by functional principal component logistic regression. *Environmetrics*, **16**, 95–107.
FAN, J., ZHANG, J.T. (2000). Two-step estimation of functional linear models with applications to longitudinal data. *J. R. Statist. Soc. Ser. B*, **62**, 303–22.
FARAWAY, J.J. (1997). Regression analysis for a functional response. *Technometrics*, **39**, 254–61.
FERRATY, F., GOÏA, A., VIEU, P. (2002a). Régression non-paramétrique pour des variables aléatoires fonctionnelles mélangeantes. *C. R. Math. Acad. Sci.*, **334**, 217–20.
FERRATY, F., GOÏA, A., VIEU, P. (2002b). Functional nonparametric model for time series: a fractal approach to dimension reduction. *Test*, **11**, 317–44.
FERRATY, F., GOÏA, A., VIEU, P. (2007a). On the using of modal curves for radar waveforms classification. *Comput. Statist. Data Anal.*, **51**, 4878–90.

FERRATY, F., GOÏA, A., VIEU, P. (2007b). Nonparametric functional methods: new tools for chemometric analysis. In *Statistical Methods for Biostatistics and Related Fields* (W. Härdle, Y. Mori, P. Vieu, eds) 245–64. Springer, Berlin.

FERRATY, F., VIEU, P. (2000). Fractal dimensionality and regression estimation in semi-normed vectorial spaces. *C.R. Acad. Sci. Paris Sér. I*, **330**, 139–42.

FERRATY, F., VIEU, P. (2002). The functional nonparametric model and application to spectrometric data. *Comput. Stat.*, **17**, 545–64.

FERRATY, F., VIEU, P. (2003). Curves discrimination: a nonparametric functional approach. *Comput. Stat. Data Anal.*, **44**, 161–73.

FERRATY, F., VIEU, P. (2004). Nonparametric models for functional data, with application in regression, time-series prediction and curve discrimination. *J. Nonparametr. Stat.*, **16**, 111–25.

FERRATY, F., VIEU, P. (2006a). Functional nonparametric statistics in action. In *The Art of Semiparametrics* (W. Härdle, ed.), 112–29. Physica/Springer, Heidelberg.

FERRATY, F., VIEU, P. (2006b). *Nonparametric Functional Data Analysis: Theory and Practice*. Springer, New York.

FRIEDMAN, J.H., TUKEY, J.W. (1974). A projection pursuit algorithm for exploratory data analysis. *IEEE Trans. Comput.*, **C-23**, 881–90.

GASSER, T., HALL, P., PRESNELL, B. (1998). Nonparametric estimation of the mode of a distribution of random curves. *J. R. Stat. Soc. Ser. B*, **60**, 681–91.

GERVINI, D. (2008). Robust functional estimation using the median and spherical principal components. *Biometrika*, **95**, 587–600.

GIRARD, S. (2000). A nonlinear PCA based on manifold approximation. *Comput. Statist.*, **15**, 145–67.

GRAMBSCH, P.M., RANDALL, B.L., BOSTICK, R.M., POTTER, J.D., LOUIS, T.A. (1995). Modeling the labeling index distribution: an application of functional data analysis. *J. Amer. Statist. Assoc.*, **90**, 813–21.

HALL, P., HECKMAN, N. (2002). Estimating and depicting the structure of a distribution of random functions. *Biometrika*, **89**, 145–58.

HALL, P., HOROWITZ, J.L. (2007). Methodology and convergence rates for functional linear regression. *Ann. Statist.*, **35**, 70–91.

HALL, P., HOSSEINI-NASAB, M. (2009). Theory for high-order bounds in functional principal components analysis. *Math. Proc. Cambridge Philos. Soc.*, **146**, 225–56.

HALL, P., HOSSEINI-NASAB, M. (2006). On properties of functional principal components analysis. *J. R. Stat. Soc. Ser. B*, **68**, 109–26.

HALL, P., LEE, Y.K., PARK, B.U. (2007). A method for projecting functional data onto a low-dimensional space. *J. Comput. Graph. Statist.*, **16**, 799–812.

HALL, P., MÜLLER, H.-G., WANG, J.-L. (2006). Properties of principal component methods for functional and longitudinal data analysis. *Ann. Statist.*, **34**, 1493–517.

HALL, P., YANG, Y.-J. (2010). Ordering and selecting components in multivariate or functional-data linear regression. *J. Royal Statistical Society Series B*, **72**, 93–110.

HE, G.Z., MÜLLER, H.-G., WANG, J.-L. (2003). Functional canonical analysis for square integrable stochastic processes. *J. Multivariate Anal.*, **85**, 54–77.

HOCKING, R.R. (1976). The analysis and selection of variables in linear regression. *Biometrics*, **32**, 1–49.

HUANG, J.Z., SHEN, H., BUJA, A. (2008). Functional principal components analysis via penalized rank one approximation. *Electron. J. Stat.*, **2**, 678–95.

HUANG, J.H.Z., WU, C.O., ZHOU, L. (2002). Varying-coefficient models and basis function approximations for the analysis of repeated measurements. *Biometrika*, **89**, 111–28.

IZEM, R., MARRON, J.S. (2007). Analysis of nonlinear modes of variation for functional data. *Electron. J. Stat.*, **1**, 641–76.

JAMES, G.M., HASTIE, T.J., SUGAR, C.A. (2000). Principal component models for sparse functional data. *Biometrika*, **87**, 587–602.

KENDALL, M.G. (1957). *A Course in Multivariate Analysis*. Griffin, London.

KNEIP, A., UTIKAL, K.J. (2001). Inference for density families using functional principal component analysis. *J. Amer. Statist. Assoc.*, **96**, 519–32.

LEE, S.Y., ZHANG, W.Y., SONG, X.Y. (2002). Estimating the covariance function with functional data. *Brit. J. Math. Statist. Psych.*, **55**, 247–61.

LENG, X., MÜLLER, H.-G. (2006). Classification using functional data analysis for temporal gene expression data. *Bioinformatics*, **22**, 68–76.

LI, B. (2007). Comment: Fisher Lecture: Dimension reduction in regression. *Statistical Science*, **22**, 32–5.

LI, L., NACHTSHEIM, C.J. (2007). Comment: Fisher Lecture: Dimension reduction in regression. *Statistical Science*, **22**, 36–9.

MAS, A. (2002). Weak convergence for the covariance operators of a Hilbertian linear process. *Stochastic Process. Appl.*, **99**, 117–35.

MAS, A. (2008). Local functional principal component analysis. *Complex Anal. Oper. Theory*, **2**, 135–67.

MOSTELLER, F., TUKEY, J.W. (1977). *Data Analysis and Regression: A Second Course in Statistics*. AddisonWesley, Reading, MA.

NIANG, S. (2002). Estimation de la densité dans un espace de dimension infinie: application aux diffusions. *Comp. Rend. Acad. Sci.*, **334**, 213–16.

OCAÑA, F.A., AGUILERA, A.M., ESCABIAS, M. (2007). Computational considerations in functional principal component analysis. *Comput. Statist.*, **22**, 449–65.

OCAÑA, F.A., AGUILERA, A.M., VALDERRAMA, M.J. (1999). Functional principal components analysis by choice of norm. *J. Multivariate Anal.*, **71**, 262–76.

PFEIFFER, R.M., BURA, E., SMITH, A., RUTTER, J.L. (2002). Two approaches to mutation detection based on functional data. *Statist. Med.*, **21**, 3447–64.

PREDA, C., SAPORTA, G. (2005a). PLS regression on a stochastic process. *Comput. Statist. Data Anal.*, **48**, 149–58.

PREDA, C., SAPORTA, G. (2005b). Clusterwise PLS regression on a stochastic process. *Comput. Statist. Data Anal.*, **49**, 99–108.

RAMSAY, J.O., DALZELL, C.J. (1991). Some tools for functional data analysis. (With discussion.) *J. R. Statist. Soc. Ser. B*, **53**, 539–72.

RAMSAY, J.O., SILVERMAN, B.W. (2005). *Functional Data Analysis* (second edition). Springer, New York.

REISS, P.T., OGDEN, R.T. (2007). Functional principal component regression and functional partial least squares. *J. Amer. Statist. Assoc.*, **102**, 984–96.

RICE, J.A., SILVERMAN, B.W. (1991). Estimating the mean and covariance structure nonparametrically when the data are curves. *J. R. Statist. Soc. Ser. B*, **53**, 233–43.

SILVERMAN, B.W. (1995). Incorporating parametric effects into functional principal components analysis. *J. R. Statist. Soc. Ser. B*, **57**, 673–89.

SILVERMAN, B.W. (1996). Smoothed functional principal components analysis by choice of norm. *Ann. Statist.*, **24**, 1–24.

SPURRELL, D.J. (1963). Some metallurgical applications of principal components. *Applied Statist.*, **12**, 180–8.

STANISWALIS, J.G., LEE, J.J. (1998). Nonparametric regression analysis of longitudinal data. *J. Amer. Statist. Assoc.*, **93**, 1403–18.

WANG, K.M., GASSER, T. (1997). Alignment of curves by dynamic time warping. *Ann. Statist.*, **25**, 1251–76.

WANG, K.M., GASSER, T. (1999). Synchronizing sample curves nonparametrically. *Ann. Statist.*, **27**, 439–60.

YAMANISHI, Y., TANAKA, Y. (2005). Sensitivity analysis in functional principal component analysis. *Comput. Statist.*, **20**, 311–26.

YAO, F. (2007). Functional principal component analysis for longitudinal and survival data. *Statist. Sinica*, **17**, 965–83.

YAO, F., LEE, T.C.M. (2006). Penalized spline models for functional principal component analysis. *J. R. Stat. Soc. Ser. B*, **68**, 3–25.

YAO, F., MÜLLER, H.-G., WANG, J.-L. (2003a). Functional data analysis for sparse longitudinal data. *J. Amer. Statist. Assoc.*, **100**, 577–90.

YAO, F., MÜLLER, H.-G., CLIFFORD, A.J., DUEKER, S.R., FOLLETT, J., LIN, Y., BUCHHOLZ, B.A., VOGEL, J.S. (2003b). Shrinkage estimation for functional principal component scores with application to the population kinetics of plasma folate. *Biometrics*, **59**, 676–85.

YU, Y., LAMBERT, D. (1999). Fitting trees to functional data, with an application to time-of-day patterns. *J. Comput. Graph. Statist.*, **8**, 749–62.

ZHANG, J. (2006). Projection-pursuit based principal component analysis: a large sample theory. *J. Syst. Sci. Complex.*, **19**, 365–85.

CHAPTER 9

CURVE REGISTRATION

JAMES RAMSAY

We often note that prominent features in a set of curves, such as the pubertal growth spurt in human growth and mid-winter in temperature records, vary in timing as well as in size. Curve registration refers to methods for aligning these features by transforming their abscissa variables, often as a preliminary to applying some functional data analysis method designed to explore size and shape variation. In Section 9.1 the concepts of amplitude and phase variation are illustrated schematically and with real data. Phase variation involves a nonlinear transformation of time intrinsic to the system being observed, that is, a transformation from *system time s* to *clock time t*. The time-warping function $h_i(s)$ is defined along with its functional inverse in Section 9.2. Section 9.3 provides a decomposition of total mean squared variation into separate amplitude and phase components, along with an R^2 measure of the proportion of functional variation due to phase in a sample of curves. When applied to growth acceleration curves, we find that about 75% of the variation among these curves across a sample of 54 girls is due to phase. Landmark registration is the oldest curve registration technique, and is taken up in Section 9.4, where we also consider novel ways of defining *curve features*. New methods use the entire curve in registering to a target curve, called *continuous registration*; these are considered in Section 9.5. This approach is extended in Section 9.6 to the situation in which the target function for a curve is defined by its approximation defined by a model, such as functional principal component analysis or a functional linear model. Many recent methods are based on structured models for amplitude and phase variation combined with more statistically oriented fitting methods such as maximum likelihood or Bayesian estimation; these are discussed in Section 9.7. Software resources for registration are briefly surveyed in Section 9.8.

9.1 AMPLITUDE AND PHASE VARIATION

Figure 9.1 plots the second derivatives of functions x_i fit to 31 height measurements for the first ten girls in the Berkeley Growth Study (Tuddenham and Snyder 1954). Figure 9.1(a) presents the problem that curve registration is designed to solve: the mean curve, shown as a dashed line, is unlike any of the individual curves in that:

- the duration of the pubertal growth spurt (PGS) in the mean curve is longer than that of any single curve, and
- the slope of acceleration during puberty is not nearly as steep as even the shallowest of the individual curves.

These aberrations occur because the ten girls are not in the same phase of growth at around 10 to 12 years of age. We see that peak growth acceleration occurs around age 10.5 for many girls, but it occurred before age 8 for one girl and after age 13 for another. Similarly, the maximum pubertal growth rate occurs where the acceleration drops to zero following the maximum pubertal acceleration. The middle of the pubertal growth spurt (mid-PGS) averages 11.7 years of age for the entire Berkeley sample, but occurs before age 10 for two girls and around age 14 for another in this sub-sample of 10. That is, one girl has not yet begun her pubertal growth spurt, three others are at or just past their peak acceleration, and the rest are beyond their peak pubertal growth rate with negative acceleration.

Figure 9.1(b) shows these curves aligned by landmark registration, which is described in Section 9.4. Here the acceleration curves for all girls cross zero at mid-PGS at the same time. The average of these curves is a much more realistic representation of the typical pubertal growth spurt.

There is physiological *growth time* that unrolls at different rates from child to child relative to *clock time*. In terms of growth time, all girls experience puberty at the same age, with the peak growth rate (zero acceleration) occurring at about 11.7 years of age for the Berkeley sample. If we want a reasonable sense of amplitude variation, we must consider it within this growth-time frame of reference. Growth time itself is an elastic medium that can vary randomly from girl to girl when viewed relative to clock time, and functional variation has the potential to be *bivariate*, with variation in both the *range* and *domain* of a function.

Phase variation is illustrated in Fig. 9.2(a) as a variation in the location of curve features along the horizontal axis, as opposed to *amplitude variation*, shown in panel (b) as the size of these curves. The mean curve in Fig. 9.2(a), shown as

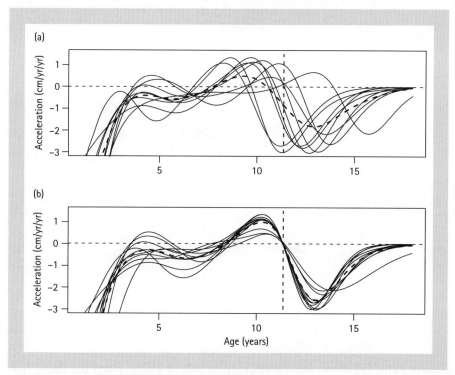

Fig. 9.1 Panel (a) contains the second derivatives of human growth curves for ten girls. The landmark-registered curves corresponding to these are shown in panel (b), where the single landmark was the crossing of zero in the middle of the mean pubertal growth spurt. The dashed line in each panel indicates the mean curve for the curves in that panel.

a dashed curve, does not resemble any of the other curves; it has less amplitude variation, but its horizontal extent is greater than that of any individual curve. The mean has, effectively, borrowed from amplitude to accommodate phase. Moreover, if we carry out a functional principal component analysis of the curves in each panel, we find in panel (a) that the first three principal components account for 56%, 39%, and 5% of the variation. On the other hand, the same analysis of the amplitude-varying curves requires a single principal component to account for 100% of the variation.

Like the mean, most statistical methods when translated into the functional domain are purely designed to model amplitude variation. This certainly includes the functional version of regression analysis, reviewed in Chapter 1 of this volume and treated in more detail in Chapters 2, 3, and 4. As we have just seen, even an exploratory tool like principal component analysis, discussed in Chapter 8 above, is likely to use more principal components in the presence of substantial phase variation than it would if a preliminary curve registration were applied. In Section 9.7 we will see how registration can be folded into the modeling process.

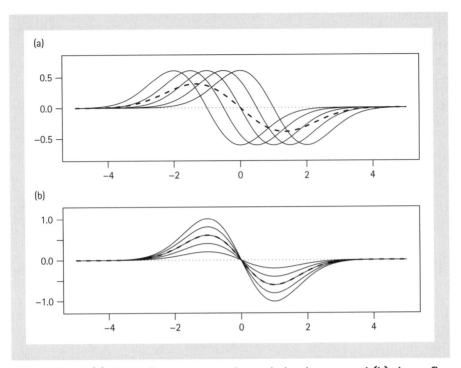

Fig. 9.2 Panel (a) shows five curves varying only in phase; panel (b) shows five curves varying only in amplitude. The dashed line in each panel indicates the mean of the five curves. The mean curve in the bottom panel is superimposed exactly on the central curve.

9.2 TIME-WARPING FUNCTIONS h

9.2.1 System and clock times

Let $x_i(t)$ be an observed or estimated function value at clock time t for functional observation i, where $i = 1, \ldots, N$. The registration problem is expressed by

$$y_i(s) = x_i[(h_i(s))] \quad \text{where} \quad t = h_i(s). \tag{9.1}$$

The *time-warping function* $h_i(s)$ transforms system time $s \in [0, S]$ to clock time $t \in [0, T_i]$, or possibly $t \in [\delta_i, T_i + \delta_i]$, for observation i. System terminal time S is the system-time value at which all the N observed systems are at the same end-state, such as, for example, when they have completed a task or achieved maximum height.

We observe clock times t and the corresponding function values $x_i(t)$, but what really defines the evolution of each system is unobserved system time s, and we need the unobserved functions y_i in order to understand how the systems work.

We lose no generality in assuming that system time begins at 0. On the other hand, clock times are often observed over the variable intervals $[0, T_i]$ and, in addition, optional *time-shift* parameters δ_i may be required to allow for periodic processes where simple translations of time $h_i(s) = s + \delta_i$ are also feasible.

We refer to the functions y_i as *registered* or *synchronized* because, relative to system time s, salient features or events in the curve will occur at the same system time for each observation. As such, the y_i's exhibit only amplitude variation. On the other hand, the unregistered functions x_i are apt to have these same features and other characteristics occurring at variable clock times t, and are therefore considered to mix phase variation with amplitude variation.

For example, we can require that $h_i(11.7) = t_i$ for all girls, where 11.7 years is the average clock time of the mid-PGS and t_i is the clock age at which the ith girl reached this event. If, at any time s, $h_i(s) < s$, we may say that the girl is growing faster than average at that clock time but slower than average if $h_i(s) > s$. This is illustrated in Fig. 9.3, where the growth acceleration curves for the earliest and latest of the first ten girls are shown in the left panels and their corresponding time-warping functions in the right panels.

It is important to appreciate that differentiation of a function can transform phase variation in the function itself to amplitude variation in the first derivative because of the chain-rule relation

$$\frac{dx_i}{ds} = Dx_i(t)Dh_i(s). \tag{9.2}$$

That is, since Dh_i is a positive functional multiplier of clock-time derivative Dx_i, it effectively modulates the amplitude variation of Dx_i. We cannot, therefore, attempt to synchronize more than one level of derivative at the same time, and it will be important to consider beforehand the order m of derivative at which phase and amplitude variation are to be disentangled. Typically this decision will depend on the derivative level at which the main causal factors defining amplitude variation are acting. In mechanical systems, for example, we would use $m = 2$ since energy input has a direct impact at the level of acceleration, and one would probably argue for $m = 2$ as well for the growth data.

Time-warping functions h_i must, of course, be strictly increasing; we cannot allow time to go backwards in either frame of reference. This implies that h_i's first derivative, Dh_i, is everywhere positive and that, consequently, we can re-express it as

$$(Dh_i)(s) = \exp[w_i(s)] \tag{9.3}$$

for some log-derivative function w_i whose value is unconstrained. The warping function h_i therefore has the expression

$$h_i(s) = C + \int_0^s \exp[w_i(u)]du = C + [D^{-1}\exp(w_i)](s). \tag{9.4}$$

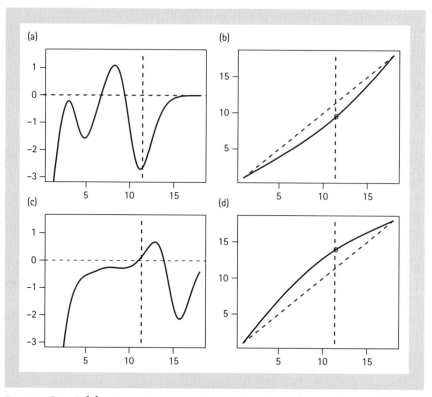

Fig. 9.3 Panel (a) shows the growth acceleration curve and panel (b) the corresponding time-warping function h(t) for the girl among the first ten in the Berkeley growth study with the earliest PGS. The corresponding plots for the girl with the latest PGS are in panels (c) and (d). The middle of the PGS is indicated by a vertical dashed line in all panels.

Time-warping functions must, because of the chain rule, also be smooth, in the sense of being differentiable up to order m for the curves being registered, implying that the differentiability of w_i must be at least $m - 1$.

9.2.2 The deformation function d

The time-warping functions are expected to vary from case to case, and in general we expect them to vary around the null-warping function $h(s) = s$. For plotting and many other purposes, it is easier to inspect the deformation function

$$d(s) = h(s) = s \qquad (9.5)$$

since it naturally varies around 0, with end boundary conditions $d(0) = d(S) = 0$. It directly reflects in clock-time units the extent to which clock time is running early ($d(s) < 0$) or late ($d(s) > 0$) relative to system time. Also, since we frequently use computing algorithms that build up warping functions iteratively, their corresponding deformation functions can simply be added to provide the final or total deformation and, by adding t, also to provide the final warping function.

9.2.3 The functional inverse warping function g

Since each warping function h_i is strictly increasing, the functional inverse h_i^{-1} defining values $s = h_i^{-1}(t)$ exists. The notation h_i^{-1} makes equations hard to parse, and we therefore replace it by g throughout this paper. That is,

$$g[h(s)] = s \text{ and } h[g(t)] = t \text{ or } (g \circ h)(s) = s \text{ and } (h \circ g)(t) = t. \quad (9.6)$$

Applying the chain rule to the relation $g[h(s)] = s$, we have $Dg(t) = 1/Dh(s)$, and therefore the inverse warping function satisfies the ordinary differential equation

$$(Dg)(t) = \exp[-w(s)] = \exp\{-w[g(t)]\}. \quad (9.7)$$

Although there is an agreeable antisymmetry in (9.3) and (9.7):

$$(Dh) = \exp[w]$$
$$(Dg) = \exp[-w(g)],$$

in fact the two differential equations have quite different structures. Equation (9.3) is a first-order *linear* differential equation with a homogenous component $Dh = 0$ and a forcing or nonhomogeneous component $\exp(w)$, while (9.7) is an unforced *nonlinear* differential equation since the function appears on both sides. This indicates that we cannot in general compute the functional inverse g by merely changing the sign of w in (9.4), although experience indicates that in practice this gets us very close to the functional inverse, provided that Dh is not too far from 1 everywhere.

These two differential equations are of more than just theoretical interest. Although there are ways of approximating the indefinite integral in (9.4), its numerical approximation by applying well-developed differential equation approximation algorithms to (9.3) yields fast and accurate results in most cases.

Moreover, once w_i has been estimated by methods that will be treated below, we have the differential equation for y_i

$$(Dx_i)(s) = [Dy_i \exp(-w_i)](s), \quad (9.8)$$

as well as the functional equation

$$y_i(s) = [y_i(g_i)](t) = [x_i(h_i)](s) = \{x_i[h_i(g_i)]\}(t) = x_i(t). \quad (9.9)$$

This tells us that the system-time function y can be estimated by interpolating or smoothing the relationship between $x_i(t)$ and $g_i(t)$ evaluated at a fine mesh of t-values.

9.2.4 Parametric and nonparametric families for warping functions

While in general we will prefer to have the kind of control over estimated warping functions that is achieved by working with high-dimensional basis function expansions for w_i combined with a roughness penalty, a simple parametric family remains useful for either illustrating registration or for applications requiring only simple phase variation. A one-parameter family for $h : [0, T] \to [0, S]$ and its functional inverse is, for $\beta \neq 0$,

$$t = h(s|\beta) = T\frac{e^{\beta s} - 1}{e^{\beta S} - 1}$$

$$s = g(t) = \log[(e^{\beta S} - 1)t/T + 1]/\beta. \tag{9.10}$$

In the limit $\beta \to 0$, these relations become $h(s) = Ts/S$ and $g(t) = St/T$.

Figure 9.4 displays a set of unregistered functions $\sin(4\pi t)$ where $s \in [0, 1]$ has been warped in this way with $\beta = -1, -0.5, 0, 0.5,$ and 1. We will use this and related examples at several points in this chapter. Figure 9.4(b) also shows as dotted

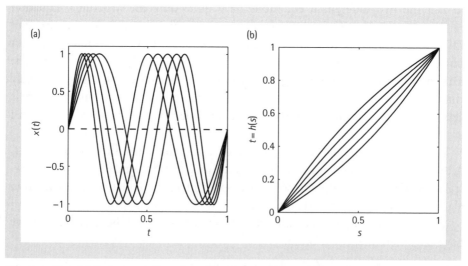

Fig. 9.4 Panel (a) shows $\sin(4\pi t)$ as functions of the warping function values $h_i(s)$ shown in panel (b) as solid lines. Also in panel (b) are the inverse functions $h_i(s)$, shown as dotted lines. The warping functions are defined by (9.10) for $\beta = -1, -0.5, 0, 0.5,$ and 1.

lines the inverse functions g, and the slight discrepancy between warping functions reflected about the diagonal line and the true inverse functions is a warning that h_i and g_i are not, in general, exactly reflections of each other.

Maximal flexibility is obtained by expanding w in terms of a set of basis function ϕ_k as

$$w(t) = \sum_{k}^{K} c_k \phi_k(t) = c'\phi(t). \tag{9.11}$$

We will return to this approach below in considering computational strategies.

An intermediate solution is to work with an expansion using positive coefficients and I-splines defined in Ramsay (1988):

$$h(s) = T \frac{\sum_k e^{\beta_k} I_k(s|\zeta)}{\sum_k e^{\beta_k} I_k(S|\zeta)}. \tag{9.12}$$

I-splines are indefinite integrals of M-splines, and as such rise monotonically from 0 to 1 over their range, and M-splines in turn are proportional to B-splines, but normalized so that their integrals, rather than their sum at any point, are equal to 1. Vector ζ contains a suitable knot sequence. A special case of this approach is the use of piecewise linear interpolating of warping function values over a sufficiently fine mesh of system-time values (Tang and Müller 2008).

Alternatively, strictly increasing functions may be achieved by replacing the I-splines in (9.12) by the far more popular B-splines, and forcing the coefficients β_k to be a strictly increasing sequence. Telesca and Inoue (2008) use this strategy.

9.3 A DECOMPOSITION INTO AMPLITUDE AND PHASE SUMS OF SQUARES

Kneip and Ramsay (2008) have quantified the amount of these two types of variation by comparing results for a sample of N functional observations before and after registration. Let the population or sample means of the unregistered and registered samples be $\mu = Ex$ and $\eta = Ey$, respectively.

The *total mean square error* is defined as

$$MSE_{\text{total}} = E \int [x_i(t) - \mu(t)]^2 dt. \tag{9.13}$$

We define the constant C_R as

$$C_R = \frac{E \int x^2(s)\,dt}{E \int y^2(s)\,ds}$$

$$= 1 + \frac{E \int \{Dh(t) - E[Dh](t)\}\{y^2(t) - E[y^2](t)\}\,dt}{E \int y^2(s)\,ds}, \quad (9.14)$$

where the expectation can either be an average over a sample of observed functions or an expectation over a measure space of functions.

The structure of C_R indicates that $C_R - 1$ is related to the covariation between the deformation functions Dh and the squared registered functions y^2. When these two sets of functions are independent, the number of the ratio is 0 and $C_R = 1$. Alternatively, we see in the first expression that $C_R = 1$ implies that the expected squared Euclidean norm of the functions is preserved under registration. Now it can easily be shown by a change of variable that $C_R = 1$ exactly for simple translations such as we see in Fig. 9.2, and if we plot the squared function values for each curve, we see that they are just translations of each other. In practice, however, C_R is usually very close to 1 even for nonconstant warping functions h and, moreover, $\int x_i(t)\,dt$ and $\int y_i(t)\,dt$ tend to be nearly equal for each i. This means that the registration process tends to shift function mass *laterally*, so that $\int x^2(t)dt$ is conserved. For example, for the sines de-registered by the one-parameter warping functions in Fig. 9.4, the base 10 logarithm of C_R ranges from -0.0017 for $\beta = -1$ to 0 for $\beta = 1$.

The measures of amplitude and phase mean square error are, respectively,

$$MSE_{amp} = C_R E \int [y(t) - \eta(t)]^2\,dt$$

$$MSE_{phase} = C_R \int \eta^2(t)\,dt - \int \mu^2(t)\,dt. \quad (9.15)$$

Kneip and Ramsay (2008) show that, defined in this way, $MSE_{total} = MSE_{amp} + MSE_{phase}$.

The interpretation of this decomposition is as follows. If we have registered our functions well, then the registered functions y will have higher and sharper peaks and valleys, since the main effect of mixing phase variation with amplitude variation is to smear variation over a wider range of t-values, as we saw in Figs. 9.1 and 9.2. Consequently, the first term in MSE_{phase} will exceed the second and is a measure of how much phase variation has been removed from the y's by registration. On the other hand, MSE_{amp} is now a measure of pure amplitude variation to the extent that the registration has been successful. The decomposition does depend on the success of the registration step, however, since it is possible in principle for MSE_{phase} to be negative.

From this decomposition we can get a useful squared multiple correlation index of the proportion of the total variation due to phase:

$$R^2 = \frac{MSE_{\text{phase}}}{MSE_{\text{total}}}. \tag{9.16}$$

Figure 9.5 shows the unregistered sine curves in Fig. 9.4 with some amplitude variation added by multiplying each curve by a Gaussian random number with mean 1 and standard deviation 0.5. The figure also shows the registered functions. The amplitude phase decomposition yields 0.049 and 0.143 for the mean squares for amplitude and phase, respectively, $R^2 = 0.74$ and $C_R = 1.001$.

Is the amplitude/phase decomposition uniquely defined? More often than not, the answer is no. There is usually some variation that can be accounted for either by the use of a sufficiently aggressive warping function, or simply as amplitude

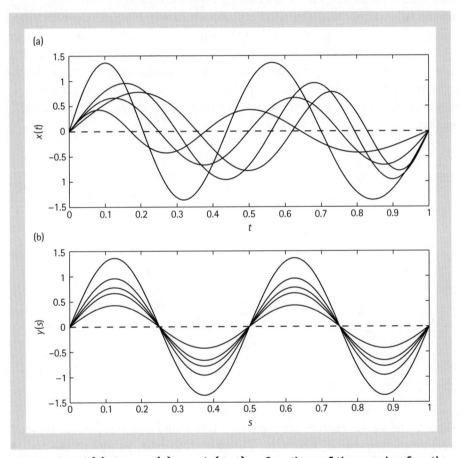

Fig. 9.5 Panel (a) shows $x_i(t) = c_i \sin(4\pi t)$ as functions of the warping function values $h_i(s)$ shown in Fig. 9.4(b), where c_i have a normal distribution with mean 1 and standard deviation 0.5. Panel (b) shows the registered functions $y_i(s)$.

variation. For example, Chicago and Montreal get their weather from the same source for much of the year, namely the powerful weather-generation systems in the Gulf of Mexico. Montreal is substantially colder in midwinter, and this is amplitude variation by any standard. But it is also colder in mid-May, and should we attribute this to Montreal's spring coming later or being colder? It all depends partly on what one means by spring. The gardening season is often defined by the number of days for which the temperature exceeds 5 °C (50 °F), and by that standard, spring does come earlier in Chicago. On the other hand, someone traveling from Montreal to Chicago in mid-May is more likely to say that it will be warmer in Chicago, suggesting that the amplitude variation is involved.

This issue of amplitude and phase identification is illustrated in Fig. 9.6. Churchill, Manitoba, situated on the western shore of Hudson's Bay, is one of the most northerly weather stations in the continental climate zone. The target curve indicated by the dotted line is the mean temperature for the entire zone, and is substantially warmer than Churchill over the entire year. If we register with only a constant time shift, we get the solid curve in panel (a), which is shifted to the left by about 10 days due to the impact of the Bay on Churchill's climate. If we also allow nonlinear time warping, then the summer season is lengthened so that the registered curve can come as close as possible to the target. For most purposes, we would regard the right registered curved as an inadmissible distortion of Churchill's weather.

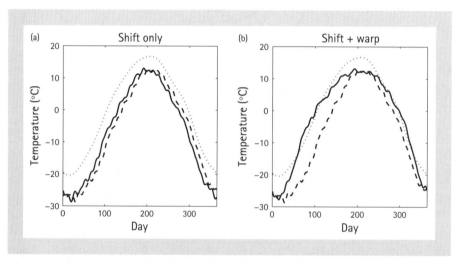

Fig. 9.6 The panels contain the registered (solid line) and unregistered (dashed line) mean temperature curves for Churchill, Manitoba, as well as the target curve (dotted line) defined by the registration procedure in Section 9.6. Panel (a) is for registration with only constant shift, and (b) for constant shift plus a nonlinear time warping.

9.4 Landmark registration

The simplest curve alignment procedure is *landmark registration,* a term introduced in Bookstein (1991), a reference which is itself a landmark in the registration literature. A landmark is a feature with a location that is clearly identifiable in all curves. Landmarks may be the locations of minima, maxima, or crossings of zero, and we see three such landmarks in each curve in Fig. 9.2. We align the curves by transforming t for each curve so that landmark locations are the same for all curves. For the warped sine functions in Fig. 9.4, we may use as features the three times at which each curve crosses the zero line, as well as the four times for the tops of the peaks and the bottoms of the valleys. But, of course, we know that the actual warping functions in this case are so simple that choosing a single feature, such as the second zero-crossing, would be sufficient to estimate them. Features of this nature are often more obvious at the level of some derivative, such as the PGS peak for the growth curves in the first derivative or the zero crossing in the second derivative. The second derivative curve also has a clear peak before the PGS zero-crossing, and a clear valley afterwards. These features can serve as additional landmarks, whereas the first derivative curve reveals only the PGS peak as an unambiguous feature in all curves.

For Fig. 9.1(b), we used a single landmark, t_i being the age for girl i at which her acceleration curve crossed 0 with a negative slope during the pubertal growth spurt. Let s_0 be a system time specified for the middle of the *average* pubertal growth spurt, such as 11.7 years of age for the Berkeley Growth Study girls. Then we specify time-warping functions h_i by fitting a smooth function to the three points $(1, 1)$, (s_0, t_i), and $(18, 18)$. This function should be as differentiable as the curves themselves, and in this case could be simply the unique parabola passing through the three points $(1, 1)$, (s_0, t_i) and $(18, 18)$, which is what is shown in Fig. 9.3(b) and (d).

Since at time $h_i(s_0)$ girl i is in the middle of her pubertal growth spurt, and since in her registered time $g_i[h_i(s_0)] = s_0$, she and all the other girls will experience puberty at time s_0 in terms of system or growth time. In particular, if $h_i(s_0) < s_0$ for a girl i reaching puberty early, then aligning function $g_i(t)$ effectively slows down or stretches out her clock time so as to conform with growth time.

Figure 9.1(b) displays the same ten female growth acceleration curves after registering to the middle of the pubertal growth spurt. We see that the curves are now exactly aligned at the mean PGS (pubertal growth spurt) age, but that there is still some misalignment for the maximum and minimum acceleration ages. Our eye is now drawn to the curve for girl seven, whose acceleration minimum is substantially later than the others and who has still not reached zero acceleration by age 18. The long period of near-zero acceleration for girl four prior to puberty also stands out as unusual. The mean curve is now much more satisfactory as a summary of the typical shape of growth acceleration curves, and in particular is nicely placed in the middle of the curves for the entire PGS period.

After landmark registration of all 54 growth curves for the Berkeley data, the amplitude/phase decomposition was computed for ages ranging from 3 to 18 years, since the variation in second derivative estimates in the first two or three years are far larger than afterwards. The amplitude and phase mean squares in (9.15) are 1.75 and 4.95, respectively, $C_R = 0.99$, and $R^2 = 0.74$.

Landmarks may be defined in many ways besides as crossings, peaks, and valleys. James (2007) chooses functional templates, called *probes* in Ramsay and Silverman (2005), to define features as pre-specified patterns of variation that are defined by locations. These can then be aligned as indicated above, and they also tend to avoid the over-registration at specific points associated with crossing, peak, and valley alignment, since features defined in this way involve a range of curve values. Bigot (2006) describes a sophisticated automatic landmark registration method using zero-crossings of the wavelet transforms of a set of curves, features that he called *structural intensities*. Muñoz Maldonado, Staniswalis, Irwin and Byers (2002) use landmark registration to improve the power of a functional two-sample test.

9.5 Continuous registration

We may need registration methods that use the entire curves rather than their values at specified points. A number of such methods have been developed, and the problem continues to be actively researched. Landmark registration is usually a good first step, but we need a more refined registration process if landmarks are not visible in all curves. For example, many but not all female growth acceleration curves have at least one peak prior to the pubertal growth spurt that might be considered a landmark. Even when landmarks are clear, identifying their timing may involve tedious interactive graphical procedures, and we might prefer a fully automatic method. Finally, as we saw in Fig. 9.1, landmark registration using just a few landmarks can still leave aspects of the curves unregistered at other locations, such as at the peak and valley preceding and following the zero-crossing for the growth acceleration curves.

Most continuous registration methods attempt to align a sample of curves to a target curve x_0, and the usual choice for this target is the mean function $\bar{x}$. Of course, we have already seen in Fig. 9.1(a) that the mean curve can be badly thrown off by the presence of substantial phase variation. But if it nevertheless retains variation that tends to be seen in most curves, it represents an acceptable starting target for a process that can be iterated. That is, one begins by taking x_0 as the mean of the unregistered functions, and then a method for continuous registration, such as is detailed below, is applied. The mean of these registered curves can then be used as a new target, and is likely to have improved variation. The registration is repeated, and so on. In practice, it is rare to find that more than a single iteration of the process yields substantial improvement in the mean registered function, which is often referred to as the *structural mean* of the sample. The iterative process is itself an example of what is referred to as a *Procrustes iteration* in the psychometric literature.

In fact, the structural mean or any other target function x_0 is, in effect, a sort of feature, and differs from a single peak, valley, or crossing point in that it is distributed across the entire curve rather than being positioned at a specific point. In this sense, landmark registration and continuous registration are not so far apart in concept.

What fitting criterion should one optimize with respect to the parameters defining a warping function h_i? Least squares fitting has been repeatedly noted to yield poor results (Ramsay and Silverman 2005), and this is unsurprising since comparing an unregistered function to a target at any fixed point t may be comparing two curves in very different states of evolution. Least squares and, in fact, all classical measures of fit, are essentially tuned to pure amplitude variation, and inevitably do badly when phase variation is substantial.

Ramsay and Li (1996) suggest an alternative criterion. If an arbitrary sample registered curve $x[h(s)]$ and target curve $x_0(t)$ differ only in terms of amplitude variation, then their values will tend to be proportional to one another across the range of t-values. That is, if we were to plot the values of the registered curve against the target curve, we would see something approaching a straight line tending to pass through the origin, although not necessarily at a 45° angle with respect to the axes of the plot. If this is true, then a principal component analysis of the following order-two matrix $C(h)$ of integrated products of these values should reveal essentially one component, and the smallest eigenvalue should be near 0:

$$C(h) = \begin{bmatrix} \int \{x_0(t)\}^2 \, dt & \int x_0(t) x[h(t)] \, dt \\ \int x_0(t) x[h(t)] \, dt & \int \{x[h(t)]\}^2 \, dt \end{bmatrix}. \tag{9.17}$$

According to this rationale, then, estimating h so as to minimize the smallest eigenvalue of $C(h)$ should do the trick, and is closely related to maximizing the correlation between two vectors of curve values. If the curves are multivariate, such as coordinates of a moving point, then what is minimized is the sum of the smallest eigenvalues across the components of the curve vectors.

It can be important to control the amount of time-warping, often in order to avoid the problem that we saw in Fig. 9.6. A basis for expanding the functions $w(t)$ in (9.4) can be combined with a roughness penalty to estimate strictly monotone

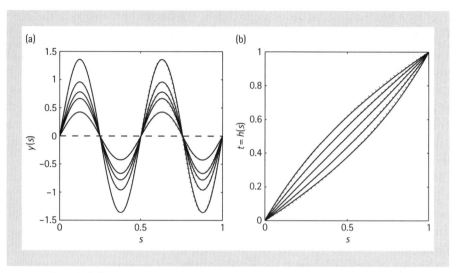

Fig. 9.7 Panel (a) shows the registered functions resulting from continuously registering the functions in Fig. 9.5(a) as solid lines, and the actual registered functions used to generate them as dotted lines. Panel (b) shows the estimated warping function as solid lines and the actual warping functions as dotted lines.

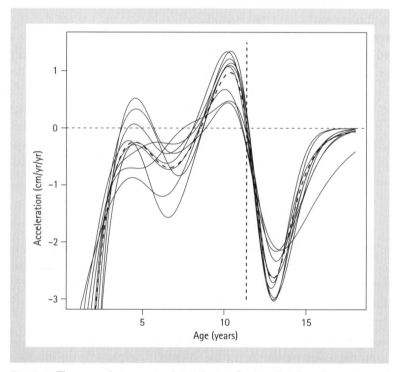

Fig. 9.8 The continuous registration of the landmark-registered height acceleration curves in Fig. 9.1. The vertical dashed line indicates the target landmark age used in the landmark registration.

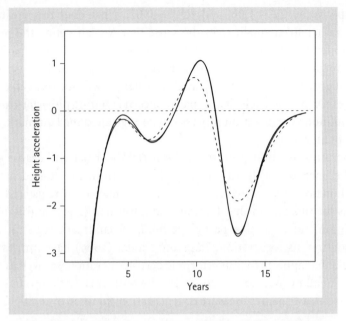

Fig. 9.9 The mean of the continuously registered acceleration curves is shown as a heavy solid line, while that of the landmark-registered curves is a light solid line. The light dashed line is the mean of the unregistered curves.

warping functions. Because the continuous registration process requires iterative numerical optimization techniques, starting values for the coefficients defining the functions w must be supplied, and zeros usually suffice.

Figure 9.7 shows the result of a single iteration of this process, using 13 B-splines to expand w_i for each curve. Although there is some small failure to get the warping functions right, for most purposes the registration would be considered to be entirely satisfactory. The estimates of the warping functions were correct to within graphing accuracy after a second iteration, but the improvement in the registered functions was negligible.

Figure 9.8 shows that the continuously registered height acceleration curves are now aligned over the entire PGS relative to the landmark–registered curves, although we have sacrificed a small amount of alignment of the zero-crossings of these curves. Figure 9.9 shows the impacts of the two types of registrations, and we see that, while both registrations provide average curves with maximum and minimum values much more typical of the individual curves, as well as the width of the PGS, the final continuously registered mean curve does a better job in the mid-spurt period centered on five years of age.

The amplitude and phase mean squares in (9.15) comparing the landmark-registered curves to the continuously registered curves over [3,18] are 1.60 and

−0.10, respectively, $C_R = 1.00$, and $R^2 = -0.07$. What does this mean? "Registered" is a rather fuzzy qualifier in the sense that we can define the registration process in different ways and obtain different answers. A careful comparison of the two figures might suggest that the landmark registration process has over-registered the pubertal growth spurt at the expense of earlier growth spurts visible in several of the curves. Or, alternatively, if our main concern is getting pubertal growth right, then the continuous registration process has de-registered the landmark-registered curves by about 6%.

The Procrustes iteration process has been rightly criticized because the initial mean of the unregistered functions may be so far from the desired mean of the registered functions that convergence to a suboptimal set of registered functions results. It is for this reason that Ramsay and Silverman (2005) and Ramsay et al. (2009) recommend at least a partial preliminary landmark registration before applying continuous registration. Tang and Müller (2009) have proposed a one-stage process that appears to outperform Procrustes iterations, as well as the methods of Gervini and Gasser (2004, 2005), for the problems that they consider. They use a penalized least squares measure or registration summed over all possible *pairs* of curves.

9.6 CONTINUOUS REGISTRATION TO A MODEL

Kneip and Ramsay (2008) consider the possibility of folding the registration step into a model-fitting procedure, rather than pre-registering the data before fitting a model. For example, one may contemplate a principal components analysis to explore the data, for which the model is

$$x_i(t) = \mu(t) + \sum_{\ell}^{L} f_{i\ell} \xi_\ell(t) + \epsilon_i(t),$$

the L eigenfunctions ξ_ℓ being mutually orthogonal and having mean zero. It is a relatively trivial modification to suggest

$$y_i(s) = x_i[h_i(s)] = \mu(s) + \sum_{\ell}^{L} f_{i\ell} \xi_\ell(s) + \epsilon_i(s), \quad (9.18)$$

where the warping functions h_i are required to optimize the relationship between the registered functions y_i and their approximations $\hat{y}_i(s) = \mu(s) + \sum_\ell f_{i\ell} \xi_\ell(s)$.

In this case, we may consider the eigenfunctions ξ_ℓ to be L "features" or patterns of variation in the data that generalize the concept presented in Section 9.5 of the mean function as a feature. Moreover, it can be hoped that registration will also improve the fit to the model, in this case defined as the proportion of the variance in the data accounted for by L principal components.

Although Kneip and Ramsay (2008) propose a somewhat complex iterative process involving alternating between principal component analysis and small registration steps, it is possible to proceed more simply by adapting the Procrustes iteration process to this case. After initializing the process with the unregistered functions, an iteration proceeds in these steps:

1. A functional principal component analysis is applied to the functions after either the initialization or their registered versions after the previous iteration.
2. The approximation on the right side of (9.18) is computed, where the population mean μ is replaced by the mean of the functions being analyzed.
3. The continuous registration procedure is applied, where the target x_0 for the ith function is that function's approximation.

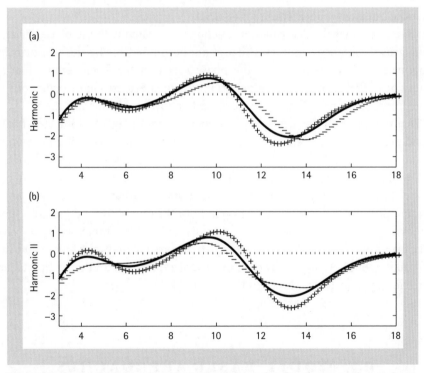

Fig. 9.10 Two principal components or harmonics for the growth acceleration curves resulting from combining registration and principal component analysis. The mean of the registered curves is shown as a solid line, and the effects of adding each harmonic to the mean is indicated by "+", and the effects of subtracting by "−".

4. The registered functions replace the functions used for the PCA in this step, in preparation for the next iteration.
5. Convergence tests are applied if desired.

The deformation functions $d_i(s) = h_i(s) - t$ can be accumulated over these iterations to provide the total deformation function.

It will be evident that this approach may use some of the phase variation to define the model, and especially with a relatively "greedy" analysis like PCA, and therefore move some of the phase variation into the model. However, this in itself may be useful, since we can see in the model how amplitude and a certain amount of phase variation interact.

Figure 9.10 plots two functional principal components as perturbations of the mean for the growth acceleration curves evaluated over the age interval [3,18] and registered in this way. We see that the first principal component, accounting for 66% of the variation, represents an early growth spurt combined with no change in the childhood "mid-spurt", while the second indicates a late growth spurt combined with positive amplitude variation in the "mid-spurt", and accounts for 21% of the variation. Before registration, two components account for 78.5% of the total variation, but afterwards account for 88.1%.

Figure 9.11 shows the deformation functions associated with the phase variation in the growth acceleration curves that is not accounted for by the two principal components. Although much of the phase variation has been removed in the vicinity of the mean time for the pubertal growth spurt, strong deformations are still required (of as much as two years) for ages above and below this time.

The distribution of the principal component scores in Fig. 9.12 indicates that the second principal component is primarily required by two clusters of girls, 6 on the left and 3 on the right, where there are large negative scores on the second component or harmonic. The "−" curve in Fig. 9.10(b) indicates that these girls will tend to have earlier growth spurts than their neighbors immediately above them, and at the same time weak amplitude variation in both the mid-spurt and pubertal spurt.

Combining registration and cluster analysis can lead to simpler and more compact clusters of curves, and two recent approaches are Liu and Yang (2009) and Tang and Müller (2009). An especially exciting area of application for this work is to gene expression trajectories, as shown in the second of these papers.

9.7 Model-based approaches to registration

Landmark and continuous registration are examples of algorithmic data analysis in the sense that a task is defined, in this case curve alignment, and methods are

CURVE REGISTRATION   255

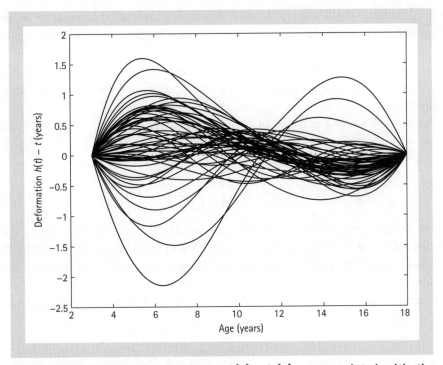

Fig. 9.11 The deformation functions $d_i(s) = h_i(s) - t$ associated with the registration to two principal components.

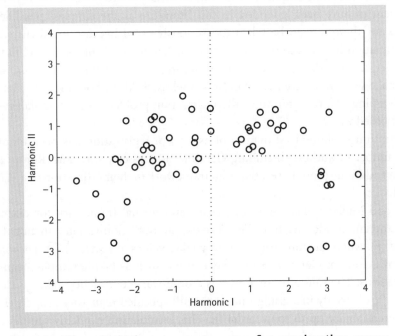

Fig. 9.12 The principal component scores for acceleration curves registered to two principal components.

evolved without much attempt to provide either a structured account or model of the sources of variation involved, or a stochastic framework within which both the model and the estimation procedure are framed. An example of an algorithmic approach is the earliest use of the term "time-warping" by Sakoe and Chiba (1978), where a dynamic programing was used to register discrete sequences of phonemes.

Possibly the earliest example of registration in the statistical literature is also based on a specific model for inter-curve variation. Let μ be a function and let the registered function y_i be defined as

$$y_i(s) = \beta_{i0} + \beta_{i1}\mu\left(\frac{s - a_{i0}}{a_{i1}}\right) + \epsilon_i(s), \qquad (9.19)$$

where rescaling parameters a_{i1} and β_{i1} are usually assumed to be positive. This model, called *self-modeling nonlinear regression* and often abbreviated as SEMOR, was first explored by Lawton, Sylvestre and Maggio (1972) and later taken up by a number of authors, including Gervini and Gasser (2004), Kneip and Gasser (1988, 1992) and Telesca and Inoue (2008). We can re-express this model in terms of (9.1) by defining

$$x_i(t) = \beta_{i0} + \beta_{i1}\mu(t)$$
$$h_i(s) = \frac{s - a_{i0}}{a_{i1}}. \qquad (9.20)$$

The models for both amplitude and phase variation are linear transformations, and seem to have been intended in this literature as the simplest possible surrogates for these, not to be taken too seriously—and because the focus was on the estimation by parametric and nonparametric methods of the kernel function μ. In fact, as we indicated above, incorporating a time-shift parameter a_{01} would only make sense if μ were periodic. Nevertheless, SEMOR both precedes dynamic time-warping as a recognition of the registration problem, and formulates it in a model-based way.

Maximum likelihood estimation of aspects of registration has been employed by a number of authors, including Gervini and Gasser (2005) and Rønn (2001); Telesca and Inoue (2008) use a Bayesian framework in their estimation of a SEMOR model.

Liu and Müller (2004) offer an elegant and formal framework for discussing phase and amplitude variation. They propose an implicit bivariate functional variable (h, y), where the random function y takes values in a space $\mathcal{Y}$ of *synchronized* functions closed under *convex linear operations* such as taking sample means and expectations. The random function h is a time-warping function taking values in a space $\mathcal{H}$ of strictly increasing functions with specified regularity properties such as differentiability up to some order m. This formulation explicitly models phase variation in terms of h, while the variation in y can be defined as pure amplitude variation.

What is observed, of course, is a random function x taking values in an *unsynchronized* function space $\mathcal{X}$ where convex linear operations do not, in general, preserve certain features such as the number of peaks and valleys in each x_i. The two spaces are related by the mapping

$$G : (h, y) \to x \mid y[g(t)] = x(t) \tag{9.21}$$

where, as above, $g = h^{-1}$.

The spirit of SEMOR and the registration to a model in Section 9.6 can be incorporated into this framework by postulating a trivariate functional random variable (h, p, y), where y takes values in a synchronized and low-dimensional function space having a simple structure such as (9.18), and random *amplitude modulation function* p adds amplitude variation to the low-dimensional core of synchronized functions y through

$$GP : (h, p, y) \to x \mid p[g(t)]y[g(t)] = x(t). \tag{9.22}$$

It will often be appropriate to assume that $p(s) > 0$ for all s.

9.8 SOFTWARE AND OTHER RESOURCES

It can be a long time before a promising registration method is encapsulated in accessible software so that it can be widely applied. This is a shame, since much is usually learned through applications about both limitations and fruitful extensions of the technique. On the other hand, setting in reliable code with adequate checking and good documentation in high-level languages like MATLAB and R is no small task, and the value of this contribution is too often overlooked in university promotion processes. Things are, however, definitely improving, and the number of new packages in R appearing each year continues to increase.

Ramsay and Silverman (2002, 2005) and Ramsay et al. (2009) have developed a set of functions in both MATLAB and R that contains many registration tools. These, along with sample data and sample analyses, may be downloaded as either the R package fda or from the web site www.functionaldata.org.

MATLAB functions for dynamic time-warping are available as contributed toolboxes from www.mathworks.com. For other software resources, one has to search author websites or search engines. Useful phrases for searches are "curve registration" (statistics), "curve alignment" (biology), and "time warping" (engineering).

References

Bigot, J. (2006). Landmark-based registration of curves via the continuous wavelet transform. *J. Comp. Graph. Stat.*, **15**, 542–64.

Bookstein, F. L. (1991). *Morphometric Tools for Landmark Data: Geometry and Biology.* Cambridge University Press, Cambridge.

Gervini, D., Gasser, T. (2004). Self-modeling warping functions. *J. Roy. Stat. Soc., Ser. B*, **66**, 959–71.

Gervini, D., Gasser, T. (2005). Nonparametric maximum likelihood estimation of the structural mean of a sample of curves. *Biometrika*, **92**, 801–20.

James, G. (2007). Curve alignment by moments. *Ann. Appl. Stat.*, **1**, 480–501.

Kneip, A., Gasser, T. (1988). Convergence and consistency properties for self-modeling nonlinear regression. *Ann. Stat.*, **16**, 82–113.

Kneip, A., Gasser, T. (1992). Statistical tools to analyze data representing a sample of curves. *Ann. Stat.*, **20**, 1266–305.

Kneip, A., Ramsay, J.O. (2008). Combining registration and fitting for functional models. *J. Amer. Stat. Assoc.*, **103**, 1155–65.

Lawton, W.H., Sylvestre, E.A., Maggio, M.S. (1972). Self-modeling nonlinear regression. *Technometrics*, **14**, 513–32.

Liu, X., Müller, H.-G. (2004). Functional convex averaging and synchronization for time-warped random curves. *J. Amer. Stat. Assoc.*, **99**, 687–99.

Liu, X., Yang, M.C.K. (2009). Simultaneous curve registration and clustering for functional data. *Comp. Stat. & Data Anal.*, **4**, 1361–76.

Muñoz Maldonado, Y., Staniswalis, J.G., Irwin, L.N., Byers, D. (2002). A similarity analysis of curves. *The Canadian Journal of Statistics*, **30**, 373–81.

Ramsay, J.O. (1988). Monotone regression splines in action. *Statistical Science*, 3(4), 425–41.

Ramsay, J.O., Li, X. (1996). Curve registration. *J. Roy. Stat. Soc., Ser. B*, **60**, 351–63.

Ramsay, J.O., Silverman, B.W. (2002). *Applied Functional Data Analysis* (second edition). Springer, New York.

Ramsay, J.O., Silverman, B.W. (2005). *Functional Data Analysis* (second edition). Springer, New York.

Ramsay, J.O., Hooker, G., Graves, S. (2009). *Functional Data Analysis with R and Matlab*. Springer, New York.

Rønn, B. (2001). Nonparametric maximum likelihood estimation for shifted curves. *J. Roy. Stat. Soc., Ser. B*, **63**, 243–59.

Sakoe, H., Chiba, S. (1978). Dynamic programming algorithm optimization for spoken word recognition. *IEEE Trans. ASSP-26*, **1**, 43–9.

Tang, R., Müller, H.-G. (2008). Pairwise curve synchronization for functional data. *Biometrika*, **95**, 875–89.

Tang, R., Müller, H.G. (2009). Time-synchronized clustering of gene expression trajectories. *Biostatistics*, **10**, 32–45.

Telesca, D., Inoue, L.Y.T. (2008). Bayesian hierarchical curve registration. *J. Amer. Stat. Assoc.*, **103**, 328–39.

Tuddenham, R., Snyder, M. (1954). *Physical Growth of California Boys and Girls from Birth to Eighteen Years*. University of California Press, Berkeley.

CHAPTER 10

CLASSIFICATION METHODS FOR FUNCTIONAL DATA

AMPARO BAÍLLO
ANTONIO CUEVAS
RICARDO FRAIMAN

THE aim of this chapter is to provide a survey of the literature concerning classification of functional data. Roughly half of the survey is devoted to supervised classification and the rest to the unsupervised case. Whereas most available methods (k-NN, kernel, K-means, ...) are adaptations to the infinite-dimensional setting of other well-known methodologies from the classical multivariate framework, others (such as those based on random projections) have been especially developed for functional problems. Some practical issues, in particular real-data examples and simulations, are also reviewed. Some selected proofs are sketched.

10.1 INTRODUCTION

In modern statistical terminology the word "classification" is used with two principal meanings: while "unsupervised classification" roughly stands for "clustering", "supervised classification" is used as a synonym for the more classical name "discriminant analysis".

Both meanings correspond, almost exactly, with those given by the *Oxford English Dictionary* under the entry "classify": first, *arrange (a group) in classes according to shared characteristics*; second, *assign to a particular class or category*.

On the meaning of unsupervised classification

The basic aim of unsupervised classification (sometimes called "unsupervised learning") techniques is to partition a (usually large) data sample $X_1, \ldots, X_n$ into a number k of classes or clusters. These are defined in such a way that the members of each class are "alike" or "similar to each other" in a sense specified by the clustering algorithm used. There are several algorithms widely known in the literature and implemented in the standard software packages. The number of clusters k has to be given in advance but most clustering procedures include some guidelines for the choice of k. Unsupervised classification is typically used when a large amount of data is available and the presence of some internal structure of hidden data groups is suspected. In a way, the clustering methods are designed to reveal such structure through the final partition into clusters. This will typically aid in the understanding of the problem at hand.

On the meaning of supervised classification

The supervised classification methodology corresponds to the second meaning of "classify" mentioned above. This methodology applies when k populations (or classes) $P_1, \ldots, P_k$ are given and clearly defined in advance. The observable feature for all the individuals in these populations is a random variable X. The available data consist of a "training sample" $\{(X_i, Y_i), 1 \leq i \leq n\}$, where X_i, $i = 1, \ldots, n$, are independent replications of X measured on n randomly chosen individuals, and Y_i are the corresponding values of an indicator variable, such that $Y_i = j$ whenever the ith individual belongs to the population P_j. The distribution of X is assumed to be different in each population considered; unless otherwise stated we also denote the conditional distribution $X|Y = j$ by P_j and assume that all these distributions are different. The final aim is to classify a newly appearing observation X into one of the populations P_j. That is, we want to predict the corresponding value Y using the information provided by the training sample.

The term "supervised" accounts for the fact that the elements of the training sample are supposed to be classified with no error, through some non-statistical procedure (maybe implemented with the help of some expert advisor). We will mainly focus on the binary case $k = 2$, which has a special practical importance.

In this case it is customary to denote the populations (as well as the corresponding distributions of X on them) by P_0 and P_1. Binary supervised classification typically arises, for example, as an automatic diagnosis tool in medical studies, where P_0 and P_1 could correspond to the group of "healthy" and "ill" individuals, respectively.

Some general references

The classical books by Devroye et al. (1996), Duda et al. (2000), and Hastie et al. (2001) provide a broad coverage of these topics, mostly confined to the standard multivariate case where the variable X takes values in the *feature space* $\mathcal{F} = \mathbb{R}^d$.

We will consider here the situation where $\mathcal{F}$ is an infinite-dimensional space as, for example, $\mathcal{F} = L^2[a, b]$, which would lead to a typical problem in functional data analysis, as, for example, those considered in the monographs by Ramsay and Silverman (2005) and Ferraty and Vieu (2006). The infinite-dimensional variables will be denoted by calligraphic letters such as $\mathcal{X}$. In most cases of practical interest the infinite-dimensional feature space can be identified with a functional space. For this reason we will use the expression "functional classification" as nearly synonymous with "infinite-dimensional classification," though clearly the latter has a broader meaning.

Our aim here is to summarize the main ideas and techniques used so far in the classification of functional data, as well as to provide a survey of the existing literature on the subject. The chapter is organized as follows. Sections 10.2 and 10.3 are devoted respectively to supervised and unsupervised classification for functional data. Some examples and practical aspects are considered in Section 10.4.

10.2 Supervised classification

We will show here that the formal statement of the supervised classification problem in the infinite-dimensional case largely coincides with that of discriminant analysis in the classical multivariate case. However, some crucial differences arise when it comes to implementation in the functional case of the simplest and most popular multivariate discrimination methodologies, namely Fisher's (linear) discrimination procedure and the nonparametric method based on nearest neighbors. Not surprisingly, our conclusion is that the discrimination problem is much harder in the infinite-dimensional setting as it gives rise to some additional difficulties, both theoretical and computational.

10.2.1 Some general background and notation

The optimal classifier and plug-in rules

We concentrate on the binary problem. So we assume that the available information is given by a training sample $(\mathcal{X}_i, Y_i)$, $1 \leq i \leq n$. Here $\mathcal{X}_i$, $i = 1, \ldots, n$, are independent observations of the feature variable $\mathcal{X}$, taking values in the functional space $\mathcal{F}$, and the Y_i's are the corresponding values of the indicator variable Y, which takes values 0 or 1 depending on whether the individual belongs to P_0 or P_1, respectively.

Then the problem is to find a "classifier" (or "classification rule") $g : \mathcal{F} \to \{0, 1\}$ that minimizes the classification error (often called "risk") $P(g(\mathcal{X}) \neq Y)$. It can be shown (see, for example, Devroye et al. 1996, p. 11) that the optimal classification rule (or "Bayes rule") is

$$g^*(x) = 1_{\{\eta(x) \geq 1/2\}}, \tag{10.1}$$

where 1_A denotes the indicator function of the set A and $\eta(x) = \mathbb{E}(Y|\mathcal{X} = x)$.

Since the exact expression of $\eta(x)$ is seldom available, (10.1) cannot be directly used to obtain the optimal classifier. In practice, the information provided by the training sample $\mathcal{D}_n = ((\mathcal{X}_i, Y_i), 1 \leq i \leq n)$ is used to construct classifiers $g_n(x) = g_n(x; \mathcal{D}_n)$, whose conditional error $L_n = P(g_n(\mathcal{X}) \neq Y|\mathcal{D}_n)$ is as close as possible to the Bayes error $L^* = P(g^*(\mathcal{X}) \neq Y)$.

A sequence of classifiers $\{g_n\}$ is said to be *weakly consistent* if $L_n \longrightarrow L^*$ in probability or, equivalently, $\mathbb{E}(L_n) \longrightarrow L^*$ as $n \to \infty$. The definition of strong consistency is obtained by just imposing $L_n \longrightarrow L^*$ "almost surely" (a.s.).

To a large extent, the supervised classification theory deals with the construction of classifiers suitable for different situations. Expression (10.1) suggests an obvious general procedure for constructing classifiers by simply replacing the regression function $\eta(x)$ with appropriate data-based estimators. This is the the so-called *plug-in methodology*.

Empirical risk and empirical minimization rules

Again, as the distribution of $(\mathcal{X}, Y)$ is unknown in general, the exact value of the risk for a given classifier g is also unknown. It can be estimated by the "empirical risk"

$$\hat{L}_n = \hat{L}_n(g) = \frac{1}{n} \sum_{i=1}^{n} 1_{\{g(\mathcal{X}_i) \neq Y_i\}}. \tag{10.2}$$

This expression provides a natural way (alternative to the plug-in methodology) for constructing classification rules. It is based on the idea of selecting a class $\mathcal{C}$ of rules (with a simple structure or any other interesting property) and solving the minimization problem

$$g_n^* = \mathrm{argmin}_{g \in \mathcal{C}} \hat{L}_n(g). \tag{10.3}$$

It is often argued that this risk-minimization methodology provides better results than the plug-in procedure. The reason for this is that risk-minimization takes a more direct approach to supervised classification, since it does not require the estimation of the regression function $\eta(x)$, which is in fact a harder problem. However, this is still a controversial issue; see Audibert and Tsybakov (2007).

10.2.2 Linear discrimination rules

A short overview of Fisher's methodology

In the multivariate case, where the feature space is $\mathcal{F} \subset \mathbb{R}^d$, Fisher's linear classifier is still the most popular discrimination method among users of such methods. Let us briefly recall the main ideas behind Fisher's multivariate classification rule in order to see why it cannot be straightforwardly translated to the functional case. Fisher's linear rule typically applies to the homoscedastic case in which P_0 and P_1 have the same covariance matrix Σ but different vectors of means $\boldsymbol{\mu}_0$ and $\boldsymbol{\mu}_1$. The rule is based on a linear transformation $\mathbf{x} \mapsto \boldsymbol{\beta}'\mathbf{x}$ in such a way that $\mathbf{x}$ is assigned to P_0 whenever $\boldsymbol{\beta}'\mathbf{x}$ is closer to $\boldsymbol{\beta}'\boldsymbol{\mu}_0$ than to $\boldsymbol{\beta}'\boldsymbol{\mu}_1$. Such a linear transformation is chosen by maximizing (in $\boldsymbol{\beta}$) the separation between the projected means $\boldsymbol{\beta}'\boldsymbol{\mu}_0$ and $\boldsymbol{\beta}'\boldsymbol{\mu}_1$. In order to have a finite optimum, this maximization is subject to the restriction that the common variance $\boldsymbol{\beta}'\Sigma\boldsymbol{\beta}$ of $\boldsymbol{\beta}'\mathbf{X}$ is 1. This amounts to maximizing in $\boldsymbol{\beta}$ the following ratio of the "between-class variance" to the "within-class variance",

$$\frac{(\boldsymbol{\beta}'\boldsymbol{\mu}_0 - \boldsymbol{\beta}'\boldsymbol{\mu}_1)^2}{\boldsymbol{\beta}'\Sigma\boldsymbol{\beta}}.$$

The resulting classifier is in turn equivalent to assigning an observation $\mathbf{x}$ to the population P_i whose mean $\boldsymbol{\mu}_i$ is closest to $\mathbf{x}$ in the Mahalanobis distance. That is, $\mathbf{x}$ is assigned to P_0 whenever

$$(\mathbf{x} - \boldsymbol{\mu}_0)^t \Sigma^{-1}(\mathbf{x} - \boldsymbol{\mu}_0) < (\mathbf{x} - \boldsymbol{\mu}_1)^t \Sigma^{-1}(\mathbf{x} - \boldsymbol{\mu}_1).$$

The optimal solution in $\boldsymbol{\beta}$ depends on Σ^{-1}, which is one of the crucial difficulties in extending the rule to functional data.

On the extension of the linear rule to functional data

Indeed, the functional counterpart of the above setting would go as follows: assume that $\mathcal{F} \subset L^2[a, b]$ for a closed finite interval $[a, b] \subset \mathbb{R}$. Denote by $\langle f, g \rangle = \int_a^b f(t) g(t) \, dt$, the standard inner product in $L^2[a, b]$. A linear classifier for the functional measurement $x \in \mathcal{F}$ would be obtained by projecting the infinite-dimensional x onto the real line and comparing such a projection with those corresponding to the mean functions $\mu_j(t) = \mathbb{E}(\mathcal{X}(t)|Y = j)$, $j = 0, 1$. The projection "direction" β would be selected as the maximizer of the distance between the projected class means $\langle \beta, \mu_0 \rangle$ and $\langle \beta, \mu_1 \rangle$. The optimization is subject to the restriction $\int_a^b \beta(t)\langle \beta, V(t, \cdot)\rangle dt = 1$, where $V(s, t) = \text{Cov}(\mathcal{X}(t), \mathcal{X}(s)|Y = j)$ for $j = 0, 1$. As in the finite-dimensional case this would lead to maximizing

$$\frac{\text{Var}(\mathbb{E}(\langle \beta, \mathcal{X}\rangle|Y))}{\mathbb{E}(\text{Var}(\langle \beta, \mathcal{X}\rangle|Y))}.$$

Unfortunately, the analogy with the finite-dimensional case must stop here since the covariance operator associated with the kernel $V(s, t)$ is not in general invertible, so the above maximization problem has no solution.

Remarks on the literature on functional linear discrimination

Thus, we must content ourselves with finding some approximate solution. Two popular ways to achieve this are regularization and filtering. The first method can be applied when all the functional observations $\mathcal{X}$ are discretized by evaluating them on the same nodes $(\mathcal{X}(t_1), \ldots, \mathcal{X}(t_N))$ and the integral inner products may be approximated by sums. Then, the components of these discretized high-dimensional data $\mathcal{X}(t_i), \mathcal{X}(t_j)$ are typically highly correlated when t_i and t_j are close together. Consequently, the corresponding covariance matrix is nearly singular and thus solving the problem becomes numerically unfeasible. This difficulty can be overcome via regularization, for instance by adding a penalization (diagonal) matrix to the original (nearly singular) covariance matrix. See Hastie *et al.* (1995) and Yu *et al.* (1999) for details. In Li and Yu (2008) linear discriminant analysis over sections of curves is combined with support vector machines techniques in the so-called functional segment discriminant analysis.

In the filtering approach we choose a subspace $\mathcal{F}_1$ of $\mathcal{F}$ spanned by a finite functional basis. The coefficients of the projections on $\mathcal{F}_1$ of the functional data in terms of this basis (given by the inner products of the data with the elements of the basis) are multivariate observations to which the classical discriminant rules may be applied: see, for example, Hall *et al.* (2001). One of the difficulties here lies in choosing an appropriate basis for the infinite-dimensional data under study.

Finally James and Hastie (2001) propose a parametric model to fit irregularly sampled fragments of curves. This allows one to extend Fisher's linear discrimination rule to a type of functional data which does not admit regularization or filtering. See Chapter 11 of this volume for more details concerning sparseness in functional data analysis.

10.2.3 k-NN rules

General ideas

The k nearest neighbors (k-NN) method is based on a very simple idea which can be expressed equally well in both the multivariate and the functional settings. Thus, assume that $\mathcal{F}$ is a metric space. In order to classify a datum x, simply look at the k training data closest to x (in the metric D of $\mathcal{F}$) and assign x to P_0 when the majority of these k data belong to P_0. In the case of a tie, some randomized decision procedure can be used. Here $k = k_n$ is a given integer which plays the role of a smoothing parameter.

It is worth noting that the k-NN procedure can be seen as a particular case of the plug-in principle mentioned above. Indeed, the k-NN classifier can be obtained by replacing the unknown regression function $\eta(x)$ in (10.1) with the regression estimator

$$\eta_n(x) = \frac{1}{k} \sum_{i=1}^{n} 1_{\{\mathcal{X}_i \in k(x)\}} Y_i, \qquad (10.4)$$

where "$\mathcal{X}_i \in k(x)$" means that $\mathcal{X}_i$ is one of the k nearest neighbours of x. More concretely, if the pairs $(\mathcal{X}_i, Y_i)_{1 \leq i \leq n}$ are re-indexed as $(\mathcal{X}_{(i)}, Y_{(i)})_{1 \leq i \leq n}$ so that the $\mathcal{X}_{(i)}$'s are arranged in increasing distance from x, $D(x, \mathcal{X}_{(1)}) \leq D(x, \mathcal{X}_{(2)}) \leq \cdots \leq D(x, \mathcal{X}_{(n)})$, then $k(x) = \{\mathcal{X}_{(i)}, 1 \leq i \leq k\}$. This leads to the k-NN classifier $g_n(x) = 1_{\{\eta_n(x) \geq 1/2\}}$. Therefore, the k-NN classifier relies on the k-NN estimator (10.4) of the regression function $\eta(x) = P(Y = 1 | \mathcal{X} = x)$; see Burba et al. (2009) for a recent reference.

In practice, for a given training dataset, the value of k can be chosen by minimizing in k the following cross-validated version of $\hat{L}_n$:

$$\tilde{L}_n = \frac{1}{n} \sum_{i=1}^{n} 1_{\{g_{ni}(\mathcal{X}_i) \neq Y_i\}}, \qquad (10.5)$$

where g_{ni} denotes the *leave-one-out* k-NN rule based on the original sample of size n in which the ith observation $(\mathcal{X}_i, Y_i)$ has been deleted.

The ideas behind k-NN classifiers apply exactly along the same lines in the case of finite- or infinite-dimensional (functional) data. For the latter, an appropriate functional distance D must be chosen. This is equivalent to specifying the feature metric space $\mathcal{F}$ where the functional data are supposed to take values. Typical choices are $\mathcal{F} = L^2[a, b]$, endowed with the usual L^2-metric, and $\mathcal{F} = C[a, b]$, the space of continuous functions endowed with the supremum metric. As usual, the functional distances between two functional data must be approximated via discretization, but this can be seen as a numerical device which keeps intact the functional motivation for the method.

On consistency and lack of consistency

An important difference between the finite- and the infinite-dimensional situations arises with regard to consistency. As a consequence of a celebrated result by Stone (1977), the finite-dimensional k-NN classifiers are *universally (weakly) consistent* provided that $k = k_n \to \infty$ and $k/n \to 0$, as $n \to \infty$. This means that, for these k-NN classifiers, $L_n \to L^*$, in probability, as $n \to \infty$, *with no restriction at all on the underlying distribution* of (X_i, Y_i). In the infinite-dimensional case this result is no longer true. This has been pointed out by Cérou and Guyader (2006), who have studied the consistency of the k-NN classifier when $\mathcal{F}$ is a metric space. These authors have obtained the following sufficient condition for weak consistency:

Theorem 10.1 *Denote by L_n and L^*, respectively, the error (conditional on the data) associated with the k-NN classifier defined above and the Bayes (optimal) error for the problem under consideration. Assume that $(\mathcal{F}, D)$ is separable and the following Besicovich condition holds:*

$$\lim_{\delta \to 0} \frac{1}{P_\mathcal{X}(B(\mathcal{X}, \delta))} \int_{B(\mathcal{X}, \delta)} |\eta(z) - \eta(x)| dP_x(z) = 0 \quad \text{in probability,} \qquad (10.6)$$

where $B(x, \delta) := \{z \in \mathcal{F} : D(x, z) \leq \delta\}$ is the closed ball with center x and radius δ and $P_\mathcal{X}$ is the distribution of $\mathcal{X}$.

Then the k-NN classifier is weakly consistent, that is, $\mathbb{E}(L_n) \to L^*$ as $n \to \infty$, provided that $k \to \infty$ and $k/n \to 0$.

A sketch of the proof is given in Section 10.5.1. Condition (10.6) is clearly reminiscent of the conclusion of the classical finite-dimensional Lebesgue differentiation theorem. It can be seen (see again Cérou and Guyader 2006) that the following more convenient property ensures the Besicovich condition (10.6):

For every $\epsilon > 0$ and for $P_\mathcal{X}$-almost every $x \in \mathcal{F}$,

$$\lim_{\delta \to 0} P_\mathcal{X}\{z \in \mathcal{F} : |\eta(z) - \eta(x)| > \epsilon | D(x, z) < \delta\} = 0.$$

Consequently, (10.6) is also implied by the continuity ($P_\mathcal{X}$-a.e.) of the regression function $\eta(x) = P(Y = 1 | \mathcal{X} = x)$. This can be used to establish the consistency of the k-NN rules for a broad family of discrimination problems. Assume, for example, that the data from the populations P_0 and P_1 are continuous functions coming, respectively, from Gaussian processes with distributions ν_0 and ν_1 and covariance functions (sometimes called "triangular") of type $\Gamma(s, t) = u(\min(s, t)) v(\max(s, t))$. Varberg (1961) and Jørsboe (1968) have proved that, under some additional conditions, such processes fulfill $\nu_0 \ll \nu_1$ (i.e., ν_0 is absolutely continuous with respect to ν_1) and, more importantly, the explicit expression of the corresponding Radon–Nikodym derivative $d\nu_0/d\nu_1$ can be obtained. Baíllo et al. (2010) give an outline of the applicability of such results to the functional discrimination problem. In particular, the continuity of the Radon–Nikodym derivative $d\nu_0/d\nu_1$ can be used to prove the continuity of $\eta(x)$ and, consequently, the consistency of the k-NN classifier. Moreover, the relative simplicity of dv_0/dv_1 can also be employed to develop alternative plug-in estimators of the optimal classification rule (10.1).

On the possible role of k-NN as a reference method

As for the practical performance of the functional k-NN classifiers, the results available are still quite limited (see, for example, Cuevas et al. 2007). Nevertheless, they suggest that these classification rules could represent a sort of standard default choice in the functional setting (much as Fisher's linear rule does in finite-dimensional problems), providing a reasonable balance between ease of implementation, interpretability, and performance. These practical aspects will be considered in more detail in Sections 10.4.1 and 10.4.2.

10.2.4 Kernel rules

Apart from the k-NN rule there exists another simple classification procedure, the moving window rule, which is based on the majority vote of the data surrounding the point $\mathcal{X}$ to be classified. In this case a positive smoothing parameter $h = h_n$ is fixed and the datum $\mathcal{X}$ is assigned to population P_0 if the majority of the data within distance h of $\mathcal{X}$ belong to P_0. In general the "distance" will be specified by a metric D in the feature functional space $\mathcal{F}$.

Specifically, the moving window rule assigns $\mathcal{X}$ to P_0 if

$$\sum_{i=1}^{n} 1_{\{Y_i=0,\ D(\mathcal{X}_i, \mathcal{X}) \leq h\}} > \sum_{i=1}^{n} 1_{\{Y_i=1,\ D(\mathcal{X}_i, \mathcal{X}) \leq h\}} \tag{10.7}$$

and to P_1 otherwise (see Abraham et al. 2006).

A smoother and more general version of this discrimination rule is given by the kernel classifier

$$g_n(\mathcal{X}) = \begin{cases} 0 \text{ if } \sum_{i=1}^{n} 1_{\{Y_i=0\}} K\left(\frac{D(\mathcal{X}_i, \mathcal{X})}{h}\right) > \sum_{i=1}^{n} 1_{\{Y_i=1\}} K\left(\frac{D(\mathcal{X}_i, \mathcal{X})}{h}\right), \\ 1 \text{ otherwise} \end{cases}$$

where the kernel $K : [0, \infty) \to [0, \infty)$ is a nonincreasing known function. Some popular choices for K are the Gaussian kernel $K(x) = e^{-x^2}$, the Epanechnikov kernel $K(x) = (1 - x^2) 1_{[0,1]}$, and the uniform kernel $K(x) = 1_{[0,1]}$. When using the uniform kernel, the kernel classification rule actually turns into the moving window rule. But if we wish to give more weight to the sample points $\mathcal{X}_i$ that are closer to $\mathcal{X}$, we would usually choose a smoother K.

Observe that the kernel rule may be derived as a particular case of the plug-in methodology, when the regression function η in (10.1) is replaced by the kernel regression estimator (see Chapter 4)

$$\eta_n(x) = \frac{\sum_{i=1}^{n} Y_i K\left(\frac{D(\mathcal{X}_i, \mathcal{X})}{h}\right)}{\sum_{i=1}^{n} K\left(\frac{D(\mathcal{X}_i, \mathcal{X})}{h}\right)}.$$

Hence, as in the *k*-NN case, the kernel classifier is another example of the plug-in methodology mentioned in Section 10.2.1.

In the finite-dimensional context it has been proved (see, for example, Devroye *et al.* 1996, p. 149) that the kernel classification rule is strongly universally consistent if the kernel function is regular and if $h \to 0$ and $nh^d \to \infty$ when $n \to \infty$, where d is the dimension of X. The study of general sufficient conditions for consistency in the case of an infinite-dimensional $\mathcal{X}$ is an interesting research problem (see Ferraty and Vieu (2006), p. 124). For example, Abraham *et al.* (2006) prove that universal consistency does not hold for the moving window rule in the infinite-dimensional context. Moreover, these authors also offer the following consistency result, that relies on a "differentiability condition" similar to (10.6):

Theorem 10.2 *Given any set $\mathcal{G} \subset \mathcal{F}$, and any $\epsilon > 0$, let $\mathcal{N}(\epsilon, \mathcal{G})$ denote the ϵ-covering number defined by*

$$\mathcal{N}(\epsilon, \mathcal{G}) = \inf\left\{N \geq 1 : \exists x_1, \ldots, x_N \in \mathcal{F} \text{ with } \mathcal{G} \subset \bigcup_{i=1}^{N} B^\circ(x_i, \epsilon)\right\},$$

where $B^\circ(x, \epsilon)$ denotes the open ball of center x and radius ϵ (i.e., $\mathcal{N}(\epsilon, \mathcal{G})$ is the minimum number of open balls of radius ϵ which are required in order to cover $\mathcal{G}$).
Assume that

(i) There exists a sequence $\{\mathcal{F}_k\}$ of totally bounded sets (i.e. $\mathcal{N}_k(\epsilon) := \mathcal{N}(\epsilon, \mathcal{F}_k) < \infty$) such that for all k, $\mathcal{F}_k \subset \mathcal{F}_{k+1}$ and $P_\mathcal{X}\left(\bigcup_{k=1}^{\infty} \mathcal{F}_k\right) = 1$.

(ii) $\lim_{\delta \to 0} \left[\int_{B(\chi, \delta)} \eta(z) dP_\mathcal{X}(z)\right] / P_\mathcal{X}(B(\chi, \delta)) = \eta(\chi)$, in probability.

(iii) $h = h_n \to 0$ and $\mathcal{N}_k(h/2)/n \to 0$, as $n \to \infty$.

Then the "moving window rule" defined in (10.7) is weakly consistent.

From a more practical point of view there remains to be addressed the issues of how to select the bandwidth h and the metric D. In fact (as discussed in Ferraty and Vieu 2003 and Chapter 3 in Ferraty and Vieu 2006), the use of a semi-metric, instead of a proper metric, can be advantageous in some cases. Regarding the choice of such a semi-metric, Ferraty and Vieu (2006, p. 28), describe three semi-metrics which are well-adapted for functional data. The first one is based on functional principal components (see Chapter 8 for an overview of functional principal component analysis). The second is inspired by the multivariate partial least squares regression approach and thus is appropriate only when an additional response variable is observed, as is the case in the discrimination problem. These two semi-metrics are suitable for rough functional data. The third semi-metric, which depends on the distance between the derivatives of the curves, is clearly only suitable for smooth functional data.

Concerning the smoothing parameter h, a simple and natural idea would be to find the h that minimizes the cross-validated version (10.5) of $\hat{L}_n$, as when choosing the optimal k for the k-NN rule. However, the optimization task is typically harder for h than for k, since this last parameter can only take a finite number of values.

10.2.5 Classification based on partial least squares

The definition and meaning of PLS

Partial Least Squares (PLS) is an increasingly popular dimension-reduction technique for multivariate regression problems with predictor X and a response Y. The idea behind this method is similar to that of principal components: to project the data along directions of high variability. The main difference is that PLS also takes into account the response variable Y when defining the projection directions. More precisely, let X be a p-dimensional random vector of explanatory variables with covariance matrix Σ_x. Let Y be the q-dimensional response vector with covariance matrix Σ_y.

Denote by Σ_{xy} the $p \times q$ matrix of covariances between the variables of X and those of Y. We define the first pair of PLS directions as the unit vectors $\mathbf{a}_1 \in \mathbb{R}^p$, $\mathbf{b}_1 \in \mathbb{R}^q$ which maximize with respect to $\mathbf{a}$ and $\mathbf{b}$ the expression

$$\frac{(\text{cov}(\mathbf{a}'X, \mathbf{b}'Y))^2}{(\mathbf{a}'\mathbf{a})(\mathbf{b}'\mathbf{b})}.$$

It can be shown that $\mathbf{a}_1$ is the eigenvector of $\Sigma_{xy}\Sigma_{yx}$ corresponding to the largest eigenvalue of this matrix. The direction $\mathbf{b}_1$ fulfills $\mathbf{b}_1 = \Sigma_{yx}\mathbf{a}_1$.

The remaining PLS directions can be found in a similar way by imposing the additional condition that the direction $\mathbf{a}_{k+1}$ is orthogonal to the $\mathbf{a}_1, \ldots, \mathbf{a}_k$ already found. We thus obtain

$$(\mathbf{a}_{k+1}, \mathbf{b}_{k+1}) = \underset{\substack{\mathbf{a} \in \mathbb{R}^p, \mathbf{b} \in \mathbb{R}^q \\ \mathbf{a}'A = 0}}{\arg \max} \frac{(\text{cov}(\mathbf{a}'X, \mathbf{b}'Y))^2}{(\mathbf{a}'\mathbf{a})(\mathbf{b}'\mathbf{b})},$$

where A is the $p \times k$ matrix $A = [\mathbf{a}_1, \ldots, \mathbf{a}_k]$. It can be shown that $\mathbf{a}_{k+1}$ is the eigenvector of $\Sigma_{xy}\Sigma_{yx}$ corresponding to the $(k+1)$th largest eigenvalue and $\mathbf{b}_{k+1} = \Sigma_{yx}\mathbf{a}_{k+1}$. Of course, in practice, the covariance population matrices involved in these calculations must be estimated from the sample data.

Applications to binary classification; the functional case

In the binary classification problem the variable Y takes values in $\{0, 1\}$, which allows for some simplification in the above expressions. For example, the matrix $\Sigma_{xy}\Sigma_{yx}$ can be estimated by

$$S_{xy}S_{yx} = \sum_{i=0}^{1} \frac{1}{(n-1)^2} n_i^2 (\bar{x}_i - \bar{x})(\bar{x}_i - \bar{x})',$$

where $\bar{x} = (n_1\bar{x}_1 + n_0\bar{x}_0)/(n_0 + n_1)$ and, for $i = 0, 1$, $\bar{x}_i$ denotes the vector of means estimated from the n_i sample elements belonging to the population P_i.

The general idea of PLS classification is to reduce the dimension of the data via PLS and then to use any standard classification method (for example, Fisher's linear procedure) on the projected data of lower dimension. See Barker and Rayens (2003), and Liu and Rayens (2007) for further details and references.

The case where the predictor variable $\mathcal{X}$ is functional has been considered in Preda *et al.* (2007), where Fisher's linear method is used after applying the PLS dimension-reduction technique to a discretized version of the functional data of type $X = (\mathcal{X}(t_1), \mathcal{X}(t_2), \ldots, \mathcal{X}(t_N))$. One important issue here is the choice of the number of PLS directions to project these data on to. This can be chosen by minimizing the cross-validated classification error in the training dataset. The practical performance of PLS classification is quite satisfactory according to the results of this study. However, the less sophisticated k-NN method also shows good behaviour.

For a practical use of PLS and k-NN in a functional classification problem involving magnetic resonance spectra, see Barba *et al.* (2008). An application of the PLS methodology to the logit functional regression problem (closely related to binary classification) can be found in Escabias *et al.* (2007).

10.2.6 Classification based on reproducing kernels

The concept of reproducing kernels; some basic notions

The theory of reproducing kernels, which comes from mathematical analysis, has many important ramifications in different fields of theoretical and applied mathematics. The book by Berlinet and Thomas-Agnan (2004) provides a comprehensive account of its statistical applications. Here we will just outline a few of them, concerning the use of reproducing kernels for constructing functional classifiers in a binary classification problem.

A function $\kappa : \mathcal{F} \times \mathcal{F} \longrightarrow \mathbb{R}$ is said to be a reproducing kernel of a real Hilbert space $(\mathcal{H}, \langle \cdot, \cdot \rangle)$ of real functions defined on $\mathcal{F}$ if

(a) For all $x \in \mathcal{F}$, $\kappa(., x) \in \mathcal{H}$.
(b) For all $x \in \mathcal{F}$ and $\varphi \in \mathcal{H}$, $\langle \varphi, \kappa(\cdot, x) \rangle = \varphi(x)$ (this is the so-called "reproducing property").

From (a) and (b) it follows that $\kappa(x_1, x_2) = \langle \kappa(\cdot, x_1), \kappa(\cdot, x_2) \rangle$. A Hilbert space of functions for which there exists a reproducing kernel is called a reproducing kernel Hilbert space (RKHS).

It can be shown (see Theorem 10.1, Lemma 2, and Theorem 10.3 in Berlinet and Thomas-Agnan 2004) that the continuity on $\mathcal{H}$ of the evaluation functionals e_x defined by $g \mapsto e_x(g) = g(x)$ is a necessary and sufficient condition for $\mathcal{H}$ to be a RKHS. Also, κ is a reproducing kernel for some (unique) Hilbert space $\mathcal{H}$ if and only if κ is a positive type function, that is, the matrix $(\kappa(x_i, x_j))_{1 \le i, j \le n}$ is nonnegative definite for any $n \in \mathbb{N}$ and $(x_1, \ldots, x_n) \in \mathcal{F}^n$. In this case the subspace $\mathcal{H}_0$ of $\mathcal{H}$ spanned by the functions $\{\kappa(., x)\}_{x \in \mathcal{F}}$ is dense in $\mathcal{H}$.

The rationale behind many practical uses of reproducing kernels in statistics is summarized by Persi Diaconis (in the preface of Berlinet and Thomas-Agnan 2004) in the following words:

The idea is to use a kernel $K(x, y)$ to embed the observed points $\mathcal{X}_1, \ldots, \mathcal{X}_n$ into a Hilbert space by declaring that the distance between $\mathcal{X}_i$ and $\mathcal{X}_j$ is $K(\mathcal{X}_i, \mathcal{X}_j)$. This distance assigned, standard tools of Multivariate Analysis (Principal Components, Canonical Correlations, Discriminant Analysis) can be applied.

There is a clear link between reproducing kernels and the theory of second-order stochastic processes since the covariance functions $\Gamma(s, t)$ of such processes are known to be functions of positive type. This fact allows for the use of reproducing kernel methods in some problems of filtering or best linear prediction. However, the most interesting aspect of the RKHS to be considered here is the ease of computation for obtaining explicit solutions of some constrained functional optimization problems. This is particularly important in the spline methodology of nonparametric curve estimation (see Chapter 3 in Berlinet and Thomas-Agnan 2004).

Applications to supervised classification

In this framework one could consider using a plug-in classifier obtained by replacing the regression function η in the Bayes rule (10.1) with the function $\hat{\eta} \in \mathcal{H}_\kappa$ (the RKHS associated with the kernel κ), minimizing the "regularized empirical risk"

$$\frac{1}{n} \sum_{i=1}^{n} \mathcal{C}(\mathcal{X}_i, Y_i, \hat{\eta}(\mathcal{X}_i)) + J(\hat{\eta}),$$

where $\mathcal{C}$ is a loss function convex with respect to the third argument and $J(\hat{\eta})$ is a penalty term. Typical choices are $\mathcal{C}(\mathcal{X}, Y, \hat{\eta}(\mathcal{X})) = (Y - \hat{\eta}(\mathcal{X}))^2$ and $J(\hat{\eta}) = \lambda \|\hat{\eta}\|_\kappa^2$, for a regularization parameter $\lambda > 0$. In the classification problems where $Y \in \{0, 1\}$ is a categorical response, another possible choice for $\mathcal{C}$ is the logistic loss function $\mathcal{C}(\mathcal{X}, Y, \hat{\eta}(\mathcal{X})) = -Y \log \hat{\eta}(\mathcal{X}) + \log(1/(1 - \hat{\eta}(\mathcal{X})))$.

This approach has been considered in the literature from slightly different points of view (see Preda 2007 and references therein). In particular, it has been shown that the optimal solution $\hat{\eta}$ exists, is unique, and has a representation of the form

$$\hat{\eta}(x) = \sum_{i=1}^{n} a_i \kappa(x, \mathcal{X}_i),$$

with $a_i \in \mathbb{R}$. A standard choice for κ is the Gaussian kernel $\kappa(x_1, x_2) = \exp\left(-\|x_1 - x_2\|_2^2/\sigma_\kappa^2\right)$, where $\sigma_\kappa > 0$ is a fixed parameter.

The RKHS approach is more of a general methodology than a uniquely defined method. Even when the kernel κ and the loss $\mathcal{C}$ have been chosen, there still remain some tuning parameters to be determined (typically by a cross-validation procedure). So far we have only limited experience in the use of these procedures for functional supervised discrimination and the results are still inconclusive.

10.2.7 Classification methods based on depth measures

On notions of data depth

The notion of depth is especially important in multivariate analysis. If we are able to decide how deep a datum is in a population, then we have a criterion to define the median as the deepest value in the population. The corresponding empirical definitions of depth, based on the sample data, would allow us to measure how deep a datum is in a data cloud, so providing different multivariate notions of sampling medians.

The application of the same ideas to detect outliers or to define α-trimmed means as the average of the $100(1 - \alpha)\%$ deepest observations is also very clear.

In the real line the depth of a datum x with respect to a population defined by a distribution function F is easily evaluated by the quantities $F(x)(1 - F(x-))$ or $\min\{F(x), 1 - F(x)\}$ which lead, in several respects, to equivalent results. In the multivariate case there are several popular ways to measure depth (simplicial depth, Tukey's depth, etc.). An account of them, even concise, is beyond the scope of this work. We refer the reader, for example, to Zuo and Serfling (2000).

The use of data depth for functional classification

Suppose we have a measurement $D(P_i, x)$ of the depth of the datum x in the population P_i, for $i = 0, 1$. Suppose also we have empirical versions $D_{ni}(x)$ of these depth measures, defined in terms of the training sample. A simple depth-based classifier for an incoming value x would be $g(x) = 1_{\{D_{n1}(x) > D_{n0}(x)\}}$, which amounts to assigning x to the population in which we estimate it is most deeply placed. For a study on the relationship between classification and depth measures in a multivariate setup see Ghosh and Chaudhuri (2005).

A recent proposal that applies depth tools to functional classification is given in Cuevas and Fraiman (2009), relying on previous ideas of Fraiman and Muniz

(2001) and Cuesta-Albertos et al. (2007). These authors suggest evaluating the depth of a function $x \in \mathcal{F}$ (a real separable Banach space) with respect to a probability measure P on $\mathcal{F}$ as the average

$$D(P, x) = \int D(P_f, f(x)) \, dQ(f), \qquad (10.8)$$

where $f : \mathcal{F} \to \mathbb{R}$ denotes a linear continuous functional (so f is an element of the dual $\mathcal{F}^*$ of $\mathcal{F}$), $P_f = P_{f(\mathcal{X})}$ is the univariate distribution of the random variable $f(\mathcal{X})$, $\mathcal{X}$ being a random element in $\mathcal{F}$ with distribution P, $D(P_f, f(x))$ is the univariate depth $D(P_f, f(x)) = \mathbb{P}(f(\mathcal{X}) \leq x)\mathbb{P}(f(\mathcal{X}) \geq x)$, and Q stands for a given probability measure on the dual space $\mathcal{F}^*$ (or on the unit ball in $\mathcal{F}^*$).

Thus the idea is to randomly choose a direction f along which to project the data on the real line, to evaluate the depth of the univariate projections obtained in this way (by using an empirical version of $D(P_f, f(x))$) and, finally, to calculate the average of such depths. The measure Q can be chosen, in the finite-dimensional case, as the uniform distribution on the unit sphere or (in the functional case) as a Gaussian measure.

The properties of this depth measure as well as its applicability to different statistical problems (including supervised classification) are considered in Cuevas and Fraiman (2009). Some simulations are given in a previous paper by Cuevas et al. (2007), where other versions of this idea are also proposed: for example, the method can be used by simultaneously projecting the functions in the sample and their derivatives. In this way we get a bivariate sample of projections (from the function and its derivative) and tackle the problem in an analogous fashion, this time by using a depth measure in $\mathbb{R}^2$.

Another proposal for a functional depth, with applications to classification, can be found in López-Pintado and Romo (2006).

The h-modal depth: another look at the kernel rule

Let us finally mention that the kernel rules considered above can also be motivated in terms of *modal depth*, as indicated in Cuevas et al. (2007). According to this point of view one can define the population *h-modal depth* of a datum x as given by the function

$$f_h(x) = \mathbb{E}(K_h(\|x - \mathcal{X}\|)), \qquad (10.9)$$

where $\|\cdot\|$ denotes a suitable norm (for example the L_2-norm in the functional case), K is a specified kernel function (for example the standard Gaussian), $K_h(t) = h^{-1}K(t/h)$, and h a fixed tuning parameter. Consequently, one can define the *h-mode* of $\mathcal{X}$ as the deepest value of x obtained by maximizing (in x) the function (10.9). Of course, the empirical version of (10.9) given by

$$\hat{f}_h(x) = \frac{1}{n} \sum_{i=1}^{n} K_h \left(\| z - \mathcal{X}_i \| \right) \qquad (10.10)$$

would be used in practice.

This notion is clearly reminiscent of the multivariate depth concepts based on likelihood (see, for example, Fraiman and Meloche 1999 and references therein) but in the functional case, one must bear in mind that there is no real underlying density to be estimated by (10.10) by letting $h \to 0$ slowly enough, as is done in the well-known finite-dimensional kernel estimates. This is associated with the lack of a simple translation-invariant dominating measure which could play the role that Lebesgue measure does in the Euclidean space $\mathbb{R}^d$. Still, the expression (10.9) can be informally seen as a measure of the depth of x, as it indicates "how surrounded" the function x is in the sample space. The bandwidth h is in principle fixed but there are different possibilities for selecting it. In Cuevas et al. (2007) it is chosen as the 20th percentile in the L^2 distances between the functions in the training sample. Observe that different choices of h lead to h-modal depths of noncomparable magnitude, since, for instance, $\max_x f_h(x)$ (where the maximum is taken on the support of f_h) will typically increase as h decreases. However, the use of a re-scaled version of f_h such as $(f_h(z) - \min_x f_h(x))/(\max_x f_h(x) - \min_x f_h(x))$ could circumvent this difficulty.

10.2.8 Other classification methods

Some general ideas on support vector machines

The so-called *support vector machines* (SVM) methodology has become very popular in recent years. It has its origins in work by Vapnik in the late seventies; see, for example, Vapnik (1998). A tutorial is given in Burgess (1998). We do not attempt to present the SVM theory here, but only to roughly outline some relevant ideas. If the training data (X_i, Y_i) are linearly separable in the sense that those with $Y_i = 0$ lie in a different halfspace to those with $Y_i = 1$, the discrimination problem can be reduced to that of finding the "best separating hyperplane". This turns out to be an optimization problem which essentially depends on the X_i's only through the inner products $\langle X_i, X_j \rangle$. The method can be also adapted (Cortes and Vapnik 1995) to the cases where the training data are not exactly separable by a hyperplane. The highly nonlinear cases can be treated by defining an appropriate transformation $x \mapsto \Phi(x)$ which takes the data into a Hilbert space $(\mathcal{H}, \langle \cdot \rangle_\mathcal{H})$ (usually of higher dimension than the original x's space) where the transformed data are linearly (or "almost linearly") separable. This space can be endowed with the inner product $K(x_1, x_2) = \langle \Phi(x_1), \Phi(x_2) \rangle_\mathcal{H}$. The function K is a reproducing kernel; see Lemma 1 in Berlinet and Thomas-Agnan (2004). The theoretical properties of the resulting classifiers can be studied in terms of the theory of structural risk minimization developed by Vapnik: see, for example, Section 6 of Burgess (1998) for more details and references.

For a recent use of these ideas in the setting of supervised functional classification the reader is referred to Rossi and Villa (2006).

On neural networks in functional classification

The methodology of *classification neural networks* has also been considered in functional problems. A recent reference is Ferré and Villa (2006), where a multilayer perceptron for functional inputs is studied from a theoretical and practical point of view.

10.3 UNSUPERVISED CLASSIFICATION

As mentioned in the introduction, unsupervised classification is largely identified with cluster analysis although, in fact, other subjects could be included under this title; see, for example, Hastie *et al.* (2001). In any case, here we will only consider clustering methods for functional data.

As is well known, cluster analysis deals with the problem of identifying groupings, relatively isolated from each other and constituted by similar points. In other words, the aim is to partition the data into dissimilar groups of similar items which is, in some sense, a harder problem than supervised classification. In the unsupervised case, we want to find homogeneous groups within a large data cloud, but there is no training sample to serve as a guide. Furthermore, in most real-data applications, the number of groups is unknown.

In the finite-dimensional case, there are two main avenues for cluster analysis: K-mean type methods and hierarchical methods. In the first case there is a particular population target in mind which motivates the clustering algorithm: it is given by the population K-means, defined as a natural extension of the notion of mean (understood as the "mass center" of the distribution) to the collection of the K clusters. The main ideas behind K-means clustering are presented in the next section, where we also show that the method can be adapted in a quite natural way to problems with infinite-dimensional data.

The hierarchical methods rely on the use of a matrix of distances between the sample points. They are not usually based on the sampling approximation of an optimization problem defined in population terms. However, the resulting clusters can sometimes be considered as sampling approximations to the population clusters defined, according to Hartigan's (1975) notion, as the connected components of the level set $\{g \geq c\}$, where g is the underlying data density and c is a given constant. In principle, the distance-based hierarchical algorithms can be adapted to the functional clustering problems. However, the geometrical motivation based on connected components is obviously lost in these infinite-dimensional problems, as no reference measure (and therefore no underlying density) is usually available in these cases.

Choosing an adequate criterion; the number of clusters

In the case of supervised classification there is an obvious aim, which is to minimize the misclassification rate over the test sample. For consistent classification rules this leads to an asymptotic minimization of the "true" (population) misclassification rate. The situation is a bit different in the unsupervised case where there is no unique widely accepted criterion for deciding what is a "satisfactory" clustering criterion. While, as mentioned above, the popular K-means clustering procedure is motivated in terms of a "reasonable" objective function, we cannot say that this function represents an indisputable target for all clustering problems.

A recent attempt to define such a general target (which moreover provides a guide for the choice of the number of clusters) has been proposed in Fraiman et al. (2008). Given a random element $\mathcal{X}$ in $\mathcal{F}$, a dispersion measure $\mathcal{D}$ for which $\mathcal{D}(\mathcal{X})$ is finite, and a family $\{f_K\}$ of partition methods $f_K : \mathcal{F} \to \{1, \ldots, K\}$, let us first define $W_{j,K}$ to be the conditional distributions $\mathcal{X} | f_K(\mathcal{X}) = j$, $j = 1, \ldots, K$ of the random element $\mathcal{X}$ given the partition sets $f_K(\mathcal{X}) = j$. A population cluster method is "good" within the family of methods given by $\{f_K\}$ if the following three quantities are small enough:

$$\frac{\sum_{j=1}^{K} \mathcal{D}(W_{j,K})}{\mathcal{D}(X)}, \quad \frac{\sum_{j=1}^{K} \mathcal{D}(W_{j,K})}{\sum_{j=1}^{K-1} \mathcal{D}(W_{j,K-1})}, \quad \frac{\sum_{j=1}^{K+1} \mathcal{D}(W_{j,K+1})}{\sum_{j=1}^{K} \mathcal{D}(W_{j,K})} - 1. \tag{10.11}$$

The first condition requires that the dispersions within the clusters are small relative to the global dispersion, the second and third account for the fact that we are looking for the "correct" number of clusters K. Clearly the first condition is not sufficient, since we can decrease the dispersion of the $W_{j,K}$ variables by increasing the number of clusters.

For a fixed number K of clusters, an optimality criterion (which could play an analogous role to that of the Bayes error in supervised classification) could be the infimum over all possible partitions $G_1, \ldots, G_K$ of size K of $\sum_{j=1}^{K} \mathcal{D}(W_{j,K}^*)/\mathcal{D}(\mathcal{X})$, where $W_{j,K}^*$ stands for the conditional distribution $\mathcal{X} | \mathcal{X} \in G_j$. All the quantities in equation (10.11) can be estimated by simply replacing the true underlying distribution by the empirical distribution.

In the following section we will summarize the infinite-dimensional clustering methods that are available, especially those inspired by the K-means methodology.

The mathematical foundations of this theory are more developed than those corresponding to supervised functional classification (although this is not the case with real-data applications). For this reason we shall consider some technical aspects, including some proofs, in more detail in this section.

10.3.1 K-means for functional data: the objective function

The extension of K-means clustering methods to infinite-dimensional settings has been considered by various authors, for example Pollard (1981) and Cuesta-Albertos and Matrán (1988).

For a fixed number of clusters K, we look for functions $f : \mathcal{F} \to \{1, \ldots, K\}$ which determine to which cluster each single point belongs. We denote the space partition by $G_k = f^{-1}(k)$, $k = 1, \ldots, K$. We assume $\bigcup_{k=1}^{K} G_k = \mathcal{F}$ and $G_i \cap G_j = \emptyset$ for $i \neq j$.

Let us begin by defining the population K-mean parameter (or K-cluster centers). If $\mathcal{F}$ is a normed space, and $\mathcal{X}$ is an $\mathcal{F}$-valued random element with distribution P, let us denote

$$I_K(P, \mathbf{h}) = I_K(\mathcal{X}, \mathbf{h}) = \mathbb{E}\left(\min_{i=1,\ldots,K} \|\mathcal{X} - h_i\|^2\right) \qquad (10.12)$$

for a vector of centers $\mathbf{h} = (h_1, \ldots, h_K) \in \mathcal{F}^K$. This will be the objective function we want to minimize. Let us also define the "optimal clustering risk" $I_K(P)$ associated with the objective function (10.12) as

$$I_K(P) = I_K(\mathcal{X}) = \inf_{\mathbf{h} \in \mathcal{F}^K} I_K(P, \mathbf{h}). \qquad (10.13)$$

$H_K(P)$ will stand for any set with K elements $\{h_1^P, \ldots, h_K^P\} \subset \mathcal{F}$ for which equality in (10.13) is reached. The elements of $H_K(P)$ are the centers which define the K-means clusters: each observation of $\mathcal{F}$ is assigned to the nearest cluster center and the ties are decided at random.

For instance, if we consider $K = 2$ then $c_1, c_2 \in \mathcal{F}$ are the optimal cluster centers if they are the values of h_1 and h_2 which minimize the expectation $\mathbb{E}\left(\min(\|\mathcal{X} - h_1\|^2, \|\mathcal{X} - h_2\|^2)\right)$, the set G_1 is given by $G_1 = \{x \in \mathcal{F} : \|x - c_1\| \leq \|x - c_2\|\}$, and $G_2 = G_1^c$.

In order to estimate the parameters, let us assume that we have a sample of random elements $\mathcal{X}_1, \ldots, \mathcal{X}_n$ with the same distribution as $\mathcal{X}$, taking values on $\mathcal{F}$. For $n \in \mathbb{N}$, let P_n be the empirical probability distribution on the space $\mathcal{F}$ corresponding to this sample. Natural plug-in estimates of the set $H_K(P)$ and the value $I_K(P)$ are $H_K(P_n)$ and $I_K(P_n)$, respectively. The empirical objective function, sometimes called the *empirical squared norm criterion*, is given by

$$I_K(P_n, \mathbf{h}_n) = \frac{1}{n} \sum_{i=1}^{n} \min_{k=1,\ldots,K} \|\mathcal{X}_i - h_k\|^2. \tag{10.14}$$

The K-means clustering prescribes the criterion for partitioning the sample $\mathcal{X}_1, \ldots, \mathcal{X}_n$ into K groups, by minimizing the empirical squared norm (10.14) over all possible choices of cluster centers $\mathbf{h}$. Associated with each empirically optimal center h_{nj}^0 is the convex polyhedron $G_{j,n}$ of all points in $\mathcal{F}$ closer to h_{nj}^0 than to any other center, called the *Voronoi cell* of h_{nj}^0 (ties are broken randomly). Each $\mathcal{X}_i$ is assigned to its nearest center, and each center $h_{n1}^0, \ldots, h_{nK}^0$ is just the average of those $\mathcal{X}_i$'s falling into its Voronoi cell.

First we will focus on the results given in Biau et al. (2008) on K-means for data from Hilbert spaces. Afterwards we describe the impartial trimmed K-means procedure, a robust extension of K-means, given in Cuesta-Albertos and Fraiman (2006), for data on uniformly convex Banach spaces.

10.3.2 K-means for data in a Hilbert space: some theoretical results

Throughout this section we will assume that $\mathcal{X}$ is a random element in a Hilbert space $\mathcal{F}$ that satisfies $\mathbb{E}(\|\mathcal{X}\|^2) < \infty$. Since the "absolute" empirical minimizer $\mathbf{h}_n^0$ of the criterion defined in (10.14) is not always easy to find, we can consider approximate versions of it, as follows. Given $\delta_n \geq 0$, $\mathbf{h}_n \in \mathcal{F}^k$ is called a δ_n-minimizer of the empirical objective function (10.14) (or a vector of δ_n-optimal empirical centers) if

$$I_K(P_n, \mathbf{h}_n) \leq I_K(P_n) + \delta_n.$$

Of course, for $\delta_n = 0$, we get $\mathbf{h}_n = \mathbf{h}_n^0$. The following result, due to Biau et al. (2008), states the consistency of $I_K(P, \mathbf{h}_n)$ towards $I_K(P)$, as $n \to \infty$.

Theorem 10.3 *Let $\mathbf{h}_n$ be a δ_n-minimizer of the empirical objective function. If $\lim_{n \to \infty} \delta_n = 0$, then*

(i) $\lim_{n \to \infty} I_K(P, \mathbf{h}_n) = I_K(P)$ *a.s.*
(ii) $\lim_{n \to \infty} \mathbb{E}(I_K(P, \mathbf{h}_n)) = I_K(P).$

In the finite-dimensional case, statement (i) has been proved in Pollard (1982); see also Pollard (1981) and Abaya and Wise (1984).

This result does not, however, provide information on what sample size is required in order to ensure that $I_K(P_n, \mathbf{h}_n)$ is close to the optimal value $I_K(P)$ for a δ_n-optimal vector $\mathbf{h}_n$ of empirical centers.

A possible way of tackling this problem is given by the following inequality (see Devroye et al. 1996, Chapter 8),

$$\mathbb{E}\left(I_K(P, \mathbf{h}_n) - I_K(P)\right) = \mathbb{E}\left((I_K(P, \mathbf{h}_n) - I_K(P_n, \mathbf{h}_n)) + (I_K(P_n, \mathbf{h}_n) - I_K(P))\right)$$

$$\leq \mathbb{E}\left(\sup_{\mathbf{h} \in \mathcal{F}^K} (I_K(P_n, \mathbf{h}) - I_K(P, \mathbf{h}))\right)$$

$$+ \sup_{\mathbf{h} \in \mathcal{F}^K} \mathbb{E}\left(I_K(P_n, \mathbf{h}) - I_K(P, \mathbf{h})\right) + \delta_n.$$

Thus, it will suffice to control the convergence rate of the uniform deviation

$$\mathbb{E}\left(\sup_{\mathbf{h} \in \mathcal{F}^K} (I_K(P_n, \mathbf{h}) - I_K(P, \mathbf{h}))\right).$$

In the finite-dimensional case this can be done using standard empirical processes techniques (see, for instance, Pollard 1981 and Antos et al. 2005). However, as pointed out in Biau et al. (2008), in infinite-dimensional Hilbert spaces these techniques yield suboptimal bounds.

The following theorem gives a sharp result in this direction using Rademacher averages. The proof can be found in Biau et al. (2008). A sketch of this proof is given in Section 10.5.2.

For any $R \geq 0$, let $\mathcal{P}(R)$ denote the set of probability measures on $\mathcal{F}$ supported on the closed ball $B(0, R)$ of radius R centered at the origin.

Theorem 10.4 *Assume $P \in \mathcal{P}(R)$. For any δ_n-minimizer $\mathbf{h}_n$ of the empirical objective function, we have*

$$\mathbb{E}(I_K(P, \mathbf{h}_n)) - I_K(P) \leq \frac{8K R \sqrt{\mathbb{E}(\|\mathcal{X}\|^2)} + 4K R^2}{\sqrt{n}} + \delta_n,$$

and, consequently,

$$\mathbb{E}(I_K(P, \mathbf{h}_n)) - I_K(P) \leq \frac{12K R^2}{\sqrt{n}} + \delta_n.$$

As a corollary, it can be shown that for $P \in \mathcal{P}(R)$ and for any $M > 0$,

$$\mathbb{P}\left\{I_K(P, \mathbf{h}_n) - I_K(P) \leq \frac{12K R^2 + 4R^2\sqrt{2M}}{\sqrt{n}} + \delta_n\right\} \geq 1 - e^{-M}.$$

As stated in the following theorem (see also Biau et al. 2008) the requirement $P \in \mathcal{P}(R)$ can be removed:

Theorem 10.5 *For any $M > 0$, there exist positive constants $C(P)$ and $N_0 = N_0(P, K, M)$ such that, for all $n \geq N_0$,*

$$\mathbb{P}\left\{I_K(P, \mathbf{h}_n) - I_K(P) \leq C(P)\frac{K + \sqrt{M}}{\sqrt{n}} + \delta_n\right\} \geq 1 - 2e^{-M}.$$

The proof can be found in Biau et al. (2008).

10.3.3 Impartial trimmed K-means for functional data

The notion of impartial trimming was introduced in Gordaliza (1991) as a robust alternative technique for the multidimensional location problem. This idea was also adapted by Cuesta-Albertos et al. (1997) to the field of cluster analysis as a robust method in the spirit of K-means, but one that is less sensitive than K-means to the presence of a small group of outliers. This is achieved through the notion of impartial trimmed K-means (ITKM in what follows). Roughly speaking, the idea is to drop a small proportion of the data before starting the search for the centers of the groups. The deletion step is "impartial" in the sense that the deleted points are self-determined by the sample.

In the rest of this section we are interested in the application of this method to a quite general feature space which includes the case of functional data or more general data structures (see, for instance, Locantore et al. 1999). Specifically we will assume here that the data belong to a Banach space $(\mathcal{F}, \|\cdot\|)$. We now summarize the applicability of the ITKM ideas in such a general framework following, essentially, the approach in Cuesta-Albertos and Fraiman (2007). See also García-Escudero and Gordaliza (2005) for related ideas.

Let us first note that, if our target is to robustify the K-means procedure, we could imagine employing this method in a generalized version which results from replacing the quadratic function in (10.12) by a continuous and strictly increasing score function $\Phi : \mathbb{R}^+ \to \mathbb{R}^+$. If $\mathcal{X}$ is an $\mathcal{F}$-valued random element with distribution P and $K \in \mathbb{N}$, let us denote

$$I_K^\Phi(P) = I_K^\Phi(\mathcal{X}) = \inf_{h_1,\ldots,h_K \in \mathcal{F}} \mathbb{E}\left(\Phi\left[\min_{i=1,\ldots,K} \|\mathcal{X} - h_i\|\right]\right). \tag{10.15}$$

Then we could follow the same steps as for K-means, that is, we could define the empirical version of (10.15) and estimate the optimal K-centers (i.e. the minimizers of (10.15)) with their empirical counterparts.

However, this will not solve the problem of the lack of robustness in K-means. Indeed, in the case $K > 1$, assume that we replace a single point in the sample by another point x_0 located far away from the data cloud. It can be seen that, for any score function Φ, the outlying point x_0 will be one of the "optimal centers" provided that it is located far enough from the remaining data. This shows that, if the sample size is n, the estimate of the optimal centers would "break down" by just replacing a proportion of n^{-1} points in the sample.

The idea of ITKM is to choose a (usually small) number $\alpha \in (0, 1)$, and then change the function to be minimized a little in order to allow a proportion α of the data points in the sample to be dropped.

To be precise, let $\mathcal{P}_{\alpha,n}$ be the family of all measurable functions $\tau : \mathcal{F} \to [0, 1]$ such that $\int \tau(x) d P_n(x) \geq 1 - \alpha$. Thus, if $\tau \in \mathcal{P}_{\alpha,n}$, then τ trims at most a proportion α of points in the sample. However, τ does not necessarily trim complete points (giving weight 0 or 1 to all data points), but may give them a weight in (0, 1). This condition is required to obtain an exact α-trimming level and to show the existence of optimal trimming functions, exactly in the same way as randomized tests are required to obtain uniformly most powerful tests in the Neyman–Pearson theory.

Trimming functions can also be defined in "population terms," that is, not only for sample distributions. In general, given a probability distribution P, $\mathcal{P}_\alpha$ will denote the family of all measurable functions $\tau : \mathcal{F} \to [0, 1]$ such that $\int \tau(x) \, dP(x) \geq 1 - \alpha$.

Now, given $K \geq 1$, a distribution P on $\mathcal{F}$, $\tau \in \mathcal{P}_\alpha$, and $h_1, \ldots, h_K \in \mathcal{F}$, define

$$I_K^{\Phi,\alpha}(\tau, \mathbf{h}, P) = I(\tau, \mathbf{h}, P) = \mathbb{E}\left(\Phi\left[\min_{i=1,\ldots,K} \|\mathcal{X} - h_i\|\right] \tau(\mathcal{X})\right).$$

Let $H(P)$ be any set of centers $H(P) = \{h_1, \ldots, h_K\}$ in which this infimum is reached and

$$I(P) = \inf_{\tau \in \mathcal{P}_\alpha} \inf_{h_1, \ldots, h_K \in \mathcal{F}} I(\tau, \mathbf{h}, P). \tag{10.16}$$

We omit the superscripts Φ, α, and the subscript K since they will remain fixed throughout.

The empirical version is defined in the natural way as follows:

$$I(P_n) = \inf_{\tau \in \mathcal{P}_{\alpha,n}} \inf_{h_1, \ldots, h_K \in \mathcal{F}} I(\tau, \mathbf{h}, P_n), \tag{10.17}$$

and $H(P_n)$ is any set of centers in which the infimum in (10.17) is reached.

Functions in $\mathcal{P}_\alpha$ are often called *α-trimming functions*. We will denote by τ_α any trimming function in which the infimum in (10.16) is reached and we will refer to it as an *optimal α-trimming function*. The analogous notions of *empirical α-trimming functions* (which are the functions in $\mathcal{P}_{\alpha,n}$) and *optimal empirical α-trimming functions* (which are the minimizers of (10.17) denoted by $\tau_{n,\alpha}$) are defined from the corresponding empirical concepts.

In general terms, the purpose of the ITKM methodology is to obtain the set of empirical centers $H(P_n)$, viewed as an estimator of its population counterpart $H(P)$. The main underlying idea is that in the ITKM procedure we are allowed to get rid of a proportion α of the points in $\mathcal{F}$ while searching for the centers. The points to be trimmed are those for which the value of $I(\tau, \mathbf{h}, P)$ is as small as possible.

The effective application of the ITKM method, with the corresponding calculation of the ITKM centers for a given dataset, is in general quite an involved problem from the computational point of view. A well-known difficulty is that, even in the one-dimensional, non-trimmed K-means case (with $K \geq 2$), the only effective algorithm for computing the set $H(P_n)$ needs to check all possible partitions of the sample obtained with $(K - 1)$ hyperplanes. The situation is even worse if we try to apply trimming because, in this case, it is also required to check all possible trimmings. To circumvent this, several procedures have been proposed (see, for instance, García-Escudero *et al.* (2003) where, in fact, the absolute minimum in (10.17) is not provided, but instead a stationary point). Thus, the consistency of those algorithms is not guaranteed unless the objective function contains only one stationary point.

Following the idea proposed in Cuesta-Albertos and Fraiman (2006) for the case $K = 1$, we can restrict the search for the set of centers $H(P_n)$ to the family of all possible subsets of the sample with cardinality K. Thus, it is only required to make the search in a family of $\binom{n}{K}$ possible candidates. This idea, in fact, reduces to replacing $\mathcal{F}$ in the second infimum in equation (10.17) by $\mathcal{S}_n = \{\mathcal{X}_1, \ldots, \mathcal{X}_n\}$, i.e. to minimizing

$$\widehat{I}(P_n) = \inf_{\tau \in \mathcal{P}_{a,n}} \inf_{h_1, \ldots, h_K \in \mathcal{S}_n} I(\tau, \mathbf{h}, P_n). \tag{10.18}$$

The existence of the corresponding optimum $\widehat{H}(P_n)$ as well as its asymptotic properties are stated below. Note that the use of $\widehat{H}(P_n)$ instead of $H(P_n)$ might entail a loss of efficiency. However, the smaller computational cost required for the calculation of $\widehat{H}(P_n)$ suggests the possibility of using $\widehat{H}(P_n)$ as a starting point in the search for the infimum $H(P_n)$ in (10.17).

Existence and asymptotic results

The material in this section is mostly based on the results in Cuesta-Albertos and Fraiman (2007), where all the proofs can be found. An important property of the optimal trimming functions is that they are, essentially, the indicator functions of a union of K balls with the same radius. Also, their existence and the consistency of the empirical versions are guaranteed under general conditions. All these theoretical properties are summarized in the next theorem. However, in order to formally establish these properties we need to handle a notion of convergence of sets which is defined as follows. Given $K \geq 1$, let $H_n = \{h_1^n, \ldots, h_K^n\}$ be a sequence of subsets of $\mathcal{F}$ with cardinality K. We say that the sequence $\{H_n\}$ converges to the set $H = \{h_1, \ldots, h_m\}$ with $m \leq K$ if there exists a labeling such that if we denote $H_n = \{h_{i_1}^n, \ldots, h_{i_K}^n\}$, then, for every $j = 1, \ldots, m$, we have that $\lim_n h_{i_j}^n = h_j$ and, if $j > m$ then $\lim_n \|h_{i_j}^n\| = \infty$. Of course, when the sequence $\{H_n\}$ is random this definition can be used, in the obvious way, to define weak or strong convergence of $\{H_n\}$ towards the limit H.

Theorem 10.6

(i) *Structure of the trimming functions. Let P be a Borel probability on $\mathcal{F}$ and let $G = \{g_1, \ldots, g_m\} \subset \mathcal{F}$. Let us define*

$$r_P(G) := \inf\{r > 0 : P[\cup_{i \leq m} B(g_i, r)] \geq 1 - \alpha\}.$$

Let $\tau_G \in \mathcal{P}_a$ such that

$$\int \tau_G(y) d P(y) = 1 - a \text{ and } I_{\cup_{i \leq m} B^o(g_i, r_P(G))} \leq \tau_G \leq I_{\cup_{i \leq m} B(g_i, r_P(G))}. \tag{10.19}$$

Then, we have that for every $\tau \in \mathcal{P}_a$,

$$\int \inf_{i=1,\dots,m} \Phi(\|x - g_i\|) \tau_G(x) d P(x) \leq \int \inf_{i=1,\dots,m} \Phi(\|x - g_i\|) \tau(x) d P(x).$$

(ii) *Existence.* Let P be a Borel probability on $\mathcal{F}$, $a \in [0, 1)$ and $K \geq 1$. There exists $H(P) = \{h_1^P, \dots, h_K^P\} \subset \mathcal{F}$ such that

$$I(P) = \mathbb{E}\left(\Phi\left[\inf_{i=1,\dots,K} \|\mathcal{X} - h_i^P\|\right] \tau_{H(P)}(\mathcal{X})\right), \tag{10.20}$$

where $\tau_{H(P)}$ denotes any function in $\mathcal{P}_a$ which satisfies (10.19) for $G = H(P)$.

(iii) *Consistency.* Let $a \in (0, 1)$. Let P be a Borel probability on $\mathcal{F}$ such that the set $H(P)$ is unique. Then the sequence $\{H(P_n)\}$ converges (in norm) to $H(P)$ almost surely (a.s.). If $H(P)$ is unique and is contained in the support of P then the sequence $\{\widehat{H}(P_n)\}$ of approximate solutions converges a.s. to $H(P)$.

The uniqueness assumption is not always guaranteed even if $a = 0$. This assumption has been discussed very often in the K-means literature (see, for instance, Cuesta-Albertos and Fraiman 2007). Some results are known in the case where $\mathcal{F} = \mathbb{R}$ and $a = 0$ (see Zoppé 1994 and Li and Flury 1995), but, to our knowledge, there is no satisfactory result yet, even in the case $\mathcal{F} = \mathbb{R}^2$, $K = 2$, and $a = 0$.

However, if the uniqueness assumption is omitted, the proofs of all consistency results work (with obvious modifications) to show that, with probability one, the sequence $\{\widehat{H}(P_n)\}$ is sequentially compact and its adherence values are trimmed K-means of P.

10.4 SOME EXAMPLES AND PRACTICAL ASPECTS

This section is devoted to the practice of functional data classification. In Section 10.4.1 we will describe several real-data examples that are used in the literature to illustrate the classification techniques described in previous sections. Also for illustration purposes, in Section 10.4.2 we will give a concise account of the simulation studies carried out in different works.

10.4.1 On real-data examples

Usually, the development of any relevant statistical idea needs to be tested by means of its application to one or several real-data examples. Some particular datasets attain a special status as standard references, revisited each time that a variant of some statistical methodology is proposed within a specific field. In the classical multivariate classification theory the *Iris data set*, first considered by R.A. Fisher in his original proposals for linear classification, has been re-analyzed many times since, coinciding with almost every new proposal for low-dimensional discrimination or clustering. The real-data examples are particularly important when dealing with classification, since sometimes the correct classification of the data is known in advance. Thus they become a sort of test for the new proposals under study. For these reasons we believe that a brief overview of some popular functional datasets, used for classification in different works, may be of interest.

The Berkeley growth data

In the Berkeley Growth Study (Tuddenham and Snyder 1954) the heights of 54 girls and 39 boys were measured at 31 time points between the ages of 1 and 18 years. The data may be downloaded, for example, from the website www.psych.mcgill.ca/faculty/ramsay/ramsay.html; some of them are displayed in Chapter 9. Ramsay and Silverman (2002) use these data to highlight the importance of studying not only the original height records, but also their corresponding velocity and acceleration curves. Chiou and Li (2007) cluster these same data by growth patterns to check if the resulting clusters reflect gender differences. The Berkeley growth curves were also used in Cuevas *et al.* (2007) to compare supervised classification techniques based on different notions of depth for functional data. Another clustering study was performed by Tokushige *et al.* (2007) on the growth data of the girls.

The electrocardiogram (ECG) data

The data, from the MIT-BIH Arrhythmia Database (see Goldberger *et al.* 2000 and www.physionet.org), are available at the web page www.cs.ucr.edu/~wli. Each observation in the ECG dataset is a curve containing the measurements recorded by one electrode during one heartbeat. A group of cardiologists have assigned a label of normal or abnormal to each ECG record. Of the 2026 observations in the dataset, 520 were identified as abnormal and 1506 as normal. The ECG data were used by Wei and Keogh (2006) to test a semi-supervised classification technique (based on the 1-NN rule) for time series. In this same work other interesting examples of functional data sets are considered and they are also available at the web page cited above.

Cell data

These data come from research in experimental cardiology conducted by Dr David García-Dorado at the Vall d'Hebron Hospital (Barcelona, Spain). The variable under study is the mitochondrial calcium overload (MCO), a measure of the mitochondrial calcium ion Ca2+ levels. This variable was measured every 10 seconds during an hour in isolated mouse cardiac cells. During mitochardial ischemia high levels of MCO are related to a better protection against the ischemia process. Thus, it is interesting to study different procedures that might increase the MCO level. The cell data consist of two samples of the MCO evolution for 45 cells in a control group and 44 cells in a treatment group. The drug used in this last group is called Cariporide and, as a selective blocker of the ionic exchange, it was expected to induce an MCO increase. See Ruiz-Meana *et al.* (2003) for details on the biochemical and medical aspects of this example.

Cuevas *et al.* (2004) checked whether there were any significant differences between the control and treatment groups via an ANOVA (analysis of variance) test for functional data. Cuevas *et al.* (2006) carried out an exploratory analysis of these cell data and compute several location estimators as well as their bootstrap confidence bands.

The Tecator dataset

These well-known data correspond to 215 pieces of finely chopped pure meat with different moisture, fat, and protein contents. The three contents, measured as percentages, were determined by analytic chemistry. Also, for each piece of meat, a 100-channel spectrum of absorbances (a curve) was obtained using a Tecator Infratec Food and Feed Analyzer working in the wavelength range 850–1050 nm by the near-infrared (NIR) transmission principle. The absorbance is $-\log 10$ of the transmittance measured by the spectrometer. The data are accessible at http://lib.stat.cmu.edu/datasets/tecator.

The Tecator dataset was first studied by Borggaard and Thodberg (1992) using neural networks. Ferraty and Vieu (2006, Chapter 8), use these spectrometric curves in the context of supervised learning with functional data. They separate the sample into two populations depending on whether the fat content is smaller than 20% or not, and check the performance of the kernel classification procedure described in Section 10.2.4. Dabo-Niang *et al.* (2006) and Ferraty and Vieu (2006, Chapter 9) also use the Tecator data to illustrate the use of functional unsupervised classification techniques.

The phoneme data

These are speech-recognition data, downloadable at www-stat.stanford.edu/tibs/ElemStatLearn/.

They consist of 4509 digitized speech frames of 32 msec duration from the TIMIT database. Each speech frame, given by 512 samples at a 16kHz sampling rate, corresponds to one of the following five phonemes: "sh" as in "s<u>h</u>e", "dcl" as in "<u>d</u>ark", "iy" as in "sh<u>e</u>", "aa" as in "d<u>a</u>rk", and "ao" as in "w<u>a</u>ter". A log-periodogram was computed for each speech frame, so that the final phoneme data are 4509 log-periodograms of length 256, with known class (phoneme) memberships.

Hastie et al. (1995) use the phoneme data to illustrate a penalized version of linear discriminant analysis (see Section 10.2.2). The kernel discrimination procedure (Section 10.2.4) is also applied to the speech data in Chapter 8 of Ferraty and Vieu (2006). In an example with these phoneme data, Ferré and Villa (2006) compare the performance of classification based on functional sliced inverse regression with other discrimination techniques.

The medflies data

The observations (available at http://anson.ucdavis.edu/~mueller) consist of the number of eggs laid daily by each of 1000 medflies (Mediterranean fruit flies, *Ceratitis capitata*) until time of death. For each medfly in the sample its lifespan and egg-laying trajectory were recorded (see Carey et al. 1998). One of the objectives of this research was to explore the relationship between longevity and lifetime reproduction, since a high degree of reproduction is suspected to shorten the life of the reproducing individual.

From the original collection of 1000 medflies, Müller and Stadtmüller (2005) select flies which lived past 34 days, yielding a sample of 534 medflies. These authors focus on the problem of predicting whether a fly is short- or long-lived after a 30-day period of egg-laying is observed (the period starts at the fifth living day of the medfly). A fly is considered as long-lived if the remaining lifetime past 30 days is 14 days or longer, otherwise it is short-lived. Of the $n = 534$ flies, 256 were short-lived and 278 were long-lived. Classification in Müller and Stadtmüller (2005) is carried out using a functional logistic regression model, where a smoothed version of the reproductive curve is the predictor and the longevity status of the fly is the response.

Gene expression data

The *Drosophila melanogaster* gene expression profile data (Arbeitman et al. 2002) contain gene expression patterns for nearly one-third of all genes of this common fruit fly during its life cycle. The genes have been separated according to their biological functions.

Chiou and Li (2007) use a subset of 77 gene data (downloadable at www.blackwellpublishing.com/rss) to illustrate the k-centres functional clustering method. The functional gene expression data employed by these authors exclude the observations at adulthood and consist of 30 sequential time points from the embryonic period, 10 observations from the larval, and 18 time points from the pupal period. Specifically, they choose 21 transient early zygotic genes, 23 muscle-specific genes, and 33 eye-specific genes and compare the clustering results to the known gene categories.

A similar experiment, where the resulting gene clusters are compared with the biological functions of the genes, was conducted by Ma and Zhong (2008). This time the functional data are gene expression changes of wild-type budding yeast (*Saccharomyces cerevisiae*) under aerobic and anaerobic conditions; the data had been obtained by Lai *et al.* (2006).

Food industry data

Abraham *et al.* (2003) consider the evolution of pH along time in the process of converting milk into a certain type of cheese. Since the production process may take several months and acidification is a key factor for good quality, clustering these pH curves would help determine the quality of cheese prior to the end of the production process. The data consist of $n = 148$ observations of pH evolution between 5800 and 70000 seconds. Here the cheese manufacturer previously chose $K = 3$ clusters and Abraham *et al.* (2003) applied a clustering procedure based on dimension reduction of the curves (via B-splines fitting) and a K-means algorithm.

10.4.2 Some comments on simulation studies

Cuevas *et al.* (2007) compare functional classification methods based on different notions of data depth (see Section 10.2.7) such as the h-modal depth, the integrated depth proposed by Fraiman and Muniz (2001), or depths based on random projections. The comparisons are carried out via a simulation study with two models. The first one corresponds to populations P_0 and P_1 with smooth mean functions. Specifically, $\mathcal{X}|Y = i$ is a Gaussian process on $[0, 1]$ with mean $m_i(t) = 30(1 - t)^{1.1^i} t^{1.1^{1-i}}$ and covariance function $\Gamma_i(s, t) = 0.2 \exp(-|s - t|/0.3)$, for $i = 0, 1$. In the second model (Cuevas *et al.* 2007) the trajectories in both populations correspond to Gaussian processes with the same covariance structure. The mean functions are close together in terms of the supremum distance but they have very different "shapes" (one of them being much wigglier than the other one). The idea is to check the performance of some classifiers that use information on the derivatives.

Preda *et al.* (2007) carry out a simulation study to compare the performance of PLS classification, k-NN, discrimination based on RKHS, and Fisher's linear rule applied to data whose dimension was reduced via principal components. The model used by these authors is taken from Ferraty and Vieu (2003), who in turn use a functional version of the simulated waveform data of Breiman *et al.* (1984). Ferraty and Vieu (2003) apply the kernel classification rule to data from three different populations, P_0, P_1, and P_2. The trajectories of $X(t)|Y = j$ for $j = 0, 1, 2$ are given by $Uh_j(t) + (1 - U)h_k(t)$, where $t \in [1, 21]$, U is uniformly distributed on $[0, 1]$, $k = k(j)$ denotes the integer part of $(j + 1)/2$, $h_1(t) = \max(6 - |t - 11|, 0)$, $h_2(t) = h_1(t - 4)$ and $h_3(t) = h_1(t + 4)$ are shifted triangular waveforms. These curves are observed at 101 equidistant points, $t_0 = 0 < \cdots < t_{101} = 21$. The error $\epsilon(t)$ is then generated at these grid points in such a way that the $\epsilon(t_i)$ are independent standard normals. This model leads to very irregular trajectories in the three populations P_0, P_1, and P_2.

It is clear that these studies cannot be conclusive since much more research is required in these practical aspects of functional classification. Therefore, the conclusions should be understood just as partial hints to be completed in further studies. In general terms, the k-NN classifier is the global winner under model 1 of those considered in Cuevas et al. (2007), but the method based on random projections also shows good behaviour. Under the more irregular model 2, the classifiers based on the random projections methodology (suitably modified in order to take into account the information provided by the derivatives) are the winners. In the study by Preda et al. (2007), the PLS method and the k-NN are the overall winners with similar error rates. This suggests again, as noted at the end of Section 10.2.3, the possibility of considering k-NN as a "default method" with a reasonable equilibrium between implementation simplicity, interpretability, and performance.

Shin (2008) proposes an extension of Fisher's linear rule to data which are realizations of second-order stochastic processes. The simulated data from P_i are of type

$$\mu_i(t) + \sum_{j=1}^{30} j^{-1/2} U_j \sqrt{2} \cos(j\pi t) \quad \text{for } i = 0, 1,$$

where $t \in [0, 1]$, the U_j are i.i.d. $N(0, 1)$ variables, $\mu_0(t) = 3\sqrt{2}\cos(\pi t) + \sqrt{2}\cos(2\pi t)$, and $\mu_1(t) = \sqrt{2}\cos(2\pi t)$.

While the above references are rather focused on supervised classification, there are other interesting recent proposals in the field of functional clustering. All of them include simulation results or examples:

Chiou and Li (2007) study a so-called *k-centres functional clustering methodology*, inspired by the K-means procedure.

Tokushige et al. (2007) adapt to the functional framework two previous variants of the K-means method called *crisp k-means clustering algorithms* (MacQueen 1967) and *fuzzy k-means algorithms* (Ruspini 1969).

Another flexible penalized functional data clustering method has been proposed by Ma and Zhong (2008), motivated by an analysis of temporal gene expression data.

Let us finally mention the paper by James and Sugar (2003) where a model-based procedure is developed for clustering functional data. It is particularly useful when the sample curves are observed at sparse and irregular time points. The basic idea is to model each individual curve by a spline of order p and use spline coefficients for clustering the curves into G groups.

10.5 SKETCHES OF SOME SELECTED PROOFS

For illustration purposes we will present a sketch of the proofs of two results, selected from among those stated above: Theorem 10.1 (from Cérou and Guyader 2006), concerning the consistency of k-NN infinite-dimensional classification rules, and Theorem 10.4 (from Biau et al. 2008) on the effect of using approximate k-means for clustering. These examples have been chosen by taking into account several criteria: (a) they both provide relevant results with a clear interpretation; (b) the proofs include some important typical tools which are, in some sense, representative; and (c) both results are quite recent and may not yet be well known.

10.5.1 Sketch of the proof of Theorem 10.1

We follow here the proof given in Cérou and Guyader (2006, Theorem 2). Our aim is to convey some basic ideas, so the more involved technical details will be omitted.

Step 1. In view of the structure of the optimal classifier (10.1) we only need to prove that

$$\lim_{n \to \infty} \mathbb{E}|(\eta_n - \eta)(\mathcal{X})| = 0, \tag{10.21}$$

where η_n is the k-NN estimator (10.4) of the regression function $\eta(x) = \mathbb{E}(Y|\mathcal{X} = x)$.

Now, introduce the auxiliary estimator

$$\tilde{\eta}_n(x) = \frac{1}{k} \sum_{i=1}^{k} \eta(\mathcal{X}_{(i)}(x)),$$

where $\mathcal{X}_{(i)}(x)$ represents the ith element of the sample, when arranged in increasing order of distance from x. We have

$$\mathbb{E}|(\eta_n - \eta)(\mathcal{X})| \leq \mathbb{E}|(\eta_n - \tilde{\eta}_n)(\mathcal{X})| + \mathbb{E}|(\tilde{\eta}_n - \eta)(\mathcal{X})|.$$

Step 2. From the Cauchy–Schwartz inequality we have that, denoting $\mathcal{X}_{(i)} = \mathcal{X}_{(i)}(\mathcal{X})$, the first term of the right-hand side can be bounded by

$$\mathbb{E}|(\eta_n - \tilde{\eta}_n)(\mathcal{X})| \leq \left\{ \frac{1}{k^2} \sum_{1 \leq i,j \leq k} \mathbb{E}[(Y_{(i)} - \eta(\mathcal{X}_{(i)}))(Y_{(j)} - \eta(\mathcal{X}_{(j)}))] \right\}^{1/2}.$$

Denote by $\mathbb{E}(\mathcal{S}_{ij})$ the expectation of the i, j term in the above sum. By using the iterated expectation $\mathbb{E}(\mathcal{S}_{ij}) = \mathbb{E}\left(\mathbb{E}(\mathcal{S}_{ij}|\mathcal{X}_1, \ldots, \mathcal{X}_n, \mathcal{X}) \right)$ we can show that, for $i \neq j$, $\mathbb{E}(\mathcal{S}_{ij}) = 0$ since, conditionally to $\mathcal{X}_1, \ldots, \mathcal{X}_n, \mathcal{X}$, the variables $Y_{(i)} - \eta(\mathcal{X}_{(i)})$ are i.i.d. Bernoulli with zero mean. We thus get

$$\mathbb{E}|(\eta_n - \tilde{\eta}_n)(\mathcal{X})| \leq \left\{ \frac{1}{k^2} \sum_{i=1}^{k} \mathbb{E}[(Y_{(i)} - \eta(\mathcal{X}_{(i)}))^2] \right\}^{1/2}$$

and since $|Y_{(i)} - \eta(\mathcal{X}_{(i)})| \leq 1$ we conclude that $\lim_{n \to \infty} \mathbb{E}|(\eta_n - \tilde{\eta}_n)(\mathcal{X})| = 0$.

Step 3. The convergence to zero of the term $\mathbb{E}|(\tilde{\eta}_n - \eta)(\mathcal{X})|$ is the most delicate step in the proof. It involves some technical reasonings that will not be considered in detail. Let us only mention that, using the conditioning trick again together with a technical lemma (Lemma 2 in Cérou and Guyader 2006), we finally get the following bound:

$$\mathbb{E}|(\tilde{\eta}_n - \eta)(\mathcal{X})| \leq 2\mathbb{E}\left[\frac{1}{P_{\mathcal{X}}(B(\mathcal{X}, D_{(k+1)}))} \int_{B(\mathcal{X}, D_{(k+1)})} |\eta - \eta(\mathcal{X})| d P_{\mathcal{X}} \right],$$

where $D_{(j)} = D(\mathcal{X}_{(j)}, \mathcal{X})$. Since the random variables inside the square brackets are bounded by 1, the proof will be complete if we show that they converge in probability to zero. To this end, take $\epsilon > 0$ and note that for every $\delta_0 > 0$,

$$\mathbb{P}\left\{ \frac{1}{P_{\mathcal{X}}(B(\mathcal{X}, D_{(k+1)}))} \int_{B(\mathcal{X}, D_{(k+1)})} |\eta - \eta(\mathcal{X})| d P_{\mathcal{X}} > \epsilon \right\}$$

$$\leq \mathbb{P}\{D_{(k+1)} \geq \delta_0\} + \sup_{0 \leq \delta \leq \delta_0} \mathbb{P}\left\{ \frac{1}{P_{\mathcal{X}}(B(\mathcal{X}, \delta))} \int_{B(\mathcal{X}, \delta)} |\eta - \eta(\mathcal{X})| d P_{\mathcal{X}} > \epsilon \right\}.$$

Now, the second term goes to zero from the Besicovich condition. The convergence of the first term follows from the fact (whose proof can be found in Devroye *et al.* 1996) that $\lim_{n \to \infty} D(\mathcal{X}_{(k)}, \mathcal{X}) = 0$, with probability 1, whenever $k/n \to 0$.

10.5.2 Sketch of the proof of Theorem 10.4

We summarize here the proof of Theorem 2.1 in Biau *et al.* (2008).

As a consequence of the assumption $P \in \mathcal{P}(R)$ it suffices to consider only cluster centers in the (closed and convex) ball $B(0, R)$. Further, since $P_n(B(0, R)) = 1$ a.s., we only need to search for empirical centers in $B(0, R)$.

Let l_h be the real-valued map defined by

$$l_h(x) = -2\langle x, h\rangle + \|h\|^2, \quad x, h \in \mathcal{F},$$

where $\langle \cdot, \cdot \rangle$ stands for the inner product in $\mathcal{F}$. For $\mathbf{h} \in B(0, R)^K$ we have that

$$I_K(P, \mathbf{h}) = \mathbb{E}(\|\mathcal{X}\|^2) + \mathbb{E}\left(\min_{j=1,\ldots,K} l_{h_j}(\mathcal{X})\right).$$

So, minimizing $I_K(P, \mathbf{h})$ is equivalent to minimizing

$$W(P, \mathbf{h}) = \mathbb{E}\left(\min_{j=1,\ldots,K} l_{h_j}(\mathcal{X})\right).$$

Similarly, minimizing $I_K(P_n, \mathbf{h})$ is equivalent to minimizing

$$W(P_n, \mathbf{h}) = \frac{1}{n}\sum_{i=1}^{n} \min_{j=1,\ldots,K} l_{h_j}(\mathcal{X}_i).$$

Moreover, for any δ_n-minimizer $\mathbf{h}_n$ of the empirical objective function,

$$I_K(P, \mathbf{h}_n) - \inf_{\mathbf{h} \in B(0,R)^K} I_K(P, \mathbf{h}) = W(P, \mathbf{h}_n) - \inf_{\mathbf{h} \in B(0,R)^K} W(P, \mathbf{h}), \quad (10.22)$$

and

$$\mathbb{E}(W(P, \mathbf{h}_n)) - \inf_{\mathbf{h} \in B(0,R)^K} W(P, \mathbf{h})$$

$$\leq \mathbb{E}\left(\sup_{\mathbf{h} \in B(0,R)^K} (W(P_n, \mathbf{h}) - W(P, \mathbf{h}))\right)$$

$$+ \sup_{\mathbf{h} \in B(0,R)^K} \mathbb{E}\left(W(P, \mathbf{h}) - W(P_n, \mathbf{h})\right) + \delta_n. \quad (10.23)$$

Now, in order to control the right-hand side of this inequality, it suffices to bound the maximal deviation

$$\mathbb{E}\left(\sup_{\mathbf{h} \in B(0,R)^K} (W(P_n, \mathbf{h}) - W(P, \mathbf{h}))\right),$$

since the corresponding bound will also hold for the second term.

Let $\sigma_1, \ldots, \sigma_n$ be i.i.d. Rademacher random variables (that is, $\mathbb{P}(\sigma_i = 1) = \mathbb{P}(\sigma_i = -1) = 1/2)$), independent of the $\mathcal{X}_i$'s. Let $\mathcal{G}$ be a class of real-valued functions defined on the space $\mathcal{F}$. The Rademacher averages of $\mathcal{G}$ are defined by

$$R_n(\mathcal{G}) = \mathbb{E}\left(\sup_{g \in \mathcal{G}} \frac{1}{n}\sum_{i=1}^{n} \sigma_i g(\mathcal{X}_i)\right).$$

Finally, the proof of Theorem 10.4 is a consequence of (10.22), (10.23), and the following lemma.

Lemma 10.1 *The following three statements hold:*

(i)
$$\mathbb{E}\left(\sup_{h\in B(0,R)^K} (W(P_n, h) - W(P, h))\right)$$
$$\leq 2\mathbb{E}\left(\sup_{h\in B(0,R)^K} \frac{1}{n}\sum_{i=1}^n \sigma_i \min_{j=1,\ldots,K} l_{h_j}(\mathcal{X}_i)\right).$$

(ii)
$$\mathbb{E}\left(\sup_{h\in B(0,R)^K} \frac{1}{n}\sum_{i=1}^n \sigma_i \min_{j=1,\ldots,K} l_{h_j}(\mathcal{X}_i)\right)$$
$$\leq 2K\left(\mathbb{E}\left(\sup_{h\in B(0,R)} \frac{1}{n}\sum_{i=1}^n \sigma_i \langle \mathcal{X}_i, h\rangle\right) + \frac{R^2}{2\sqrt{n}}\right).$$

(iii)
$$\mathbb{E}\left(\sup_{h\in B(0,R)} \frac{1}{n}\sum_{i=1}^n \sigma_i \langle \mathcal{X}_i, h\rangle\right) \leq R\sqrt{\frac{\mathbb{E}(\|\mathcal{X}\|^2)}{n}}.$$

Sketch of the proof of Lemma 10.1

Step 1. Let $\mathcal{X}_1', \ldots, \mathcal{X}_n'$ be an independent copy of $\mathcal{X}_1, \ldots, \mathcal{X}_n$, independent of the σ_i's. Then, by a standard symmetrization argument, we have

$$\mathbb{E}\left(\sup_{h\in B(0,R)^K} (W(P_n, h) - W(P, h))\right)$$
$$\leq \mathbb{E}\left(\sup_{h\in B(0,R)^K} \frac{1}{n}\sum_{i=1}^n \sigma_i \left[\min_{j=1,\ldots,K} l_{h_j}(\mathcal{X}_i) - \min_{j=1,\ldots,K} l_{h_j}(\mathcal{X}_i')\right]\right)$$
$$\leq 2\mathbb{E}\left(\sup_{h\in B(0,R)^K} \frac{1}{n}\sum_{i=1}^n \sigma_i \min_{j=1,\ldots,K} l_{h_j}(\mathcal{X}_i)\right).$$

Step 2. Now make use of the following properties of Rademacher averages.

(R1) $R_n(|\mathcal{G}|) \leq R_n(\mathcal{G})$, where $|\mathcal{G}| := \{|g| : g \in \mathcal{G}\}$.
(R2) $R_n(\mathcal{G}_1 \oplus \mathcal{G}_2) \leq R_n(\mathcal{G}_1) + R_n(\mathcal{G}_2)$, where $\mathcal{G}_1 \oplus \mathcal{G}_2 = \{g_1 + g_2 : g_1 \in \mathcal{G}_1, g_2 \in \mathcal{G}_2\}$.

The proof proceeds by induction on K. For $K = 1$ we have

$$\mathbb{E}\left(\sup_{h \in B(0,R)} \frac{1}{n} \sum_{i=1}^{n} \sigma_i[-2\langle \mathcal{X}_i, h \rangle + \|h\|^2]\right) \leq 2\mathbb{E}\left(\sup_{h \in B(0,R)} \frac{1}{n} \sum_{i=1}^{n} \sigma_i \langle \mathcal{X}_i, h \rangle\right) + \frac{R^2}{\sqrt{n}}.$$

For $K = 2$, we obtain

$$\mathbb{E}\left(\sup_{(h_1,h_2) \in B(0,R)^2} \frac{1}{n} \sum_{i=1}^{n} \sigma_i \min_{j=1,2} l_{h_j}(\mathcal{X}_i)\right)$$

$$= \mathbb{E}\left(\sup_{(h_1,h_2) \in B(0,R)^2} \frac{1}{2n} \sum_{i=1}^{n} \sigma_i [l_{h_1}(\mathcal{X}_i) + l_{h_2}(\mathcal{X}_i) - |l_{h_1}(\mathcal{X}_i) - l_{h_2}(\mathcal{X}_i)|]\right)$$

$$\leq 2\mathbb{E}\left(\sup_{h \in B(0,R)} \frac{1}{n} \sum_{i=1}^{n} \sigma_i l_h(\mathcal{X}_i)\right)$$

$$\leq 4\left(\mathbb{E}\left(\sup_{h \in B(0,R)} \frac{1}{n} \sum_{i=1}^{n} \sigma_i \langle \mathcal{X}_i, h \rangle\right) + \frac{R^2}{2\sqrt{n}}\right),$$

where we use properties (R1) and (R2) and the result for $K = 1$. The recurrence rule follows using the arguments presented for $K = 1$, $K = 2$, and the fact that for any $(z_1, \ldots, z_K) \in \mathbb{R}^K$,

$$\min(z_1, \ldots, z_k) = \min(\min(z_1, \ldots, z_{[K/2]}), \min(z_{[K/2]+1}, \ldots, z_K)).$$

Step 3.

$$\mathbb{E}\left(\sup_{h \in B(0,R)} \frac{1}{n} \sum_{i=1}^{n} \sigma_i \langle \mathcal{X}_i, h \rangle\right) = \mathbb{E}\left(\sup_{h \in B(0,R)} \frac{1}{n} \left\langle \sum_{i=1}^{n} \sigma_i \mathcal{X}_i, h \right\rangle\right)$$

$$= \frac{R}{n} \mathbb{E}\left(\left\|\sum_{i=1}^{n} \sigma_i \mathcal{X}_i\right\|\right) \leq \frac{R}{n} \sqrt{\mathbb{E}\left(\left\|\sum_{i=1}^{n} \sigma_i \mathcal{X}_i\right\|^2\right)} = R\sqrt{\frac{\mathbb{E}(\|\mathcal{X}\|^2)}{n}}.$$

References

ABAYA, E.A., WISE, G.L. (1984). Convergence of vector quantizers with applications to optimal quantization. *SIAM J. Appl. Math.*, 44, 183–9.

ABRAHAM, C., BIAU, G., CADRE, B. (2006). On the kernel rule for function classification. *Ann. Inst. Stat. Math.*, 58, 619–33.

ABRAHAM, C., CORNILLON, P.A., MATZNER-LOBER, E., MOLINARI, N. (2003). Unsupervised curve clustering using B-splines. *Scand. J. Statist.*, 30, 581–95.

ANTOS, A., GYÖRFI, L., GYÖRGY, A. (2005). Improved convergence rates in empirical vector quantizer design. *IEEE Trans. Inform. Theory*, **51**, 4013–22.

ARBEITMAN, M.N., FURLONG, E.E.M., IMAM, F., JOHNSON, E., NULL, B.H., BAKER, B.S., KRASNOW, M.A., SCOTT, M.P., DAVIS, R.W., WHITE, K.P. (2002). Gene expression during the life cycle of Drosophila melanogaster. *Science*, **297**, 2270–4.

AUDIBERT, J.Y., TSYBAKOV, A.B. (2007). Fast learning rates for plug-in classifiers. *Ann. Statist.*, **35**, 608–33.

BAÍLLO, A., CUESTA-ALBERTOS, J.A. and CUEVAS, A. (2010). Supervised classification for a family of Gaussian functional models. Submitted manuscript, available at arxiv.org/abs/1004.5031.

BARBA, I., DE LEN, G., MARTÍN, E., CUEVAS, A., AGUADE, S., CANDELL-RIERA, J., BARRABÉS, J.A., GARCÍA-DORADO, D. (2008). Nuclear magnetic resonance-based metabolomics predicts exercise-induced ischemia in patients with suspected coronary artery disease. *Magn. Reson. Med.*, **25**, 60(1), 27–32.

BARKER M., RAYENS, W. (2003). Partial least squares for discrimination. *J. Chemom.*, **17**, 166–73.

BERLINET, A., THOMAS-AGNAN, C. (2004). *Reproducing Kernel Hilbert Spaces in Probability and Statistics*. Kluwer, Dordrecht.

BIAU, G., DEVROYE, L., LUGOSI, G. (2008). On the performance of clustering in Hilbert spaces. *IEEE Trans. Inform. Theory*, **54**, 781–90.

BORGGAARD, C., THODBERG, H.H. (1992). Optimal minimal neural interpretation of spectra. *Anal. Chem.*, **64**, 545–51.

BREIMAN, L., FRIEDMAN, J., OLSHEN, R., STONE, C. (1984). *Classification and Regression Trees*. Wadsworth, Belmont, CA.

BURBA, F., FERRATY, F., VIEU, P. (2009). k-nearest neighbors method in functional nonparametric regression. *J. Nonparametr. Stat.*, **21**, 453–69.

BURGESS, C.J.C. (1998). A tutorial on support vector machines for pattern recognition. *Data Mining and Knowledge Discovery*, **2**, 121–67.

CAREY, J.R., LIEDO, P., MÜLLER, H.G., WANG, J.L., CHIOU, J.M. (1998). Relationship of age patterns of fecundity to mortality, longevity, and lifetime reproduction in a large cohort of Mediterranean fruit fly females. *J. Gerontol. Ser. A*, **53**, 245–51.

CÉROU, F., GUYADER, A. (2006). Nearest neighbor classification in infinite dimension. *ESAIM Probab. Stat.*, **10**, 340–55.

CHIOU, J.-M., LI, P.-L. (2007). Functional clustering and identifying substructures of longitudinal data. *J. Roy. Statist. Soc. Ser. B*, **69**, 679–99.

CORTES, C., VAPNIK, V. (1995). Support vector networks. *Mach. Learn.*, **20**, 273–97.

CUESTA-ALBERTOS, J.A., FRAIMAN, R. (2006). Impartial trimmed means for functional data. In *Data Depth: Robust Multivariate Statistical Analysis, Computational Geometry and Applications* (R. Liu, R. Serfling and D. Souvaine eds), 121–46. American Mathematical Society in DIMACS Series, **72**.

CUESTA-ALBERTOS, J.A., FRAIMAN, R. (2007). Impartial trimmed k-means for functional data. *Comput. Statist. Data Anal.*, **51**, 4864–77.

CUESTA-ALBERTOS, J.A., FRAIMAN, R., RANSFORD, T. (2007). A sharp form of the Cramer–Wold theorem. *J. Theoret. Probab.*, **20**, 201–9.

CUESTA-ALBERTOS, J.A., GORDALIZA, A., MATRÁN, C. (1997). Trimmed k-means: an attempt to robustify quantizers. *Ann. Statist.*, **25**(2), 553–76.

CUESTA-ALBERTOS, J.A., MATRÁN, C. (1988). The strong law of large numbers for k-means and best possible nets of Banach valued random variables. *Probab. Theory Related Fields*, 78, 523–34.

CUEVAS, A., FEBRERO, M., FRAIMAN, R. (2004). An anova test for functional data. *Comput. Statist. Data Anal.*, 47, 111–22.

CUEVAS, A., FEBRERO, M., FRAIMAN, R. (2006). On the use of the bootstrap for estimating functions with functional data. *Comput. Statist. Data Anal.*, 51, 1063–74.

CUEVAS, A., FEBRERO, M., FRAIMAN, R. (2007). Robust estimation and classification for functional data via projection-based depth notions. *Comput. Statist.*, 22, 481–96.

CUEVAS, A., FRAIMAN, R. (2009). On depth measures and dual statistics. A methodology for dealing with general data. *J. Multivariate Anal.*, 100, 753–66.

DABO-NIANG, S., FERRATY, F., VIEU, P. (2006). Mode estimation for functional random variable and its application for curves classification. *Far East J. Theor. Stat.*, 18, 93–119.

DEVROYE, L., GYÖRFI, L., LUGOSI, G. (1996). *A Probabilistic Theory of Pattern Recognition*. Springer, New York.

DUDA, R.O., HART, P.E., STORK, D.G. (2000). *Pattern Classification* (second edition). Wiley, New York.

ESCABIAS, M., AGUILERA, A.M., VALDERRAMA, M.J. (2007). Functional PLS logit regression model. *Comput. Statist. Data Anal.*, 51, 4891–902.

FERRATY, F., VIEU, P. (2003). Curves discrimination: a nonparametric functional approach. *Comput. Statist. Data Anal.*, 44, 161–73.

FERRATY, F., VIEU, P. (2006). *Nonparametric Functional Data Analysis: Theory and Practice*. Springer, Berlin.

FERRÉ, L., VILLA, N. (2006). Multilayer perceptron with functional inputs: an inverse regression approach. *Scand. J. Statist.*, 33, 807–23.

FRAIMAN, R., JUSTEL, A., SVARC, M. (2008). Selection of variables for cluster analysis and classification rules. *J. Amer. Stat. Assoc.*, 103, 1294–303.

FRAIMAN, R., MELOCHE, J. (1999). Multivariate L-estimation. *Test*, 8, 255–317.

FRAIMAN, R., MUNIZ, G. (2001). Trimmed means for functional data. *Test*, 10, 419–40.

GARCÍA-ESCUDERO, L.A., GORDALIZA, A. (2005). A proposal for robust curve clustering. *J. Classification*, 22, 185–201.

GARCÍA-ESCUDERO, L.A., GORDALIZA, A., MATRÁN, C. (2003). Trimming tools in exploratory data analysis. *J. Computat. Graph. Statist.*, 12, 434–59.

GOLDBERGER, A., AMARAL, L., GLASS, L., HAUSDORFF, J., IVANOV, P., MARK, R., MIETUS, J., MOODY, G., PENG, C., HE, S. (2000). PhysioBank, PhysioToolkit, and PhysioNet: components of a new research resource for complex physiologic signals. *Circulation*, 101, 215–20.

GORDALIZA, A. (1991). Best approximations to random variables based on trimming procedures. *J. Approx. Theory*, 64(2), 162–80.

GHOSH, A.K., CHAUDHURI, P. (2005). On maximal depth and related classifiers. *Scand. J. Statist.*, 32, 327–50.

HALL, P., POSKITT, D.S., PRESNELL, B. (2001). A functional dataŰanalytic approach to signal discrimination. *Technometrics*, 43, 1–9.

HARTIGAN, J.A. (1975). *Clustering Algorithms*. John Wiley, New York.

HASTIE, T., BUJA, A., TIBSHIRANI, R. (1995). Penalized discriminant analysis. *Ann. Statist.*, 23, 73–102.

HASTIE, T., TIBSHIRANI, R., FRIEDMAN, J. (2001). *The Elements of Statistical Learning.* Springer, New York.

JAMES, G.M., HASTIE, T.J. (2001). Functional linear discriminant analysis for irregularly sampled curves. *J. Roy. Statist. Soc. Ser. B*, **63**, 533–50.

JAMES, G.M., SUGAR, C.A. (2003). Clustering for sparsely sampled functional data. *J. Amer. Stat. Assoc.*, **98**, 397–408.

JØRSBOE, O.G. (1968). *Equivalence or Singularity of Gaussian Measures on Function Spaces.* Various Publications Series, No. 4. Matematisk Institut, Aarhus Universitet, Aarhus.

LAI, L.C., KOSORUKOFF, A.L., BURKE, P., KWAST, K.E. (2006). Metabolic state-dependent remodeling of the transcriptome in response to anoxia and subsequent reoxygenation in saccharomyces cerevisiae. *Eukaryot. Cell*, **5**, 1468–89.

LI, L., FLURY, B. (1995). Uniqueness of principal points for univariate distributions. *Statist. Probab. Letters*, **25**, 323–7.

LI, B., YU, Q. (2008). Classification of functional data: a segmentation approach. *Comput. Statist. Data Anal.*, **52**, 4790–800.

LIU, Y., RAYENS, W. (2007). PLS and dimension reduction for classification. *Comput. Statist.*, **22**, 189–208.

LOCANTORE, N., MARRON, J.S., SIMPSON, D.G., TRIPOLI, N., ZHANG, J.T., COHEN, K.L. (1999). Robust principal components for functional data. *Test.*, **8**, 1–73.

LÓPEZ-PINTADO, S., ROMO, J. (2006). Depth based classification for functional data. *DIMACS Series in Discrete Mathematics* **72**, 103–20.

MA, P., ZHONG, W. (2008). Penalized clustering of large-scale functional data with multiple covariates. *J. Amer. Statist. Assoc.*, **103**, 625–36.

MACQUEEN, J.B. (1967). Some methods for classification and analysis of multivariate observations. In *Proc. 5th Berkeley Symp. Mathematical Statistics and Probability*, Vol. 1, 281–97. University of California Press.

MÜLLER, H.-G., STADTMÜLLER, U. (2005). Generalized functional linear models. *Ann. Stat.*, **33**, 774–805.

POLLARD, D. (1981). Strong consistency of k-means clustering. *Ann. Statist.*, **9**, 135–40.

POLLARD, D. (1982). Quantization and the method of k-means. *IEEE Trans. Inform. Theory*, **28**, 199–205.

PREDA, C. (2007). Regression models for functional data by reproducing kernel Hilbert spaces methods. *J. Statist. Plann. Inference*, **137**, 829–40.

PREDA, C., SAPORTA, G., LÉVÉDER, C. (2007). PLS classification of functional data. *Comput. Statist.*, **22**, 223–35.

RAMSAY, J.O., SILVERMAN, B.W. (2002). *Applied Functional Data Analysis. Methods and Case Studies.* Springer, New York.

RAMSAY, J.O., SILVERMAN, B.W. (2005). *Functional Data Analysis* (second edition). Springer, New York.

ROSSI, F., VILLA, N. (2006). Support vector machine for functional data classification. *Neurocomputing*, **69**, 730–42.

RUIZ-MEANA, M., GARCÍA-DORADO, D., PINA, P., INSERTE, J., AGULLÓ, L., SOLER-SOLER, J. (2003). Cariporide preserves mitochondrial proton gradient and delays ATP depletion in cardiomyocites during ischemic conditions. *Amer. J. Physiol. Heart Circulatory Physiol.*, **285**, 999–1006.

RUSPINI, E.H. (1969). Hierarchical grouping to optimize an objective function. *J. Amer. Statist. Assoc.*, **58**, 236–44.

SHIN, H. (2008). An extension of Fisher's discriminant analysis for stochastic processes. *J. Multivariate Anal.*, **99**, 1191–216.

STONE, C.J. (1977). Consistent nonparametric regression. *The Annals of Statistics*, **5**, 595–645.

TOKUSHIGE, S., YADOHISA, H., INADA, K. (2007). Crisp and fuzzy k-means clustering algorithms for multivariate functional data. *Comput. Statist.*, **22**, 1–16.

TUDDENHAM, R.D., SNYDER, M.M. (1954). Physical growth of California boys and girls from birth to eighteen years. *University of California Publications in Child Development* **1**, 183–364.

VAPNIK, V. (1998). *Statistical Learning Theory*. Wiley, New York.

VARBERG, D.E. (1961). On equivalence of Gaussian measures. *Pacific J. Math.*, **11**, 751–62.

WEI, L., KEOGH, E. (2006). Semi-supervised time series classification. In *Proc. of the 12th ACM SIGKDD International Conference on Knowledge Discovery and Data Mining (KDD 2006)*, 748–53. Philadelphia, USA.

YU, B., OSTLAND, I.M., GONG, P., PU, R. (1999). Penalized discriminant analysis of in situ hyperspectral data for conifer species recognition. *IEEE Trans. Geosci. Remote Sens.*, **37**, 2569–77.

ZOPPÉ, A. (1994). On uniqueness and symmetry of self-consistent point of univariate continuous distributions. *J. Classification*, **14**, 147–58.

ZUO, Y., SERFLING, R. (2000). General notions of statistical depth function. *The Annals of Statistics*, **28**, 461–82.

CHAPTER 11

SPARSENESS AND FUNCTIONAL DATA ANALYSIS

GARETH JAMES

11.1 INTRODUCTION

IT is often assumed in functional data analysis (FDA) that the curve or function has been measured over all points or, more realistically, over a densely sampled grid. In this chapter we deal with two FDA settings where sparsity becomes important. We first consider the situation where, rather than densely observed functions, we only have measurements at a relatively sparse set of points, which is in practice often the case. In this situation alternative methods of analysis are required.

Figure 11.1 provides an example of two data sets that fall into the "sparsely observed" category. The "growth" data, illustrated in Fig. 11.1(a), consists of measurements of spinal bone mineral density for 280 males and females taken at various ages and is a subset of the data presented in Bachrach et al. (1999). Even though, in aggregate, there are 860 observations taken over a period of almost two decades, there are only 2–4 measurements for each individual covering no more than a few years. The "nephropathy" data, illustrated in Fig. 11.1(b), shows percentage changes in glomular filtration rate (GFR) over a six-year period, for a group of patients with membranous nephropathy, an autoimmune disease of the kidney. GFR is a

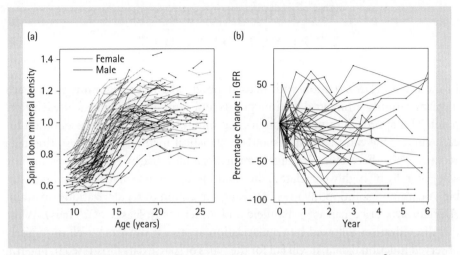

Fig. 11.1 (a) Measurements of spinal bone mineral density (g/cm^2) for males (black) and females (grey) at various ages, n = 280; (b) percentage change in GFR scores for 39 patients with membranous nephropathy.

standard measure of kidney function. Both of these datasets have sparsely observed curves, so more standard FDA techniques cannot be applied. We will use both data sets to illustrate alternative approaches that have been developed for dealing with sparse functional data. In Section 11.2 we discuss different classes of methods that can be applied to functional data, such as basis functions, mixed effects models and local smoothing, and examine the relative merits of each approach for sparsely observed data. Then in Section 11.3 we briefly outline several specific methodologies that have been developed for dealing with sparse functional data in the principal components, clustering, classification, and regression settings.

Section 11.4 considers the second situation where sparsity plays a key role. Here we examine two approaches that utilize variable selection ideas from the high-dimensional regression literature, where most of the regression coefficients are assumed to be zero, to perform functional regressions. The first approach uses variable selection methods to identify a small set of time points that are jointly most predictive for the response, Y. A local linear estimator is then used to form predictions for Y based on the observed values of the predictor, $X(t)$, at the selected time points. The second approach performs an interpretable functional linear regression by assuming sparsity in the derivatives of $\beta(t)$. For example, a simple piecewise linear estimate for $\beta(t)$ can be produced provided $\beta''(t)$ is zero at "most" time points. Variable selection methods are then used to identify the locations where $\beta''(t)$ is nonzero.

11.2 Different approaches to functional data

In the classic FDA situation with densely observed curves, a common approach involves forming a fine grid of time points and sampling the curves at each time point. This approach results in a high-, but finite-, dimensional representation for each curve. One can then apply standard statistical methods to the resulting data vectors. Since the vectors are high dimensional and nearby observations in time will generally be highly correlated, the resulting covariance matrices can be unstable. A common solution involves imposing some kind of regularization constraint (DiPillo 1976; Friedman 1989; Banfield and Raftery 1993; Hastie et al. 1995). While this discretization followed by regularization approach can work well on densely observed functional data, it will fail for the types of sparse data illustrated in Fig. 11.1. For this data, each individual has been observed at different time points. Hence, any fixed grid that is formed will involve many missing observations for each curve. For such data a new approach becomes necessary. In Sections 11.2.1, 11.2.2, and 11.2.3 we discuss three possible alternatives. Throughout this section we assume that one observes N functions, $Y_1(t), \ldots, Y_N(t)$ with the ith function observed at time points $t_{i1}, \ldots, t_{in_i}$ and $Y_{ij} = Y_i(t_{ij})$.

11.2.1 Basis functions

A natural way to model the smooth shape of each curve observed over time is to choose a finite p-dimensional basis, $s_\ell(t)$, $\ell = 1, \ldots, p$. One can then assume

$$Y_i(t) = \sum_{l=1}^{p} s_\ell(t)\eta_{i\ell} = \mathbf{s}(t)^T \boldsymbol{\eta}_i \quad (11.1)$$

where $\mathbf{s}(t) = [s_1(t), s_2(t), \ldots, s_p(t)]^T$ and $\boldsymbol{\eta}_i$ represents the basis coefficients for the ith curve. Common examples for $s_\ell(t)$ include spline and Fourier bases. Once the basis and dimension, p, have been chosen the $\boldsymbol{\eta}_i$'s can be estimated by applying standard least squares, separately to each curve, using the linear model (11.1).

Once the $\boldsymbol{\eta}_i$'s have been estimated there are two common approaches to analyzing the data. The first involves using the estimated $\boldsymbol{\eta}_i$ and (11.1) to provide an estimate of the entire curve, $Y_i(t)$. One can then treat the estimated curve as if it had been observed and apply the discretization approach previously discussed. Alternatively, one can treat the p-dimensional estimated $\boldsymbol{\eta}_i$'s as the observed data and apply standard statistical methods. A related approach to the basis method involves fitting a separate smoothing spline to each curve by finding $g_i(t)$ to minimize

$$\sum_{j=1}^{n_i}(Y_i(t_{ij}) - g_i(t_{ij}))^2 + \lambda \int (g_i''(t))^2 \, dt.$$

Once the g_i's have been estimated the discretization approach can be applied.

The basis and smoothing methods both have the advantage over the straight discretization approach that they can be applied to data where the curves have not all been measured at the same time points. For example, they can be used on the nephropathy data. However, the basis method also has several problems. First, when the curves are measured at different time points, it is easy to show that the variance of the estimated basis coefficients, $\widehat{\boldsymbol{\eta}}_i$, is different for each individual. Hence, any statistical analysis that utilizes the $\widehat{\boldsymbol{\eta}}_i$'s should place more weight on the more accurately estimated basis coefficients. However, it is often not obvious how one should treat the $\widehat{\boldsymbol{\eta}}_i$'s differently. Even more importantly, for extremely sparse data sets, many of the basis coefficients will have infinite variance, making it impossible to produce reasonable estimates. For example, in the growth data there are so few observations that it is not possible to fit a separate curve for each individual using any reasonable basis. In this case the basis approach will fail.

11.2.2 Mixed-effects methods

The main reason that the basis approach fails for the growth data is that it only uses the information from a particular curve to estimate the basis coefficients for that curve. For the growth data there are few observations per curve but a large number of curves, so in total we still have a lot of information. Hence, a potentially superior method would be to somehow utilize the information from all curves to estimate the coefficients for each curve. A natural way to achieve this goal is to utilize a mixed-effects framework. Mixed-effects models have been widely used in the analysis of functional data; see for instance Shi et al. (1996), Brumback and Rice (1998), and Rice and Wu (2001) for some early examples.

Denote by $\boldsymbol{\beta}$ an unknown but fixed vector of spline coefficients, let $\boldsymbol{\gamma}_i$ be a random vector of spline coefficients for each curve with population covariance matrix $\boldsymbol{\Gamma}$, and let $\epsilon_i(t)$ be random noise with mean zero and variance σ^2. The resulting mixed-effects model has the form

$$Y_i(t) = \mathbf{s}(t)^T \boldsymbol{\beta} + \mathbf{s}(t)^T \boldsymbol{\gamma}_i + \epsilon_i(t) \quad i = 1, \ldots, N. \tag{11.2}$$

In practice $Y_i(t)$ is only observed at a finite set of time points. Let $\mathbf{Y}_i$ be the vector consisting of the n_i observed values, let $\mathbf{S}_i$ be the corresponding n_i-by-p spline basis matrix evaluated at these time points, and let $\boldsymbol{\epsilon}_i$ be the corresponding random noise vector with covariance matrix $\sigma^2 \mathbf{I}$. The mixed-effects model then becomes

$$Y_i = S_i\beta + S_i\gamma_i + \epsilon_i \quad i = 1, \ldots, N. \tag{11.3}$$

The fixed-effects term, $S_i\beta$, models the mean curve for the population and the random-effects term, $S_i\gamma_i$, allows for individual variation. The principal patterns of variation about the mean curve are referred to as functional principal component curves.

A general approach to fitting mixed-effects models of this form uses the Expectation Maximization (EM) algorithm to estimate β, Γ and σ^2 (Laird and Ware 1982). Given these estimates, predictions are obtained for the γ_i's using the "best linear unbiased prediction" (Henderson 1950). For (11.3) above, the best linear unbiased prediction for γ_i is

$$\hat{\gamma}_i = \left(\hat{\Gamma}^{-1}/\hat{\sigma}^2 + S_i^T S_i\right)^{-1} S_i^T (Y_i - S_i\hat{\beta}).$$

Once the $\hat{\gamma}_i$'s have been computed one can then either form predictions for each $Y_i(t)$ using (11.2) and utilize the discretization approach, or else simply treat the $\hat{\gamma}_i$'s as the observed p-dimensional data.

The mixed-effects method has many advantages over the basis approach. First, it estimates the curve $Y_i(t)$ using all the observed data points rather than just those from the ith individual. This means that the mixed-effects method can be applied when there are insufficient data from each individual curve to use the basis method. For example, James et al. (2000) successfully use a mixed-effects model to fit the growth data. Secondly, this method uses maximum likelihood to estimate the parameters. Thus it automatically assigns the correct weight to each observation and the resulting estimators have all the usual asymptotic optimality properties.

11.2.3 Local smoothing

The mixed-effects approach is not the only method for building strength across functions by incorporating information from all the curves. An alternative approach, that does not require specifying basis functions, involves using local smoothing techniques to estimate the mean and covariance functions of the curves. The smoothing approach makes use of the Karhunen–Loeve expansion, which states that any smooth curve can be decomposed as $Y(t) = \mu(t) + \sum_{k=1}^{\infty} \xi_k \phi_k(t)$, where ξ_k is the kth principal component score and $\phi_k(t)$ is the kth principal component function. Hence a natural approximation for $Y(t)$ is given by

$$\hat{Y}_i(t) = \hat{\mu}(t) + \sum_{k=1}^{K} \hat{\xi}_{ik} \hat{\phi}_k(t), \tag{11.4}$$

where K represents the number of principal components used in the estimation. Lower values of K correspond to more regularization. Next we outline the method for estimating $\hat{\mu}(t)$, $\hat{\xi}_k$, and $\hat{\phi}_k(t)$.

Let $Y_i(t) = \mu(t) + X_i(t) + \epsilon_i(t)$ where $\mu(t) = EY_i(t)$ and $\text{Var}(\epsilon_i) = \sigma^2$. Further, let $G(s, t) = \text{cov}(X(s), X(t))$, the covariance function of $X(t)$. To implement this approach one first pools all the observations for $Y_1(t), \ldots, Y_N(t)$ and uses a kernel method such as a local linear smoother to estimate $\mu(t)$. Next, the estimated mean is subtracted from the curves and the "raw" covariance estimates

$$\widehat{G}_i(t_{ij}, t_{il}) = (Y_{ij} - \hat{\mu}(t_{ij}))(Y_{il} - \hat{\mu}(t_{il}))$$

are computed. A local smoother is then applied to all the $\widehat{G}_i(t_{ij}, t_{il})$ points where $j \neq l$ to produce $\widehat{G}(s, t)$, an estimate for $G(s, t)$. From the estimated covariance function one can compute the corresponding eigenvalues $\hat{\lambda}_k$ and eigenfunctions $\hat{\phi}_k(t)$ via

$$\int \widehat{G}(s, t)\hat{\phi}_k(s)\, ds = \hat{\lambda}_k \hat{\phi}_k(t) \tag{11.5}$$

where $\int \hat{\phi}_k(t)^2\, dt = 1$ and $\int \hat{\phi}_k(t)\hat{\phi}_m(t)\, dt = 0$ for $m < k$. Equation (11.5) can be estimated by evaluating $\widehat{G}(s, t)$ over a fine grid and then computing the standard eigenvectors and eigenvalues.

The final step in implementing (11.4) requires the estimation of $\hat{\xi}_{ik}$. Yao et al. (2005a) recommend using an approach they call PACE (Principal Analysis by Conditional Estimation), which involves computing the conditional expectation $E[\xi_{ik}|\mathbf{Y}_i]$ where $\mathbf{Y}_i = (Y_{i1}, \ldots, Y_{in_i})^T$. They show that

$$E[\xi_{ik}|\mathbf{Y}_i] = \lambda_k \boldsymbol{\phi}_{ik}^T \boldsymbol{\Sigma}_i^{-1}(\mathbf{Y}_i - \boldsymbol{\mu}_i), \tag{11.6}$$

where $\boldsymbol{\phi}_{ik}$ and $\boldsymbol{\mu}_i$ are vectors corresponding to the principal component function and mean function evaluated at $t_{i1}, \ldots, t_{in_i}$, and $(\boldsymbol{\Sigma}_i)_{jl} = G(t_{ij}, t_{il}) + \sigma^2 \delta_{jl}$ where $\delta_{jl} = 1$ if $j = l$ and 0 otherwise. Plugging in the estimates of these various parameters gives $\hat{\xi}_{ik}$. Once the $\hat{\xi}_{ik}$'s have been estimated, one can either compute the $Y_i(t)$'s using (11.4) and apply the discretization approach or else treat the $\hat{\xi}_{ik}$'s as a K-dimensional representation of the curves. See Yao et al. (2005a) for an application of this approach to estimate functional principal components.

Both the basis function and local smoothing approaches require the choice of tuning parameters. For basis functions one must select the dimension of the basis and for local smoothing the bandwidth. As with all problems involving tuning parameters there are a number of possible approaches that can be used such as AIC (Akaike Information Criterion), BIC (Bayesian Information Criterion), or cross-validation. In practice, depending on the sparsity level of the data, visual inspection is often used to select a reasonable tradeoff between fit to the data and smoothness of the curves.

11.3 APPLICATIONS OF FDA TO SPARSE DATA

This section illustrates some applications of the functional data analysis paradigm to sparse data situations. In particular we discuss functional principal components, functional clustering, functional classification, and functional regression methods.

11.3.1 Functional principal component analysis

A number of papers have been written on the problem of computing functional principal components (FPCA). Ramsay and Silverman (2005) discuss FPCA for densely sampled curves (see also Chapter 8 of this volume). For sparsely sampled data a few references include Shi et al. (1996), Staniswalis and Lee (1998), James et al. (2000), Rice and Wu (2001), and Yao et al. (2005a). The approaches taken in these papers tend to be either to use the mixed-effects framework of Section 11.2.2 or the local smoother method from Section 11.2.3. For example the approach of Yao et al. (2005a) first uses a local smoother to estimate the covariance function, $G(s, t)$. The covariance function is then discretized over a fine grid from which the eigenvectors and eigenvalues are computed. Finally, the principal component scores, ξ_{ik}, are computed using the PACE estimate given by (11.6).

Here we outline an alternative approach to that of Yao et al. (2005a), which utilizes the mixed-effects framework discussed in Section 11.2.2. The mixed-effects approach to functional principal components has been used in several papers (Shi et al. 1996; James et al. 2000; Rice and Wu 2001). Let $Y_i(t)$ be the measurement at time t for the ith individual or curve. Let $\mu(t)$ be the overall mean function, let f_j be the jth principal component function and set $f = (f_1, f_2, \ldots, f_K)^T$. To estimate K principal component curves we first define a general additive model

$$Y_i(t) = \mu(t) + \sum_{j=1}^{K} f_j(t)\alpha_{ij} + \epsilon_i(t)$$

$$= \mu(t) + f(t)^T \alpha_i + \epsilon_i(t) \quad i = 1, \ldots, N,$$

subject to the orthogonality constraint $\int f_j f_l = 0$ for $j \neq l$ and 1 otherwise. The random vector α_i gives the relative weights on the principal component functions for the ith individual, and $\epsilon_i(t)$ is random measurement error. The α_i's and ϵ_i's are all assumed to have mean zero. The α_i's are taken to have a common covariance matrix, Σ, and the measurement errors are assumed uncorrelated with a constant variance of σ^2.

In order to fit this model when the data are measured at only a finite number of time points it is necessary to place some restrictions on the form of the mean and principal component curves, so one represents μ and f using a finite-dimensional basis, such as spline functions (Silverman 1985; Green and Silverman 1994). Let s(t) be a p-dimensional orthogonal basis. Let Θ and θ_μ be, respectively, a p-by-k matrix and a p-dimensional vector of spline coefficients. Then $\mu(t) = s(t)^T \theta_\mu$, and $f(t)^T = s(t)^T \Theta$. The resulting restricted model has the form

$$Y_i(t) = s(t)^T \theta_\mu + s(t)^T \Theta \alpha_i + \epsilon_i(t), \quad i = 1, \ldots, N,$$

$$\epsilon_i(t) \sim N(0, \sigma^2), \quad \alpha_i \sim N(0, D),$$

subject to

$$\Theta^T \Theta = I, \quad \int s(t)^T s(t) \, dt = 1, \quad \int\int s(t)^T s(s) \, dt \, ds = 0. \tag{11.7}$$

The equations in (11.7) impose orthogonality constraints on the principal component curves. Note that, if one does not assume a special structure for the covariance matrix of the α_i's, Θ and Σ are confounded. Thus we restrict the covariance matrix to be diagonal and denote it by D.

For each individual i, let $t_{i1}, t_{i2}, \ldots, t_{in_i}$ be the time points at which measurements are available. Then $Y_i = (Y_i(t_{i1}), \ldots, Y_i(t_{in_i}))^T$, and $S_i = (s(t_{i1}), \ldots, s(t_{in_i}))^T$. Note that S_i is the basis matrix for the ith individual. To approximate the orthogonality condition in (11.7) we choose $s(\cdot)$ so that $S^T S = I$, where S is the basis matrix evaluated on a fine grid of time points. For instance, for the growth data the time interval was divided into 172 periods of 1/10th of a year each.

The final model can then be written as

$$Y_i = S_i \theta_\mu + S_i \Theta \alpha_i + \epsilon_i, \quad i = 1, \ldots, N, \tag{11.8}$$

$$\Theta^T \Theta = I, \quad \epsilon_i \sim N(0, \sigma^2 I), \quad \alpha_i \sim N(0, D).$$

James et al. (2000) provide an EM fitting procedure for this model and also suggest methods for choosing p and K. They also note that (11.8) can be interpreted as a mixed-effects model with a rank constraint on the covariance matrix. Peng and Paul (2009) propose an alternative geometric-based fitting procedure which is somewhat superior to the EM approach.

Figure 11.2 provides an illustration of this approach on the growth data. The plot gives 80% and 90% confidence intervals for the mean function and the first two principal components. The confidence intervals were produced using a parametric bootstrap approach (James et al. 2000). Despite the sparsity of the data, the intervals for the mean function are relatively narrow, with some widening in the right tail where there are few observations. The confidence intervals for the first principal component are much wider, particularly in the right tail. The large dip in the confidence band in this region occurs because approximately 20% of the bootstrap principal component curves exhibited an inverted U shape. There appear to be two distinctly different possible shapes for this component. Interestingly, given the

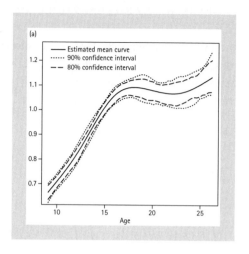

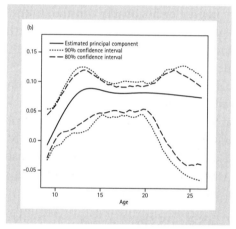

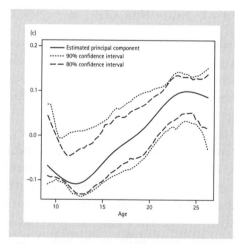

Fig. 11.2 80% and 90% pointwise confidence intervals for the mean function and both principal components for the spinal bone data.

variability of the first component, the intervals for the second component follow the general shape of the estimated curve quite tightly.

11.3.2 Functional clustering

Here we outline an approach to clustering of sparsely sampled functional data proposed in James and Sugar (2003). We develop a functional clustering model utilizing the mixed-effects framework discussed in Section 11.2.2. We describe how the model can be used to obtain low-dimensional plots of curve datasets, enabling one to visually assess clustering. Discriminant functions to identify the regions of greatest separation between clusters are then developed. Finally, an approach is presented for using the clustering model to estimate the entire curve for an individual.

A functional clustering model

Let $g_i(t)$ be the true value for the ith individual or curve at time t, and let $\mathbf{g}_i$, $\mathbf{Y}_i$, and $\boldsymbol{\epsilon}_i$ be, respectively, the corresponding vectors of true values, observed values, and measurement errors at times $t_{i1}, \ldots, t_{in_i}$. Then $\mathbf{Y}_i = \mathbf{g}_i + \boldsymbol{\epsilon}_i$ for $i = 1, \ldots, N$, where N is the number of individuals. The measurement errors are assumed to have mean zero and to be uncorrelated with each other and $\mathbf{g}_i$. It is necessary to impose some structure on the individual curves so, as with the functional PCA approach, one can model $g_i(t)$ using basis functions. Let $g_i(t) = \mathbf{s}(t)^T \boldsymbol{\eta}_i$, where $\mathbf{s}(t)$ is a p-dimensional spline basis vector and $\boldsymbol{\eta}_i$ is a vector of spline coefficients. The $\boldsymbol{\eta}_i$'s are modeled using a Gaussian distribution,

$$\boldsymbol{\eta}_i = \boldsymbol{\mu}_{z_i} + \boldsymbol{\gamma}_i, \quad \boldsymbol{\gamma} \sim N(0, \boldsymbol{\Gamma}),$$

where $z_i \in \{1, \ldots, G\}$ denotes the unknown cluster membership and G represents the number of clusters. The z_i's can either be modeled as missing data using the "mixture likelihood" approach or as unknown parameters using the "classification likelihood" approach.

There is a further parameterization of the cluster means that proves useful for producing low-dimensional representations of the curves. Note that $\boldsymbol{\mu}_k$ can be rewritten as

$$\boldsymbol{\mu}_k = \boldsymbol{\lambda}_0 + \boldsymbol{\Lambda} \boldsymbol{\alpha}_k, \tag{11.9}$$

where $\boldsymbol{\lambda}_0$ and $\boldsymbol{\alpha}_k$ are respectively p- and h-dimensional vectors, and $\boldsymbol{\Lambda}$ is a $p \times h$ matrix with $h \leq \min(p, G-1)$. When $h = G - 1$, (11.9) involves no loss of generality while $h < G - 1$ implies that the means lie in a restricted subspace. With this formulation, the functional clustering model (FCM) can be written as

$$Y_i = S_i(\lambda_0 + \Lambda \alpha_{z_i} + \gamma_i) + \epsilon_i, \quad i = 1, \ldots N, \tag{11.10}$$

$$\epsilon_i \sim N(0, \sigma^2 I), \quad \gamma_i \sim N(0, \Gamma),$$

where $S_i = (s(t_{i1}), \ldots, s(t_{in_i}))^T$ is the basis matrix for the ith curve. Note that λ_0, Λ, and α_k are confounded if no constraints are imposed. Therefore we require that

$$\sum_k \alpha_k = 0 \tag{11.11}$$

and

$$\Lambda^T S^T \Sigma^{-1} S \Lambda = I, \tag{11.12}$$

where S is the basis matrix evaluated over a fine lattice of time points that encompasses the full range of the data, and $\Sigma = \sigma^2 I + S\Gamma S^T$. The restriction in (11.11) means that $s(t)^T \lambda_0$ may be interpreted as the overall mean curve. There are many possible constraints that could be placed on Λ. The reason for the particular form used in (11.12) will become apparent below.

Fitting the functional clustering model involves estimating λ_0, Λ, α_k, Γ, and σ^2. This is achieved by maximizing the likelihood function using an EM algorithm treating the cluster memberships, z_i, as missing data. The E-step can be implemented by noting that under (11.10), conditional on the ith curve belonging to the kth cluster,

$$Y_i \sim N(S_i(\lambda_0 + \Lambda \alpha_k), \Sigma_i), \tag{11.13}$$

where $\Sigma_i = \sigma^2 I + S_i \Gamma S_i^T$. Depending on whether the classification or mixture likelihood approach is used, curves are first either assigned to a cluster (classification) or assigned a probability of belonging to a cluster (mixture). Then the parameters are estimated, given the current assignments, and the process is repeated. Further details of the algorithm are provided in James and Sugar (2003).

Low-dimensional graphical representations

One of the chief difficulties in high-dimensional clustering is visualization of the data. Plotting functional data is easier because of the continuity of the dimensions. However, it can still be hard to see the clusters since variations in shape and the location of time-points make it difficult to assess the relative distances between curves. These problems are exacerbated when the curves are fragmentary, as in Fig. 11.1(a). In this section we illustrate a set of graphical tools for use with functional data. The method is based on projecting the curves into a low-dimensional space so that they can be plotted as points, making it much easier to detect the presence of clusters.

Figure 11.3(a) shows the growth data projected onto a one-dimensional space. The horizontal axis represents the projected curve, $\hat{\alpha}_i$, while the vertical axis gives the average age of observation for each individual. Points to the left of zero are assigned to cluster 1 and the remainder to cluster 2. Squares represent males and circles females. The dotted lines at α_1 and α_2 correspond to the projected cluster

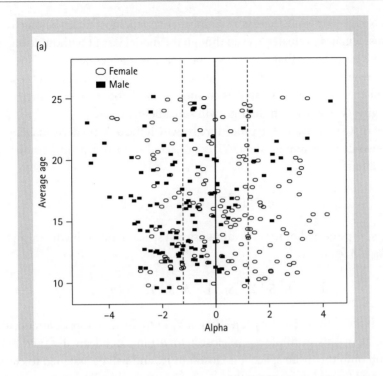

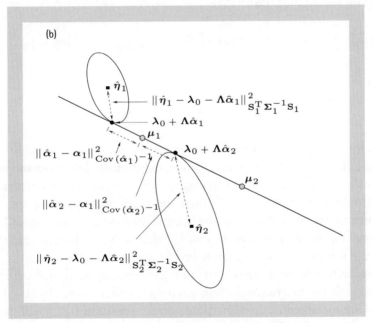

Fig. 11.3 (a) A linear discriminant plot for the bone mineral density data. (b) An illustration of the decomposition of the distance between two curves and two cluster centers.

centers. Notice that while there is a significant overlap, most males belong to cluster 1 and most females to cluster 2, even though the model was fit without using gender labels. The plot shows that the clustering separates the genders most strongly for those younger than 16 years. In fact, 74% of such individuals matched the majority gender of their cluster compared with only 57% of those older than 16. This is because girls typically begin their growth spurt before boys.

Figure 11.3(b) illustrates the procedure by which the $\widehat{\alpha}_i$'s are derived using a two-cluster, two-curve example. First, Y_i is projected onto the p-dimensional spline basis to obtain

$$\hat{\eta}_i = \left(S_i^T \Sigma_i^{-1} S_i\right)^{-1} S_i^T \Sigma_i^{-1} Y_i. \tag{11.14}$$

Second, $\hat{\eta}_i$ is projected onto the h-dimensional space spanned by the means, μ_k, to obtain $\lambda_0 + \Lambda \widehat{\alpha}_i$, where

$$\widehat{\alpha}_i = \left(\Lambda^T S_i^T \Sigma_i^{-1} S_i \Lambda\right)^{-1} \Lambda^T S_i^T \Sigma_i^{-1} S_i (\hat{\eta}_i - \lambda_0). \tag{11.15}$$

Thus, $\widehat{\alpha}_i$ is the h-dimensional projection of Y_i onto the mean space after centering. Notice that in this example $\hat{\eta}_2$ is closest to μ_2 in Euclidean distance but after projection onto the mean space it is closest to μ_1 and will be assigned to cluster 1.

James and Sugar (2003) prove that

$$\arg\max_k P(z_{ik} = 1|Y_i) = \arg\min_k \left(||\widehat{\alpha}_i - \alpha_k||^2_{\operatorname{Cov}(\widehat{\alpha}_i)^{-1}} - 2 \log \pi_k \right), \tag{11.16}$$

where

$$\operatorname{Cov}(\widehat{\alpha}_i) = \left(\Lambda^T S_i^T \Sigma_i^{-1} S_i \Lambda\right)^{-1} \tag{11.17}$$

and π_k is the prior probability that an observation belongs to the kth cluster. From (11.16) and Bayes' rule we note that cluster assignments based on the $\widehat{\alpha}_i$'s will minimize the expected number of misassignments. Thus no clustering information is lost through the projection of Y_i onto the lower-dimensional space. We call the $\widehat{\alpha}_i$'s functional linear discriminants because they are exact analogs of the low-dimensional representations used to visualize data in linear discriminant analysis. In the finite-dimensional setting the linear discriminants all have identity covariance, so separation between classes can be assessed visually using the Euclidean distance metric. In the functional clustering setting $\operatorname{Cov}(\widehat{\alpha}_i)$ is given by (11.17). When all curves are measured at the same time points, constraint (11.12) will guarantee $\operatorname{Cov}(\widehat{\alpha}_i) = I$ for all i, again allowing the Euclidean metric to be used. When curves are measured at different time points it is not possible to impose a constraint that will simultaneously cause $\operatorname{Cov}(\widehat{\alpha}_i) = I$ for all i. However, when the cluster means lie in a one-dimensional subspace ($h = 1$), assuming equal priors, (11.16) simplifies to

$$\arg\min_k \frac{1}{\text{Var}(\widehat{a}_i)}(\widehat{a}_i - a_k)^2 = \arg\min_k (\widehat{a}_i - a_k)^2,$$

which yields the same assignments as if the $\widehat{a}_i$'s all had the same variance. In this situation it is useful to plot the functional linear discriminants versus their standard deviations to indicate not only to which cluster each point belongs but also the level of accuracy with which it has been observed. Note that for a two-cluster model h must be 1. However, it will often be reasonable to assume the means lie approximately in one dimension even when there are more than two clusters.

Linear discriminant plots have other useful features. Note that the functional linear discriminant for a curve observed over the entire grid of time points used to form S will have identity covariance. Thus, the Euclidean distance between the α_k's gives the number of standard deviations separating the cluster means for a fully observed curve. The degree to which the variance for an individual curve is greater than 1 indicates how much discriminatory power has been lost due to taking observations at only a subset of time points. This has implications for experimental design in that it suggests how to achieve minimum variance, and hence optimal cluster separation, with a fixed number of time points. For instance, the cluster means in Fig. 11.3(a) are 2.4 standard deviations apart, indicating that the groups can be fairly well separated if curves are measured at all time points. The overlap between the two groups is due to the extreme sparsity of sampling, resulting in the $\widehat{\alpha}_i$'s having standard deviations up to 2.05.

Plots for the membranous nephropathy data, given in Fig. 11.4, provide an example in which the differing covariances of the $\widehat{\alpha}_i$'s must be taken into account more carefully. Nephrologists' experiences suggest that patients with this disease fall into three groups, either faring well, deteriorating gradually, or collapsing quickly. Hence we fit a three-cluster model whose mean curves are shown in Fig. 11.4(a). With three clusters the means must lie in a plane. Figure 11.4(b) shows a two-dimensional linear discriminant plot with solid circles indicating cluster centers. To circumvent the problem caused by the unequal covariances, we use different symbols for members of different clusters. Note that while most patients fall in the cluster corresponding to their closest cluster in Euclidean distance, there are several that do not. In this example the cluster centers lie essentially on a straight line so it is sufficient to fit a one-dimensional model ($h = 1$). The corresponding plots are shown in Figs 11.4(c) and (d). The basic shapes of the mean curves are reassuringly similar to those in 11.4(a), but are physiologically more sensible in the right tail. Figure 11.4(d) plots one dimensional $\widehat{\alpha}_i$'s versus their standard deviations. We see that the cluster on the right is very tight while the other two are not as well separated. Figures 11.4(e) and (f) show, respectively, the overall mean curve, $s(t)^T \lambda_0$ and the function $s(t)^T \Lambda$. The latter, when multiplied by α_k, gives the distance between $\mu_k(t)$ and the overall mean curve. From Fig. 11.4(e) we see that on average the patients showed a decline in renal function. The primary distinction lies in the

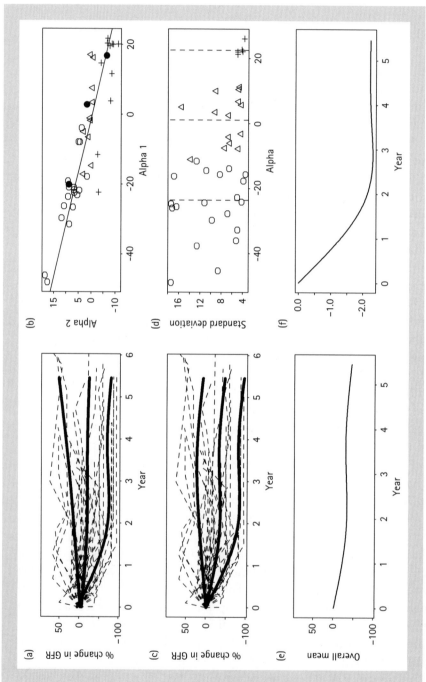

Fig. 11.4 Assessment plots for the membranous nephropathy data. The cluster mean curves and linear discriminant plots for a fit with $h = 2$ are shown in (a) and (b). The equivalent plots for a fit with $h = 1$ are given in (c) and (d). Finally, (e) shows the overall mean curve and (f) the characteristic pattern of deviations about the overall mean.

speed of the deterioration. For example, the fact that Fig. 11.4(f) shows a sharp decline in the first two years indicates that patients in the third cluster, which has a highly positive α_3, experience a much sharper initial drop than average. In fact all patients in cluster 3 eventually required dialysis.

Discriminant functions

In this section we present a set of curves that identify the dimensions, or equivalently time points, of maximum discrimination between clusters. Intuitively, the dimensions with largest average separation relative to their variability will provide the greatest discrimination. Average separation can be determined by examining $S\Lambda$, while variability is calculated using the covariance matrix, $\Sigma = S\Gamma S^T + \sigma^2 I$. These two quantities can work in opposite directions, making it difficult to identify the regions of greatest discrimination. Consider, for example, Fig. 11.5 which illustrates the covariance and correlation functions for the growth data. From Fig. 11.5(a) it is clear that the relationship between a person's bone mineral density before and after puberty is weak, but the measurements after puberty are strongly correlated with each other. Figure 11.5(b) has a sharp peak in the early puberty years, corresponding to the period of greatest variability. However, this is also the period of greatest distance between the cluster mean curves.

The dimensions of maximum discrimination must also be the ones that are most important in determining cluster assignment. When observations are made at all time points, the spline basis matrix is S, and equations (11.15) and (11.16) imply that curves should be assigned based solely on the Euclidean distance between $\widehat{\alpha} = \Lambda^T S^T \Sigma^{-1}(Y - S\lambda_0)$ and the α_k's. Thus

$$\Lambda^T S^T \Sigma^{-1}$$

gives the optimal weights to apply to each dimension for determining cluster membership. Dimensions with low weights contain little information about cluster

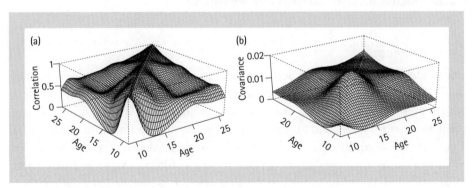

Fig. 11.5 Estimated (a) correlation and (b) covariance of $g_i(t_1)$ with $g_i(t_2)$ for the spinal bone data.

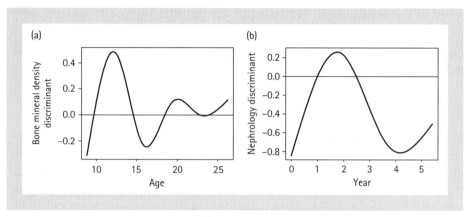

Fig. 11.6 Discriminant curves for (a) the growth data and (b) the nephropathy data with $h = 1$.

membership and therefore do little to distinguish among groups, while dimensions with large weights have high discriminatory power. Notice that this set of weights fits with the intuitive notion that dimensions with high discrimination should have large average separation, $S\Lambda$, relative to their variability, Σ. James and Sugar (2003) draw connections between this function and the classical linear discriminant function for a Gaussian mixture.

When the α_k's are one dimensional, $\Lambda^T S^T \Sigma^{-1}$ is a vector and the weights can be plotted as a single curve, as illustrated by Fig. 11.6 for the growth and nephropathy datasets. For the growth data the highest absolute weights occur in the puberty years, confirming our earlier interpretation from the linear discriminant plot, Fig. 11.3(a). For the nephropathy data most of the discrimination between clusters occurs in the early and late stages of disease. The difference between patients in the later time periods is not surprising. However, the discriminatory power of the early periods is encouraging since one of the primary goals of this study was to predict disease progression based on entry characteristics.

Curve estimation

The model at the beginning of Section 11.3.2 can also be used to accurately predict unobserved portions of $g_i(t)$, the true curve for the ith individual. When using a basis representation, a natural estimate for $g_i(t)$ is $\hat{g}_i(t) = s(t)^T \hat{\eta}_i$, where $\hat{\eta}_i$ is a prediction for η_i. James and Sugar (2003) show that the optimal estimate for η_i is $E(\eta_i|Y_i)$. This quantity takes on slightly different values depending on whether the unknown cluster membership, z_i, is taken to be a random variable or a parameter to be estimated. However, in the simpler case where z_i is modeled as a parameter,

$$E(\eta_i|Y_i) = \lambda_0 + \Lambda\alpha_{z_i} + \left(\sigma^2\Gamma^{-1} + S_i^T S_i\right)^{-1} S_i^T \left(Y_i - S_i\left(\lambda_0 + \Lambda\alpha_{z_i}\right)\right), \quad (11.18)$$

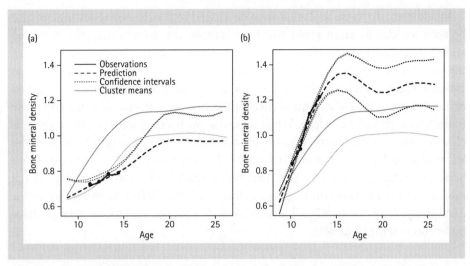

Fig. 11.7 Curve estimates, confidence intervals, and prediction intervals for two subjects from the growth data.

where $z_i = \arg\max_k f(y|z_{ik} = 1)$. In general, using (11.18) to form predictions produces significant improvements over the basis approach from Section 11.2.1 when σ^2 is very large, the components of Γ are very small, or $S_i^T S_i$ is close to singular. In fact, when $S_i^T S_i$ is singular the basis approach breaks down completely, while the functional clustering method can still produce reliable predictions.

Figure 11.7 illustrates this approach for two subjects from the growth data. For each plot, the two solid grey lines give the cluster mean curves, the solid black curve fragment gives the observed values for a single individual, and the dashed line gives the corresponding prediction. The dotted lines represent 95% confidence and prediction intervals. See James and Sugar (2003) for details on producing the intervals. Note that the confidence interval provides bounds for the underlying function $g_i(t)$ while the prediction interval bounds the observed value of $g_i(t)$. As usual, the prediction interval is produced by adding σ^2 to the variance used in the confidence interval.

11.3.3 Extensions to functional classification

The functional clustering model can easily be extended to the functional classification setting where one wishes to classify a curve, $Y_i(t)$, into one of G possible classes. In this setting James and Hastie (2001) use (11.10) to model the observed curves, except that now z_i is assumed known for the data used to train the model. This removes one step from the functional clustering EM fitting procedure, namely where z_i is estimated. In all other respects the clustering and classification models are fit in an identical fashion.

In addition, the same tools that were developed in the clustering setting can also be used for classification problems. For example, the growth data also recorded an individual's ethnicity. We fit the functional classification model to a subset of the data, females of Asian or African American descent, using ethnicity as the class variable. Figure 11.8(a) graphs the corresponding linear discriminants, as described earlier. Points to the right of the vertical center line are classified as Asian and those to the left as African American. While there is some overlap between the groups it is clear that most of the observations on the right are Asian (circles) while those on the left tend to be African American (triangles). There is strong evidence of a difference between the two groups. This is further highlighted in Fig. 11.8(b) which plots the raw curves for the two ethnic groups, along with the estimated mean curves for each group. One can clearly see that African Americans tend to, on average, have higher bone mineral density.

11.3.4 Functional regression

One of the most useful tools in FDA is that of functional regression. This setting can correspond to either functional predictors or functional responses or both. See Ramsay and Silverman (2002) and Muller and Stadtmuller (2005) for numerous specific applications. One commonly studied problem involves data with functional predictors but scalar responses. Ramsay and Silverman (2005) discuss this scenario; other papers have also been written on the topic, both for continuous and categorical responses, and for linear and nonlinear models (Hastie and Mallows 1993; James and Hastie 2001; Ferraty and Vieu 2002; James 2002; Cardot *et al.* 2003; Ferraty and Vieu 2003; Muller and Stadtmuller 2005; James and Silverman 2005; Cardot *et al.* 2007; Crambes *et al.* 2009). The alternative situation where the response is functional has also been well studied. A sampling of papers examining this situation includes Fahrmeir and Tutz (1994), Liang and Zeger (1986), Faraway (1997), Hoover *et al.* (1998), Wu *et al.* (1998), Fan and Zhang (2000), and Lin and Ying (2001). Chapters 1–6 of this volume give good overviews of this topic.

Here we briefly examine an approach by Yao *et al.* (2005b), specifically designed for performing functional regressions where the predictor and response have both been sparsely sampled. Assume we observe a set of predictor curves, X_i, and a corresponding set of response curves, Y_i. The predictors are observed at times $s_{i1}, \ldots, s_{il_i}$ and the responses at times $t_{i1}, \ldots, t_{in_i}$. A standard method for modeling the relationship between X and Y is to use the linear regression model

$$E[Y(t)|X] = \mu_Y(t) + \int \beta(s, t) X^c(s) \, ds, \qquad (11.19)$$

where X^c is the centered version of X. In this model, $\beta(s, t)$ is a two-dimensional coefficient function that must be estimated from the data. When X and Y are both

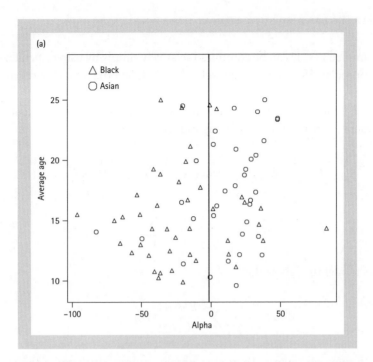

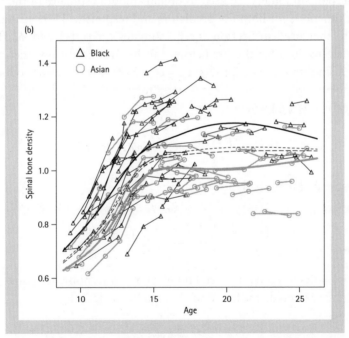

Fig. 11.8 (a) Linear discriminants for African Americans and Asians. The vertical line gives the classification boundary. (b) A plot of the raw data for African Americans and Asians. The two solid lines represent the means for African Americans and Asians, while the dotted lines are for two other ethnic groups, Hispanics and Whites.

densely sampled $\beta(s, t)$ can be computed relatively simply by using a penalized least squares approach. However, for sparsely observed data the least squares method is not feasible.

Yao et al. (2005b) avoid this problem by first assuming that the curves have measurement error using the following model

$$U_{il} = X_i(s_{il}) + \epsilon_{il} = \mu_X(s_{il}) + \sum_{m=1}^{\infty} \zeta_{im}\psi(s_{il}) + \epsilon_{il} \qquad (11.20)$$

$$V_{ij} = Y_i(t_{ij}) + e_{il} = \mu_Y(t_{ij}) + \sum_{k=1}^{\infty} \xi_{ik}\phi(t_{ik}) + e_{ik}. \qquad (11.21)$$

Using this model they show that

$$\beta(s, t) = \sum_{k=1}^{\infty} \sum_{m=1}^{\infty} \frac{E[\zeta_m \xi_k]}{E[\zeta_m^2]} \psi_m(s)\phi_k(t).$$

Next, scatterplot smoothers are used to obtain smooth estimates of the mean and covariance functions for both the predictors and responses. Then, using a similar approach to that used in Section 11.2.3, estimates for $\psi_m(s)$ and $\phi_k(t)$, as well as ρ_m, the eigenvalues for the predictors, can be produced from the covariance functions. A smooth estimate for $C(s, t) = \text{cov}(X(s), Y(t))$ is also produced using a smoothing procedure. Then ρ_m can be used to estimate $E\left[\zeta_m^2\right]$, and $\hat{\sigma}_{km}$ provides an estimate for $E[\zeta_m \xi_k]$ where

$$\hat{\sigma}_{km} = \iint \hat{\psi}_m(s)\widehat{C}(s, t)\hat{\phi}(t)\, ds\, dt.$$

The final estimate for $\beta(s, t)$ is given by

$$\hat{\beta}(s, t) = \sum_{k=1}^{K} \sum_{m=1}^{M} \frac{\hat{\sigma}_{km}}{\hat{\rho}_m} \hat{\psi}_m(s)\hat{\phi}_k(t).$$

Once this model has been fit, predictions for a new response function, Y^*, given a partially observed predictor function, X^*, can be made using

$$Y^*(t) = \hat{\mu}_Y(t) + \sum_{k=1}^{K} \sum_{m=1}^{M} \frac{\hat{\sigma}_{km}}{\hat{\rho}_m} \hat{\zeta}_m^* \hat{\phi}_k(t).$$

The only quantity here that needs to be computed is $\hat{\zeta}_m^*$. However, $\hat{\zeta}_m^*$ can be computed using an analogous approach to the PACE method of Section 11.2.3. Details are provided in Yao et al. (2005b).

11.4 VARIABLE SELECTION IN A FUNCTIONAL SETTING

In this section we examine another sense in which sparsity can be important in the regression setting. Here we return to the more traditional FDA paradigm where it is assumed that the functions in question have been observed over a dense grid of time points. In this setting we outline two methods that have been proposed for performing regressions involving a functional predictor, $X(t)$, and a scalar response, Y. Both methods use ideas from the high-dimensional regression literature, where it is often assumed that the coefficient vector, β, is sparse in the sense of containing few nonzero values.

11.4.1 Selecting influential design points

Ferraty et al. (2010) suggest a method, called a "sequential algorithm for selecting design automatically" (SASDA), for taking a function predictor, $X(t)$, observed at a fine grid of time points, $t_1, \ldots, t_r$, and selecting a small subset of points that are most predictive of the response, Y. They refer to these points as the "most predictive design points". The selected design points are then used to produce a local linear regression for predicting the response.

Ferraty et al. begin with the general functional nonparametric model

$$Y = g(X) + \epsilon. \tag{11.22}$$

If one was to interpret this model using a standard nonparametric approach then (11.22) would become $Y = g(X(t_1), \ldots, X(t_r)) + \epsilon$. This approach is not practical because $X(t_1), \ldots, X(t_r)$ are both high dimensional and highly correlated. Instead, g is approximated using a local linear estimator, $\hat{g}_h$. Let $S(\mathbf{t}, h)$ be the cross-validated sum of squares error between Y_i and the prediction, $\hat{g}_h(X_i)$, constructed using the time points, $\mathbf{t}$. Then an iterative algorithm is proposed for selecting the optimal design points for $\mathbf{t}$. First $\mathbf{t}$ is initialized to the empty set. Then, the single time point t_s that gives the lowest value for $S(\mathbf{t}, h)$ is selected. The algorithm continues, at each step selecting the time point resulting in the greatest reduction in $S(\mathbf{t}, h)$, conditional on the currently chosen points. This procedure continues until there are no new points that cause a large enough reduction in $S(\mathbf{t}, h)$.

Ferraty et al. demonstrate SASDA on two real-world datasets and show that it can have higher predictive accuracy in comparison to other standard functional linear, or nonlinear, regression methods. Another advantage of SASDA is that it generally selects a small set of time points that jointly contain the relevant information in terms of predicting Y. This makes the final model much easier to interpret because the relationship between the response and predictor can be summarized in terms of these few time points.

Several extensions of SASDA are also considered. Ferraty et al. note that their approach can be computationally expensive. They suggest using a method such as the Lasso (Tibshirani 1996) to perform an initial dimension reduction, and then to implement SASDA on the time points that Lasso selects. This seems to result in a small reduction in prediction accuracy but a significant improvement in computational efficiency. Another extension that is examined involves combining SASDA with other functional regression methods. Specifically, one first fits SASDA and then uses functional linear, or nonlinear, regression methods to predict the leave-one-out residual between the response and the SASDA prediction. The final predicted response is then the sum of the SASDA fit plus the predicted residual. This combined approach appears to give further improvements in prediction accuracy.

11.4.2 Functional linear regression that's interpretable

James et al. (2009) develop a method, which they call "Functional Linear Regression That's Interpretable" (FLiRTI), that produces accurate, but also highly interpretable, estimates for the coefficient function, $\beta(t)$. The key to their procedure is to reformulate the problem as a form of variable selection. In particular they divide the time period up into a fine grid of points and then use variable selection methods to determine whether the dth derivative of $\beta(t)$ is zero or not at each of the grid points. The implicit assumption is that the dth derivative will be zero at most grid points, i.e. it will be sparse. By choosing appropriate derivatives they can produce a large range of highly interpretable $\beta(t)$ curves.

James et al. start with the standard functional linear regression model

$$Y_i = \beta_0 + \int X_i(t)\beta(t)\,dt + \epsilon_i, \quad i = 1, \ldots, n.$$

Then, modeling $\beta(t)$ using a p-dimensional basis function, $\beta(t) = B(t)^T \eta$, they arrive at

$$Y_i = \beta_0 + \mathbf{X}_i^T \eta + \epsilon_i, \tag{11.23}$$

where $\mathbf{X}_i = \int X_i(t)B(t)dt$. Estimating η presents difficulties because a high-dimensional basis is used, so $p > n$. Hence FLiRTI models $\beta(t)$ assuming that one or more of its derivatives are sparse, i.e. $\beta^{(d)}(t) = 0$ over large regions of t for one or more values of $d = 0, 1, 2, \ldots$ This approach has the advantage of both constraining η enough to allow one to fit (11.23) as well as producing a highly interpretable estimate for $\beta(t)$. For example, constraining the first derivative produces an estimate for $\beta(t)$ that is constant over large regions. The fact that the effect of $X(t)$

on Y is constant over most time points makes the relationship between predictor and response simple to understand.

Let $A = [D^d\mathbf{B}(t_1), D^d\mathbf{B}(t_2), \ldots, D^d\mathbf{B}(t_p)]^T$ where $t_1, t_2, \ldots, t_p$ represent a grid of p evenly spaced points and D^d is the dth finite difference operator, i.e.

$$D\mathbf{B}(t_j) = p\left[\mathbf{B}(t_j) - \mathbf{B}(t_{j-1})\right],$$

$$D^2\mathbf{B}(t_j) = p^2\left[\mathbf{B}(t_j) - 2\mathbf{B}(t_{j-1}) + \mathbf{B}(t_{j-2})\right],$$

etc. Then, if

$$\boldsymbol{\gamma} = A\boldsymbol{\eta}, \tag{11.24}$$

γ_j provides an approximation to $\beta^{(d)}(t_j)$ and hence, enforcing sparsity in $\boldsymbol{\gamma}$ constrains $\beta^{(d)}(t_j)$ to be zero at most time points. For example, one may believe that $\beta^{(2)}(t) = 0$ over many regions of t, i.e. $\beta(t)$ is exactly linear over large regions of t. In this situation we would let

$$A = [D^2\mathbf{B}(t_1), D^2\mathbf{B}(t_2), \ldots, D^2\mathbf{B}(t_p)]^T, \tag{11.25}$$

which implies $\gamma_j = p^2[\mathbf{B}(t_j)^T\boldsymbol{\eta} - 2\mathbf{B}(t_{j-1})^T\boldsymbol{\eta} + \mathbf{B}(t_{j-2})^T\boldsymbol{\eta}]$. Hence, provided p is large, so t is sampled on a fine grid, $\gamma_j \approx \beta^{(2)}(t_j)$. In this case enforcing sparsity in the γ_j's will produce an estimate for $\beta(t)$ that is linear except at the time points corresponding to nonzero values of γ_j.

If A is constructed using a single derivative, as in (11.25), then one can always choose a grid of p different time points, $t_1, t_2, \ldots, t_p$, such that A is a square p-by-p invertible matrix. In this case $\boldsymbol{\eta} = A^{-1}\boldsymbol{\gamma}$ so (11.23) and (11.24) can be combined to produce the FLiRTI model

$$\mathbf{Y} = V\boldsymbol{\gamma} + \boldsymbol{\epsilon},$$

where $V = [\mathbf{1}|XA^{-1}]$, $\mathbf{1}$ is a vector of ones and β_0 has been incorporated into $\boldsymbol{\gamma}$. James et al. then use high-dimensional regression methods such as the Lasso (Tibshirani 1996) and the Dantzig Selector (Candes and Tao 2007) to estimate $\boldsymbol{\gamma}$. A final estimate for $\beta(t)$ is produced using $\hat{\beta}(t) = \mathbf{B}(t)^T\hat{\boldsymbol{\eta}} = \mathbf{B}(t)^T A^{-1}\hat{\boldsymbol{\gamma}}$. James et al. show that FLiRTI performs well on simulated and real-world datasets. In addition, they present non-asymptotic theoretical bounds on the estimation error.

References

Bachrach, L.K., Hastie, T.J., Wang, M.C., Narasimhan, B., Marcus, R. (1999). Bone mineral acquisition in healthy Asian, Hispanic, Black and Caucasian youth; a longitudinal study. *Journal of Clinical Endocrinology & Metabolism*, 84, 4702–12.

Banfield, J.D., Raftery, A.E. (1993). Model-based Gaussian and non-gaussian clustering. *Biometrics*, 49, 803–21.

BRUMBACK, B., RICE, J. (1998). Smoothing spline models for the analysis of nested and crossed samples of curves. *Journal of the American Statistical Association*, 93, 961–76.

CANDES, E., TAO, T. (2007). The Dantzig selector: statistical estimation when p is much larger than n. *Annals of Statistics*, 35, 2313–51.

CARDOT, H., FERRATY, F., SARDA, P. (2003) Spline estimators for the functional linear model. *Statist. Sinica*, 13, 571–91.

CARDOT, H., MAS, A., SARDA, P. (2007). CLT in functional linear regression models. *Probab. Theory Related Fields*, 138, 325–561.

CRAMBES, C., KNEIP, A., SARDA, P. (2009). Smoothing splines estimators for functional linear regression. *Ann. Statist.*, 37, 35–72.

DIPILLO, P.J. (1976). The application of bias to discriminant analysis. *Communications in Statistics, Part A – Theory and Methods*, A5, 843–54.

FAHRMEIR, L., TUTZ, G. (1994). *Multivariate Statistical Modeling Based on Generalized Linear Models*. Springer, New York.

FAN, J., ZHANG, J. (2000). Two-step estimation of functional linear models with applications to longitudinal data. *Journal of the Royal Statistical Society, Series B*, 62, 303–22.

FARAWAY, J. (1997). Regression analysis for a functional response. *Technometrics*, 39, 254–61.

FERRATY, F., HALL, P., VIEU, P. (2010). Most-predictive design points for functional data predictors. Biometrika (to appear).

FERRATY, F., VIEU, P. (2002). The functional nonparametric model and applications to spectrometric data. *Computational Statistics*, 17, 545–64.

FERRATY, F., VIEU, P. (2003). Curves discrimination: a nonparametric functional approach. *Computational Statistics and Data Analysis*, 44, 161–73.

FRIEDMAN, J.H. (1989). Regularized discriminant analysis. *Journal of the American Statistical Association*, 84, 165–75.

GREEN, P.J., SILVERMAN, B.W. (1994). *Nonparametric Regression and Generalized Linear Models: A Roughness Penalty Approach*. Chapman and Hall, London.

HASTIE, T.J., BUJA, A., TIBSHIRANI, R.J. (1995). Penalized discriminant analysis. *Annals of Statistics*, 23, 73–102.

HASTIE, T., MALLOWS, C. (1993). Comment on "a statistical view of some chemometrics regression tools". *Technometrics*, 35, 140–3.

HENDERSON, C.R. (1950). Estimation of genetic parameters (abstract). *Annals of Mathematical Statistics*, 21, 309–10.

HOOVER, D.R., RICE, J.A., WU, C.O., YANG, L.P. (1998). Nonparametric smoothing estimates of time-varying coefficient models with longitudinal data. *Biometrika*, 85, 809–22.

JAMES, G.M. (2002). Generalized linear models with functional predictors. *Journal of the Royal Statistical Society, Series B*, 64, 411–32.

JAMES, G.M., HASTIE, T.J. (2001). Functional linear discriminant analysis for irregularly sampled curves. *Journal of the Royal Statistical Society, Series B*, 63, 533–50.

JAMES, G.M., HASTIE, T.J., SUGAR, C.A. (2000). Principal component models for sparse functional data. *Biometrika*, 87, 587–602.

JAMES, G.M., SILVERMAN, B.W. (2005). Functional adaptive model estimation. *Journal of the American Statistical Association*, 100, 565–76.

JAMES, G.M., SUGAR, C.A. (2003). Clustering for sparsely sampled functional data. *Journal of the American Statistical Association*, 98, 397–408.

JAMES, G.M., WANG, J., ZHU, J. (2009). Functional Linear Regression That's Interpretable. *Annals of Statistics*, 37, 2083–108.

LAIRD, N.M., WARE, J.H. (1982). Random-effects models for longitudinal data. *Biometrics*, 38, 963–74.

LIANG, K.Y., ZEGER, S.L. (1986). Longitudinal data analysis using generalized linear models. *Biometrika*, 73, 13–22.

LIN, D.Y., YING, Z. (2001). Semiparametric and nonparametric regression analysis of longitudinal data. *Journal of the American Statistical Association*, 96, 103–13.

MÜLLER, H.G., STADTMÜLLER, U. (2005). Generalized functional linear models. *Annals of Statistics*, 33, 774–805.

PENG, J., PAUL, D. (2009). "A Geometric Approach to Maximum Likelihood Estimation of the Functional Principal Components From Sparse Longitudinal Data." *Journal of Computational and Graphical Statistics*, 18, 995–1015.

RAMSAY, J.O., SILVERMAN, B.W. (2002). *Applied Functional Data Analysis*. Springer, New York.

RAMSAY, J.O., SILVERMAN, B.W. (2005). *Functional Data Analysis* (second edition). Springer, New York.

RICE, J.A., WU, C.O. (2001). Nonparametric mixed effects models for unequally sampled noisy curves. *Biometrics*, 57, 253–9.

SHI, M., WEISS, R., TAYLOR, J. (1996). An analysis of pediatric CD4 counts for acquired immune deficiency syndrome using flexible random curves. *Applied Statistics*, 45, 151–64.

SILVERMAN, B.W. (1985). Some aspects of the spline smoothing approach to non-parametric regression curve fitting (with discussion). *R. Statist. Soc. B*, 47, 1–52.

STANISWALIS, J.G., LEE, J.J. (1998). Nonparametric regression analysis of longitudinal data. *Journal of the American Statistical Association*, 93, 1403–18.

TIBSHIRANI, R. (1996). Regression shrinkage and selection via the lasso. *Roy. Statist. Soc. Ser. B*, 58, 267–88.

WU, C.O., CHIANG, C.T., HOOVER, D.R. (1998). Asymptotic confidence regions for kernel smoothing of a varying-coefficient model with longitudinal data. *Journal of the American Statistical Association*, 93, 1388–1402.

YAO, F., MULLER, H., WANG, J. (2005a). Functional data analysis for sparse longitudinal data. *Journal of the American Statistical Association*, 100, 577–90.

YAO, F., MULLER, H., WANG, J. (2005b). Functional linear regression analysis for longitudinal data. *Annals of Statistics*, 33, 2873–903.

PART III

TOWARDS A STOCHASTIC BACKGROUND IN INFINITE-DIMENSIONAL SPACES

CHAPTER 12

VECTOR INTEGRATION AND STOCHASTIC INTEGRATION IN BANACH SPACES

NICOLAE DINCULEANU

12.1 Introduction

THIS chapter is devoted to the theory of integration with respect to vector measures with finite semivariation and its applications. This theory reduces to integration with respect to vector measures with *finite variation* which, in turn, reduces to the Bochner integral with respect to a positive measure. The Bochner integral, itself, is based upon the classical integral of real-valued functions with respect to a positive measure. The above presentation is a description, in reverse order, of the four stages in the development of this integration theory. We shall present these four stages in their natural order.

Among the many approaches to the *classical integral* we have chosen one which seems to be relatively simple and is presented, for example, by Rudin (1973). Any one of these approaches leads to a vector space $L^1(\mu)$ of integrable functions, equipped with a seminorm $||f||_1$, under which it is complete and in which the Lebesgue convergence theorem is valid.

We shall impose the same requirements on any kind of integration theory. An integration theory satisfying these requirements is called a "satisfactory" theory. Here, "satisfactory" refers to the possibility of using the integral in a wide range of applications. The other three stages of the integration theory yield satisfactory integrals.

The Bochner integrability of vector-valued functions f reduces to the classical integrability of $|f|$. The Bochner integral is obtained by extending by continuity the integral of step functions.

There are other types of integrals for vector-valued functions, for example the Pettis integral. These integrals are not satisfactory in the above sense.

Integrability with respect to a vector measure m with *finite variation* $|m|$ is defined as Bochner integrability with respect to the positive measure $|m|$. The integral is again obtained by extending by continuity the integral of step functions. This stage of the integration theory is presented in detail by Dinculeanu (1967).

Finally, integrability with respect to a measure with *finite semivariation* reduces to integrability with respect to a family of measures m_z with finite variation. The integral $\int f\,dm$ is defined then as the linear operation $z \mapsto \int f\,dm_z$. This last stage is very important. In fact, the most interesting measures do not have finite variation, but may have finite semivaration.

Among the applications of the last stage of integration theory we mention: the integral representation of linear operations on L^p-spaces, the Riesz representation theorem, and the Stieltjes integral with respect to a vector-valued function with *finite semivariation*. The most important application is the stochastic integral in Banach spaces, with respect to summable processes; in particular with respect to square-integrable martingales and with respect to processes with integrable variation or integrable semivariation.

For a detailed presentation of vector integration and its applications, the reader is referred to Dinculeanu (2000).

In this chapter we give complete statements of definitions and theorems, but no proofs of theorems are given. For proofs, the reader will be referred to the literature.

12.2 Preliminaries

In this section we establish the notation that will be used throughout the chapter.

12.2.1 Banach spaces

Throughout the chapter E, F, G, D will denote Banach spaces. For any Banach space M the norm of an element $x \in M$ is denoted by $|x|$, the dual of M is denoted by M^*, and the unit ball of M by M_1. The duality between M and M^* is denoted by $\langle x, x^*\rangle$, $\langle x^*, x\rangle$, x^*x, or even xx^*. $L(F, G)$ is the space of bounded linear operations from F to G. We write $E \subset L(F, G)$ to mean that E is continuously embedded into $L(F, G)$, i.e. $|xy| \leq |x||y|$ for $x \in E$ and $y \in F$. For example, $E = L(\mathbb{R}, E)$, $E \subset L(E^*, \mathbb{R}) = E^{**}$; if E is a Hilbert space, $E = L(E, \mathbb{R})$.

We write $c_0 \subseteq M$ to mean that M contains a subspace which is isomorphic to the Banach space c_0.

If M is a Banach space, a subspace $Z \subset M^*$ is said to be *norming* for M, if for every $x \in M$ we have

$$|x| = \sup\{|\langle x, z\rangle| : z \in Z_1\}.$$

12.2.2 Measurable functions

Throughout the chapter, S is a nonempty set, $\mathcal{P}, \mathcal{R}, \mathcal{A}, \mathcal{D}, \mathcal{S}, \pm$ are respectively a semiring, a ring, an algebra, a δ-ring, a σ-ring, and the σ-algebra of subsets of S.

For any class $\mathcal{C}$ of subsets S we denote by $r(\mathcal{C})$, $a(\mathcal{C})$, $\delta r(\mathcal{C})$, $\sigma r(\mathcal{C})$, $\sigma a(\mathcal{C})$, respectively the ring, the algebra, the δ-ring, the σ-ring, and the σ-algebra generated by $\mathcal{C}$.

If $\mathcal{R}$ is a ring, we denote by $\mathcal{S}_F(\mathcal{R})$ the vector space of $\mathcal{R}$-step functions $f : S \to F$ of the form $f = \sum_{i=1}^{n} \chi_{A_i} x_i$, with $A_i \in \mathcal{R}$ and $x_i \in F$. The sets A_i can be chosen to be mutually disjoint. Then

$$|f| = \sum_{i=1}^{n} \chi_{A_i} |x_i|.$$

For any function $f : S \to F$ or $\overline{\mathbb{R}}$, we denote by $|f|$ the function defined by

$$|f|(s) = |f(s)| \quad \text{for } s \in \mathcal{S}.$$

We emphasize that a positive $\mathcal{R}$-step function takes only finite values.

Measurability is defined with respect to a σ-algebra. Let Σ be a σ-algebra of subsets of $\mathcal{S}$.

Definition 12.1 *A function $f : S \to \mathcal{F}$ or $\overline{\mathbb{R}}$ is said to be Σ-measurable if there is a sequence (f_n) of Σ-step functions $f_n : S \to \mathcal{F}$ such that $f_n \to f$, pointwise.*

The Σ-step functions are Σ-measurable. It follows that the set of Σ-measurable functions is a vector space. If f is Σ-measurable, then $|f|$ is also Σ-measurable.

In the above definition we can choose the sequence (f_n) with additional properties.

Theorem 12.1 Let $f : S \to \mathcal{F}$ or $\overline{\mathbb{R}}$ be a Σ-measurable function. Then there is a sequence (f_n) of Σ-step functions $f_n : S \to \mathcal{F}$ or $\mathbb{R}$ such that $f_n \to f$ pointwise and $|f_n| \le |f|$ for each n.

If f is positive (with values in $[0, +\infty]$), the sequence (f_n) can be chosen to be increasing.

If f is real-valued and bounded, the sequence (f_n) can be chosen to be uniformly convergent.

The following theorem gives a characterization of Σ-measurability.

Theorem 12.2 A function $f : S \to \mathcal{F}$ or $\overline{\mathbb{R}}$ is Σ-measurable if it has separable range and $f^{-1}(B) \in \Sigma$ for every Borel set $B \subset F$ or $B \subset \overline{\mathbb{R}}$. Σ-measurability is preserved by pointwise convergence:

Theorem 12.3 If (f_n) is a sequence of F or $\overline{\mathbb{R}}$-valued, Σ-measurable functions converging pointwise to a function f, then f is also Σ-measurable.

12.2.3 The integral of step functions

Let $\mathcal{R}$ be a ring of subsets of S and $m : \mathcal{R} \to \mathcal{E} \subset \mathcal{L}(\mathcal{F}, \mathcal{G})$ be an additive measure. For any F-valued, $\mathcal{R}$-step function

$$f = \sum_{i=1}^{n} \chi_{A_i} x_i \quad \text{with } A_i \in \mathcal{R} \text{ and } x_i \in \mathcal{F},$$

we define the integral $\int f\, dm$ by the equality

$$\int f\, dm = \sum_{i=1}^{n} m(A_i) x_i \in G.$$

The definition of the integral is independent of the particular representation of f as a step function. In fact, one can prove that if $\sum_{i=1}^{n} \chi_{A_i} x_i \equiv 0$, then $\sum_{i=1}^{n} m(A_i) x_i = 0$. Consequently, if

$$f = \sum_{i=1}^{m} \chi_{A_i} x_i = \sum_{j=1}^{m} \chi_{B_j} y_j$$

with $A_i, B_j \in \mathcal{R}$ and $x_i, y_j \in F$, then

$$\sum_{i=1}^{n} m(A_i) x_i = \sum_{j=1}^{m} m(B_j) y_j,$$

and $\int f\, dm$ is defined unambiguously.

We have the following immediate properties of the integral of $\mathcal{R}$-step functions:

1. $\int (f+g)\, dm = \int f\, dm + \int g\, dm$.
2. $\int cf\, dm = c \int f\, dm$, for $c \in \mathbb{R}$.
3. If μ is a positive, finite, additive measure on $\mathcal{R}$ and if $f : S \to F$ is an $\mathcal{R}$-step function, then $\int f\, d\mu \in F$ and

$$\left| \int f\, d\mu \right| \leq \int |f|\, d\mu.$$

If we want to extend the integral to a larger class of functions, we have to impose additional conditions on $\mathcal{R}$ and m, for example, that $\mathcal{R}$ is a δ-ring or a σ-algebra and m is σ-additive and with finite semivariation. In particular, m can have finite variation or can be positive.

There are four main stages in the development of the integral $\int f\, dm$:

(1) The *classical integral*, with $m \geq 0$ and f real-valued.
(2) The *Bochner integral*, with $m \geq 0$ and f vector-valued.
(3) The integral $\int f\, dm$, where m is a vector measure with *finite variation* and f is *vector-valued*.
(4) The integral $\int f\, dm$, where m is a vector measure with *finite semivariation* and f is *vector-valued*.

The most important stage is the first one. The other stages can be performed to the extent that they can be reduced to the first one.

In the following paragraphs we shall succinctly present each of the above stages, except (1), for which we refer the reader to Rudin (1973).

A set $A \subset S$ is μ-negligible if $\mu(B) = 0$ for every set $B \in \Sigma$ with $B \subset A$.

A property $P(s)$ defined for all points of S is said to be true μ-a.e. if the set of points $s \in S$ for which $P(s)$ is false is μ-negligible.

12.2.4 Measurability with respect to a positive measure

The framework for this section is a measure space (S, Σ, μ), consisting of a non-empty set S, a σ-algebra Σ of subsets of S, and a positive, σ-additive measure, with finite or infinite values, $\mu : \Sigma \to [0, +\infty]$, such that $\mu(\phi) = 0$.

A function $f : S \to F$ or $\overline{\mathbb{R}}$ is said to be μ-*negligible* if $f(s) = 0$, μ-a.e. Then let $\Sigma(\mu)$ be the σ-algebra of subsets $A \subset S$ for which there is a set $A' \in \Sigma$ such that $A \triangle A'$ is μ-negligible.

Measurability with respect to the σ-algebra $\Sigma(\mu)$ is called μ-measurability. The $\Sigma(\mu)$-step functions are also called μ-*measurable step functions*.

A function $f : S \to F$ or $\overline{\mathbb{R}}$ is called μ-*measurable* if it is $\Sigma(\mu)$-measurable, i.e. if there is a sequence (f_n) of F or $\mathbb{R}$-valued μ-measurable step functions such that $f_n \to f$ pointwise.

A function $f : S \to F$ or $\overline{\mathbb{R}}$ is μ-measurable if there is a Σ-measurable function $g : S \to F$ or $\overline{\mathbb{R}}$ such that $f = g$, μ-a.e. If $f_1 = f_2$, μ-a.e. and if f_1 is μ-measurable, then f_2 is μ-measurable. It follows that f is μ-measurable if there is a sequence (f_n) of Σ-step functions such that $f_n \to f$, μ-a.e.

12.3 THE BOCHNER INTEGRAL

The framework for this section is a measure space (S, Σ, μ) and a Banach space F. The definition of the Bochner integral is very similar to the classical integral, but with some differences.

12.3.1 Bochner integrability

For every F-valued μ-measurable function f defined on S we set $||f||_1 = \int |f| d\mu$.

Definition 12.2 *We say that an F-valued function f defined μ-a.e. on S is Bochner μ-integrable if f is μ-measurable and $||f||_1 < \infty$, i.e. $\int |f| d\mu < \infty$.*

If $||f_n - f||_1 \to 0$, we say that $f_n \to f$ in the mean.

Definition 12.3 *We denote by $L_F^1(\mu)$ or $L_F^1(\Sigma, \mu)$ or $L_F^1(S, \Sigma, \mu)$ the set of all Bochner μ-integrable functions $f : S \to F$, defined everywhere on S.*

The Bochner-integrable functions have the following properties:

1. A step function $f = \sum_{i=1}^n \chi_{A_i} x_i$ with $A_i \in \Sigma(\mu)$ and $x_i \in F$ is μ-integrable if $A_i \in \Sigma(\mu)$ and $\mu(A_i) < \infty$ for each i.
2. If $f \in L_F^1(\mu)$, then $|f| \in L^1(\mu)$.
3. $L_F^1(\mu)$ is a vector space.
4. The mapping $f \mapsto ||f||_1$ is a seminorm on $L_F^1(\mu)$. The topology defined by this seminorm is called the topology of convergence in the mean.
5. If $f \in L_F^1(\mu)$ and $g : S \to F$ is μ-measurable and satisfies $|g| \leq |f|$, μ-a.e., then $g \in L_F^1(\mu)$.

We can prove Lebesgue's theorem before defining the Bochner integral; at this stage one can conclude only convergence in the mean.

Theorem 12.4 (Lebesgue's Dominated Convergence Theorem) *Let (f_n) be a sequence from $L_F^1(\mu)$, $f : S \to F$ a function, and $g \in L^1(\mu)$. Assume that*

(a) $f_n \to f$, μ-a.e.
(b) $|f_n| \leq g$, μ-a.e., for each n.

Then $f \in L_F^1(\mu)$ and $f_n \to f$ in the mean.

For the proof, we apply Fatou's lemma to the sequences $(|f_n - f| + 2g)$ and $(2g - |f_n - f|)$. After we have defined the Bochner integral, we can add the further conclusion that $\int f_n \, d\mu \to \int f \, d\mu$.

Further properties:

6. The set $\mathcal{S}_F(\Sigma)$ of F-valued, Σ-step functions is dense in $L_F^1(\mu)$. In fact, for each $f \in L_F^1(\mu)$ there is a sequence (f_n) of Σ-step functions such that $f_n \to f$ pointwise, μ-a.e. and $|f_n| \le |f|$, μ-a.e. for every n; then we apply Lebesgue's theorem.

7. $F_F^1(\mu)$ is complete.

12.3.2 The Bochner integral

For a Σ-step function $f : S \to F$ of the form $f = \sum_{i=1}^n \chi_{A_i} x_i$ with $A_i \in \Sigma$, $\mu(A_i) < \infty$, and $x_i \in F$, we define the integral $\int f \, d\mu$ by the equality

$$\int f \, d\mu = \sum_{i=1}^n \mu(A_i) x_i.$$

Taking the sets A_i to be mutually disjoint we have

$$|f| = \sum_{i=1}^n \chi_{A_i} |x_i|,$$

hence

$$\int |f| \, d\mu = \sum_{i=1}^n \mu(A_i) |x_i|.$$

It follows that

$$\left| \int f \, d\mu \right| \le \int |f| \, d\mu = \|f\|_1.$$

This inequality shows that the linear mapping $L : \mathcal{S}_F(\pm) \to F$ defined by

$$L(f) = \int f \, d\mu \quad \text{for } \in \mathcal{S}_F(\pm)$$

is continuous for the seminorm $\|f\|_1$. Since the space $\mathcal{S}_F(\pm)$ is dense in $L_F^1(\mu)$ for the topology defined by the seminorm $\|f\|_1$, one can extend L to a continuous linear mapping $L^* : L_F^1(\mu) \to F$, and we still have

$$|L^*(f)| \le \|f\|_1 \quad \text{for } f \in L_F^1(\mu).$$

For each function $f \in L_F^1(\mu)$, the value $L^*(f) \in F$ is denoted by $\int f \, d\mu$ and is called the *Bochner integral* of f with respect to μ.

The Bochner integral has the following properties:

1. $\int (f+g)\, d\mu = \int f\, d\mu + \int g\, d\mu$.
2. $\int cf\, d\mu = c\int f\, d\mu$, for $c \in \mathbb{R}$.
3. $|\int f\, d\mu| \le \int |f|\, d\mu = \|f\|_1$.
4. If $f_n \to f$ in $L_F^1(\mu)$, then $\int f_n\, d\mu \to \int f\, d\mu$.
   This property allows us to add $\int f_n\, d\mu \to \int f\, d\mu$ to the conclusion of Lebesgue's therorem.
5. If $f \in L_F^1(\mu)$, the set function $m : \Sigma(\mu) \to F$ defined by

$$m(A) = \int_A f\, d\mu \text{ for } A \in \Sigma(\mu)$$

is a σ-additive measure.

Definition 12.4 Let $1 \le p < \infty$. We denote by $L_F^p(\mu)$ the set of μ-measurable functions $f : S \to F$ such that $|f|^p \in L^1(\mu)$, i.e. such that $\int |f|^p\, d\mu < \infty$. For $f \in L_F^p(\mu)$, we define

$$\|f\|_p = \left(\int |f|^p\, d\mu \right)^{\frac{1}{p}}.$$

$L_F^p(\mu)$ is a vector space and the mapping $f \mapsto \|f\|_p$ is a seminorm on it. The topology defined by this seminorm is called the topology of convergence in the mean of order p.

$L_F^p(\mu)$ is complete and $\mathcal{S}_F(\pm)$ is dense in $L_F^p(\mu)$.

Definition 12.5 We denote by $L_F^\infty(\mu)$ the set of μ-measurable functions $f : S \to F$ such that

$$\|f\|_\infty = \inf\{a : 0 \le a \le +\infty, |f(s)| \le a, \mu\text{-a.e.}\} < \infty.$$

$L_F^\infty(\mu)$ is a vector space and $\|f\|_\infty$ is a seminorm on it, under which it is complete. We have $f_n \to f$ in $L_F^\infty(\mu)$ iff there is a μ-negligible set $N \subset S$ such that $f_n \to f$ uniformly on $S \setminus N$.

If $F = \mathbb{R}$, the set $\mathcal{S}(\pm)$ of real-valued, Σ-step functions is dense in $L^\infty(\mu) := L_\mathbb{R}^\infty(\mu)$.

Theorem 12.5 (Lebesgue) *Let $1 \le p < \infty$, let (f_n) be a sequence from $L_F^p(\mu)$, $f : S \to F$ a function, and $g \in L^p(\mu)$, such that*

(a) $f_n \to f, \mu$-a.e.;
(b) $|f| \le g, \mu$-a.e. for each n.

Then $f \in L_F^p(\mu)$ and $f_n \to f$ in $L_F^p(\mu)$. If $p = 1$, then $\int f_n\, d\mu \to \int f\, d\mu$.

Finally, we state Hölder's inequality:

Theorem 12.6 *Let $p \leq q \leq +\infty$ be conjugate numbers, i.e. $1/p + 1/q = 1$. Let E, F, G be Banach spaces such that $E \subset L(F, G)$. Then:*

(a) *If $f \in L_E^p(\mu)$ and $g \in L_G^1(\mu)$, then $fg \in L_G^1(\mu)$ and we have*

$$\left| \int fg \, d\mu \right| \leq \|f\|_p \|g\|_q.$$

Here, the product function $fg : S \to F$ is defined by

$$(fg)(s) = f(s)g(s) \quad \text{for } s \in S.$$

(b) *Let $\mathcal{R}$ be a ring generating the δ-ring Σ. Then for every $f \in L_E^p(\mu)$ we have*

$$\|f\|_p = \sup \int |fg| \, d\mu = \sup \int |f| |g| \, d\mu,$$

the supremum being taken for $g \in \mathcal{S}_{\mathcal{F}}(\mathcal{R})$ with $\|f\|_q \leq 1$.

(c) *If $G = \mathcal{R}$ and $f \in L_E^p(\mu)$, then*

$$\|f\|_p = \sup \left| \int fg \, d\mu \right| \quad \text{for } g \in \mathcal{S}_{\mathcal{F}}(\mathcal{R}) \text{ with } \|f\|_q \leq \infty.$$

12.4 INTEGRATION WITH RESPECT TO MEASURES WITH FINITE VARIATION

In this section we present the third stage in the development of the integral. We study measures with finite variation, and integration with respect to such measures. Integrability with respect to a measure with finite variation reduces to Bochner integrability with respect to the variation measure. For a more detailed account, the reader is referred to Dinculeanu (1967, 2000).

12.4.1 Measures with finite variation

The framework for this section is a ring $\mathcal{R}$ of subsets of S, a Banach space E, and an additive measure $m : \mathcal{R} \to E$.

Definition 12.6 *For every set $A \subset S$, the* variation *of m on A is a number, finite or $+\infty$, denoted by $\mathrm{var}(m, A)$ or $\overline{m}(A)$, and defined by*

$$\mathrm{var}(m, A) = \sup \sum_{i \in I} |m(A_i)|,$$

the supremum being taken for all finite families $(A_i)_{i \in I}$ of disjoint sets from $\mathcal{R}$ contained in A.

If $A \in \sigma a(\mathcal{R})$, we write $|m|(A) = \mathrm{var}(m, A)$. We say that m has *finite* (respectively, *bounded*) *variation* if $\mathrm{var}(m, A) < \infty$ for every $A \in \mathcal{R}$ (respectively, $\mathrm{var}(m, S) < \infty$). The variation of m has the following properties:

1. $|m(A)| \leq |m|(A)$, for $A \in \mathcal{R}$.
2. If $A \subset B$, then $\mathrm{var}(m, A) \leq \mathrm{var}(m, B)$.
3. $\mathrm{var}(m, A) = 0$ if $m(B) = 0$ for every $B \in \mathcal{R}$ with $B \subset A$.
4. $\overline{m}$ is superadditive: for an arbitrary family $(A_i)_{i \in I}$ of disjoint subsets of S, we have

$$\overline{m}\left(\bigcup_{i \in I} A_i\right) \geq \sum_{i \in I} \overline{m}(A_i).$$

5. $\mathrm{var}(m + n, A) \leq \mathrm{var}(m, A) + \mathrm{var}(n, A)$
6. $\mathrm{var}(am, A) = |a|\mathrm{var}(m, A)$, for $a \in \mathbb{R}$.
7. The variation $|m| : \sigma a(\mathcal{R}) \to [0, +\infty]$ is additive. If m is σ-additive on $\mathcal{R}$, then $|m|$ is σ-additive on $\mathcal{R}$.

Let $\mu : \mathcal{R} \to \mathbb{R}$ be a real-valued, additive measure. Then for every set $A \subset S$ we have

$$\sup\{|\mu(B)| : B \in \mathcal{R}, B \subset A\} \leq \overline{\mu}(A) \leq 2\sup\{|\mu(B)| : B \in \mathcal{R}, B \subset A\}.$$

If μ is complex-valued, we replace 2 by 4 in the above inequality.

A real valued, σ-additive measure $\mu : \mathcal{S} \to \mathbb{R}$ on a δ-ring $\mathcal{D}$ has finite variation; if $\mathcal{D}$ is a σ-ring, μ has bounded variation.

12.4.2 Integration with respect to a measure with finite variation

The framework for this section and those that follow consists of a δ-ring $\mathcal{D}$ of subsets of S, three Banach spaces $E \subset L(F, G)$, and a σ-additive measure $m : \mathcal{D} \to E$ with finite variation $|m|$.

Then $|m|$ is σ-additive on the σ-algebra $\Sigma = \sigma a(\mathcal{D})$ generated by $\mathcal{D}$, possibly with infinite values. Integrability with respect to m of vector-valued functions $f : S \to F$ is reduced to Bochner integrability of f with respect to the positive measure $|m|$.

Definition 12.7 We say that a set $A \subset S$ is m-negligible (respectively, m-measurable) if A is $|m|$-negligible (respectively, $|m|$-measurable). We define $\Sigma(m) = \Sigma(|m|)$. We say a property P is true m-a.e. if it is true $|m|$-a.e.

We say that an F or $\mathbb{R}$-valued function defined m-a.e. on S is m-negligible (respectively, m-measurable, m-integrable) if it has the same property with respect to $|m|$.

For $1 \le p \le +\infty$ we denote

$$L_F^p(m) = L_F^p(|m|)$$

and endow $L_F^p(m)$ with the seminorm of $L_F^p(|m|)$:

$$\|f\|_p = \||f|\|_p = \left(\int |f|^p \, d|m|\right)^{\frac{1}{p}},$$

for $f \in L_F^p(m)$, if $1 \le p < \infty$ and

$$\|f\|_\infty = \||f|\|_\infty$$

The vector space $L_F^p(m)$ is complete for the seminorm $\|f\|_p$.

If $1 \le p < \infty$, the set $\mathcal{S}_\mathcal{F}(\mathcal{D})$ of $\mathcal{D}$-step functions $f: S \to F$ is dense in $L_F^p(m)$. However, if F is infinite dimensional, $\mathcal{S}_\mathcal{F}(\pm)$ is no longer dense in $L_F^\infty(m)$. The Lebesgue convergence theorems are valid in $L_F^p(m)$.

The following assertions are equivalent for an m-measurable function $f: S \to F$:

(a) f is m-integrable;
(b) f is $|m|$-integrable;
(c) $|f|$ is $|m|$-integrable.

We can now define the integral $\int f \, dm$ for functions $f \in L_F^1(m)$. For a $\mathcal{D}$-step function $f = \sum_{i=1}^n \chi_{A_i} x_i$ with $A_i \in \mathcal{D}$ and $x_i \in F$, we define the integral

$$\int f \, dm = \sum_{i=1}^n m(A_i) x_i \in G.$$

If the sets A_i are mutually disjoint, then

$$|f| = \sum_{i=1}^n \chi_{A_i} |x_i|$$

and

$$\left|\int f \, dm\right| = \left|\sum m(A_i) x_i\right| \le \sum |m(A_i)| |x_i|$$
$$\le \sum |m|(A_i) |x_i| = \int |f| \, d|m| = \|f\|_1.$$

It follows that the linear mapping $L: \mathcal{S}_G(\mathcal{D}) \to \mathcal{G}$ defined by

$$L(f) = \int f \, dm \quad \text{for } f \in \mathcal{S}_F(\mathcal{D})$$

is continuous for the seminorm $\|f\|_1$. Since $\mathcal{S}_F(\mathcal{D})$ is dense in $L_F^1(m)$, the mapping L can be extended to a continuous linear mapping $L^*: L_F^1(m) \to G$.

The value $L^*(f)$ of the extension for a function $f \in L_F^1(m)$ is denoted by $\int f \, dm$ and is called the *integral of f with respect to m*. We still have

$$\left| \int f \, dm \right| \leq \int |f| \, d|m| = \|f\|_1, \text{ for } f \in L_F^1(m).$$

If $f_n \to f$ in L_F^1, then $\int f_n \, dm \to \int f \, dm$.
If $f \in L_F^1(m)$ and $A \in \Sigma(m)$, we have $f \chi_A \in L_F^1(m)$. We denote, as usual,

$$\int_A f \, dm = \int f \chi_A \, dm.$$

If $f \in L_F^1(m)$, the mapping $A \mapsto \int_A f \, dm$ from $\Sigma(m)$ into G is σ-additive and

$$\lim_{|m|(A) \to 0} \int_A |f| \, d|m| = 0.$$

12.5 SEMIVARIATION OF VECTOR MEASURES

In this section we study measures with finite semivariation; the results will be used in the next section to define the integral with respect to such measures. For the proof of the results in this section, the reader is referred to Dinculeanu (2000, Sections 12.4 and 12.7).

12.5.1 Semivariation

The framework for this section is a ring $\mathcal{R}$ of subsets of S, three Banach spaces E, F, G such that $E \subset (F, G)$ continuously, and an additive measure $m: \mathcal{R} \to E \subset L(F, G)$.

Definition 12.8 *For every set $A \subset S$, the semivariation of m on A, relative to the embedding $E \subset L(F, G)$ (or relative to the pair (F, G)), is a number, finite or $+\infty$,* denoted $\text{var}_{F,G}(m, A)$ or $\tilde{m}_{F,G}(A)$, and defined by the equality

$$\tilde{m}_{F,G}(A) = \sup \left| \sum_{i \in I} m(A_i) x_i \right|,$$

the supremum being taken for all finite families $(A_i)_{i \in I}$ of disjoint sets from $\mathcal{R}$ contained in A and all families $(x_i)_{i \in I}$ of elements of F_1.

We say m has finite (respectively, bounded) semivariation $\tilde{m}_{F,G}$ on $\mathcal{R}$ of $\tilde{m}_{F,G}(A) < \infty$ for every $A \in \mathcal{R}$ (respectively, $\tilde{m}_{F,G}(S) < \infty$).

Equivalently, the semivariation $\tilde{m}_{F,G}$ can be defined by the following equality:

$$\tilde{m}_{F,G}(A) = \sup \left| \int s \, dm \right|,$$

where the supremum is taken for all $\mathcal{R}$-step functions $s : S \to F$ with $|s| \leq \varphi_A$. The semivariation has the following properties:

1. If $A \subset B$ then $\tilde{m}_{F,G}(A) \leq \tilde{m}_{F,G}(A)$.
2. $\tilde{m}_{F,G}(A) = 0$ iff $m(B) = 0$ for every set $B \in \mathcal{R}$ with $B \subset A$.
3. $\tilde{m}_{F,G}(A) \leq \text{var}(m, A)$.
4. If the embedding $E \subset L(F, G)$ is an isometry, then

$$|m(A)| \leq \tilde{m}_{F,G}(A) \quad \text{for } A \in \mathcal{R}.$$

5. If $m, n : \mathcal{R} \to E$ are additive measures and $a \in \mathbb{R}$, then

$$\widetilde{(m+n)}_{F,G}(A) \leq \tilde{m}_{F,G}(A) + \tilde{n}_{F,G}(A),$$

$$\widetilde{(am)}_{F,G}(A) \leq |a| \tilde{m}_{F,G}(A).$$

6. The set function $\tilde{m}_{F,G}$ is finitely subadditive on $\sigma a(\mathcal{R})$. If m is σ-additive on $\mathcal{R}$, then $\tilde{m}_{F,G}$ is σ-subadditive on $\sigma a(\mathcal{R})$.
7. Assume $E \subset L(F, R)$ isometrically. Then

$$\tilde{m}_{F,\mathbb{R}}(A) = \text{var}(m, A) \quad \text{for any } A \subset S.$$

8. We have

$$\tilde{m}_{F,G} \leq \tilde{m}_{E^*,\mathbb{R}} = \overline{m}.$$

If the embedding $E \subset L(F, G)$ is an isometry, then

$$\tilde{m}_{\mathbb{R},E} \leq \tilde{m}_{F,G}.$$

12.5.2 Semivariation and norming spaces

We maintain as a framework an additive measure $m : \mathcal{R} \to E \subset L(F, G)$ defined on a ring $\mathcal{R}$, and a norming space $Z \subset G^*$ for G. For each $z \in Z$ we define the set function $m_z : \mathcal{R} \to F^*$ by the equality

$$\langle x, m_z(A) \rangle = \langle m(A)x, z \rangle \quad \text{for } x \in F \text{ and } A \in \mathcal{R}.$$

Then m_z is additive. If we consider the embedding $F^* = L(F, \mathbb{R})$, then by property 7 above we have $(\tilde{m}_z)_{F,\mathbb{R}} = \overline{m}_z$.

The semivariation $\tilde{m}_{F,G}$ can be computed by means of the variations $\overline{m}_z$:

Proposition 12.1 *For any norming space $Z \subset G^*$ for G and for every set $A \subset S$, we have*

$$\tilde{m}_{F,G}(A) = \sup_{z \in Z_1} \overline{m}_z(A).$$

If $F = \mathbb{R}$, the semivariation $\tilde{m}_{\mathbb{R},E}$ has special properties. If $m : \mathcal{R} \to E = L(\mathbb{R}, E)$ is an additive measure, then:

(a) For every set $A \subset S$ we have

$$\tilde{m}_{\mathbb{R},E}(A) \leq 2 \sup\{|m(B)| : B \in \mathcal{R}, B \subset A\}.$$

(b) m is locally bounded (respectively, bounded) iff $\tilde{m}_{\mathbb{R},E}(A) < \infty$ for every $A \in \mathcal{R}$ (respectively, $\tilde{m}_{\mathbb{R},E}(S) < \infty$).

To say that m is *locally bounded* on $\mathcal{R}$ means that for every set $A \in \mathcal{R}$, m is bounded on the ring $\mathcal{R} \cap A$.

For the proof, we use a corresponding property for the variations $\overline{m}_z$ and then we apply Proposition 12.1.

12.5.3 Semivariation of σ-additive measures

The semivariation of σ-additive measures has additional properties.

Let $m : \mathcal{D} \to E \subset L(F, G)$ be a σ-additive measure on a δ-ring $\mathcal{D}$. Then for any $z \in G^*$, the measure m_z is σ-additive on $\mathcal{D}$ and the variation $|m_z|$ is σ-additive on the σ-algebra $\Sigma = \sigma a(\mathcal{D})$ generated by $\mathcal{D}$.

Let $m : \mathcal{R} \to E \subset L(F, G)$ be an additive measure on a ring $\mathcal{R}$. Assume m can be extended to an additive measure $m' : \mathcal{D} \to E \subset L(F, G)$ on the δ-ring $\mathcal{D}$ generated by $\mathcal{R}$. If $|m_z|$ is σ-additive for every z in a norming space $Z \subset G^*$ for G, then

$$\text{svar}_{F,G}(m', A) = \text{svar}_{F,G}(m, A) \quad \text{for } A \in \mathcal{R}.$$

The σ-additive measures on a δ-ring always have finite semivariation $\tilde{m}_{\mathbb{R},E}$.

Proposition 12.2 *Let $m : \mathcal{P} \to E$ or $\overline{\mathbb{R}}_+$ be a finitely additive measure defined on a semiring $\mathcal{P}$. Then m can be extended uniquely to an additive measure $m' : \mathcal{R} \to E$ or $\overline{\mathbb{R}}_+$ on the ring $\mathcal{R} = r(\mathcal{P})$ generated by $\mathcal{P}$.*
If m is σ-additive, then m' is also σ-additive.

If $A \in \mathcal{R}$, then $A = \bigcup_{i=1}^{n} B_i$ with $B_i \in \mathcal{P}$ mutually disjoint. The measure m' is defined by

$$m'(A) = \sum_{i=1}^{n} m(B_i).$$

12.6 INTEGRATION WITH RESPECT TO A MEASURE WITH FINITE SEMIVARIATION

The framework for this section is an *additive* measure $m : \mathcal{D} \to E \subset L(F, G)$ with finite semivariation $\tilde{m}_{F,G}$ on a δ-ring and a norming space $Z \subset G^*$ for G, such that for each $z \in Z$, the measure $m_z : \mathcal{D} \to F^*$ is σ-additive. $\Sigma = \sigma a(\mathcal{D})$ is the σ-algebra generated by $\mathcal{D}$. We do not assume that m is σ-additive.

In this section we present the fourth stage in the development of the integral $\int f dm$, for functions $f : S \to F$. For this purpose we define a seminorm $\tilde{m}_{F,G}(f)$ for such functions, then the space $\mathcal{F}_{F,G}(m)$ of measurable functions f with $\tilde{m}_{F,G}(f) < \infty$, and then the integral $\int f dm \in Z^*$ for functions $f \in \mathcal{F}_{F,G}(m)$.

This is the most important part of the chapter. In fact, most interesting vector measures do not have finite variation, but may have finite semivariation. This is the case, for example, for the stochastic measure I_X associated with a summable process X, even if X is real-valued. We shall apply the integration theory of this paragraph to obtain the stochastic integral.

Some of the results are valid under additional conditions such as:

(a) $\tilde{m}_{F,G}(S) < \infty$, or
(b) $S = \bigcup S_n$, with $S_n \in \mathcal{D}$, or
(c) m is σ-additive.

Conditions (a) and (b) are satisfied if $\mathcal{D}$ is a σ-algebra. But most results are valid without imposing these restrictions and can be used in a wide range of applications, such as the integral representation of Gaussian measures or the Riesz representation theorem.

If the reader is not interested in this generality, he or she can assume from the very beginning that $\mathcal{D}$ is a σ-algebra and m is σ-additive. For the proof of the results stated in this section, the reader is referred to Dinculeanu (2000, Section 12.5).

12.6.1 Measurability with respect to a vector measure

First we define *negligible* sets and functions with respect to m.

A set $A \in \Sigma$ is said to be *m-negligible* if $m(B) = 0$ for every $B \in \Sigma$ with $B \subset A$. It follows that a set $A \in \Sigma$ is *m*-negligible iff $|m|(A) = 0$, iff $\tilde{m}_{F,G}(A) = 0$, iff A is m_z-negligible for every $z \in Z$.

A set $A \subset S$ is said to be *m-negligible* is it is contained in an *m*-negligible set $B \in \Sigma$. A countable union of *m*-negligible sets is again *m*-negligible. If $A \subset S$ is *m*-negligible then A is m_z-negligible for every $z \in Z$ and if Z is separable or if the measures $|m_z|$ with $z \in Z_1$ are uniformly σ-additive, then A is *m*-negligible.

A property valid outside an *m*-negligible set is said to be valid *m-almost everywhere* (*m*-a.e.).

A function $f : S \to F$ or $\overline{\mathbb{R}}_+$ is said to be *m-negligible* iff $f = 0$, *m*-a.e. A function f is m_z-negligible for every $z \in Z$. Conversely, if f is *m*-negligible for every $z \in Z$, then f is *m*-negligible if either f is Σ-measurable, or Z is separable, or the measures $|m_z|$ with $z \in Z_1$ are uniformly σ-additive.

Now we define measurability with respect to m.

A function $f : S \to F$ or $\overline{\mathbb{R}}_+$ is said to be *m-measurable* if it is equal *m*-a.e. to a Σ-measurable function, i.e. if there is a sequence (f_n) of D or $\mathbb{R}_+$-valued, Σ-step functions, converging to f, *m*-a.e. Moreover, we can choose the functions f_n such that $|f_n| \leq |f|$, *m*-a.e., for each n.

If f is *m*-measurable, then it is m_z-measurable for every $z \in Z$. Conversely, if f is m_z-measurable for every $z \in Z$, then f is *m*-measurable, provided that Z is separable or the measures $|m_z|$ with $z \in Z_1$ are uniformly σ-additive (Dinculeanu 2000), Proposition 5.5).

12.6.2 The seminorm $\tilde{m}_{F,G}(f)$

The alternative definition of semivariation given after Definition 12.8 is extended now for functions.

Definition 12.9 *For every function $f : S \to F$ or $\overline{\mathbb{R}}$ we define*

$$\tilde{m}_{F,G}(f) = \sup \left| \int s \, dm \right|,$$

where the supremum is taken for all Σ-step functions $s : S \to F$ or $\mathbb{R}$ with $|s| \leq |f|$.

We can compute $\tilde{m}_{F,G}(f)$ in terms of the measures $|m_z|$; compare with Proposition 12.1.

If $f : S \to F$ or $\overline{\mathbb{R}}$ is *m*-measurable, then

$$\tilde{m}_{F,G}(f) = \sup \left\{ \int |f| \, d|m_z| : z \in Z_1 \right\}.$$

If the spaces F and G are understood, we shall write $\tilde{m}$ instead of $\tilde{m}_{F,G}$.

We have the following properties of $\tilde{m}_{F,G}(f)$ for m-measurable functions f:

1. $0 \le \tilde{m}(f) \le \infty$.
2. $\tilde{m}(f) = \tilde{m}(|f|)$.
3. $\tilde{m}(f) = 0$ iff $f = 0$, m-a.e.
4. If $|f| = |g|$, m-a.e., then $\tilde{m}(f) = \tilde{m}(g)$.
5. If $|f| \le |g|$, m-a.e., then $\tilde{m}(f) \le \tilde{m}(g)$.
6. $\tilde{m}(f+g) \le \tilde{m}(f) + \tilde{m}(g)$.
7. $\tilde{m}(af) = |a|\tilde{m}(f)$, for $a \in \mathbb{R}$.

Theorem 12.7 (Monotone Convergence Theorem) *If (f_n) is an increasing sequence of positive, m-measurable functions $f_n : S \to \overline{\mathbb{R}}_+$, then*

$$\tilde{m}(\sup f_n) = \sup \tilde{m}(f_n).$$

For every sequence (f_n) of positive, m-measurable functions $f_n : S \to \overline{\mathbb{R}}_+$ we have

$$\tilde{m}\left(\sum_{n=1}^{\infty} f_n\right) \le \sum_{n=1}^{\infty} \tilde{m}(f_n).$$

Lemma 12.1 (Fatou's Lemma) *If (f_n) is a sequence of positive, m-measurable functions, then*

$$\tilde{m}\left(\liminf_{n \to \infty} f_n\right) \le \liminf_{n \to \infty} \tilde{m}(f_n).$$

If m is m-measurable and $c0$, then

$$\tilde{m}(\{|f|c\}) \le \frac{1}{c}\tilde{m}(f).$$

If $f : S \to F$ or $\overline{\mathbb{R}}$ is m-measurable and $\tilde{m}(f) < \infty$, then $f < \infty$, m-a.e.

If $f : S \to F$ or $\overline{\mathbb{R}}$ is m-measurable, then the set $\{f \ne 0\}$ is contained in the union of a sequence (A_n) from Σ with $\tilde{m}(A_n) < \infty$.

12.6.3 The space $\mathcal{F}_D(\tilde{m}_{F,G})$

Now we define a space similar to the space L^1 of integrable functions.

Definition 12.10 *We denote by $\mathcal{F}_F(\tilde{m}_{F,G})$, the set of all m-measurable functions $f : S \to F$ with $\tilde{m}_{F,G}(f) < \infty$.*

If the spaces F and G are understood, we shall write $\mathcal{F}_F(\tilde{m})$ instead of $\mathcal{F}_F(\tilde{m}_{F,G})$. The functions $f \in \mathcal{F}_F(\tilde{m})$ are called m-integrable functions.

For each $z \in Z_1$ we have $\int |f| d|m_z| \leq \tilde{m}_{F,G}(f)$, hence $\mathcal{F}_F(\tilde{m}) \subset L_F^1(m_z)$, therefore

$$\mathcal{F}_F(\tilde{m}) \subset \bigcap_{z \in Z} L_F^1(m_z).$$

If Z is a closed subspace of G^*, then we have the equality

$$\mathcal{F}_F(\tilde{m}) = \bigcap_{z \in Z} L_F^1(m_z).$$

$\mathcal{F}_F(\tilde{m})$ is a vector space and $\tilde{m}_{F,G}(f)$ is a seminorm on it. We shall consider the topology on $\mathcal{F}_F(\tilde{m})$ defined by this seminorm.

Corollary 12.1 *The space $\mathcal{F}_F(\tilde{m})$ is complete for the seminorm $\tilde{m}$.*

Remark 12.1 *The set of Σ-step functions of $\mathcal{F}_F(\tilde{m})$ is not necessarily dense on $\mathcal{F}_F(\tilde{m})$.*
If the measures (m_z) with $z \in Z_1$ are uniformly σ-additive, then the set of Σ-step functions of $\mathcal{F}_F(\tilde{m})$ is dense.

12.6.4 The integral

We can now define the integral $\int f dm$ for functions $f \in \mathcal{F}_F(\tilde{m}_{F,G})$; the integral belongs to Z^*. To simplify the notation we shall write $\mathcal{F}_{F,G}(\tilde{m})$ or $\mathcal{F}_{F,G}(m)$ instead of $\mathcal{F}_F(\tilde{m}_{F,G})$.

The construction of the integral $\int f dm$ is carried out in the following way: Let $f \in \mathcal{F}_{F,G}(m)$. Then $f \in L_F^1(|m_z|)$ for every $z \in Z$. Since $m_z : \Sigma \to F^*$ has finite variation $|m_z|$, the integral $\int f dm_z$ is defined in the sense of the third stage and $\int f dm_z \in \mathbb{R}$. The mapping $z \to \int f dm_z$ is a continuous linear functional on Z:

$$\left| \int f dm_z \right| \leq \int |f| d|m_z| \leq |z| \tilde{m}(f).$$

We denote the linear mapping $f \mapsto \int f dm_z$ by $\int f dm$ and call it the integral of f with respect to m. We have $\int f dm \in Z^*$,

$$\left\langle \int f dm, z \right\rangle = \int f dm_z,$$

and

$$\left| \int f dm \right| \leq \tilde{m}_{F,G}(f).$$

From this last inequality it follows that the mapping $f \mapsto \int f dm$ from $\mathcal{F}_{F,G}(\tilde{m})$ into Z^* is continuous for the topology of $\mathcal{F}_{F,G}(\tilde{m})$.

If we take $Z = G^*$, we have $\int f dm \in G^{**}$ for $f \in \mathcal{F}_{F,G}(\tilde{m})$.

We are particularly interested in the case when $\int f dm \in F$. This is evidently the case if G is reflexive and we take $Z = G^*$. If the measures $|m_z|$ with $z \in Z_1$ are uniformly σ-additive, then $\int f dm \in G$ for every bounded function $f \in \mathcal{F}_{F,G}(\tilde{m})$.

Assume m has finite variation $|m|$. Then m has also finite semivariation $\tilde{m}_{F,G}$. We can consider the space $L_F^1(m) = L_F^1(|m|)$ of functions $f : S \to F$ which are m-integrable, in the sense of Section 12.4. For each $z \in F_1^*$ we have $|m_z| \le |m|$, hence $\tilde{m}_{F,G}(f) \le \int |f| d|m| = \|f\|_1$, the norm in $L_F^1(m)$. It follows that $L_F^1(m) \subset \mathcal{F}_{F,G}(\tilde{m})$ and the embedding is continuous.

If $f \in L_F^1(m)$, then $\int f dm$ is the same, whether we consider f in $L_F^1(m)$ or in $\mathcal{F}_{F,G}(\tilde{m})$.

For Lebesgue's theorem, we have to restrict ourselves to the space $\mathcal{F}_F(\mathcal{B}, \tilde{m})$.

Theorem 12.8 (Lebesgue) *Assume $S \in \mathcal{D}$. Let (f_n) be a sequence from $\mathcal{F}_F(\tilde{m})$, $f \in \mathcal{F}_F(\tilde{m})$, and $g \in \mathcal{F}_{\mathbb{R}}(\tilde{m})$ a positive function. Assume that:*

(a) $f_n \to f$ in $\tilde{m}$-measure; or
(a') $f_n \to f$, m-a.e. and the measures $|m_z|$ with $z \in Z_1$ are uniformly σ–additive;
(b) $|f_n| \le g$, m-a.e. and the measures $|m_z|$ with $z \in Z_1$ are uniformly σ–additive.

Then $f \in \mathcal{F}_F(\tilde{m})$, $\tilde{m}(f_n - f) \to 0$, and $\int f_n dm \to \int f dm$.

12.7 THE STOCHASTIC INTEGRAL

The main application of the integration theory with respect to a measure with finite semivariation is the stochastic integral.

12.7.1 Notations and definitions

The reader is assumed to be familiar with the general theory of stochastic processes, as presented, for example, in Dellacherie and Meyer (1975 & 1980). We present below a few definitions and some notation that will be used in what follows.

Here, E, F, G are Banach spaces with $E \subset L(F, G)$, and $(\Omega, \mathcal{F}, P)$ is a probability space. The P-negligible (respectively, P-integrable) sets or functions are called, simply, negligible (respectively, integrable). Instead of P-a.e. we shall write a.s. (almost surely). The space $L_F^p(P)$ with $1 \le p \le \infty$ will be denoted by L_F^p. A set $M \subset \mathbb{R}_+ \times \Omega$ is called *evanescent* if it is contained in a set of the form $\mathbb{R} \times A$ with $A \subset \Omega$ negligible. $(\mathcal{F}_t)_{t \in \mathbb{R}_+}$ is a filtration, i.e. each $\mathcal{F}_t$ is a σ-algebra contained in $\mathcal{F}$ and $\mathcal{F}_s \subset \mathcal{F}_t$ if $s \le t$. We assume the filtration satisfies the usual conditions, i.e. $\mathcal{F}_t = \cap_{s>t} \mathcal{F}_s$ for every $t \ge 0$ and each $\mathcal{F}_t$ contains all the negligible sets.

A *stopping time* (or optional stopping time) is a function $T : \Omega \to \overline{\mathbb{R}}_+$ such that $\{T \leq t\} \in \mathcal{F}_t$ for every $t \geq 0$. If $S \leq T$ are two stopping times, we define the *Stochastic interval*

$$]S, T] = \{(t, \omega) \in \mathbb{R} \times \Omega : S(\omega) < t \leq T(\omega)\}.$$

Other stochastic intervals, $[S, T[$, $]S, T[$, $[S, T]$ are defined similarly. The graph of a stopping time T is $[T] := [T, T]$.

For $1 \leq p < \infty$ and from $E \subset L(F, G)$ we deduce $L_E^p \subset L(F, L_G^p)$.

$\mathcal{R}$ is the ring of subsets $\mathbb{R}_+ \times \Omega$ generated by the semiring of predictable rectangles $\{0\} \times A$ with $A \in \mathcal{F}_0$ and $(s, t] \times A$ with $A \in \mathcal{F}_s$. The σ-algebra generated by $\mathcal{R}$ is called the *predictable σ-algebra* and is denoted by $\mathcal{P}$. The predictable σ-algebra is also generated by the adapted, left-continuous, real-valued processes (adapted processes are defined below).

A $\mathcal{P}$-measurable process is called a *predictable process*.

A stopping time T is said to be *predictable* if the stochastic interval $[T, \infty)$ is predictable.

The *optional σ-algebra* $\mathcal{O}$ is the σ-algebra generated by the adapted, right-continuous processes. It is also generated by the stochastic intervals $]S, T]$ with $S \leq T$ simple stopping times. An $\mathcal{O}$-measurable process is called an *optional process*.

Other σ-algebras of interest are $\mathcal{F}_T$ and $\mathcal{F}_{T-}$, where T is a stopping time. $\mathcal{F}_T$ is the σ-algebra of the sets $A \in \mathcal{F}$ with $A \cap \{T \leq t\} \in \mathcal{F}_t$ for every $t \geq 0$. The σ-algebra $\mathcal{F}_{T-}$ is generated by the sets of the form $A \cap \{t < T\}$ with $t \geq 0$ and $A \in \mathcal{F}_t$.

A function $Y : \mathbb{R}_+ \times \Omega \to D$ with values in a Banach space D is called a process, or stochastic process.

We say Y is *adapted* if for each $t \geq 0$, the random variable Y_t is $\mathcal{F}_t$-measurable. Y is *cadlag* if for each $\omega \in \Omega$, the path $t \to Y_t(\omega)$ is right continuous and has left limits.

In what follows, $X : \mathbb{R}_+ \times \Omega \to E$ will denote a cadlag, adapted process with $X_t \in L_E^p$ for every $t \geq 0$. We consider X automatically extended on $\mathbb{R} \times \Omega$ with $Z_t = 0$ for $t < 0$. Then $X_{0-} = 0$. We also extend the filtration with $\mathcal{F}_t = \mathcal{F}_0$ for $t < 0$.

Let $H : \mathbb{R}_+ \times \Omega \to D$ be a process and T a stopping time. We denote by H_T the function defined by $H_T(\omega) + H_{T(\omega)}(\omega)$ for $\omega \in \Omega$ with $T(\omega) < \infty$, and by $H_T I_{T < \infty}$ the function equal to H_T on $\{T < \infty\}$ and equal to 0 on $\{T = \infty\}$.

The *stopped process* H^T, obtained by stopping the process H at the time T, is defined by $H_t^T = H_{t \wedge T}$. The process H^{T-} obtained by stopping the process H before T is defined by $H_t^{T-} = H_t$ if $t < T$ and $H_t^{T-} = H_{T-}$ if $t \geq T$.

A process Y is said to be a *modification* of a process Z, if for every $t \geq 0$ we have $Y_t = Z_t$ a.s., the negligible set depending on t.

A process Y is said to be *evanescent* if the set $\{Y \neq 0\}$ is evanescent.

Two processes Y and Z, with values in the same Banach space, are said to be *indistinguishable* if $Y - Z$ is evanescent.

12.7.2 The measure I_X and summable processes

Let $1 \leq p < \infty$. Let $X : \mathbb{R}_+ \times \Omega \to E \subset L(F, G)$ be an adapted, cadlag process with $X_t \in L_E^p$ for every $t \geq 0$.

We associate to X the additive measure $I_X : \mathcal{R} \to L_E^p \subset L\left(F, L_G^p\right)$ defined, first, for predictable rectangles by

$$I_X(\{0\} \times A) = 1_A X_0 \quad \text{for } A \in \mathcal{F}_0$$

and

$$I_X(]s, t] \times A) = 1_A(X_t - X_s) \quad \text{for } A \in \mathcal{F}_s,$$

and then extended by additivity to the whole ring $\mathcal{R}$.

We have

$$I_X([0, t] \times A) = 1_A X_t \quad \text{for } t \geq 0 \text{ and } A \in \mathcal{F}_0.$$

In particular,

$$I_X([0, t] \times \Omega) = X_t \quad \text{for } t \geq 0.$$

If the process X is understood we shall write I instead of I_X.

Since $L_E^p \subset L\left(F, L_G^p\right)$, we can consider the semivariation of I_X relative to the pair (F, L_G^p). To simplify the notation, we shall write $\tilde{I}$ or $(\tilde{I}_X)_{F,G}$ instead of $(\tilde{I}_X)_{F,L_G^p}$:

$$\tilde{I}_{F,G}(A) = \sup \left\| \sum_{i \in I} I_X(A_i) x_i \right\|_p \quad \text{for } A \in \mathcal{R},$$

where the supremum is taken for all finite families $(A_i)_{i \in I}$ of disjoint sets A_i from $\mathcal{R}$ contained in A and for all families $(x_i)_{i \in I}$ of elements of F_1.

Definition 12.11 *We say that the process X is p-summable relative to the pair (F, G) if I_X has a σ-additive extension $I_X : \mathcal{P} \to L_E^p$ with finite semivariation relative to (F, L_G^p).*

If $p = 1$, we say that X is *summable* relative to (F, G).

The stochastic integral $H \cdot X$ will be defined with respect to p-summable processes X. Examples of p-summable processes are:

(a) processes with integrable variation;
(b) processes with integrable semivariation, provided that $c_0 \not\subset E$ and $c_0 \not\subset G$;
(c) square integrable martingales X, in the case when E and G are Hilbert spaces.

12.7.3 The stochastic integral

Let $1 \leq p \leq \infty$ and $X : \mathbb{R}_+ \times \Omega \to E \subset L(F, G)$ be a cadlag, adapted process. Assume X is p-summable relative to (F, G). Consider the σ-additive measures

$$I_X : \mathcal{P} \to L_E^p \subset L\left(F, L_G^p\right)$$

with finite semivariation $\tilde{I}_{F,G}$ relative to $\left(F, L_G^p\right)$.

We can apply the integration theory presented in Section 12.6, replacing S, Σ, m with $\mathbb{R} \times \Omega, \mathcal{P}, I_X$ and E, F, G with L_E^p, F, L_G^p, respectively.

Let $Z \subset L_{G*}^q$ and $1/p + 1/q = 1$ be a norming space for L_G^p. For $z \in Z$, we consider the measure $(I_X)_z : \mathcal{P} \to F^*$, defined for $A \in \mathcal{P}$ and $y \in F$ by

$$\langle y, (I_X)_z(A)\rangle = \langle I_X(A) y, z\rangle = \int \langle I_X(A)(\omega)\rangle \, dP(\omega),$$

where the bracket in the integral represents the duality between G and G^*. Then

$$(\tilde{I}_X)_{F, L_G^p} = \sup\{|(I_X)_z| : z \in Z, \|z\|_q \leq 1\}.$$

If p and X are understood, we shall write $I = I_X$ and $\tilde{I}_{F,G} = \tilde{I}_{F,L_G^p}$. For a Banach space D we define $\mathcal{F}_D(\tilde{I}_{F,G}), = \mathcal{F}_D(\tilde{I}_{F,L^pG})$, the space of predictable processes $H : \mathbb{R} \times \Omega \to D$ such that

$$\tilde{I}_{F,G}(H) = \sup\left\{\int |H|d|(I_X)_z| : \|z\|_q \leq 1\right\} < \infty.$$

Then $\mathcal{F}_D(\tilde{I}_{F,G})$ is a vector space and $\tilde{I}_{F,G}$ is a seminorm on $\mathcal{F}_D(\tilde{I}_{F,G})$, under which it is complete. The simple processes are not necessarily dense in $\mathcal{F}_D(\tilde{I}_{F,G})$.

If $D = F$ we shall write $\mathcal{F}_{F,G}(X), \mathcal{F}_{F,L^pG}(X), \mathcal{F}_{F,G}(I_X)$, or $\mathcal{F}_{F,L^pG}(I_X)$ instead of $\mathcal{F}_F((\tilde{I}_X)_{F,G})$. In this case we can define the stochastic integral $H \cdot X$ for processes $H \in \mathcal{F}_{F,G}(X)$ as follows:

Let $H \in \mathcal{F}_{F,G}(X)$; then $H \in L_F^1((I_X)_z)$ for every $z \in Z$, hence the integral $\int H d(I_X)_z$ is defined and is a scalar. The mapping $z \mapsto \int Hd(I_X)_z$ is linear and continuous on Z, and is denoted by $\int H d I_X$. We have, therefore for $\int H d I_X \in Z^*$,

$$\left\langle \int H d I_X, z\right\rangle = \int H d(I_X)_z \quad \text{for } z \in Z$$

and

$$\left\|\int H d I_X\right\| \leq \tilde{I}_{F,G}(H).$$

If we take $Z = \left(L_G^p\right)^*$, then $\int H d I_X \in \left(L_G^p\right)^{**}$.

We are interested in those processes $H \in \mathcal{F}_{F,G}(X)$ for which the integral $\int_{[0,t]} H d I_X$ belongs to L_G^p for every $t \geq 0$. In this case we denote by the same symbol the equivalence class $\int_{[0,t]} H d I_X$ in L_G^p, as well as any random variable belonging to this equivalence class. In this way we obtain a process $(\int_{[0,t]} H d I_X)_{t \geq 0}$

with values in G. This process is always adapted, but it is not necessarily cadlag. This leads to the following definition:

Definition 12.12 *We denote by $L^1_{F,G}(X)$ the set of processes $H \in \mathcal{F}_{F,G}(X)$ satisfying the following two conditions:*

(a) $\int_{[0,t]} H d I_X \in L^p_G$ *for each $t \geq 0$.*
(b) *The process $\left(\int_{[0,t]} H d I_X \right)_{t \geq 0}$ has a cadlag modification.*

The processes $H \in L_{F,G}(X)$ are said to be integrable with respect to X.

If $H \in L^1_{F,G}(X)$, any cadlag modification of the process $\left(\int_{[0,t]} H d I_X \right)_{t \geq 0}$ is called the stochastic integral *of H with respect to X and is denoted by $H \cdot X$ or $\int H d X$:*

$$(H \cdot X)_t(\omega) = \left(\int H d X \right)_t (\omega) = \left(\int_{[0,t]} H d I_X \right)(\omega) \quad a.s.$$

for each $t \geq 0$.

It follows that the stochastic integral $H \cdot X$ is defined up to an evanescent process, and is a cadlag, adapted process.

Theorem 12.9 *Assume X is p-summable relative to (F, G), let $H \in L^1_{F,G}(X)$, and T be a stopping time. Then:*

(a) X^T *is p-summable relative to (F, G).*
(b) $H \in L^1_{F,G}(X^T)$.
(c) $I_{[0,T]} H \in L^1_{F,G}(X)$.
(d) $(H \cdot X)^T = H \cdot X^T = (1_{[0,T]} H) \cdot X$.

If T is predictable, then:

(a') X^{T-} *is p-summable relative to (F, G).*
(b') $H \in L^1_{F,G}(X^{T-})$.
(c') $I_{[0,T]} H \in L^1_{F,G}(X)$.
(d') $(H \cdot X)^{T-} = H \cdot X^{T-} = (1_{[0,T]} H) \cdot X$.

12.7.4 Convergence theorems

Let $1 \leq p < \infty$ and $X : \mathbb{R}_+ \times \Omega \to E \subset L(F, G)$ be a p-summable process relative to (F, G).

Theorem 12.10 *The space $L^1_{F,G}(X)$ is complete for the semivariation $\tilde{I}_{F,G}$.*

Theorem 12.11 (Lebesgue) *Let (H^n) be a sequence from $L^1_{F,G}(X)$, converging pointwise to a process H. Assume there is a positive process $\Phi \in \mathcal{F}_\mathbb{R}(\tilde{I}_{F,G})$ such that*

$$|H^n| \leq \Phi \quad \text{for each } n.$$

Assume, in addition, that Φ can be approximated in $\mathcal{F}_{\mathbb{R}}(\tilde{I}_{F,G})$ by bounded processes and that the measures $|(I_X)_z|$ are uniformly σ-additive for

$$z \in L^q_{G**}, \quad 1/p + 1/q = 1.$$

Then $H \in L^1_{F,G}(X)$, $H^n \to H$ in $L^1_{F,G}(X)$, and $(H^n \cdot X)_t \to (H \cdot X)_t$ in L^p_G for each $t \geq 0$.

12.7.5 Local summability and local integrability

Let $1 \leq p < \infty$ and $X: \mathbb{R}_+ \times \Omega \to E \subset L(F, G)$ ba a cadlag, adapted process with $X \in L^p_E$ for each $t \geq 0$. We shall now define the local summability of X and the stochastic integral $H \cdot X$.

Definition 12.13

(a) We say X is *locally p-summable relative to* (F, G) if there is an increasing sequence (T_n) of stopping times with $T_n \uparrow \infty$, such that for each n, the stopped process X^{T_n} is p-summable relative to (F, G).

(b) Assume X is locally p-summable relative to (F, G). A predictable process $H: \mathbb{R}_+ \times \Omega \to F$ is said to be *locally integrable with respect to X* if there is an increasing sequence (T_n) of stopping times with $T_n \uparrow \infty$ such that, for each n, X^{T_n} is p-summable relative to (F, G) and $I_{[0, T_n]} H$ is integral with respect to X^{T_n}.

Then for each n, the stochastic integral $(1_{[0, T_n]} H) \cdot X^{T_n}$ is defined. The following theorem confirms the existence of the pointwise limit of the above sequence of stochastic integrals.

Theorem 12.12 *Assume X is locally p-summable relative to (F, G) and let $H: \mathbb{R}_+ \times \Omega \to F$ be a predictable process, locally integrable with respect to X. Let (T_n) be an increasing sequence of stopping times with $T_n \uparrow \infty$, determining the local integrability of H with respect to X. Then the limit*

$$\lim (1_{[0, T_n]} H) \cdot X^{T_n}$$

exists pointwise, outside an evanescent set, is cadlag, adapted, and independent of the sequence (T_n).

The limit in the above theorem is called the *stochastic integral of H with respect to X* and is denoted by $H \cdot X$ or $\int H dX$. For each T_n we have

$$(H \cdot X)^{T_n} = (1_{[0, T_n]} H) \cdot X^{T_n}.$$

The stochastic integral with respect to locally p-summable processes has the main properties of the stochastic integral with respect to summable processes.

12.8 Processes with integrable variation or integrable semivariation

In this section we set out the summability of processes X wit integrable variation or integrable semivariation. The stochastic integral can be computed pathwise as a Stieltjes integral.

We shall write $\mathcal{M} = \mathcal{B}(\mathbb{R}_+) \times \mathcal{F}$. We say that a process X with values in a Banach space is *measurable* if it is measurable with respect to $\mathcal{M}$. Every process $X : \mathbb{R}_+ \times \Omega \to E$ is automatically extended with $X_t(\omega) = 0$ for $t < 0$ and $\omega \in \Omega$. A measurable process which is not adapted is called a *raw process*.

12.8.1 Processes with finite variation or semivariation

Let $X : \mathbb{R}_+ \times \Omega \to E \subset L(F, G)$ be a process.

Definition 12.14 *We say that the process X has finite variation (respectively, finite semivariation relative to (F, G)) if for every $\omega \in \Omega$, the path $t \mapsto X_t(\omega)$ has finite variation (respectively, finite semivariation relative to (F, G)) on each interval $[0, t]$, or equivalently, on each interval $]-\infty, t]$.*

We write

$$|X|_t(\omega) = \mathrm{var}(X.(\omega),]-\infty, t])$$

and

$$(\tilde{X}_{F,G})_t(\omega) = \mathrm{svar}_{F,G}(X.(\omega),]-\infty, t])$$

We say that X has p-integrable variation $|X|$ (respectively, p-integrable semivariation $\tilde{X}_{F,G}$) if $|X|_\infty \in L^1$ (respectively, $(\tilde{X}_{F,G})_\infty \in L^1$).

Proposition 12.3

(a) *Assume X is right continuous and has finite variation $|X|$. If X is measurable (respectively, optional, predictable), then so is $|X|$.*
(b) *Assume X is right continuous and has finite semivariation $\tilde{X}_{F,G}$. Assume also that either F or G is separable. If X is measurable (respectively, optional, predictable), then so is $\tilde{X}_{F,G}$.*

12.8.2 Summability of processes with integrable variation or integrable semivariation

The following theorem asserts the summability of processes with integrable variation or integrable semivariation. Moreover, in this case, the stochastic integral can be computed pathwise, as a Stieltjes integral.

Theorem 12.13 *Let $X : \mathbb{R}_+ \times \Omega \to E \subset L(F, G)$ be a cadlag, adapted process. Assume that X has integrable variation $|X|$ (respectively, $c_0 \not\subset G$ and X has integrable semivariations $\tilde{X}_{\mathbb{R},E}$ and $\tilde{X}_{F,G}$). Then:*

(a) *The measure $I_X : \mathcal{R} \to L_E^1$ can be extended to a σ-additive measure $I_X : \mathcal{P} \to (\tilde{I}_X)_{F,G}$ with finite variation $|I_X|$ (respectively, with finite semivariations $(\tilde{I}_X)_{\mathbb{R},E}$ and $(\tilde{I}_X)_{F,G}$).*
(b) *X is summable relative to (F, G).*

12.9 MARTINGALES

A martingale $M : \mathbb{R}_+ \times \Omega \to E \subset L(F, G)$ is not necessarily summable, but if it is, the stochastic integral $H \cdot X$ is again a martingale. If E and G are Hilbert spaces and if M is a square integrable martingale, then M is 2-summable.

Theorem 12.14 *Let $M : \mathbb{R}_+ \times \Omega \to E \subset L(F, G)$ be a martingale. Assume M is p-summable relative to (F, G) and let $H \in \mathcal{F}_{F,G}(M)$.*

If $\int_{[0,t]} H \, dI_M \in L_G^p$ for every $t \geq 0$, then $H \in L_{F,G}^1(M)$ and the stochastic integral $H \cdot M$ is a uniformly integrable martingale, bounded in L_G^p.

If L_G^p is reflexive, then $L_{F,G}^1(M) = \mathcal{F}_{F,G}(M)$.

Theorem 12.15 *Assume E and G are Hilbert spaces and let $M : \mathbb{R}_+ \times \Omega \to E \subset L(F, G)$ be a square integrable martingale. Then M is 2-summable relative to (F, G). We have*

$$L_{F,L^2_G}^1(M) = \mathcal{F}_{F,L^2_G}(M)$$

and for every $H \in L_{F,L^2_G}^1(M)$, the stochastic integral $H \cdot M$ is a square integrable martingale.

Remark 12.2 *Here is the classical approach to the stochastic integral of real-valued semimartingales. A semimartingale is a process of the form $X = M + A$, where M is a local martingale and A is a process with finite variation. We separately define the stochastic integrals $H \cdot M$ and $H \cdot A$ and then define the stochastic integral*

$H \cdot X = H \cdot M + H \cdot A$. The stochastic integral $H \cdot A$ is defined pathwise, as a Stieltjes integral:

$$(H \cdot A)_t(\omega) = \int_{[0,t]} H_s(\omega) \, dA_s(\omega) \quad \text{for } t \geq 0 \text{ and } \omega \in \Omega.$$

The stochasic integral $H \cdot M$ reduces to the case where M is a square integrable martingale. So, assume M is a square integrable martingale. We identify the space $\mathcal{M}^2$ of cadlag, square integrable martingales with the space $L^2(P)$, by identifying a martingale $M \in \mathcal{M}^2$ with the random variable $M_\infty \in L^2(P)$, and endow $\mathcal{M}^2$ with the norm of $L^2(P)$. For a simple process

$$H = a_0 1_0 + \sum_{i=1}^n a_i 1_{]s_i t_i]}$$

we define the stochastic integral $H \cdot M$ by

$$(H \cdot M)_t = a_0 M_0 + \sum_{i=1}^n a_i (M_{t_i \wedge t} - M_{s_i \wedge t}).$$

Then $H \cdot M \in \mathcal{M}^2$. If we consider the simple process H as an element of $L^2(\mu_{\langle M \rangle}) = L^2(\mathcal{P}, \mu_{\langle M \rangle})$, then one can prove that the mapping $H \mapsto H \cdot M$ is an isometry:

$$\|H \cdot M\|_{\mathcal{M}^2} = \|H\|_{L^2_{(\mu_{\langle M \rangle})}}.$$

Since the simple processes are dense in $L^2_{(\mu_{\langle M \rangle})}$ one can extend the above isometry to an isometry of the whole space $L^2_{(\mu_{\langle M \rangle})}$ into $\mathcal{M}^2$. The value of this extension for a process $H \in L^2_{(\mu_{\langle M \rangle})}$ is called the stochastic integral of H with respect to M.

The isometry between the spaces $\mathcal{M}^2$ and $L^2_{(\mu_{\langle M \rangle})}$, which is the first step in the classical approach to the stochastic integral with respect to a square integrable martingale, can be obtained by using the measure-theoretic approach presented in this chapter.

Bibliography

Bochner, S. (1933). Integration von Funktionen, deren Werte die Elemente eines Vectorraumes sind. *Fund. Math.*, **20**, 262–76.

Bongiorno, B., Dinculeanu, N. (2001). The Riesz representation theorem and extension of additive measures. *J. Math, Analysis and Appl.*, **261**, 706–32.

Bourbaki, N. (1952–1959). Integration, Chapitres I–VI. Hermann, Paris.

Brooks, J.K., Dinculeanu, N. (1976). Lebesgue type spaces for vector integration, linear operations, weak completeness and weak compactness. *J. Math Analysis Appl.*, **54**, 348–89.

Brooks, J.K., Dinculeanu, N. (1991). Stochastic integration in Banach spaces. *Seminar on Stoch. Proc.*, 27–115 Birkhauser, Baston.

Dellacherie, C., Meyer, P.A. (1975 & 1980). *Probabilities et Potentiel*, 2 vols. Hermann, Paris.

Dinculeanu, N. (1967). *Vector Measures*. Pergamon Press, Oxford.

Dinculeanu, N. (2000). *Vector Integration and Stochastic Integration in Banach spaces*. Wiley, New York.

Dunford, N., Schwartz, J. (1958). *Linear Operators, Part I: General Theory*. Wiley, New York.

Ionescu tulcea, A., Ionescu tulcea, C. (1969). *Topics in the Theory of Lifting*. Springer, New York.

Kluvanek, I. (1961). On the theory of vector measures I. *Mat. Fyz. Casopis*, 11, 173–91.

Kluvanek, I. (1966). On the theory of vector measures II. *Casopis*, 16, 76–81.

Kussmaul, A.U. (1977). *Stochastic Integration and Generalized Martingales*. Pitman, London.

Metievier, M. (1982). *Semimartingales*. De Gruyter, Berlin.

Metievier, M., Pellaumail, J. (1982). *Stochastic Integration*. Academic Press, New York.

Pellaumail, J. (1973). Sur l'integrale stochastique et la decomposition de Doob–Meyer. *Astérisque*, No. 9.

Rao, M.M. (1979). *Stochastic Processes and Integration*. Sijthoff and Noordhoff, Alphenaan den Rijin.

Rudin W. (1973). *Real and Complex Analysis*. McGraw-Hill, New York.

CHAPTER 13

OPERATOR GEOMETRY IN STATISTICS

KARL GUSTAFSON

An operator trigonometry developed chiefly by this author during the past 40 years has interesting applications to statistical efficiency, canonical correlations, and other statistical bounds and inequalities. The main goal of this chapter is to present the essentials of this operator trigonometry as it applies to statistics. Key elements (as originally named by this author) are operator antieigenvalues, operator antieigenvectors, and operator (maximal) turning angles. A secondary goal is to develop this operator trigonometry and some other related work by this author within a wider context of a general operator geometry of stochastic processes in Hilbert space. I will write in an informal and somewhat personal manner in order to immediately expose key ideas and underlying insights that may not be so easily discernible in the usual scientific journal publications. I also hope that makes this exposition more interesting.

13.1 Motivating example: statistical efficiency

In 1999 (Gustafson 1999a) I decided to try to apply my operator trigonometry theory to statistics. This opportunity presented itself to me early in 1999 when I was asked to referee two papers, one for the journal *Linear Algebra and its Applications* in which I have published a number of papers, and another for the journal *Metrika*, in which I have never published. I accepted both papers with minor modifications. These two papers dealt with the theory of statistical efficiency. Because I had never published in *Metrica*, I finished writing a paper on the new application of my operator trigonometry to statistical efficiency and sent it to that journal. One referee found it okay except for minor typos, but the second referee flatly rejected it. Being somewhat senior and therefore experienced with the vagaries and even "tail event" dishonesties that one sometimes encounters with referees, I perservered and published the Gustafson (1999a) preprint paper as the first four sections of Gustafson (2002). However, my 1999 preprint, in which I was the first to establish the connection of my operator trigonometry to statistics, was in circulation from 1999 onwards, and attracted some interest from within the matrix statistics community.

For more details of this story and, more importantly, my applications of my operator trigonometry to statistics, readers may access the two papers Gustafson (2002) and the later invited survey paper Gustafson (2006a). In fact these two papers should provide both a number of results and also all additional background that may be needed to augment what I present here on statistical efficiency. Readers might also look at the related papers Gustafson (2005) and (2007).

In brief now: for the well-known general linear model in statistics,

$$y = X\beta + e,$$

the statistical efficiency for comparing an ordinary least squares estimator (OLSE) $\widehat{\beta}$ and the best linear unbiased estimator (BLUE) β^* is defined as

$$\text{eff}(\widehat{\beta}) = \frac{|\text{Cov}(\beta^*)|}{|\text{Cov}(\widehat{\beta})|} = \frac{1}{|X'VX||X'V^{-1}X|}.$$

Here it is assumed that the noise e has expected value 0 and symmetric positive definite (SPD) $n \times n$ covariance matrix V with eigenvalues $\lambda_1 \geq \lambda_2 \geq \cdots \geq \lambda_n > 0$. An important lower bound, the DBWK (Durbin, Bloomfield, Watson, Knott) lower bound, for the efficiency is

$$\text{eff}(\widehat{\beta}) \geq \prod_{i=1}^{n/2} \frac{4\lambda_i \lambda_{n-i+1}}{\lambda_i + \lambda_{n-i+1}}.$$

I immediately recognized that the geometrical meaning of the DBWK lower bound is

$$\text{eff}(\widehat{\beta}) \geq \prod_{i=1}^{n/2} \mu_i^2(V) = \prod_{i=1}^{n/2} \cos^2(\phi_i),$$

where the ϕ_i are my operator-trigonometric critical turning angles and the μ_i are my antieigenvalues. In other words, statistical efficiency is inherently geometric, even trigonometric. That is the main result of Gustafson (1999a). For simplicity in the above I have assumed that the model matrix X is $n \times p$ with $p = n/2$, and that $X'X = I_p$.

What are these operator turning angles ϕ_i and the antieigenvalues μ_i? They are best first understood by simple examples. Here are three. In general (see, for example, the books Gustafson (1997) and Gustafson and Rao (1997)), I prefer the linear algebra notation A for general matrices and the operator trigonometry.

Example 13.1

$$A = \begin{bmatrix} 1 & 0 \\ 0 & 2 \end{bmatrix}$$

$$\cos \phi_1(A) = \frac{2\sqrt{\lambda_1 \lambda_2}}{\lambda_1 + \lambda_2} = \frac{2\sqrt{2}}{3} \cong 0.94281$$

$$\phi_1(A) = 19.471°$$

Example 13.2

$$A = \begin{bmatrix} 9 & 0 \\ 0 & 16 \end{bmatrix}$$

$$\cos \phi_1(A) = \frac{2\sqrt{\lambda_1 \lambda_2}}{\lambda_1 + \lambda_2} = \frac{2\sqrt{144}}{25} = 0.96$$

$$\phi_1(A) = 16.269°$$

Example 13.3

$$A = \begin{bmatrix} 1 & 0 & 0 & 0 \\ 0 & 2 & 0 & 0 \\ 0 & 0 & 10 & 0 \\ 0 & 0 & 0 & 20 \end{bmatrix}$$

$$\cos \phi_1(A) = \frac{2\sqrt{\lambda_1 \lambda_2}}{\lambda_1 + \lambda_2} = \frac{2\sqrt{20}}{21} = 0.42591771$$

$$\phi_1(A) = 64.7912347°$$

$$\cos \phi_2(A) = \frac{2\sqrt{\lambda_2 \lambda_3}}{\lambda_2 + \lambda_3} = \frac{2\sqrt{20}}{12} = 0.74535593$$

$$\phi_2(A) = 41.8103149°$$

I have taken these simple examples from the book Gustafson and Rao (1997), which provides the best reference for my operator trigonometry up to about 10 years ago. Since then I have developed it considerably for applications.

I will develop some further theoretical aspects of the operator trigonometry in the next section. For now, in this motivating application to statistical efficiency, if you happened to have a noise covariance matrix V in your statistical linear estimation problem that scaled to one of the above three examples, then your efficiency eff($\widehat{\beta}$) would be bounded above by 1 and bounded below by the DBWK lower bounds, respectively:

$$\cos^2 \phi_1 = 0.888888889 \quad \text{(Example 1)},$$
$$\cos^2 \phi_1 = 0.9216 \quad \text{(Example 2)},$$

and for Example 3,

$$\cos^2 \phi_1 \cos^2 \phi_2 = \frac{4(20)}{(21)^2} \cdot \frac{4(20)}{(12)^2} = 0.100781053.$$

Notice that these bounds are uniform for whatever regressor (model) matrix X you may have implemented in your experiment.

The geometry of the operator trigonometry is as follows. In each of the three examples above, the angle $\phi_1(A)$, which I call the operator (maximal) turning angle, is indeed the largest angle through which the operator A may turn any vector x. This will be explained further in the next section. For the first two examples, we may find the most-turned vectors, which I call the (first) antieigenvectors, according to the formula

$$x_\pm = \pm \left(\frac{\lambda_1}{\lambda_1 + \lambda_2}\right)^{1/2} x_2 + \left(\frac{\lambda_2}{\lambda_1 + \lambda_2}\right)^{1/2} x_1,$$

where x_2 is the eigenvector for the smaller eigenvalue λ_2 and x_1 is the eigenvector for the larger eigenvalue λ_1. There are always 2 antieigenvectors. Thus for the first example we have the antieigenvectors

$$x_\pm = \pm \left(\frac{2}{3}\right)^{1/2} \begin{bmatrix} 1 \\ 0 \end{bmatrix} + \left(\frac{1}{3}\right)^{1/2} \begin{bmatrix} 0 \\ 1 \end{bmatrix}$$

$$= \frac{1}{\sqrt{3}} \begin{bmatrix} \sqrt{2} \\ 1 \end{bmatrix} \quad \text{and} \quad \frac{1}{\sqrt{3}} \begin{bmatrix} -\sqrt{2} \\ 1 \end{bmatrix}.$$

For the second example the (most-turned) vectors are the two antieigenvectors

$$x_+ = \frac{1}{5} \begin{bmatrix} 4 \\ 3 \end{bmatrix} \quad \text{and} \quad x_- = \frac{1}{5} \begin{bmatrix} -4 \\ 3 \end{bmatrix}.$$

For the third example the antieigenvector formula becomes

$$x_\pm = \pm \left(\frac{\lambda_1}{\lambda_1 + \lambda_4}\right)^{1/2} x_4 + \left(\frac{\lambda_4}{\lambda_1 + \lambda_4}\right)^{1/2} x_1$$

$$= \frac{1}{\sqrt{21}} \begin{bmatrix} \pm\sqrt{20} \\ 0 \\ 0 \\ 1 \end{bmatrix}.$$

Then for this third example one may strip off the lowest and highest eigenspaces, which determine the most-turned first antieigenvector pair, and speak of the next (or interior) critical turning angle ϕ_2 determined by the submatrix

$$A_{23} = \begin{bmatrix} 2 & 0 \\ 0 & 10 \end{bmatrix},$$

from which the corresponding interior, or second antieigenvector, pair for the original matrix A is computed as above to be

$$x_\pm^2 = \frac{1}{\sqrt{6}} \begin{bmatrix} 0 \\ \pm\sqrt{5} \\ 1 \\ 0 \end{bmatrix}.$$

In general, and as above, we like to normalize all the eigenvectors to length = norm = 1, and then the antieigenvectors will also have norm 1.

To continue this example-led explanation of the connection I have established of my operator trigonometry with statistical efficiency, let me now refer to the paper Chu et al. (2005). This paper goes into nice detail about efficiency and, moreover, presents some concrete examples. I have not previously discussed what follows, so readers are referred to Chu et al. (2005) for more details. They consider the noise matrix

$$V = \begin{bmatrix} 3 & 1 & 1 & 3\rho \\ 1 & 3 & 1 & 1 \\ 1 & 1 & 3 & 1 \\ 3\rho & 1 & 1 & 3 \end{bmatrix}.$$

Thus $n = 4$ and $p = 2$. V is SPD whenever $1 > \rho > -2/3$. Let me take $\rho = 0$ here for simplicity, then work out on a preliminary basis some operator trigonometry, and then return to Chu et al.'s considerations of Watson factorized efficiencies. I have worked out the following details by hand (using a handheld calculator) and readers may easily do so too. Doing this should enhance readers' understanding of statistical efficiencies from the operator geometric point of view of this chapter.

First we need to calculate the eigenvalues of $\mathbf{V}$ with ρ taken equal to zero. Its characteristic equation becomes:

$$0 = \begin{vmatrix} (3-\lambda) & 1 & 1 & 0 \\ 1 & (3-\lambda) & 1 & 1 \\ 1 & 1 & (3-\lambda) & 1 \\ 0 & 1 & 1 & (3-\lambda) \end{vmatrix}$$

$$= (3-\lambda) \begin{vmatrix} (3-\lambda) & 1 & 1 \\ 1 & (3-\lambda) & 1 \\ 1 & 1 & (3-\lambda) \end{vmatrix} - \begin{vmatrix} 1 & 1 & 1 \\ 1 & (3-\lambda) & 1 \\ 0 & 1 & (3-\lambda) \end{vmatrix}$$

$$+ \begin{vmatrix} 1 & (3-\lambda) & 1 \\ 1 & 1 & 1 \\ 0 & 1 & (3-\lambda) \end{vmatrix}$$

$$= (3-\lambda)[(3-\lambda)^2 + 2 - 3(3-\lambda)] - [(3-\lambda)^2 + 1 - ((3-\lambda) + 1)]$$
$$+ [(3-\lambda) + 1 - ((3-\lambda)^2 + 1)]$$
$$= (3-\lambda)^4 - 5(3-\lambda)^2 + 4(3-\lambda).$$

Thus one eigenvalue is $\lambda = 3$. Replacing $3 - \lambda$ by x, we need to factor the cubic equation $x^3 - 5x + 4 = 0$ and we see that $x = 1$, hence $\lambda = 2$, will be one root. That leaves the quadratic equation $x^2 + x - 4 = 0$, from which we find the other two eigenvalues of $\mathbf{V}$ from the roots $x = \left(-1 \pm \sqrt{17}\right)/2$. Thus $\mathbf{V}$ has the four eigenvalues

$$\lambda_1 = 5.561552813$$

$$\lambda_2 = 3$$

$$\lambda_3 = 2$$

$$\lambda_4 = 1.438447187.$$

From these we could now find the critical turning angles ϕ_1 and ϕ_2, as we did in the 4×4 example above from the operator cosine formula given above. But, alternatively, let me take this opportunity to introduce the other important operator-trigonometric entity,

$$\sin \phi_1(\mathbf{A}) = \frac{\lambda_1 - \lambda_4}{\lambda_1 + \lambda_4}.$$

An important early result in Gustafson (1968a) states that, generally, $\sin^2 \phi(\mathbf{A}) + \cos^2 \phi(\mathbf{A}) = 1$ in the operator trigonometry for all strongly accretive operators on a Hilbert space. (See Gustafson (1972) for a proof). Indeed, as I have frequently pointed out recently, one really cannot speak of a trigonometry until one has both a cosine and a sine. The entity $\sin \phi(\mathbf{A})$ appears in applications perhaps more often than $\cos \phi(\mathbf{A})$ as the important application entity. I will comment further on this later in this chapter. Here we prefer $\sin \phi(A)$ because it is simpler arithmetically. Let us look at some examples.

For the noise matrix V we have

$$\frac{\lambda_1 - \lambda_4}{\lambda_1 + \lambda_4} = \frac{4.123105626}{7} = 0.589015089 = \sin \phi_1(V),$$

and hence the maximal turning angle $\phi_1(V)$ is

$$\phi_{14} = 36.08714707°.$$

Thus we immediately obtain

$$\cos \phi_{14} = 0.808122036$$
$$\cos^2 \phi_{14} = 0.653061224.$$

In the same manner, from

$$\frac{\lambda_2 - \lambda_3}{\lambda_2 + \lambda_3} = \frac{1}{5} = \sin \phi_2(V),$$

we have the second (interior) critical turning angle $\phi_2(V)$:

$$\phi_{23} = 11.53695903°$$

and

$$\cos \phi_{23} = 0.979795897$$
$$\cos^2 \phi_{23} = 0.96.$$

Hence we may compute the DBWK lower bound

$$\text{eff}(\hat{\beta}) \geq \cos^2 \phi_1(V) \cos^2 \phi_2(V)$$
$$= 0.626938775.$$

Notice that, so far, the regressor matrix X has not yet entered the picture.

Chu et al. (2005) now introduce two examples with specific partitioned $X = (X_1 : X_2)$ and factorize the efficiency. So let us continue illustrating our new operator-trigonometric connection to statistical efficiency by using their specific examples for X for the model problem

$$y = X_1\beta_1 + X_2\beta_2 + \epsilon,$$

where the noise covariance matrix is the V above. Without going into their analysis of efficiency factorization multipliers, etc., we may proceed directly to their first example:

$$X = [X_1 : X_2] = \begin{bmatrix} 1 : -1 \\ 1 : -2 \\ 1 : +2 \\ 1 : +1 \end{bmatrix}.$$

They find a factorized total efficiency

$$\text{eff}(\widehat{\boldsymbol{\beta}}) = \text{eff}(\widehat{\boldsymbol{\beta}}_1)\text{eff}(\widehat{\boldsymbol{\beta}}_2)$$

$$= \frac{400(2 + 3\rho)(1 - \rho)}{(1 + \rho)(11 + 3\rho)(7 - 6\rho)(11 - 3\rho)},$$

which in our case with $\rho = 0$ becomes

$$\text{eff}(\widehat{\boldsymbol{\beta}}; \text{given } X) = \frac{800}{7(121)} = 0.944510035,$$

indicating an estimation accuracy quite a bit better than our worst-case DBWK lower bound 0.626938775 obtained above. Their second example is

$$X = [X_1 : X_2] = \begin{bmatrix} 1 : -1 \\ 1 : +1 \\ 1 : +2 \\ 1 : -2 \end{bmatrix},$$

and they arrive at a total efficiency expression (correcting their obvious minor sign typo on p. 170, eqn. (2.16)):

$$\text{eff}(\widehat{\boldsymbol{\beta}}; \text{given } X) = \frac{12800(3\rho + 2)(1 - \rho)}{(77 - 70\rho - 3\rho^2)(343 + 414\rho - 9\rho^2)}$$

$$= \frac{25600}{(77)(343)} = 0.969293098$$

when $\rho = 0$, a bit better than with the first regressor matrix X.

13.1.1 Recapitulation and commentary

In teaching, learning, and writing, there are two approaches. Of course there are degrees and shades of each, but let us discuss just these two approaches. In academia mathematicians often prefer the "Theorem, Proof, Example" approach, which we could call the German, metaphysical way. Deduction. In practice, however, it often goes the other way. Induction. In this introductory section I have chosen the latter approach. You jump in and start swimming. In the next section I will formalize the operator trigonometry in a more deductive manner. Then in later sections we will come back to statistics and further aspects of the statistical applications of the geometry of my operator trigonometry.

After my initial Gustafson (1999a) preprint, some people in the statistics community became interested. For example, see Khattree (2003), which has considerable overlap with Gustafson (2002) but in which the viewpoints are different. Then I am pleased to say that the most noted C.R. Rao became interested in my antieigenvalue theory as it applies to several aspects of matrix statistics (see Rao (2005, 2007)). We first met at the New Zealand IWMS14 workshop in 2005, with warm resulting relations. Rao's (2007) paper takes a wider view of what my operator trigonometry might say, one that in some specifics I do not agree with; for example, see my remarks in Gustafson (2006a) and in the next section. However, his interest is of course greatly appreciated, and enhances the impact of the initial connection in Gustafson (1999a) of my theory to the statistics community.

In general here I do not have the space to develop all the background material and all the literature. I will therefore simply cite pertinent references as we go along. I will also use this device of a Recapitulation and Commentary at the end of each section to bring out some salient background material and references to the literature, or to discuss some important point that I think the literature (including my papers) does not emphasize enough.

What might be one such point to highlight here? I think it is the following. In this section I have chosen simple, low-dimensional examples. But as we will see in the next section, my operator trigonometry started out infinite dimensional, and remains infinite dimensional, even for Banach spaces. Thus, it is in principle applicable not only to infinite-dimensional statistics but also to functional data situations, the two contexts of this *Handbook*. It does not matter whether the vector space consists of vectors or functions. Of course, in either case, when one actually needs to compute, or see things more clearly, one often truncates one's operators, vectors, and functions to finite-dimensional subspaces.

13.2 Operator Trigonometry: The Essentials

I created the essentials of the operator trigonometry in the period 1967–1969. I was very fortunate that the application which motivated me brought out the two essential trigonometric entities $\cos \phi(A)$ and $\sin \phi(A)$. Here is how it happened.

Following my additive perturbation paper Gustafson (1966), I wanted to obtain a similar result for the multiplicative perturbation of contraction semigroups. Let A be the infinitesimal generator of the contraction semigroup Z_t which evolves on a Banach space X. Then A is a densely-defined differential operator whose exponential $Z_t = \exp(tA)$ propagates the initial data forward in time. We may think of Z_t as a continuous Markov process. See Kato (1980) for the operator-theoretic treatment of such semigroups. All such generators have nonpositive quadratic forms: $\text{Re}[\mathbf{Ax}, \mathbf{x}] \leq 0$ for all $\mathbf{x}$ in the domain of $\mathbf{A}$. Here $[\cdot, \cdot]$ denotes a semi-inner product for the Banach space. This means in particular that all eigenvalues of $\mathbf{A}$ have nonpositive real part so that when we exponentiate, the norm $||Z_t||$ stays less than or equal to 1. Hence the term "contraction" semigroup.

The additive perturbation result was that if we consider $\mathbf{A} + \mathbf{B}$, where $\mathbf{A} + \mathbf{B}$ stays negative, i.e. $\text{Re}[(\mathbf{A} + \mathbf{B})\mathbf{x}, \mathbf{x}] \leq 0$, then if $\mathbf{B}$ is suitably small, $\mathbf{A} + \mathbf{B}$ still generates a contraction semigroup. I used this result to obtain the multiplicative perturbation result in Gustafson (1968b): if $\mathbf{B}$ is bounded on X and is strongly accretive, i.e. for all $\mathbf{x}$ in X, $\text{Re}[\mathbf{Bx}, \mathbf{x}] \geq m||\mathbf{x}||^2$ where $m > 0$, then $\mathbf{BA}$ still generates a contraction semigroup iff $\mathbf{BA}$ is dissipative, i.e. $\text{Re}[\mathbf{BAx}, \mathbf{x}] \leq 0$ for all $\mathbf{x}$ in the domain of the original generator $\mathbf{A}$.

Focusing on the main ideas, one can say that the $\mathbf{BA}$ generator issue comes down to asking: when is a "negative" operator $\mathbf{A}$, when multiplied by a "positive" operator $\mathbf{B}$, still a "negative" operator? When $\mathbf{B}$ and $\mathbf{A}$ commute some simple answers emerge, but what about the general case? I have explored this question in Gustafson (1968c) and (1968d) and found the sufficient condition

$$\min_{\epsilon > 0} ||\epsilon \mathbf{B} - \mathbf{I}|| \leq \inf_{\substack{x \in D(A) \\ x \neq 0}} \text{Re} \frac{[(-\mathbf{A})\mathbf{x}, \mathbf{x}]}{||\mathbf{Ax}||\,||\mathbf{x}||}.$$

It is easier just to think of both $\mathbf{A}$ and $\mathbf{B}$ as strongly accretive operators, i.e. to absorb the minus sign into $\mathbf{A}$. Let us also pass to a Hilbert space and to everywhere defined, accretive, bounded operators $\mathbf{A}$ and $\mathbf{B}$. So then the above sufficient condition for $\mathbf{BA}$ to be accretive, i.e. $\text{Re}\langle\mathbf{BAx}, \mathbf{x}\rangle \geq 0$ for all vectors $\mathbf{x}$, becomes

$$\nu_1(\mathbf{B}) \equiv \min_{\epsilon > 0} ||\epsilon \mathbf{B} - \mathbf{I}|| \leq \inf_{x \neq 0} \text{Re} \frac{\langle \mathbf{Ax}, \mathbf{x}\rangle}{||\mathbf{Ax}||\,||\mathbf{x}||} \equiv \mu_1(\mathbf{A}).$$

Originally I called $\nu_1(\mathbf{B})$ the norm minimum $g_m(\mathbf{B})$. That was because the curve $g(\epsilon, \mathbf{B}) = ||\epsilon \mathbf{B} - \mathbf{I}||$ for $-\infty < \epsilon < \infty$ is known to be generally convex and differentiable almost everywhere. In Hilbert space the entity $\nu_1(\mathbf{B}) \equiv g_m(\mathbf{B})$ occurs for a unique ϵ which I denote by ϵ_m. Then the fundamental result in Gustafson (1968a) established that, in fact, $\nu_1(\mathbf{B})$ is $\sin \phi(\mathbf{B})$. We are a bit ahead of ourselves here but to continue I should point out that I also originally called (see Gustafson (1968d)) the right-hand-side entity $\mu_1(\mathbf{A})$ the (real) cosine of $\mathbf{A}$, using notations $\cos \mathbf{A}$, later $\cos \phi(\mathbf{A})$. The intuition for that terminology is obvious from the Schwarz inequality, the only technicality being if the infimum is actually attained by some vector $\mathbf{x}$. Then in my 1969 lecture (see Gustafson (1972)) at the Third Symposium on Inequalities at UCLA, 1–9 September 1969, I introduced the name *antieigenvalue* for $\mu_1(\mathbf{A})$, and *antieigenvector* for any such maximally-turned vector $\mathbf{x}$. The maximally-turned angle $\phi(\mathbf{A})$ I called the *operator angle* of $\mathbf{A}$ (Gustafson 1968d).

That is the origin of my operator trigonometry. For the purposes of this chapter, let me quickly summarize the ensuing developments. Aside from a little work with my PhD students D.K.M. Rao and M. Seddighin, I essentially left the operator trigonometry after 1972 for other pursuits. Then twenty years later in 1992 I returned to the subject, wishing to clarify certain parts of it, and I turned especially toward applications. See the three papers Gustafson (1994a, 1994b, 1995), the invited survey Gustafson (1996), and the books Gustafson (1997) and Gustafson and Rao (1997). Recently I wrote a survey (Gustafson 2006b) in which readers will find 60 citations to my work and 40 citations to related work by others. My operator trigonometry has important applications to numerical analysis and to quantum physics which I don't want to discuss here.

So what are the essential ideas that are important to statistics? Let A be an SPD $n \times n$ matrix with eigenvalues $\lambda_1 \geq \lambda_2 \geq \cdots \geq \lambda_n > 0$. Then we have

$$\mu_1(A) \equiv \cos\phi(A) = \min_{x \neq 0} \frac{\langle Ax, x \rangle}{||Ax||\,||x||} = \frac{2\sqrt{\lambda_1 \lambda_n}}{\lambda_1 + \lambda_n},$$

$$\nu_1(A) \equiv \sin\phi(A) = \min_{\epsilon > 0} ||\epsilon A - I|| = \frac{\lambda_1 - \lambda_n}{\lambda_1 + \lambda_n}.$$

From either $\cos\phi(A)$ or $\sin\phi(A)$ we have the maximum turning angle of A. The most-turned vectors, the antieigenvectors, are

$$x_\pm = \pm \left(\frac{\lambda_1}{\lambda_1 + \lambda_n}\right)^{1/2} x_n + \left(\frac{\lambda_n}{\lambda_1 + \lambda_n}\right)^{1/2} x_1.$$

Just think of these as the two extreme eigenvectors, weighted oppositely by their eigenvalues in order to achieve a "most twisted" angle when operated upon by the matrix A.

"Higher" (smaller) operator angles, corresponding (larger) antieigenvalues, and corresponding "higher" antieigenvectors are defined in the same way. See especially Gustafson (1994b, 1995), where I define these more carefully than in my casual mention in Gustafson (1972). See also the more general paper Gustafson (2000). The entities

$$\mu_k(A) = \cos\phi_k(A) = \frac{2\sqrt{\lambda_k \lambda_{n-k+1}}}{\lambda_k + \lambda_{n-k+1}}$$

$$\nu_k(A) = \sin\phi_k(A) = \frac{\lambda_k - \lambda_{n-k+1}}{\lambda_k + \lambda_{n-k+1}}$$

$$x_\pm^k = \pm \left(\frac{\lambda_k}{\lambda_k + \lambda_{n-k+1}}\right)^{1/2} x_{n-k+1} + \left(\frac{\lambda_{n-k+1}}{\lambda_k + \lambda_{n-k+1}}\right)^{1/2} x_k$$

become important in the applications to statistics, as I will make clear in the next section.

Another important entity (Gustafson (1972); see also Gustafson (1994b)) is my Euler equation satisfied by antieigenvectors:

$$2||Ax||^2||x||^2(\operatorname{Re} A)x - ||x||^2\operatorname{Re}\langle Ax, x\rangle A^*Ax - ||Ax||^2\operatorname{Re}\langle Ax, x\rangle x = 0.$$

For A an $n \times n$ SPD matrix and $||x|| = 1$, this becomes:

$$\frac{A^2 x}{\langle A^2 x, x\rangle} - \frac{2Ax}{\langle Ax, x\rangle} + x = 0.$$

The context here is to view my operator trigonometry as a spectral extension of the old Rayleigh–Ritz variational theory for eigenvalues and eigenvectors. Indeed, that is why I introduced the names antieigenvalues and antieigenvectors. It turns out that these Euler equations can be connected to related equations in variational statistics, which I will explain in the next section.

Readers may now verify how we obtained the operator angles, cosines, sines, antieigenvalues, and antieigenvectors in Examples 13.1, 13.2, and 13.3 in the preceding section. Readers could also calculate the antieigenvectors for the 4×4 covariance noise matrix V we treated there. Then it can be seen how the antieigenvectors will somehow provide the "worst" regressors, and the DBWK lower bound for this example can be obtained. I haven't worked that out here, however (see Gustafson (2007)). Indeed, let me state that there remains much unexplored terrain once one starts asking specific questions.

There are two points, however, which I have known since the beginning, but which seem to be not sufficiently appreciated by others now entering the subject. To highlight these points, let me state them as Lemma 13.1 and Lemma 13.2.

Lemma 13.1 *Let A be an $n \times n$ SPD matrix. Then $\sin\phi(A)$, the minimum value of $||\epsilon A - I||$ for $\epsilon > 0$, is attained by all vectors $x = c_1 x_1 + c_n x_n$, $|c_1|^2 + |c_n|^2 = 1$, whereas $\cos\phi(A)$ is attained only for*

$$c_1 = \pm\left(\frac{\lambda_n}{\lambda_1 + \lambda_n}\right)^{1/2} \quad \text{and} \quad c_n = \left(\frac{\lambda_1}{\lambda_1 + \lambda_n}\right)^{1/2},$$

i.e. only for the antieigenvectors.

The proof is just a careful examination of the vectors x which "work" in the following important relations:

$$\max_{||x||=1} \min_{\epsilon>0} ||(\epsilon A - I)x||^2 = \max_{||x||=1}\left[1 - \left(\operatorname{Re}\frac{\langle Ax, x\rangle}{||Ax||}\right)^2\right]$$

$$= 1 - \min_{||x||=1}\left(\frac{\operatorname{Re}\langle Ax, x\rangle}{||Ax||}\right)^2$$

$$= 1 - \cos^2\phi(A)$$

and
$$\min_{\epsilon>0} \max_{||x||=1} ||(\epsilon A - I)||^2 = \min_{\epsilon>0} ||\epsilon A - I||^2$$
$$= \sin^2 \phi(A).$$

Let me make this point crystal clear here. Take the A of Example 13.1,
$$A = \begin{bmatrix} 1 & 0 \\ 0 & 2 \end{bmatrix}.$$

Plot $||\epsilon A - I||$ versus $\epsilon \geq 0$. One immediately sees that this norm curve is very simple. It is first the line
$$\ell_1 : ||(\epsilon A - I)x_n|| = 1 - \epsilon \lambda_n, \quad \epsilon \leq \frac{2}{\lambda_1 + \lambda_n},$$

and then the line
$$\ell_2 : ||(\epsilon A - I)x_1|| = -1 + \epsilon \lambda_1, \quad \epsilon \geq \frac{2}{\lambda_1 + \lambda_n}.$$

For our Example 13.1, these are just the lines
$$\ell_1 : 1 - \epsilon \quad \text{for} \quad \epsilon \leq 2/3$$

and the line
$$\ell_2 : -1 + 2\epsilon \quad \text{for} \quad \epsilon \geq 2/3.$$

The intersection of these two lines comes at $\epsilon_m = 2/3$ and the height of that intersection is $1/3 = \sin \phi(A)$. In other words, the norm curve $||\epsilon A - I||$ is a "one-component" line to the left, a "one-component" line to the right, and only where the lines intersect does it become "two-component."

We may call Lemma 13.1 the General Two-Component Lemma. By minimizing the cosine functional $\mu_1(A)$, I already knew for SPD $n \times n$ matrices A that the antieigenvectors were the (properly weighted) linear combination of the first and last eigenvectors x_n and x_1. Then one expects the same antieigenvectors to give the sine functional $v_1(A)$ also. But Lemma 13.1 shows that the latter functional is a much more general thing. It is attained not only at A's antieigenvectors but much more widely over the whole unit sphere in the span of the first and last eigenspaces of A. I plan to develop elsewhere this concept of $v_1(A)$ as a general optimization functional for a variety of optimization situations.

I have not seen Lemma 13.1 anywhere, and indeed I have neglected to highlight it before. But Lemma 13.1 leads to a deeper result, for which we must look at the proof of my Min–Max Theorem (1968a). It is a kind of converse to Lemma 13.1, and says that we may indeed extract the antieigenvectors from the $v_1(A)$ general functional. I'm pretty sure no one but me has seen it.

Lemma 13.2 *Let A be an n × n SPD matrix. Then we can obtain A's antieigenvectors just from the norm minimum* $v_1(A) \equiv \sin\phi(A) \equiv ||\epsilon_m A - I||$ *and the Min–Max theorem proof.*

To prove Lemma 13.2, we need to dig into some details of the proof of my Min–Max theorem (Gustafson 1968a). See Gustafson (1972), Gustafson (1995), or Gustafson and Rao (1997) for these details. Because I view this Min–Max theorem as the crux of my operator trigonometry, let me first take readers through its general proof. A is a strongly accretive operator on a Hilbert space and the convex curve $||\epsilon A - I||$ has left and right derivatives for almost all $\epsilon > 0$, so its unique minimium $||\epsilon_m A - I||$ is attained at either a lower corner or at the bottom of a cup. As the former case is the situation for all nontrivial selfadjoint operators A (witness the examples I gave earlier, and Lemma 13.1), we only consider the corner case here, but still for general A. Although we do not know exact minimizing vectors, we do know that for arbitrary small $\delta > 0$ we can always find norm-one vectors x_1 and x_2 such that to the left and right of ϵ_m, respectively, $||(\epsilon A - I)x_1||^2$ and $||(\epsilon A - I)x_2||^2$ approximate to within δ the curve $||\epsilon A - I||^2$ near ϵ_m. Therefore we take $x = \xi x_1 + \eta x_2$, where ξ and η are real, and

$$1 = ||x||^2 = \xi^2 + \eta^2 + 2\eta\xi \text{Re}\langle x_1, x_2\rangle$$

and

$$||(\epsilon_m A - I)x||^2 \geq ||\epsilon_m A - I||^2 - \delta + 2\xi\eta C,$$

where

$$C = \text{Re}\{\langle(\epsilon_m A - I)x_1, (\epsilon_m A - I)x_2\rangle - (||\epsilon_m A - I||^2 - \delta)\langle x_1, x_2\rangle\}.$$

By choosing ξ and η in the appropriate quadrant we can ensure that $2\xi\eta C \geq 0$. Because $||(\epsilon_m A - I)x||^2$ must always lie below $||(\epsilon_m A - I)||^2$, and yet also above the lower bound we have just found, we have obtained an approximate minimizing vector x with $||(\epsilon_m A - I)x||^2$ within arbitrary small $\delta > 0$ just below $||\epsilon_m A - I||^2$. Since the Min–Max equality states that

$$\sup_{||x||\leq 1} \inf_{\epsilon} ||(\epsilon A - I)x||^2 = \inf_{\epsilon>0} \sup_{||x||\leq 1} ||(\epsilon A - I)x||^2,$$

we have established that the left-hand side attains the right-hand side. That establishes the Min–Max theorem because the right-hand side is a priori never less than the left-hand side.

Moreover I then checked (see Gustafson (1968a, 1972, 1995)) that $\epsilon_m(x)$ can be made arbitrarily close to $\epsilon_m(A)$, as must be possible if the pair $(\epsilon_m(x), ||(\epsilon A - I)x||^2)$ is to converge to $(\epsilon_m(A), ||(\epsilon A - I)||^2)$. My proof of this reveals, if I might say so, a rather nice use of elementary ellipses and hyperbolas. Then to go further and ask that $\epsilon_m(x)$ can be made to exactly equal $\epsilon_m(A)$, I show this to be equivalent to the relation

$$\xi^2 \left\{ \text{Re}\langle Ax_1, x_1\rangle \left(1 - \frac{\epsilon_m}{\epsilon_1}\right)\right\} + \eta^2 \left\{ \text{Re}\langle Ax_2, x_2\rangle \left(1 - \frac{\epsilon_m}{\epsilon_2}\right)\right\}$$
$$- 2\xi\eta \text{Re}\langle Ax_1, (\epsilon_m A - I)x_1\rangle = 0.$$

Here $\epsilon_1 = \text{Re}\langle Ax_1, x_1\rangle/||Ax_1||^2$ is where the curve $||(\epsilon A - I)x_1||^2$ attains its minimum $1 - (\text{Re}\langle Ax_1, x_1\rangle/||\epsilon A - I||)^2$; likewise ϵ_2 is where $||(\epsilon A - I)x_2||^2$ achieves its minimum.

Now I want to focus here on the situation of Lemma 13.2 where A is an $n \times n$ SPD matrix. Then we know just from the norm curve $||\epsilon A - I||$ that x_1 and x_2 in the Min–Max theorem proof may be taken to be any norm-one eigenvectors x_n and x_1 corresponding to the smallest and largest eigenvalues λ_n and λ_1, respectively. (See the example given after Lemma 13.1 above.) Then $\epsilon_1 = 1/\lambda_n$ and $\epsilon_2 = 1/\lambda_1$ and $\epsilon_m(A) = 2/(\lambda_1 + \lambda_n)$, and the expression above simplifies to

$$\xi^2 \left\{\lambda_n \left(1 - \frac{2\lambda_n}{\lambda_1 + \lambda_n}\right)\right\} + \eta^2 \left\{\lambda_1 \left(1 - \frac{2\lambda_1}{\lambda_1 + \lambda_n}\right)\right\} = 0,$$

or seen more trigonometrically, to the expression

$$\sin\phi(A)[\lambda_n\xi^2 - \lambda_1\eta^2] = 0.$$

Also the expression $||(\epsilon_m A - I)x||^2$ simplifies to

$$||(\epsilon_m A - I)x||^2 = \sin^2\phi(A)[\xi^2 + \eta^2].$$

We have of course used the orthogonality property $\langle x_1, x_n\rangle = 0$. Thus because $\xi^2 + \eta^2 = 1$ we see that the minimum attaining vector $x = \xi x_n + \eta x_1$ may be found from the conic system

$$\begin{cases} \xi^2 + \eta^2 = 1 \\ \lambda_1\xi^2 - \lambda_2\eta^2 = 0. \end{cases}$$

Solving these equations yields

$$\xi = \pm\left(\frac{\lambda_1}{\lambda_1 + \lambda_n}\right)^{1/2}, \quad \eta = \pm\left(\frac{\lambda_n}{\lambda_1 + \lambda_n}\right)^{1/2}.$$

These are the correct coefficients of the antieigenvectors $x_\pm$. Thus Lemma 13.2 has been proved.

We may call Lemma 13.2 the Antieigenvector Reconstruction Lemma. It was not clear to me until after I proved the Min–Max Theorem that my proof actually needed to construct either exact (in the selfadjoint A case) or approximate (in the general A, strongly accretive case) antieigenvectors. And although they do indeed give one the maximum turning angle $\phi(A)$ of A by their attainment of $\sin\phi(A)$, one doesn't really understand the maximal turning action of A until one looks at the variational quotient defining $\cos\phi(A)$.

13.2.1 Recapitulation and commentary

These two lemmas, implicit and essential to my operator trigonometry from its creation, have not been highlighted or even exposed until now. I plead guilty. But originally I was after a more general new trigonometric theory for arbitrary strongly accretive operators A. Then I left the subject in 1972 and did not really return to it until 1992 (see Gustafson (1994a, 1994b, 1995)).

The Min–Max theorem holds for all strongly accretive operators A on a Hilbert space. In the conceptually easier case of a smooth cup minimum $||\epsilon A - I||$, the x_1 and x_2 in my proof still work as before but may coincide. It is easy to construct simple, even normal, matrices, where the minimum $||\epsilon_m A - I||$ is attained by $||(\epsilon_m A - I)x||$ where x is even a single eigenvector. But there can be an even smaller $||\lambda_m A - I||$ elsewhere, at some complex λ_m. One must be careful in the nonselfadjoint case when extracting trigonometrical interpretations. Still, I believe that my Min–Max theorem proof, being inherently two (or one) component, indicates some kind of antieigenvector meanings for the approximate minimizing vectors $x = \xi x_1 + \eta x_2$ that I used in its proof, for general matrices A. This is especially true in view of Lemmas 13.1 and 13.2 above. Although we have investigated normal matrices to some extent (see the citations in Gustafson and Rao (1997) and in Gustafson (2006b)), it is safe to say that the antieigenvector theory is not much developed for general matrices.

The Lemmas 13.1 and 13.2 that I have highlighted here are important to keep in mind as we approach the statistics and inequalities context of the next section. We will find that there are many instances in which one wishes to "optimize" some statistical entity or inequality, and often this optimization may be shown to occur at some "two-component" vector $x = c_1 x_1 + c_n x_n$ which linearly combines the first and last eigenvectors. These two lemmas make clear that "many" such linear combinations will give us the $\sin\phi(A)$ optimization (minimization of $||\epsilon A - I||$ over ϵ), but only very special linear combinations (the antieigenvectors) correspond to the $\cos\phi(A)$ optimization (minimization of $\langle Ax, x\rangle/||Ax||||x||$ over x).

It is on this point that I differ from C.R. Rao's (2005, 2007) treatment of antieigenvalues and antieigenvectors. To me, the antieigenvectors (my creation, after all) must represent maximal or at least critical turnings. Vectors which optimize some other functional should not be called antieigenvectors unless it can be shown that they are in fact the antieigenvectors. We will see some other examples of this distinction in the next section. That having been said, let me hasten to add that Rao's viewpoint, which I might describe as looking at more general "departures from eigenvector conditions", creates a very interesting and more general extension of my antieigenvector theory.

13.3 STATISTICS INEQUALITIES: CANONICAL CORRELATIONS

To save time here I want to refer the reader immediately to the extensive treatment of statistics matrix inequalities and canonical correlations in the long survey paper Drury et al. (2002), and also to my papers Gustafson (2002, 2006a, 2007). Very roughly, one can say that once one knows the operator trigonometry geometry that I have described in Section 13.2, then one can give corresponding new geometrical meanings to a number of important inequalities already established within the statistics literature. We have already seen one instance of this in Section 13.1.

As a further example, already treated in Gustafson (1999a, 2002), one finds in the statistics literature some quite highly developed Lagrangian variational methods from which statistical inequalities and related conclusions are drawn. For example, the DBWK lower bound we discussed in Section 13.1 can be derived by applying such methods to the expression

$$(\text{eff}(\hat{\beta}))^{-1} = |X'V^{-1}X||X'VX|.$$

The minimization leads to an equation

$$\frac{z^2}{\langle Vx, x \rangle} - 2z + \frac{1}{\langle V^{-1}x, x \rangle} = 0$$

which invites comparison to my Euler equation for antieigenvectors which I formulated in 1969 and which I have already described above. But as I made clear in Gustafson (1999a, 2002), the optimizing vectors are not the same. Notably, the solutions to the above equation are of the form

$$x_{\pm}^{j+k} = \pm \frac{1}{\sqrt{2}} x_j + \frac{1}{\sqrt{2}} x_k.$$

In other words, the two components are equally weighted, unlike what is needed for an antieigenvector. I look quite closely at these Lagrangian-derived equations versus my Euler equation in my papers, so readers are referred there for more developments. Note that I am speaking in this example of an interesting specific instance of my Lemma 13.1 in the preceding section.

I have applied my operator trigonometry to other notable inequalities in statistics; see especially Gustafson (2006a). The inequalities include Khattree–Rao–Ando and Bartmann–Bloomfield. Certainly there are more out there.

As my next example, I want to conclude my treatment of the Hotelling correlation coefficient, which I left slightly unfinished in Gustafson (2006a). However, before I do that, let me demonstrate why what are often called canonical correlations in statistics can often be seen to be essentially my operator sines.

In their literature review Chu et al. (2005, Section 3) relate efficiency to canonical correlations via the expression (Chu et al. (2005), eqn (3.10)):

$$\text{eff}(\hat{\beta}) = \frac{|X'X|^2}{|X'VX||X'V^{-1}X|}$$

$$= |I - (X'VX)^{-1}X'V\overline{X}(\overline{X}'VX)^{-1}\overline{X}'VX|$$

$$= \prod_{i=1}^{m}\left(1 - \rho_i^2\right),$$

where $\overline{X}$ is an $n \times (n-p)$ matrix such that $X'\overline{X} = 0$ and range $R(\overline{X}) =$ null space $N(X)$. The ρ_i^2 are the eigenvalues of the four-factor operator subtracted from the identity in the above expression. These ρ_i may therefore be taken as the canonical correlations between the fitted values from OLS (ordinary least squares), i.e. the elements of the vector $X\hat{\beta}_{\text{OLS}}$, and the residuals from OLS, i.e. the elements of the vector $y - X\hat{\beta}_{\text{OLS}}$. For a more general discussion of canonical correlations, see Drury et al. (2002). These canonical correlations are called κ_i.

The point I made in Gustafson (2006a) is that if we couple the above expression for canonical correlations with the DBWK bound, we have

$$\prod_{i=1}^{n/2}(1 - \kappa_i^2) \geq \prod_{i=1}^{n/2}\frac{4\lambda_i\lambda_{n-i+1}}{(\lambda_i + \lambda_{n-i+1})^2} = \prod_{i=1}^{n/2}\cos^2\phi_i(V).$$

In other words from my operator-trigonometric point of view, the canonical correlations κ_i look like my $\sin\phi_i$ critical operator turning angles, just because all of those angles satisfy the Min–Max theorem result

$$\cos^2\phi_i(V) + \sin^2\phi_i(V) = 1.$$

Devoid of my operator-trigonometric intuition, in Drury et al. (2002) the above relation was recast into a general inequality proposition for an $n \times n$ SPD dispersion matrix A, where the product $\prod \kappa_i^2$ is called the Hotelling correlation coefficient. Their inequality is

$$\prod_{i=1}^{m}\kappa_i^2 \leq \prod_{i=1}^{m}\left(\frac{\lambda_i - \lambda_{n-i+1}}{\lambda_i + \lambda_{n-i+1}}\right)^2 \leq \prod_{i=1}^{m}\left(\frac{\lambda_i - \lambda_{n-m+i}}{\lambda_i + \lambda_{n-m+i}}\right)^2 \leq \left(\frac{\lambda_1 - \lambda_n}{\lambda_1 + \lambda_n}\right)^n.$$

The first and last inequalities are less essential than the middle inequality, which we write out in the special case of $0 < \lambda_4 < \lambda_3 < \lambda_2 < \lambda_1$:

$$\left(\frac{\lambda_1 - \lambda_4}{\lambda_1 + \lambda_4}\right)^2 \left(\frac{\lambda_2 - \lambda_3}{\lambda_2 + \lambda_3}\right)^2 \leq \left(\frac{\lambda_1 - \lambda_3}{\lambda_1 + \lambda_3}\right)^2 \left(\frac{\lambda_2 - \lambda_4}{\lambda_2 + \lambda_4}\right)^2.$$

Anyone versed in the operator trigonometry immediately notices that this is, trigonometrically (and first taking square roots, of course):

$$\sin \phi_{14} \sin \phi_{23} \leq \sin \phi_{13} \sin \phi_{24},$$

where ϕ_{14} is the operator maximal turning angle $\phi(A)$, ϕ_{23} is the next (interior) turning angle, and ϕ_{13} and ϕ_{24} are "skew" critical interior turning angles determined by the maximal turning action of A on the respective eigenspaces.

First let us prove the special case above. That inequality is easily seen to be equivalent to the inequality

$$\frac{1 - \left(\frac{\lambda_1\lambda_3 + \lambda_2\lambda_4}{\lambda_1\lambda_2 + \lambda_3\lambda_4}\right)}{1 + \left(\frac{\lambda_1\lambda_3 + \lambda_2\lambda_4}{\lambda_1\lambda_2 + \lambda_3\lambda_4}\right)} \leq \frac{1 - \left(\frac{\lambda_1\lambda_4 + \lambda_2\lambda_3}{\lambda_1\lambda_2 + \lambda_3\lambda_4}\right)}{1 + \left(\frac{\lambda_1\lambda_4 + \lambda_2\lambda_3}{\lambda_1\lambda_2 + \lambda_3\lambda_4}\right)}.$$

This expression is of the form $f(t_1) \leq f(t_2)$, where $f(t) = (1-t)/(1+t)$ is strictly decreasing for $t > -1$. Because $\lambda_1\lambda_3 + \lambda_2\lambda_4 > \lambda_1\lambda_4 + \lambda_2\lambda_3$ (because $\lambda_3 > \lambda_4$), we have a proof of the special case by working backward from the above. But here we want to see this in an operator-trigonometric geometric way. So we write the above as

$$\sin \phi(A_{\text{LHS}}) \leq \sin \phi(A_{\text{RHS}}),$$

where A_{LHS} is an imagined operator with largest eigenvalue $\lambda_1\lambda_2 + \lambda_3\lambda_4$ and smallest eigenvalue $\lambda_1\lambda_3 + \lambda_2\lambda_4$, and similarly A_{RHS} has largest eigenvalue $\lambda_1\lambda_2 + \lambda_3\lambda_4$ and smallest eigenvalue $\lambda_1\lambda_4 + \lambda_2\lambda_3$. Then, working backward from the last expression proves the Hotelling bound in the special case. The geometrical meaning of the Hotelling bound is thus that A_{LHS} has smaller turning angle than A_{RHS}.

Now let us consider the general case. But it reduces to the special case. To see this, consider the case $n = 8$, $m = 4$. Then the inequality is

$$\left(\frac{\lambda_1 - \lambda_8}{\lambda_1 + \lambda_8}\right)^2 \left(\frac{\lambda_2 - \lambda_7}{\lambda_2 + \lambda_7}\right)^2 \left(\frac{\lambda_3 - \lambda_6}{\lambda_3 + \lambda_6}\right)^2 \left(\frac{\lambda_4 - \lambda_5}{\lambda_4 + \lambda_5}\right)^2$$

$$\leq \left(\frac{\lambda_1 - \lambda_5}{\lambda_1 + \lambda_5}\right)^2 \left(\frac{\lambda_2 - \lambda_6}{\lambda_2 + \lambda_6}\right)^2 \left(\frac{\lambda_3 - \lambda_7}{\lambda_3 + \lambda_7}\right)^2 \left(\frac{\lambda_4 - \lambda_8}{\lambda_4 + \lambda_8}\right)^2.$$

Let us first select the two "outside" factors of both the LHS and RHS of this. They then satisfy the special case. Then we select the two "inside" factors of both the LHS and RHS. They also satisfy the spectral ordering needed for the special case. One may proceed from the outside in, or the inside out. In the case that m is odd, there is a factor repeated on both sides. Thus the general inequality proposition has been shown.

13.3.1 Recapitulation and commentary

I think there is much more to be done in using my operator trigonometry to put more geometry, even trigonometry, into statistics inequalities, or in order to provide geometrically inspired proofs, as I have done above. There are also related issues of so-called canonical coordinates that arise, for example, in signal-to-noise ratio problems. In addition Rao (2005, 2007) brings into the picture other related issues such as sphericity tests and homologous canonical correlations. The latter give rise to rather general Rayleigh quotients and have application to feedback control systems. As I showed in Gustafson (2002), these Rayleigh quotients can be viewed in the same context as my antieigenvector Euler equation.

A point I want to make clear here concerns the use of the Kantorovich inequality

$$\frac{x'x}{x'Vx \cdot x'V^{-1}x} \geq \frac{4\lambda_1 \lambda_n}{(\lambda_1 + \lambda_n)^2}$$

in many inequalities in statistics. (See Chu et al. (2005) and Drury et al. (2002) and citations therein.) I want to state very clearly that I did not know this inequality when I developed my operator trigonometry in 1967–69. It was first shown to me by Chandler Davis in Toronto in 1981. My $\cos \phi(A)$ definition in the special case of an SPD matrix A, namely,

$$\cos\phi(A) = \min_{x \neq 0} \frac{\langle Ax, x \rangle}{||Ax|| \, ||x||}$$

can be seen as a version of the Kantorovich inequality. In Gustafson (1997) I show how the Kantorovich inequality in another of its commonly expressed forms,

$$\max_{||x||=1}(X'AX)(X'A^{-1}X) = \frac{1}{4}\left\{\sqrt{\frac{\lambda_1}{\lambda_n}} + \sqrt{\frac{\lambda_n}{\lambda_1}}\right\}^2,$$

can lead to my $\cos \phi(A)$ expression given above. But one needs a change of variable: $\tilde{x} = \langle Ax, x \rangle^{-1/2} x$. Notice also that the right-hand side of the expression just above is

$$\frac{1}{4}\frac{(\lambda_1 + \lambda_n)^2}{\lambda_1 \lambda_n} = \frac{1}{\cos^2 \phi(A)}.$$

So the minimizing vectors are not the antieigenvectors. This is an instance of the admonition I stressed in the previous section.

Also I want to emphasize that neither Kantorovich nor Krein nor Wielandt nor anyone else saw the operator-trigonometric content of their inequalities. I have explained this in detail elsewhere, for example see Gustafson (1996). As I noted at the beginning of this chapter, I was fortunate also to need the entity $\min ||\epsilon A - I||$, which by my Min–Max Theorem I showed to be $\sin \phi(A)$. That was the big step towards the operator trigonometry.

13.4 PREDICTION THEORY: ASSOCIATION MEASURES

I would like to conclude this chapter by placing it within a somewhat wider context, and by giving some new results here. To begin that widening of context, I would like to look at the canonical correlations between the past and future of a stationary stochastic process. This possibility was pointed out to me by Mohsen Pourahmadi at my lecture at the IWFOS' 2008 conference, whose proceedings may be found in Dabo-Niang and Ferraty (2008). Such canonical correlations are treated in Pourahmadi (2001, p. 288). I knew the Helson, Szego, Lax prediction theory from earlier work I had done (see Goodrich and Gustafson (1981) and some later joint papers). But I had not thought about its canonical correlation aspect. The canonical correlation angle between past and future generalizes to the theory of the angle between any two closed subspaces in a Hilbert space, and I will also have more to say about that below. So these connections are new.

Let us consider discrete-parameter stationary stochastic processes X_t, for example as in Pourahmadi (2001), for convenience. Then one can define the canonical correlations as follows. Let $X = (X_1, \ldots, X_m)$ and $Y = (Y_1, \ldots, Y_n)$ be random vectors. Let Σ_{11} be the $m \times m$ matrix Cov(X), Σ_{22} be the $n \times n$ matrix Cov(Y), and Σ_{12} the $m \times n$ matrix Cov(X, Y). Then the first canonical correlation is

$$\rho_1 = \max_{a,b} \frac{a'\Sigma_{12}b}{\sqrt{a'\Sigma_{11}a}\sqrt{b'\Sigma_{22}b}}.$$

Here a and b are vectors of coefficients which linearly combine the given random vectors to produce this largest canonical correlation. The succeeding $\rho_2, \ldots, \rho_m$ are found in the same way with Rayleigh–Ritz orthogonality constraints relative to the previous subspaces. Without going into further detail here, let me just mention that those details are much like those one encounters in Singular Value Decomposition (SVD) and Principal Component Analysis (PCA) combined with the Schur complement applied to partitioned matrices

$$A = \begin{bmatrix} A_{11} & A_{12} \\ A_{21} & A_{22} \end{bmatrix}.$$

One could probably express all of the canonical correlation and canonical coordinates theory in those terms. But I don't have time to attempt that here.

The first point to note is that ρ_1 defines an angle by setting $\cos \theta = \rho_1$. Because ρ_1 is a maximum, this angle is a minimum. Please note that this distinguishes the canonical correlation theory from my antieigenvalue theory, where

$$\mu_1 = \cos \phi(A) = \min \frac{\langle Ax, x \rangle}{||Ax|| \, ||x||}$$

gives us a maximum angle.

Let me now jump ahead to the angle between past and future. Let X_t be a discrete-parameter stationary stochastic process, let $P = \overline{\mathrm{sp}}\{X_t, , t \leq 0\}$ be the subspace of its past (and present), and let $F = \overline{\mathrm{sp}}\{X_t, t = 1, 2, \ldots\}$ be the subspace of its (strict) future. Let the past–future first (largest) canonical correlation be

$$\rho_1 = \sup\{|\mathrm{Corr}(X, Y); X \in P, Y \in F|\}.$$

Introduce an angle θ according to $\cos \theta = \rho_1$ as above. One says that subspaces P and F are at a positive angle if $\theta > 0$, or equivalently $\rho_1 < 1$. There is a large and technical prediction theory which elaborates these notions. Let me just mention that in my experience, much of that theory can be helpfully focused by remembering the old F. and M. Riesz theory of Hardy spaces and the decomposition $L^2_+(S^1)$ and $L^2_-(S^1)$ of functions on the (complex) circle.

As pointed out in Pourahmadi (2001, p. 294), the paper Jewell et al. (1983) looks more closely at how to compute canonical correlations of past and future. This was the paper and theory I did not know. Therefore let us go to it now and establish a new connection to my operator trigonometry. Jewell et al. use the notation λ_1 instead of ρ_1 for the first canonical correlation, but I stay with ρ_1 here. In their theory, ρ_1 is the largest eigenvalue of $H^*H = I - T^*T$, where H is a Hankel and T is a Toeplitz operator with symbol $\overline{h}/h$, where h is an outer function in the Hardy space H^2. This is a fairly standard way to approach prediction theory, and for more about outer functions readers might refer to Goodrich and Gustafson (1981). If $\rho_1 < 1$, i.e. if the past and future subspaces are at a positive angle, then it is desirable to try to bound it away from 1, to get some estimation of the subspace angle θ. Using a spectral density argument Jewell et al. represent the symbol $\overline{h}/h$ in the form

$$\frac{\overline{h}}{h} = c e^{i(v - \tilde{u})},$$

where $|c| = 1$ and where $\tilde{u}$ denotes the Hilbert transform of u. Then their result is that ρ_1 is bounded above, below 1, according to

$$\rho_1 \leq \frac{\left[1 + ||e^u||^2_\infty ||e^{-u}||^2_\infty - 2||e^u||_\infty ||e^{-u}||_\infty \cos(2||v||_\infty)\right]^{1/2}}{1 + ||e^u||_\infty ||e^{-u}||_\infty}.$$

This is a very technical result, but may be felt intuitively as the way the Law of Cosines enters into the computation of the Poisson Kernel; see, for example, Gustafson (1999b, p. 185).

Now I can connect to my operator trigonometry. In the special case that the general spectral density $w = \exp(\mathbf{u} + \tilde{\mathbf{v}})$ is just $w = \exp(\mathbf{u})$ with $\mathbf{u} \in L^\infty$, or more generally when $m \leq w \leq M$, then the above bound simplifies to

$$\rho_1 \leq \frac{M-m}{M+m}.$$

Since the "eigenvalues" of the joint dispersion matrix of the past and future are just the values of w, we may take this to mean that such a matrix $\mathbf{A}$ is SPD, and the above bound becomes

$$\rho_1 \leq \frac{\lambda_{\max} - \lambda_{\min}}{\lambda_{\max} + \lambda_{\min}} = \sin\phi(\mathbf{A}).$$

When equality holds, that means the largest canonical correlation $\rho_1 = \cos\theta = \sin\phi(\mathbf{A})$, thereby connecting to my operator trigonometry. We have here an extension to prediction theory of my statement in the preceding section that the canonical correlations ρ_k correspond to my $\sin\phi_k(\mathbf{A})$ turning angles. Jewel et al. (1983) go on to show how an alternate derivation yields a best constant k:

$$k = \frac{2Mm}{M+m} = \frac{2\sqrt{Mm}}{M+m} \cdot \sqrt{Mm} = (\lambda_{\max}\lambda_{\min})^{1/2} \cos\phi(\mathbf{A}),$$

where I have gone ahead in this expression and factorized in order to show the relation to the operator trigonometry. No doubt there are many more interesting connections to be obtained here. I also recommend the interested reader look at the other related papers cited in Jewell et al. (1983), especially those related to sphericity tests in multivariate analysis.

Now I turn to another recent paper. Canonical correlations are linked to statistical association measures (e.g. correlation coefficients) and to gaps between subspaces of a Hilbert space in Dauxois et al. (2004). Without too much ado, I would like to connect those studies to my operator trigonometry. In the Dauxois et al. (2004) paper, the lattice $\mathcal{F}(H)$ of closed subspaces of a real Hilbert space H is considered. Then relative orthogonal projections onto subspaces of the form $\overline{H_k + H_3} \ominus H_3$ are considered, where H_1, H_2, and H_3 are three selected subspaces, and $k = 1$ or 2. The maximal canonical correlation coefficient between H_1 and H_2 relative to H_3 is defined as

$$\rho_1 = \rho_{123} = \text{spectral radius of } \widehat{\mathbf{A}},$$

where $\widehat{\mathbf{A}}$ is a positive selfadjoint operator defined in terms of the projections. I found the notation of Dauxois et al. (2004) a bit specialized, so for simplicity of explanation I will just use my own notation below. Also, although below I refer to their projectors, readers may also consider my projectors to be general, for possible applications elsewhere.

Let $B_{12} = (P_{13} + P_{23} - I)(P_{13} + P_{23})$, where the $P_{k,3}$ are, for example, the orthogonal projectors described above. Note that

$$\begin{aligned}B_{12} &= (P_{13} + P_{23})^2 - (P_{13} + P_{23}) \\ &= P_{13}^2 + P_{23}^2 + P_{13}P_{23} + P_{23}P_{13} - P_{13} - P_{23} \\ &= P_{13}P_{23} + P_{23}P_{13}.\end{aligned}$$

So far we have only used the fact that the P's are (allowably even oblique) projectors. So B_{12} is the anticommutator $\operatorname{Re} P_{13}P_{23}$. Next we require the subspace projectors to be the orthogonal ones. Then B_{12} is selfadjoint. Let $A \equiv A_{12} = B_{12}^2$. Then A is positive selfadjoint with spectral radius $\rho_1 \leq 1$. If A has discrete spectrum, then its largest eigenvalue is $\lambda_1 = \rho_1$, and its smaller eigenvalues can be taken to be the successively smaller canonical correlation coefficients. If A is compact, then only 0 is in its continuous spectrum, so there again one obtains discrete eigenvalue canonical correlations converging to zero.

The new link to my operator trigonometry is now apparent. If A has discrete spectra, one can use its (canonical correlation) eigenvalues to define my critical operator turning angles via the operator trigonometry $\sin \phi_i(A)$ formula. I don't give details here, but further developments surely follow.

Dauxois et al. (2004) go on to discuss subspace gaps and principal angles between subspaces. For further information on these, see Kato (1980). Please remember that these principal subspace angles are minimum, not maximum, angles between subspaces.

13.4.1 Recapitulation and commentary

I would like to elaborate here the relation of my operator trigonometry to the gap theory of angles between subspaces. The latter theory is well known and may be found in operator theory and matrix theory texts. I recently stumbled across a nice, simple presentation in Meyer (2000) so I will refer to that presentation here. For clearness I consider here only the finite-dimensional vector-space situation, even just the R^n case. When $R^n = R \oplus N$ the subspace angle θ between R and N is defined to be that of

$$\cos \theta = \max_{\substack{u \in R \\ v \in N}} \frac{v^T u}{||v||_2 ||u||_2} = \max_{u \in R, v \in N, ||u||_2 = ||v||_2 = 1} v^T u.$$

Let P be the (generally oblique) projector of R^n onto R along N. P has 2-norm

$$||P||_2 = \max_{||x||_2 = 1} ||Px||_2 = \sqrt{\lambda_{\max}(P^T P)} = \frac{1}{\sin \theta},$$

where θ is the above angle between the two subspaces.

Now let us turn to my operator angle

$$\cos\phi(\mathbf{A}) = \min_{\mathbf{x}\neq 0}\left\langle \frac{\mathbf{A}\mathbf{x}}{||\mathbf{A}\mathbf{x}||}, \frac{\mathbf{x}}{||\mathbf{x}||}\right\rangle,$$

where here I have written it in such a way as to emphasize that it measures a maximum turning by $\mathbf{A}$ of vectors normalized on the unit sphere. So we see that we are simply looking at one-dimensional subspaces sp{x} and sp{Ax}, but we are letting them run over all such subspace pairs. Let us now force a connection to the subspace angle context. Fix $\mathbf{x}$, rule out eigenvectors, so in the two-dimensional space $M = \text{sp}\{\mathbf{x}, \mathbf{Ax}\}$ we may take the oblique projector $\mathbf{P}_\mathbf{x}$ of sp{Ax} along sp{x}. Then we have an x-dependent angle $\theta_\mathbf{x}$ according to the above angle-between-subspaces theory:

$$\cos\theta_\mathbf{x} = \max_{||\mathbf{x}||=||\mathbf{A}\mathbf{x}||=1}\langle \mathbf{A}\mathbf{x}, \mathbf{x}\rangle = \frac{\langle \mathbf{A}\mathbf{x}, \mathbf{x}\rangle}{||\mathbf{A}\mathbf{x}||\,||\mathbf{x}||}.$$

But since x is fixed, there is really no maximum or minimum involved. My operator trigonometry angle $\phi(\mathbf{A})$ then takes the maximum of these angles $\theta_\mathbf{x}$, i.e. the minimum of the $\cos\theta_\mathbf{x}$, as x runs over all directions.

As I write this, I have a feeling that one could develop some further interesting geometry along these lines, for example a better study of the oblique projectors $\mathbf{P}_\mathbf{x}$, their separations, relations with the operator trigonometry, and so on. For example, trivial as a two-dimensional subspace angle may be, we could now consider, for each of our fixed x, the orthogonal projectors P and Q onto sp{Ax} and sp{x}, respectively. Then for each x our $\cos\theta_\mathbf{x} = ||PQ||_2$ and letting x run over all directions might produce some interesting dynamics between these orthogonal projectors and our oblique projectors as x converges toward the antieigenvectors of A.

To conclude this section whose goal was to expand context, note that we have here entered the realm of general stochastic processes. Let me recall a few things about general square integrable stochastic processes in a Hilbert space. The use of Hilbert-space representations for such processes goes back to Cramer (1940) and Kolmogorov (1941). It is fair to say that their approaches could now be identified with the spectral representation theorem of functional analysis. In that framework, each such stochastic process could be viewed in terms of its orthogonal projections. Such a view is certainly geometric. Now I must mention our Gustafson and Misra (1976) extension of that geometry to dynamics: every such stochastic process is unitarily equivalent to a quantum-mechanical momentum evolution. This result in turn led us to formulate a theory of time operators for statistical mechanics and wavelets. This theory is exposed in my book Gustafson (1997); see also some later papers.

So one could say that there has already been, since 1940, considerable geometry in prediction theory, time-series analysis, and regular stochastic processes in a Hilbert space. But it was based on the orthogonal-projection viewpoint, and it was not for statistics as distinguished from stochastic processes. My operator trigonometry arose from stochastic process theory in 1967, for example the Markov semigroup and other semigroup evolutions I mentioned at the beginning of this chapter. Then in the last ten years I have applied it specifically to certain problems in statistics. And whereas in the earlier spectral orthogonal-projection approach the key entities are eigenvalues and eigenvectors, my operator trigonometry now adds to that orthogonal geometry a nonorthogonal trigonometric geometry. Here the key entities are antieigenvalues and antieigenvectors.

13.5 Conclusions

In this chapter I have (in section order):

1. shown how my operator trigonometry gives new geometrical meaning to statistical efficiency;
2. reviewed the essential parts of the general operator trigonometry and its geometrical meaning;
3. shown how the operator trigonometry provides new geometry to canonical correlations and to other inequalities and bounds in statistics;
4. found new results applying my operator trigonometry to prediction theory and to association measures.

Special points, not so easily discernable in the general literature, which I have exposed here include (in section order):

1. The operator trigonometry is both finite and infinite dimensional for use in statistics and functional data analysis.
2. The functionals $\sin \phi_i(A)$ can give us the critical operator angles and the antieigenvalues, and, as shown here, even the antieigenvectors. But we need to use the variational character of the $\cos \phi_i(A)$ functional to really understand the antieigenvectors.
3. The Kantorovich inequality is (independently) related to my operator trigonometry, but one must exercise care with each of its forms, which may have different optimizing vectors.
4. My operator turning angles are maximal angles, as contrasted to the principal angles between subspaces which are minimal angles.

I have tried to write in an informative, simple, and interesting manner, minimizing technicalities and most of the general theory here.

References

Chu, K.L., Isotalo, J., Puntanen, S., Styan, G.P.H. (2005). The efficiency factorizaton multiplier for the Watson efficiency in partitioned linear models: some examples and a literature review. *Res. Lett. Inf. Math. Sci.*, 8, 165–87.

Cramer, H. (1940). On the theory of stationary random processes. *Ann. of Math.*, 41, 214–30.

Dabo-Niang, S., Ferraty, F. (2008). *Functional and operatorial statistics.* Springer, New York.

Dauxois, J., Nkiet, G.M., Romain, Y. (2004). Canonical analysis relative to a closed subspace. *Linear Algebra Appl.*, 388, 119–45.

Drury, S.W., Liu, S., Lu, C.Y., Puntanen, S., Styan, G.P.H. (2002). Some comments on several matrix inequalities with applications to canonical correlations: historical background and recent developments. *Sankhya: The Indian Journal of Statistics*, 64(A2), 453–507.

Goodrich, R.K., Gustafson, K. (1981). Weighted trigonometric approximation and inner-outer functions on higher dimensional Euclidean spaces. *J. of Approximation Theory*, 31, 368–82.

Gustafson, K. (1966). A perturbation lemma. *Bull. Amer. Math. Soc.*, 72, 334–8.

Gustafson, K. (1968a). A min-max theorem. *Notices Amer. Math. Soc.*, 15, 799.

Gustafson, K. (1968b). A note on left multiplication of semigroup generators. *Pacific J Math.*, 24, 463–5.

Gustafson, K. (1968c). Positive (noncommuting) operator products and semigroups. *Math. Zeitschrift*, 105, 160–72.

Gustafson, K. (1968d). The angle of an operator and positive operator products. *Bull. Amer. Math. Soc.*, 74, 488–92.

Gustafson, K. (1972). Antieigenvalue inequalities in operator theory. In *Inequalities III* (O. Shisha, ed.), 115–19. Academic Press, New York.

Gustafson, K. (1994a). Operator trigonometry. *Linear and Multilinear Algebra*, 37, 139–59.

Gustafson, K. (1994b). Antieigenvalues. *Linear Algebra Appl.*, 208/209, 437–54.

Gustafson, K. (1995). Matrix trigonometry. *Linear Algebra Appl.*, 217, 117–40.

Gustafson, K. (1996). Commentary on topics in the analytic theory of matrices. In *Collected Works of Helmut Wielandt 2* (B. Huppert, H. Schneider, eds), 356–67. DeGruyters, Berlin.

Gustafson, K. (1997). *Lectures on Computational Fluid Dynamics, Mathematical Physics, and Linear Algebra.* World Scientific, Singapore.

Gustafson, K. (1999a). On geometry of statistical efficiency. (preprint).

Gustafson, K. (1999b). *Partial Differential Equations and Hilbert Space Methods* (third edition, revised). Dover, New York.

Gustafson, K. (2000). An extended operator trigonometry. *Linear Algebra Appl.*, 319, 117–35.

Gustafson, K. (2002). Operator trigonometry of statistics and economics. *Linear Algebra Appl.*, 354, 141–58.

Gustafson, K. (2005). The geometry of statistical efficiency. *Res. Lett. Inf. Math. Sci.*, 8, 105–21.

Gustafson, K. (2006a). The trigonometry of matrix statistics. *International Statistical Review* 74(2), 187–202.

Gustafson, K. (2006b). Noncommutative trigonometry. *Operator Theory: Advances and Applications*, 167, 127–55.

GUSTAFSON, K. (2007). The geometry of statistical efficiency and matrix statistics. *J. of Applied Mathematics and Decision Sciences*, **2007**, doi:10.1155/2007/94515.

GUSTAFSON, K., MISRA, B. (1976). Canonical commutation relations of quantum mechanics and stochastic regularity. *Letters in Math. Phys.*, **1**, 275–80.

GUSTAFSON, K., RAO, D.K.M. (1997). *Numerical Range: the Field of Values of Linear Operators and Matrices*. Springer, Berlin.

JEWEL, N.P., BLOOMFIELD, P., BARTMANN, F.C. (1983). Canonical correlations of past and future for time series: bounds and computation. *Ann. Statist.*, **11**, 848–55.

KATO, T. (1980). *Perturbation Theory for Linear Operators*. Springer, New York.

KHATTREE, R. (2003). Antieigenvalues and antieigenvectors in statistics. *J. of Statistical Planning and Inference*, **114**, 131–44.

KOLMOGOROV, A.N. (1941). Stationary sequences in a Hilbert space. *Bull. Moscow State Univ*, **2**, 1–40.

MEYER, C. (2000). *Matrix Analysis and Applied Linear Algebra*. SIAM Publications, Philadelphia.

POURAHMADI, M. (2001). *Foundations of Time Series Analysis and Prediction Theory*. Wiley, New York.

RAO, C.R. (2005). Antieigenvalues and antisingularvalues of a matrix and application to problems in statistics. *Res. Lett. Inf. Math. Sci.*, **8**, 53–76.

RAO, C.R. (2007). Antieigenvalues and antisingularvalues of a matrix and application to problems in statistics. *Mathematical Inequalities and Applications*, **10**, 471–89.

CHAPTER 14

ON BERNSTEIN TYPE AND MAXIMAL INEQUALITIES FOR DEPENDENT BANACH-VALUED RANDOM VECTORS AND APPLICATIONS

NOUREDDINE RHOMARI

THE purpose of this chapter is to present some results on Bernstein type and maximal inequalities for partial sums of dependent random vectors taking their values in separable Hilbert or Banach spaces of finite or infinite dimension. We consider two types of measure of dependence, strong mixing coefficients (α-mixing) and absolutely regular mixing coefficients (β-mixing). These inequalities are similar to those in the dependent real case. From these inequalities we derive the strong law of large numbers and the bounded law of the iterated logarithm for absolutely regular Hilbert- or Banach-valued processes under minimal mixing conditions.

14.1 INTRODUCTION

Maximal inequalities provide an important tool for proving minimal sufficient conditions for convergence of sums of independent random variables. In the case of independent variables it is obvious to obtain such inequalities via Doob's martingale inequality, but this is not so in the dependent case.

Rio (1995a, 2000) has obtained a maximal inequality for real random variables and has derived from it a functional law of the iterated logarithm (LIL) for partial sums of stationary mixing processes under minimal mixing conditions.

Here we extend these inequalities to bounded dependent random vectors (r.v.) with values in separable Hilbert or Banach spaces, when the dependence is measured by the mixing coefficients. The inequalities extend to the dependent case the results of Pinelis and Sakhanenko (1985, Remark 2 on p. 145) and Pinelis (1990). The inequalities are practically the same as in the dependent real case; see, for example, Rio (1995a, 2000). More precisely, here we establish maximal inequalities for partial sums of dependent r.v.'s taking their values in separable Hilbert or Banach spaces with finite or infinite dimension.

We will make a distinction between the Hilbert and Banach cases since, in the latter case, the probability is evaluated for the norm of the partial sum centered at its mean, $\left\| \sum_{i=1}^{k} X_i \right\| - E \left\| \sum_{i=1}^{k} X_i \right\|$, but in the Hilbert case we estimate the probability for the norm of the partial sum $\left\| \sum_{i=1}^{k} X_i \right\|$. See Yurinskii (1976), Pinelis and Sakhanenko (1985), and Pinelis (1990). In our formulation, the exponential bound in Hilbert spaces contains a term in ε^2 but in Banach spaces this is replaced by a term such as $\left(\varepsilon - E \left\| \sum_{i=1}^{k} X_i \right\| \right)^2$. We consider two types of dependence, strong (α) and absolutely regular (β) mixing.

We conclude the chapter with some applications to the strong law of large numbers (SLLN) and the bounded law of the iterated logarithm (LIL) for Hilbertian or Banachian absolutely regular processes, under weak mixing conditions. This is the same sufficient minimal condition for Berbee's (1987) SLLN obtained in the real case; see Rio (1995a) for a detailed discussion. Rio (1995a) has generalized and extended these last results to unbounded real random variables using strong mixing, which is weaker than absolutely regular mixing; see also Rio (2000) and Shao (1993).

As a particular case, we apply these SLLN results to the estimation of the covariance operator of Hilbertian or Banachian absolutely regular (β-mixing) processes. We also prove the strong consistency in L^2 of the recursive kernel density estimate, with optimal rate, under weak absolutely regular mixing conditions: logarithmic decrease suffices for almost sure convergence and the rate is optimal when the sum of the series of mixing coefficients is finite. Some other applications can be derived, for example, parameter estimation for Hilbert or Banach autoregressive processes; see Bosq (2000) and references therein.

We remark that the works of Berbee (1987), Shao (1993), and Rio (1995a) also deal with Marcinkiewicz–Zygmund's SLLN for real random variables under absolutely regular and strong mixing. Furthermore, Shao (1995) considers the same questions under ρ-mixing.

The rest of the chapter is organized as follows. In the following paragraphs we set out some definitions and notation. We then give the maximal inequalities: first in the strong mixing case, followed by the absolutely regular mixing case. The third part is devoted to applications: the SLLN, the bounded LIL, the recursive estimation of probability density, and lastly the covariance operator estimations. Proofs are deferred to the end of the chapter.

14.1.1 Notation and definitions

Let $\mathbf{X} = (X_t, t \in \mathbb{N})$ be a process defined on a probability space $(\Omega, \mathcal{A}, P)$ and taking values in a measurable space $\mathcal{X}$. We define the strong mixing coefficient α, introduced by Rosenblatt (1956), and the absolutely regular mixing coefficient β of Rozanov and Volkonskii (1959) between two sub σ-algebras $\mathcal{F}$ and $\mathcal{G}$ of $\mathcal{A}$, respectively, by:

$$\alpha(\mathcal{F}, \mathcal{G}) = \sup\{|P(AB) - P(A)P(B)|; \ A \in \mathcal{F}, B \in \mathcal{G}\}.$$

$$\beta(\mathcal{F}, \mathcal{G}) = \sup\left\{\frac{1}{2}\sum_{i=1}^{I}\sum_{j=1}^{J}|P(A_i B_j) - P(A_i)P(B_j)|\right\},$$

where the supremum is taken over all finite partitions $(A_i), (B_j)$ of Ω with $A_i \in \mathcal{F}$ and $B_j \in \mathcal{G}$. For two r.v.'s X and Y, we will write $\alpha(X, Y) = \alpha(\sigma(X), \sigma(Y))$, where $\sigma(X)$ is a σ-algebra generated by X. For a process $\mathbf{X}$, the coefficients $\alpha(\cdot)$ and $\beta(\cdot)$ are defined for $l \in \mathbb{N}$ as:

$$\alpha(l) = \sup_{j \in \mathbb{N}} \alpha(\mathcal{F}_j, \mathcal{G}_{j+l}) \quad \text{and} \quad \beta(l) = \sup_{j \in \mathbb{N}} \beta(\mathcal{F}_j, \mathcal{G}_{j+l}),$$

where the $\mathcal{F}_j$ are the σ-algebras generated by $(X_0, \ldots, X_j)$ and the $\mathcal{G}_i$ by $(X_t, t \geq i)$.

It is well known that $2\alpha(\cdot) \leq \beta(\cdot)$; for properties of mixing conditions and for examples the reader may consult Bradley (1986, 2005) or Doukhan (1994) and the references therein. Bradley (2005) is an excellent survey and includes a wide-ranging discussion on different aspects of mixing. We also define the following coefficients:

$$\alpha_{\triangleleft,1}(l) = \sup_{j \in \mathbb{N}} \alpha(\mathcal{F}_j, X_{j+l}) \quad \text{and} \quad \beta_{\triangleleft,1}(l) = \sup_{j \in \mathbb{N}} \beta(\mathcal{F}_j, X_{j+l});$$

$$\alpha_{1,1}(l) = \sup_{j \in \mathbb{N}} \alpha(X_j, X_{j+l}) \quad \text{and} \quad \beta_{1,1}(l) = \sup_{j \in \mathbb{N}} \beta(X_j, X_{j+l}).$$

We will use the convention throughout this chapter that, given n random vectors $Z_1, \ldots, Z_n$, we put $Z_t = 0$, $K_t(Z_t) = 0, \ldots$, for $t > n$.

14.2 Maximal inequalities

In what follows, $\mathcal{X}$ will be a separable Banach or Hilbert space endowed with its norm $\|\cdot\|$. We note, for an integer p,

- if $1 \leq p \leq n$, $q = \left[\frac{n}{p}\right]_e + 1$, and
- if $1 \leq p \leq \frac{n}{2}$, $r = \left[\frac{n}{2p}\right]_e + 1$,

where $[x]_e$ denotes the integer part of x. We remark that

$$pq \leq n(1 + p/n) < pq(1 + 1/q)$$

and

$$pr \leq n(1/2 + p/n) < pr(1 + 1/r).$$

14.2.1 The α-mixing case

The finite-dimensional framework and some Hilbertian transformations

Let $X_1, \ldots, X_n$ be n r.v.'s of $\mathbb{R}^d$ endowed with the sup norm noted as $|\cdot|$, i.e. $|x| = \max_{1 \leq i \leq d} |x_i|$ for $x = (x_1, \ldots, x_d) \in \mathbb{R}^d$.

Theorem 14.1 *If $\forall t$, $E(X_t) = 0$, and for $1 \leq p \leq n/2$, $1 \leq i \leq 2r$,*

$$|X_{(i-1)p+1} + \cdots + X_{ip}| \leq M(p) \text{ a.s.,}$$

and
$$E|X_{(i-1)p+1} + \cdots + X_{ip}|^2 \le \sigma_i^2(p),$$
then for all $0 < \theta < 1$, $\delta > 1$, and $\varepsilon > 0$

$$P\left(\max_{1 \le k \le n} \left|\sum_{t=1}^{k} X_t\right| \ge \varepsilon + \left[\frac{p}{2}\right]_e M\right) \le 4 \exp\left(-\frac{(1-\theta)^2 \varepsilon^2}{4[2\tilde{\sigma}_r^2(p) + (1-\theta)\varepsilon M(p)/3]}\right)$$
$$+ C_b\left(\frac{n}{p} + 2\right) \alpha(p),$$

where

$$\tilde{\sigma}_r^2(p) = \max\left(\sum_{i=1}^{r} \sigma_{2i}^2(p), \sum_{i=1}^{r} \sigma_{2i-1}^2(p)\right),$$

$$C_b = 2\sqrt{2}\, 3^{d/2} \left\{\frac{\delta+1}{\delta-1} \max\left(\frac{2(\delta-1)r M(p)}{\theta \varepsilon}, 1\right)\right\}^{d/2}$$

and $M := M(1)$.

Remark 14.1

1. For example, for $\delta = 1 + \frac{\theta \varepsilon}{2r M(p)}$, $C_b = 2\sqrt{2}\, 3^{d/2} \left(1 + \frac{2r M(p)}{\theta \varepsilon}\right)^{d/2}$.
2. If $\sigma_i^2(p) \le p\sigma^2$ then we have, for all $0 < \theta < 1$,

$$P\left(\max_{1 \le k \le n} \left|\sum_{t=1}^{k} X_t\right| \ge n\varepsilon + \left[\frac{p}{2}\right]_e M\right)$$
$$\le 4 \exp\left(-\frac{n(1-\theta)^2 \varepsilon^2}{4[(1 + 2p/n)\sigma^2 + (1-\theta)pM\varepsilon/3]}\right)$$
$$+ 2\sqrt{2}\, 3^{d/2} \left(\frac{2M + \theta\varepsilon}{\theta\varepsilon}\right)^{d/2} \left(\frac{n}{p} + 2\right) \alpha(p).$$

3. If the process is strictly stationary, $C_b = 2\sqrt{2}\, 3^{d/2} \left\{\max\left(\frac{2r M(p)}{\theta\varepsilon}, 1\right)\right\}^{d/2}$.

In what follows, we will be interested in maximal inequalities when the r.v. are transformed by functionals with values in $\mathcal{X}$. Consider a family of functions $K_{t,n} : (\mathbb{R}^d, |\cdot|) \longrightarrow (\mathcal{X}, \|\cdot\|)$ for t, n positive integers, fulfilling:

$$\|K_{t,n}(y) - K_{t,n}(z)\| \le L_{t,n} |y - z| \quad \text{for all } y, z \text{ in } \mathbb{R}^d, \tag{K}$$

where $L_{t,n} > 0$ for $t, n \ge 1$.

Example 14.1 The recursive and nonrecursive kernel estimators of a probability density are based on such transformations. Let $K(\cdot)$ be a Lipschitzian function from $\mathbb{R}^d$ and let $(h_{t,n})$ be a sequence of positive numbers. Given $X_1, \ldots, X_n$, r.v.'s with common density f, the kernel estimate (recursive or not) of f, based on $K(\cdot)$, is

$$\hat{f}_n(x) = \frac{1}{n} \sum_{t=1}^{n} \frac{1}{h_{t,n}^d} K((x - X_t)/h_{t,n}) \quad \text{for } x \in \mathbb{R}^d.$$

When $h_{t,n} = h_n$ we may find the kernel estimate (Parzen–Rosenblatt estimate), but $h_{t,n} = h_t$ corresponds to a recursive estimator (Wolverton and Wagner 1969; Yamato 1971).

Considering f and $\hat{f}_n$ as elements of some functional space $\mathcal{X}$, one can write $\hat{f}_n = (1/n) \sum_{t=1}^{n} K_{t,n}(X_t)$, where

$$K_{t,n} : \mathbb{R}^d \longrightarrow \mathcal{X}, \quad u \longmapsto K_{t,n}(u) \equiv \left\{ x \longmapsto \left(1/h_{t,n}^d\right) K((x-u)/h_{t,n}) \right\}.$$

since $K(\cdot)$ is Lipschitzian, it is obvious that the family $(K_{t,n})$ satisfies condition (K).

When we approximate the (X_t) by independent r.v.'s (coupling), this approximation is independent of x and is valid for all x. The error, uniform with respect to x, when approximating the $\left((1/h_{t,n}^d) K((x - X_t)/h_{t,n})\right)$, i.e. when approximating $(K_{t,n}(X_t))$ by independent r.v.'s, is simply within a multiplicative factor of the approximation error of (X_t) when K is Lipschitzian.

Throughout this chapter we adopt the following notation: for a positive integer p,

$$K_{j,i,n}(\cdot) = K_{j+(i-1)p,n}(\cdot) \quad \text{and} \quad X_{j,t} = X_{j+(t-1)p},$$

with the convention that $K_{t,n}(\cdot) = 0$ and $X_t = 0$ whenever $t > n$.

And for simplicity we omit the index n when there is no ambiguity:

$$K_{t,n} = K_t, \quad K_{j,i,n} = K_{j,i}, \quad L_{t,n} = L_t, \ldots$$

In the case of a Hilbert space $\mathcal{X}$, we obtain

Theorem 14.2 *Under condition* (K), $EK_t(X_t) = 0$, $\|K_t(X_t)\| \leq \tilde{K}$ *a.s., for* $1 \leq p \leq n$

$$\max_{1 \leq j \leq p} \sum_{i=1}^{q} E\|K_{j,i}(X_{j,i})\|^2 \leq q\sigma^2,$$

and if $\exists \gamma > 0$ such that $\|X_t\|_\gamma \leq M_\gamma$, then for all $0 < \theta < 1$, $\delta > 1$, and $\varepsilon > 0$,

$$P\left(\max_{1 \leq k \leq n} \left\| \sum_{t=1}^{k} K_t(X_t) \right\| \geq \varepsilon + \left[\frac{p}{2}\right]_e \tilde{K} \right)$$

$$\leq 2p \exp\left(-\frac{(1-\theta)^2 \varepsilon^2}{2p[(n+p)\sigma^2 + (1-\theta)\varepsilon \tilde{K}/3]} \right) + C_K(\gamma)\, n\, [\alpha_{\triangleleft,1}(p)]^{2\gamma/(2\gamma+d)},$$

where

$$C_K(\gamma) = \left(2\sqrt{2}\, 3^{d/2} + \mathbb{1}_{\gamma<\infty}\right) \left\{\frac{\delta+1}{\delta-1} \max\left(\frac{L(\delta-1)M_\gamma n}{\theta\varepsilon}, 1\right)\right\}^{d\gamma/(2\gamma+d)}$$

and $L = L(n) = \max_{1 \leq t \leq n} L_t$.

In particular, if the (X_t) are almost surely bounded then we have

$$P\left(\max_{1 \leq k \leq n} \left\|\sum_{t=1}^{k} K_t(X_t)\right\| \geq \varepsilon + \left[\frac{p}{2}\right]_e \tilde{K}\right)$$

$$\leq 2p \exp\left(-\frac{(1-\theta)^2 \varepsilon^2}{2p[(n+p)\sigma^2 + (1-\theta)\varepsilon\tilde{K}/3]}\right) + C_K(\infty)\, n\, a_{\triangleleft,1}(p).$$

Remark 14.2 *In the case when $Y_t = K_t(X_t) - E K_t(X_t)$, the assumption (K) must relate only to the function $K_t(\cdot)$. This remark remains valid for all the remainder of this chapter.*

Remark 14.3 *The constants in the above theorem can be chosen as follows:*

- *if we take $\delta = 1 + \theta\varepsilon/(L M_\gamma n)$, then*

$$C_K(\gamma) = \left(2\sqrt{2}\, 3^{d/2} + \mathbb{1}_{\gamma<\infty}\right) \left(1 + \frac{L M_\gamma n}{\theta\varepsilon}\right)^{d\gamma/(2\gamma+d)}.$$

- *in the case of a strictly stationary process,*

$$C_K(\gamma) = \left(2\sqrt{2}\, 3^{d/2} + \mathbb{1}_{\gamma<\infty}\right) \left\{\max\left(\frac{L M_\gamma n}{\theta\varepsilon}, 1\right)\right\}^{d\gamma/(2\gamma+d)}.$$

All the parameters γ, L, K, and σ, as well as the observations, can depend on the number of observations n; so the previous results and those which follow will also apply to triangular arrays.

When the K_t are chosen to be the identity in $\mathbb{R}^d$, Theorem 14.2 gives, for the partial sum of random vectors in $\mathbb{R}^d$, another bound different from that given by Theorem 14.1. It is simpler, in certain cases, to apply this last inequality rather than that of Theorem 14.1; estimating $E\|X_t\|^2$ is simpler than calculating $E\|X_{(i-1)p+1} + \cdots + X_{ip}\|^2$.

The Banach-space case

We assume now that $\mathcal{X}$ is a separable Banach space and put

$$e_j^*(q) = E\left\|\sum_{i=1}^{q} K_{j,i}\left(X_{j,i}^*\right)\right\| \quad j = 1, \ldots, p,$$

where $\left(X_{j,i}^*, i = 1, \ldots, q\right)$ are independent r.v. such that $X_{j,i}^* \stackrel{\text{law}}{=} X_{j,i} = X_{j+(i-1)p}$. We also assume that

$$\|K_t(X_t)\| \leq \tilde{K} \quad \text{a.s.,}$$

and let (z_t) be a sequence of vectors in $\mathcal{X}$. We now give the inequality for the partial sum of the transformed $K_t(X_t)$.

Theorem 14.3 *Under the conditions* (K) *and* $E K_t(X_t) = 0$, $\|K_t(X_t) - z_t\| \leq \tilde{K}_t$ *a.s., if* $\exists \gamma > 0$ *such that* $\|X_t\|_\gamma \leq M_\gamma$, *then for all* $1 \leq p \leq n$, $0 < \theta < 1$, $\delta > 1$, *and* $\varepsilon > p \max_j e_j^*(q)/(1-\theta)$,

$$P\left(\max_{1 \leq k \leq n} \left\|\sum_{t=1}^k K_t(X_t)\right\| \geq \varepsilon + \left[\frac{p}{2}\right]_e \tilde{K}\right)$$

$$\leq p \max_{1 \leq j \leq p} \exp\left(-\frac{[(1-\theta)\varepsilon - pe_j^*(q)]^2}{2p^2\left[\tilde{K}_{j,1}^2 + \cdots + \tilde{K}_{j,q}^2\right]}\right) + C_K(\gamma)\, n\, [\alpha_{\triangleleft,1}(p)]^{2\gamma/(2\gamma+d)},$$

with $C_K(\gamma)$ *as in Theorem 14.2.*

If $\max_j e_j^*(q) \leq e^*(q)$ and $\tilde{K}_t \leq \tilde{K}$, then the inequality of Theorem 14.3 becomes

$$P\left(\max_{1 \leq k \leq n} \left\|\sum_{t=1}^k K_t(X_t)\right\| \geq \varepsilon + \left[\frac{p}{2}\right]_e \tilde{K}\right)$$

$$\leq p \exp\left(-\frac{[(1-\theta)\varepsilon - pe^*(q)]^2}{2p(n+p)\tilde{K}^2}\right) + C_K(\gamma)\, n\, [\alpha_{\triangleleft,1}(p)]^{2\gamma/(2\gamma+d)}.$$

Theorem 14.4 *Under* (K), $E K_t(X_t) = 0$, $\|K_t(X_t) - z_t\| \leq \tilde{K}$ *a.s., for* $1 \leq p \leq n$, $\max_{1 \leq j \leq p} \sum_{i=1}^q E\|K_{j,i}(X_{j,i}) - z_{j,i}\|^2 \leq q\sigma^2$, *if* $\exists \gamma > 0$ *such that* $\|X_t\|_\gamma \leq M_\gamma$, *and* $\max_j e_j^*(q) \leq e^*(q)$, *then for* $0 < \theta < 1$, $\delta > 1$, *and* $\varepsilon > pe^*(q)/(1-\theta)$, *we obtain*

$$P\left(\max_{1 \leq k \leq n} \left\|\sum_{t=1}^k K_t(X_t)\right\| \geq \varepsilon + \left[\frac{p}{2}\right]_e \tilde{K}\right)$$

$$\leq p \exp\left(-\frac{[(1-\theta)\varepsilon - pe^*(q)]^2}{2p[(n+p)\sigma^2 + \{(1-\theta)\varepsilon - pe^*(q)\}\tilde{K}/3]}\right)$$

$$+ C_K(\gamma)\, n\, [\alpha_{\triangleleft,1}(p)]^{2\gamma/(2\gamma+d)},$$

with $C_K(\gamma)$ *as in Theorem 14.2.*

Remark 14.4 *As noted by Yurinskii (1976), for independent observations such an inequality does not have any interest if* $p \max_j e_j^*(q)/n$ *is not bounded; this will depend on the geometry of the Banach space.*

14.2.2 The β-mixing case

We now assume that dependence is measured by the absolutely regular mixing coefficients. In this case we consider a general framework for the r.v. Let $Y_1, \ldots, Y_n$ be centered r.v., bounded, $\|Y_t\| \leq M$ a.s., with values in a separable Hilbert or Banach space $\mathcal{Y}$.

The Hilbertian framework

Firstly we present the results when $\mathcal{Y}$ is a separable Hilbert space.

Theorem 14.5 *If $\forall t$, $E(Y_t) = 0$, and for $1 \leq p \leq n/2$, $1 \leq i \leq 2r$, $\|Y_{(i-1)p+1} + \cdots + Y_{ip}\| \leq M(p)$ a.s., and $E\|Y_{(i-1)p+1} + \cdots + Y_{ip}\|^2 \leq \sigma_i^2(p)$, then for all $\varepsilon > 0$ we have:*

$$P\left(\max_{1 \leq k \leq n} \left\|\sum_{t=1}^{k} Y_t\right\| \geq \varepsilon + \left[\frac{p}{2}\right]_e M\right)$$

$$\leq 4\exp\left(-\frac{\varepsilon^2}{4[2\tilde{\sigma}_r^2(p) + \varepsilon M(p)/3]}\right) + \left(\frac{n}{p} + 2\right)\beta(p),$$

where $\tilde{\sigma}_r^2(p) = \max\left(\sum_{i=1}^{r} \sigma_{2i}^2(p), \sum_{i=1}^{r} \sigma_{2i-1}^2(p)\right)$.

If $\sigma_i^2(p) \leq p\sigma^2$ then

$$P\left(\max_{1 \leq k \leq n} \left\|\sum_{t=1}^{k} Y_t\right\| \geq n\varepsilon + \left[\frac{p}{2}\right]_e M\right)$$

$$\leq 4\exp\left(-\frac{n\varepsilon^2}{4[(1+2p/n)\sigma^2 + pM\varepsilon/3]}\right) + \left(\frac{n}{p} + 2\right)\beta(p).$$

Using the covariance inequality of Hilbertian r.v. (Dehling and Philipp 1982, Lemma 2.2), we deduce a bound of $\sigma^2 \leq M^2\left(1 + 10\sum_{i=1}^{p-1}\beta_{1,1}(i)\right)$.

Remark 14.5 *We also have, if $\forall t$, $E(Y_t) = 0$ and for $1 \leq p \leq n$, $\max_{1 \leq j \leq p} \sum_{i=1}^{q} E\|Y_{j+(i-1)p}\|^2 \leq q\sigma^2$, for $\varepsilon > 0$:*

$$P\left(\max_{1 \leq k \leq n} \left\|\sum_{t=1}^{k} Y_t\right\| \geq \varepsilon + \left[\frac{p}{2}\right]_e M\right)$$

$$\leq 2p\exp\left(-\frac{\varepsilon^2}{2p[(n+p)\sigma^2 + \varepsilon M/3]}\right) + n\,\beta(p). \qquad (14.1)$$

This is another bound, different from that obtained by Theorem 14.5, which, although it is less precise, is useful in some cases: estimating $E\|Y_t\|^2$ is simpler than the evaluation of $E\|Y_{(i-1)p+1} + \cdots + Y_{ip}\|^2$.

The Banachian framework

We now assume that $\mathcal{Y}$ is a separable Banach space; let (z_t) be a sequence of vectors in $\mathcal{Y}$. We then obtain the following theorem:

Theorem 14.6 *If $\forall t$, $E(Y_t) = 0$, for $1 \leq p \leq n/2$ and $1 \leq i \leq 2r$, $\|Y_{(i-1)p+1} + \cdots + Y_{ip} - z_i\| \leq M_i(p)$ a.s., and $\max\left(\sum_{i=1}^{r} M_{2i}^2(p), \sum_{i=1}^{r} M_{2i-1}^2(p)\right) \leq \tilde{M}^2(p)$, then for $\varepsilon > 2e^*$,*

$$P\left(\max_{1 \leq k \leq n} \left\|\sum_{t=1}^{k} Y_t\right\| \geq \varepsilon + \left[\frac{p}{2}\right]_e M\right)$$

$$\leq 2\exp\left(-\frac{(\varepsilon - 2e^*)^2}{8\tilde{M}^2(p)}\right) + \left(\frac{n}{p} + 2\right)\beta(p),$$

where $e^* \geq \max\left(E\left\|\sum_{i=1}^{r} S_{2i}^*\right\|, E\left\|\sum_{i=1}^{r} S_{2i-1}^*\right\|\right)$ and $(S_i^*, i = 1, \ldots, 2r)$ are independent random vectors such that $S_i^* \stackrel{law}{=} Y_{(i-1)p+1} + \cdots + Y_{ip}$.

Theorem 14.7 *If $\forall t$, $E(Y_t) = 0$, for $1 \leq p \leq n/2$ and $1 \leq i \leq 2r$, $\|Y_{(i-1)p+1} + \cdots + Y_{ip} - z_i\| \leq M(p)$ a.s., and $E\|Y_{(i-1)p+1} + \cdots + Y_{ip} - z_i\|^2 \leq \sigma_i^2(p)$, then for $\varepsilon > 2e^*$,*

$$P\left(\max_{1 \leq k \leq n} \left\|\sum_{t=1}^{k} Y_t\right\| \geq \varepsilon + \left[\frac{p}{2}\right]_e M\right)$$

$$\leq 2\exp\left(-\frac{(\varepsilon - 2e^*)^2}{4\left[2\tilde{\sigma}_r^2(p) + (\varepsilon - 2e^*)M(p)/3\right]}\right) + \left(\frac{n}{p} + 2\right)\beta(p),$$

where e^* is that of Theorem 14.6 and $\tilde{\sigma}_r^2(p) = \max\left(\sum_{i=1}^{r} \sigma_{2i}^2(p), \sum_{i=1}^{r} \sigma_{2i-1}^2(p)\right)$.

Remark 14.6 *But if, for $1 \leq p \leq n$, $\|Y_t - z_t\| \leq M_1$ a.s., and*

$$\max_{1 \leq j \leq p} \sum_{i=1}^{q} E\|Y_{j+(i-1)p} - z_{j,i}\|^2 \leq q\sigma^2,$$

then for all $\varepsilon > pe^*(q)$,

$$P\left(\max_{1 \leq k \leq n} \left\|\sum_{t=1}^{k} Y_t\right\| \geq \varepsilon + \left[\frac{p}{2}\right]_e M\right)$$

$$\leq p\exp\left(-\frac{(\varepsilon - pe^*(q))^2}{2p[(n+p)\sigma^2 + (\varepsilon - pe^*(q))M_1/3]}\right) + n\beta_{\triangleleft,1}(p), \quad (14.2)$$

where $e_j^*(q) = E\left\|\sum_{i=1}^{q} Y_{j,i}^*\right\|$ and $\left(Y_{j,i}^*, i = 1, \ldots, q\right)$ are independent random vectors such that $Y_{j,i}^* \stackrel{law}{=} Y_{j,i} = Y_{j+(i-1)p}$, and $e^*(q) = \max_j e_j^*(q)$.

If $\|Y_t - z_t\| \le \tilde{M}_t$ a.s, we can replace the denominator of the exponential argument by $2p^2 \max_{1 \le j \le p} (\tilde{M}_{j,1}^2 + \cdots + \tilde{M}_{j,q}^2)$, as in Theorem 14.3, i.e. we will have

$$P\left(\max_{1 \le k \le n} \left\|\sum_{t=1}^{k} Y_t\right\| \ge \varepsilon + \left[\frac{p}{2}\right]_e M\right)$$

$$\le p \exp\left(-\frac{(\varepsilon - pe^*(q))^2}{2p^2 \max_{1 \le j \le p} (\tilde{M}_{j,1}^2 + \cdots + \tilde{M}_{j,q}^2)}\right) + n\,\beta_{\triangleleft,1}(p). \qquad (14.3)$$

Remark 14.7 Under β-mixing, Theorem 14.5 obtained for Hilbert spaces gives the same bound as in the finite-dimensional Theorem 14.1. To compare the Hilbert and Banach cases obtained in Theorem 14.6, the bound of this latter case is a function of $(\varepsilon - 2e^*)^2$, instead of ε^2 in the Hilbert-space case, where e^* only depends on $E\left\|\sum_{i=1}^{k} X_i\right\|$.

Remark 14.8 Similar inequalities are valid for continuous processes and discrete or continuous random fields taking their values in separable Hilbert or Banach spaces (by exchanging the symbol Σ for the symbol $\int$ for continuous time); constants are slightly better (i.e. smaller) in the case of continuous time.

14.3 Applications

Here we present some applications in the absolutely regular mixing (β-mixing) case. Let $Y_1, \ldots, Y_n$ be centered r.v.'s and bounded, $\|Y_t\| \le M$ a.s., with values in $\mathcal{Y}$; put $S_n = \sum_{i=1}^{n} Y_i$.

14.3.1 The m-dependence case

Assume that the (Y_t) are $(p-1)$-dependent (i.e. for all $i < j$, $(Y_t, t \le i)$ and $(Y_t, t \ge j)$ are independent when $j - i > p - 1$); then $\alpha(p) = \beta(p) = 0$. In this case the previous bounds are actually exponential. For example, when (Y_t) are in a Hilbert space and $\|Y_i\| \le M$ a.s., we have

Corollary 14.1 *If $\forall t$, $E(Y_t) = 0$, $\|Y_{(i-1)p+1} + \cdots + Y_{ip}\| \le M_p$ and*

$$\max_{j=0,1} \sum_{i=1}^{r} E\|Y_{1+(2i-j-1)p} + \cdots + Y_{(2i-j)p}\|^2 \le \sigma^2,$$

then for all $n \geq 2p$ and $\varepsilon > 0$,

$$P\left(\max_{1\leq k\leq n}\left\|\sum_{t=1}^{k}Y_t\right\| \geq \varepsilon + \left[\frac{p}{2}\right]_e M\right) \leq 4\exp\left(-\frac{\varepsilon^2}{4(2\sigma^2 + \varepsilon M_p/3)}\right).$$

And in the Banach case we obtain

Corollary 14.2 *If $\forall t$, $E(Y_t) = 0$, $\|Y_{(i-1)p+1} + \cdots + Y_{ip} - z_{j,i}\| \leq M_p$ and*

$$\max_{j=0,1}\sum_{i=1}^{r} E\|Y_{1+(2i-j-1)p} + \cdots + Y_{(2i-j)p} - z_{j,i}\|^2 \leq \sigma^2,$$

then for all $n \geq 2p$ and $\varepsilon > 2e^*$,

$$P\left(\max_{1\leq k\leq n}\left\|\sum_{t=1}^{k}Y_t\right\| \geq \varepsilon + \left[\frac{p}{2}\right]_e M\right) \leq 2\exp\left(-\frac{(\varepsilon - 2e^*)^2}{4[2\sigma^2 + (\varepsilon - 2e^*)M_p/3]}\right),$$

where e^* is defined in Theorem 14.6.

14.3.2 The strong law of large numbers and bounded LIL

Let $(Y_t)_t$ be a centered and bounded absolutely regular process (not necessarily stationary). We distinguish the cases when the r.v.'s are in Hilbert or in Banach spaces. Let us write $\tilde{\beta}_{1,1}(p) = 1 + 10\sum_{i=1}^{p-1}\beta_{1,1}(i)$; $\log_2 x = \log\log x$, for $x > e$, and $\log_2^a x = (\log\log x)^a$, $a \in \mathbb{R}$.

Corollary 14.3 *If $\mathcal{Y}$ is a Hilbert space and $(Y_i)_i$ are bounded then:*

1. $\max_{1\leq k\leq n}\|S_k/n\| \longrightarrow 0$ *a.s. when there exists $p = p_n \leq n/2$ such that $(n/p_n)_n$ is not decreasing, $p\log\log n/n \to 0$, and $\sum_n \beta(p)/p < \infty$.*
2. *If there exists $p = p_n \leq n/2$ such that $\sum_n \beta(p)/p < \infty$, then:*
    (i) *If $\frac{p^2 \log\log n}{n\tilde{\beta}_{1,1}(p)} \to 0$ then*

    $$\limsup_n \frac{1}{\sqrt{\tilde{\beta}_{1,1}(p)n\log\log n}}\max_{1\leq k\leq n}\|S_k\| \leq 2M \text{ a.s.}$$

    (ii) *If $\limsup_n \frac{p^2 \log\log n}{n\tilde{\beta}_{1,1}(p)} \leq c_0^2$ then*

    $$\limsup_n \frac{1}{\sqrt{\tilde{\beta}_{1,1}(p)n\log\log n}}\max_{1\leq k\leq n}\|S_k\| \leq 2M\left(c_0/3 + \sqrt{c_0/9 + 1}\right) \text{ a.s.}$$

    (iii) *If $\lim_n \frac{p^2 \log\log n}{n\tilde{\beta}_{1,1}(p)} \to \infty$, with $(n/p_n)_n \nearrow$, then*

    $$\limsup_n \frac{1}{p\log\log n}\max_{1\leq k\leq n}\|S_k\| \leq 4M/3 \text{ a.s.}$$

3. In particular, if $\beta(n) = O\left(n^{-a}\log^{-b} n \log_2^{-c} n\right)$ with $(a > 1$ and $b, c \in \mathbb{R})$ or $(a = 1, b > 1,$ and $c \in \mathbb{R})$ or $(a = b = 1$ and $c > 2)$, then

$$\limsup_n \frac{1}{\sqrt{n \log\log n}} \max_{1 \leq k \leq n} \|S_k\| \leq 2M\sqrt{1 + 10 \sum_i \beta_{1,1}(i)} \quad \text{a.s.}$$

In 2.(i) and 3. above, $M = \min\left(\max_i \|Y_i\|_{2,\infty}, \max_i \|Y_i - EY_i\|_{2,\infty}\right)$, where $\|Y\|_{2,\infty} = \operatorname{ess\,sup} \|Y\|_2$.

Example 14.2 If $\beta(n) = O\left(\log^{-1} n \log_2^{-c} n\right)$, with $c > 2$, then $\max_{1 \leq k \leq n} \|S_k/n\| \longrightarrow 0$, a.s.; we choose $p_n = \left[\eta_n n \log_2^{-1} n\right]_e$ with $\eta_n \to 0$ such that $\liminf_n \eta_n \log_2^{\delta} n > 0$ and $0 < \delta < c - 1$.

Remark 14.9

1. If, in addition, the process is strictly stationary, in 3 above we can replace $M\sqrt{1 + 10 \sum_i \beta_{1,1}(i)}$ by $\sqrt{\sum_{i \in \mathbb{Z}} E\langle Y_0, Y_i\rangle}$, which is absolutely convergent to a positive number $\sigma^2 > 0$, since X_0 is not a.s. a constant.
2. If there exists $0 < \kappa < 1$ such that $\sum_n \beta^\kappa(n) < \infty$, then another bound may be given for S_n:

$$\limsup_n \frac{1}{\sqrt{n \log\log n}} \max_{1 \leq k \leq n} \|S_k\|$$

$$\leq 2 \min\left(M\sqrt{1 + 10 \sum_i \beta_{1,1}(i)}, \max(E\|Y_i\|^{2/(1-\kappa)})^{(1-\kappa)/2}\sqrt{1 + 15 \sum_i \beta_{1,1}^\kappa(i)}\right) \quad \text{a.s.}$$

3. In Corollary 14.3, the bound M may depend on n (see, for example, the next section on density estimation); with truncations, we can extend 1. of this corollary to unbounded r.v.'s.

These results improve on those in Rhomari (2002a, 2002b).

Now let us establish the SLLN in Banach spaces: $\mathcal{Y}$ is a separable Banach space of type τ, $1 < \tau \leq 2$ (see the appendix below for definitions, and for details see Ledoux and Talagrand (1991)).

Corollary 14.4 ($\mathcal{Y}$ *is a Banach space of type* τ) We assume that (Y_i) is bounded, and that there exists $p = p_n \leq n/2$ such that $(n/p_n)_n \nearrow$ and $\sum_n \beta(p)/p < \infty$. Then

$$\left\|\frac{S_n}{n}\right\| = O\left(\left(\frac{p \log\log n}{n}\right)^{1/2} + \left(\frac{p}{n}\right)^{1-1/\tau}\right) \quad \text{a.s.}$$

So, if $p \log\log n / n \to 0$ we obtain $S_n/n \longrightarrow 0$, a.s. In particular, if $\beta_{\triangleleft,1}(n) = O(\varrho^n)$, $0 < \varrho < 1$, then

$$\left\|\frac{S_n}{n}\right\| = O\left(\left(\frac{\log n \log\log n}{n}\right)^{1/2} + \left(\frac{\log n}{n}\right)^{1-1/\tau}\right) \quad \text{a.s.}$$

If $\|Y_i\| \leq M$, for all $A > 2M$, and for n large enough, we have

$$\left\|\frac{S_n}{n}\right\| \leq A\sqrt{\frac{p \log \log n}{n}} + \left(\frac{c_\tau}{2}\right)^{1/\tau} \frac{A}{M} \max_i (E\|Y_i\|^\tau)^{1/\tau} \left(\frac{p}{n}\right)^{1-1/\tau} + \frac{Ap_n}{4n} \quad \text{a.s.,}$$

where the constant c_τ is that induced by the type τ of the Banach space under consideration.

And if $\beta_{\triangleleft,1}(n) = O(\varrho^n)$, $0 < \varrho < 1$, then we choose $p = C \log n$ with $C > 2/\log(1/\varrho)$.

Remark 14.10 *This bound may be improved if we can finely bound $E\|Y_{(i-1)p+1} + \cdots + Y_{ip}\|^\tau$. This is possible, for example, in the case where one can approximate the random variables of a Banach space by random variables from a Hilbert space (for example, subspaces of L^s, $s \geq 1$); using this idea, in work on recursive estimation in L^1 (Rhomari 2002b), we have proved the a.s. convergence of the density estimate under similar conditions to those of Theorem 14.8, and with similar rates to those in Corollary 14.6.*

Remark 14.11 *If $\mathcal{Y}$ is a separable Banach space and (Y_i) are r.v.'s such that there exists $\sigma > 0$ satisfying: $E\left\|\sum_{i=1+k}^{p+k} Y_i\right\|^2 \leq \sigma^2 p$ for all $k \geq 0$ and $p \geq 1$, then we get similar bounds to those of Corollary 14.3, i.e. the bounded LIL.*

14.3.3 Covariance operator estimation of Hilbert- or Banach-valued r.v.

In this section we briefly present the a.s. convergence and the bounded LIL of the empirical estimate of the covariance operator and of its eigenvalues, in the case of a Hilbert space and for absolutely regular processes. On covariance operators, see Bosq (2000); for more detail on their properties and characteristics see, for example, Ledoux and Talagrand (1991) or Vakhania et al. (1987) and the references therein.

Let $(Y, Z) \in \mathcal{Y} \times \mathcal{Z}$ be r.v.'s and let C_Y be a covariance operator of Y. Then

if $\mathcal{Y}$ is a Hilbert space, $\quad C_Y : v \in \mathcal{Y} \longmapsto E(\langle Y, v \rangle Y) \in \mathcal{Y},$

if $\mathcal{Y}$ is a Banach space, $\quad C_Y : v^* \in \mathcal{Y}^* \longmapsto E(v^*(Y)Y) \in \mathcal{Y},$

where $\mathcal{Y}^*$ is a dual of $\mathcal{Y}$. Let $D_{Y,Z}$ a cross covariance operator of (Y, Z), then

if $\mathcal{Y}$ is a Hilbert space, $\quad D_{Y,Z} : v \in \mathcal{Y} \longmapsto E(\langle Y, v \rangle Z) \in \mathcal{Z},$

if $\mathcal{Y}$ is a Banach space, $\quad D_{Y,Z} : v^* \in \mathcal{Y}^* \longmapsto E(v^*(Y)Z) \in \mathcal{Z}.$

The natural estimators, based on (Y_i, Z_i) for $i = 1, \ldots, n$ and n observations of (Y, Z), when $\mathcal{Y}$ is a Hilbert space, are defined on $\mathcal{Y}$ by:

$$C_n(\cdot) = (1/n) \sum_{i=1}^{n} \langle Y_i, \cdot \rangle Y_i = (1/n) \sum_{i=1}^{n} Y_i \otimes Y_i(\cdot),$$

$$D_n(\cdot) = (1/n) \sum_{i=1}^{n} \langle Y_i, \cdot \rangle Z_i = (1/n) \sum_{i=1}^{n} Y_i \otimes Z_i(\cdot).$$

But if $\mathcal{Y}$ is a Banach space, for $y^* \in \mathcal{Y}^*$, the dual of $\mathcal{Y}$,

$$C_n(y^*) = (1/n) \sum_{i=1}^{n} y^*(Y_i) Y_i = (1/n) \sum_{i=1}^{n} Y_i \otimes Y_i(y^*),$$

$$D_n(y^*) = (1/n) \sum_{i=1}^{n} y^*(Y_i) Z_i = (1/n) \sum_{i=1}^{n} Y_i \otimes Z_i(y^*).$$

Notation: for $y \in \mathcal{Y}$, a Hilbert space, and for $z \in \mathcal{Z}$, $y \otimes z : v \in \mathcal{Y} \mapsto \langle y, v \rangle z \in \mathcal{Z}$. And if $y \in \mathcal{Y}$, a Banach space, $y \otimes z : v^* \in \mathcal{Y}^* \mapsto v^*(y) z \in \mathcal{Z}$. $C_Y, C_n \in \mathcal{L}_b(\mathcal{Y}^*, \mathcal{Y})$ and $D_{Y,Z}, D_n \in \mathcal{L}_b(\mathcal{Y}^*, \mathcal{Z})$; when $\mathcal{Y}$ is a Hilbert space, $\mathcal{Y}^* \equiv \mathcal{Y}$.

If $((Y_i, Z_i))_i$ are β-mixing, as in Corollaries 14.3 and 14.4, then C_n and D_n are consistent with the bounds cited in these corollaries. In the framework of Hilbertian space, the norm being considered is that of Hilbert–Schmidt, which raises the linear norm of compact bounded operators (cf., e.g., Bosq 2000). From this we can deduce the a.s. convergence of the empirical estimator of the associated eigenvalues. These results improve on those in Bosq (2000) given for bounded ARH (autoregressive Hilbertian) processes which are exponentially β-mixing under certain conditions, cf. Allam and Mourid (2001).

For both the Hilbertian framework and for Banach spaces, the convergence is also valid with nuclear norms (or simply with uniform norms associated to bounded linear operators). As an example, we state the following corollary dealing with C_n when $\mathcal{Y}$ is a Hilbert space. The other estimates are also direct applications either of Corollary 14.3 or Corollary 14.4.

Denote by $\lambda_1 \geq \lambda_2 \geq \cdots \geq 0$ the eigenvalues of C_Y and by $\lambda_{1,n} \geq \lambda_{2,n} \geq \cdots \geq \lambda_{n,n} \geq 0 = \lambda_{n+1,n} = \lambda_{n+2,n} = \cdots$ those of C_n. The range of C_n is of finite dimension, at most n. Here $\|\cdot\|_\mathcal{S}$ is a Hilbert–Schmidt norm and $\beta(\cdot)$ are the absolutely regular coefficients associated to the process (Y_i); $\widetilde{\beta}_{1,1}(p) = 1 + 10 \sum_{i=1}^{p-1} \beta_{1,1}(i)$.

Corollary 14.5 ($\mathcal{Y}$ *a Hilbert space*) *Assume that* $(Y_i)_i$ *is bounded by* M, *and that there exists* $p = p_n \leq n/2$ *such that* $\sum_n \beta(p)/p < \infty$. *Then*

1. *If* $(n/p_n)_n$ *is nondecreasing and* $p \log \log n / n \to 0$, *we have*

$$\|C_n - C_Y\|_\mathcal{S} \longrightarrow 0 \quad \text{and} \quad \sup_{i \geq 1} |\lambda_{i,n} - \lambda_i| \longrightarrow 0 \quad \text{a.s.} \quad \text{when } n \to \infty.$$

2. Furthermore,
   (i) if $\limsup_n \frac{p^2 \log\log n}{n\bar{\beta}_{1,1}(p)} \leq c_0^2$ then

$$\limsup_n \sqrt{\frac{n}{\bar{\beta}_{1,1}(p) \log\log n}} \sup_{i \geq 1} |\lambda_{i,n} - \lambda_i|$$

$$\leq \limsup_n \sqrt{\frac{n}{\bar{\beta}_{1,1}(p) \log\log n}} \|C_n - C_Y\|_{\mathcal{S}}$$

$$\leq 2M^2 \left(c_0/3 + \sqrt{c_0/9 + 1}\right) \quad \text{a.s.;}$$

   (ii) if $\lim_n \frac{p^2 \log\log n}{n\bar{\beta}_{1,1}(p)} \to \infty$ with $(n/p_n)_n \nearrow$, then

$$\limsup_n \frac{n}{p \log\log n} \sup_{i \geq 1} |\lambda_{i,n} - \lambda_i|$$

$$\leq \limsup_n \frac{n}{p \log\log n} \|C_n - C_Y\|_{\mathcal{S}} \leq 4M^2/3 \quad \text{a.s.}$$

3. In particular, if $\beta(n) = O\left(n^{-a} \log^{-b} n \log_2^{-c} n\right)$ with $(a > 1$ and $b, c \in \mathbb{R})$ or $(a = 1, b > 1,$ and $c \in \mathbb{R})$ or $(a = b = 1$ and $c > 2)$, then

$$\limsup_n \sqrt{\frac{n}{\log\log n}} \sup_{i \geq 1} |\lambda_{i,n} - \lambda_i| \leq \limsup_n \sqrt{\frac{n}{\log\log n}} \|C_n - C_Y\|_{\mathcal{S}}$$

$$\leq 2M^2 \sqrt{1 + 10 \sum_i \beta_{1,1}(i)} \quad \text{a.s.}$$

This follows from Corollary 14.3, since $\|Y_i \otimes Y_i\|_{\mathcal{S}} = \|Y_i\|^2 \leq M^2$ and $\sup_{i \geq 1} |\lambda_{i,n} - \lambda_i| \leq \|C_n - C_Y\|_{\mathcal{S}}$ (see, for example, Bosq (2000, pp. 96 and 103–4)).

The ARH and ARB (autoregressive Banach) processes fulfil the mixing assumptions, under some conditions on the error process and the coefficient operators of the models; cf. Allam and Mourid (2001).

Remark 14.12 *Therefore the empirical covariance operator is strongly consistent when $\beta(n) = O\left(\log^{-1} n \log_2^{-c} n\right)$, with $c > 2$ (see Example 14.2). This improves on the results in Rhomari (2002a, 2002b). For bounded ARH processes, Bosq (2000) has obtained the bound $O\left(\sqrt{\log n/n}\right)$.*

14.3.4 Recursive kernel density estimates in L^2

The case of β-mixing

As an application of Corollary 14.3 we give the a.s. convergence, under weak mixing conditions (a logarithmic decay suffices), of the recursive kernel estimate of probability density in L^2 for any d-dimensional processes whose marginal density is in L^2. The rate of convergence is nearly optimal under the mixing in 3. below (K in L^2). With the notation of Example 14.1 and $h_{t,n} = h_t \searrow$ we have

$$\| \hat{f}_n - E \hat{f}_n \|_2^2 = O\left(\frac{\log\log n}{n h_n^d}\right) \text{ a.s.}$$

More precisely, let $K(\cdot)$ be a function on $\mathbb{R}^d$, belonging to L^2, and (h_n) be a sequence of positive numbers. Given $X_1, \ldots, X_n$, r.v.'s in $\mathbb{R}^d$, with common density f in L^2, we consider the recursive kernel estimate of f based on $K(\cdot)$, defined by

$$\hat{f}_n(x) = \frac{1}{n}\sum_{t=1}^{n} \frac{1}{h_t^d} K\left(\frac{x - X_t}{h_t}\right) \quad \text{for } x \in \mathbb{R}^d.$$

Denote $\underline{h}_n = \min_{1 \le i \le n} h_i$ and assume, for the sake of simplicity, that $n\underline{h}_n \nearrow \infty$.

Then, if the (X_i) are β-mixing, sufficient conditions which are very close to the independent case, we obtain the a.s. convergence in L^2 (Theorem 14.8) of the recursive kernel estimate of probability density. Then we specify an upper bound for the L^2-error (Corollary 14.6). Recall that $\tilde{\beta}_{1,1}(p) = 1 + 10 \sum_{i=1}^{p-1} \beta_{1,1}(i)$.

Theorem 14.8 *If there exists $p = p_n \le n/2$ such that $\sum_n \beta(p)/p < \infty$, then*

$$\| \hat{f}_n - E \hat{f}_n \|_2^2 = \int_{\mathbb{R}^d} |\hat{f}_n(x) - E \hat{f}_n(x)|^2 \, dx \longrightarrow 0 \quad \text{a.s.,}$$

under one of the following conditions:

(i) $\limsup_n \frac{p_n \underline{h}_n^{d/2}}{\tilde{\beta}_{1,1}(p_n)} < \infty$, $\frac{n\underline{h}_n^d}{\tilde{\beta}_{1,1}(p_n)} \nearrow$, and $\frac{n\underline{h}_n^d}{\tilde{\beta}_{1,1}(p_n)\log\log n} \to \infty$;

(ii) $\lim_n \frac{p_n \underline{h}_n^{d/2}}{\tilde{\beta}_{1,1}(p_n)} \to \infty$, $\frac{n\underline{h}_n^{d/2}}{p_n} \nearrow$, and $\frac{n\underline{h}_n^{d/2}}{p_n\log\log n} \to \infty$.

If in addition $\int K(x)\,dx = 1$ and $h_n \to 0$, then $\| \hat{f}_n - f \|_2 \longrightarrow 0$ a.s.

If $\tilde{\beta}_{1,1}(\infty) = 1 + 10\sum_{i=1}^{\infty} \beta_{1,1}(i) < \infty$, we find the well-known optimal condition of a.s. convergence obtained for i.i.d. (independent and identically distributed) random variables.

Corollary 14.6

1. *If there exists $p = p_n \le n/2$ such that $\sum_n \beta(p)/p < \infty$, then:*

   (i) *if $\frac{p_n^2 \log_2 n}{n\tilde{\beta}_{1,1}(p_n)} \to 0$, then*

   $$\limsup_n \sqrt{\frac{n\underline{h}_n^d}{\tilde{\beta}_{1,1}(p_n)\log_2 n}} \| \hat{f}_n - E\hat{f}_n \|_2 \le 2\|K\|_2 \quad a.s;$$

   (ii) *if $\limsup_n \frac{p_n^2 \log_2 n}{n\tilde{\beta}_{1,1}(p_n)} \le c_0^2$, then*

   $$\limsup_n \sqrt{\frac{n\underline{h}_n^d}{\tilde{\beta}_{1,1}(p_n)\log_2 n}} \| \hat{f}_n - E\hat{f}_n \|_2 \le 2\|K\|_2 (2c_0/3 + \sqrt{4c_0^2/9 + 1}) \quad a.s;$$

(iii) if $\frac{p_n^2 \log_2 n}{n\bar{\beta}_{1,1}(p_n)} \to \infty$ and $(n/p_n)_n \nearrow$, then

$$\limsup_n \frac{nh_n^{d/2}}{p_n \log_2 n} \|\hat{f}_n - E\hat{f}_n\|_2 \leq 8\|K\|_2/3 \text{ a.s.}$$

2. In particular, under the mixing used in 3. of Corollary 14.3, with ($a > 1$ and $b, c \in \mathbb{R}$) or ($a = 1, b > 1$, and $c \in \mathbb{R}$) or ($a = b = 1$ and $c > 2$), then

$$\limsup_n \sqrt{\frac{nh_n^d}{\log\log n}} \|\hat{f}_n - E\hat{f}_n\|_2 \leq 2\|K\|_2 \sqrt{1 + 10\sum_i \beta_{1,1}(i)} \text{ a.s.}$$

Remark 14.13 (Bias) We only give the bound of the random part, because the bias is purely deterministic, so its study is the same as in the independent case. The bias is $f - E\hat{f}_n = \frac{1}{n}\sum_{t=1}^{n}(f - f * K_{h_t})$; here $*$ designates the convolution operator, and $K_{h_t}(u) = (1/h_t^d)K(u/h_t)$. Its behavior in L^2 reflects the properties of approximate identity in L^2. The recursive estimate is asymptotically unbiased in L^2, since $\int K(x)\,dx = 1$, $f \in L^2$, and $h_n \to 0$ when $n \to \infty$ (see, for example, Wheeden and Zygmund (1977)).

On the other hand, a precise evaluation (rate) of bias requires regularity conditions on f; its study is the same as in the independent framework, that is, it is deterministic (see, for example, Wegman and Davies (1979)).

Remark 14.14 Even if h_n is not decreasing we can replace $\underline{h}_n$ by h_n in the conditions of Theorem 14.8 above, by means of some monotonicity assumptions:

- in (i), if $\frac{n}{\bar{\beta}_{1,1}(p_n)} \nearrow$ then $\frac{nh_n^d}{\bar{\beta}_{1,1}(p_n)\log\log n} \to \infty \iff \frac{nh_n^d}{\bar{\beta}_{1,1}(p_n)\log\log n} \to \infty$;
- in (ii), if $\frac{p_n}{\bar{\beta}_{1,1}(p_n)} \nearrow$ then $\frac{p_n h_n^{d/2}}{\bar{\beta}_{1,1}(p_n)} \to \infty \iff \frac{p_n h_n^{d/2}}{\bar{\beta}_{1,1}(p_n)} \to \infty$;
- in (ii), if $\frac{n}{p_n} \nearrow$ then $\frac{nh_n^{d/2}}{p_n \log\log n} \to \infty \iff \frac{nh_n^{d/2}}{p_n \log\log n} \to \infty$.

These equivalences are derived from the following simple proposition (see, for example, Devroye and Györfi (1985, Lemma 3 on p. 200)):
Let $u_n, v_n \geq 0$ and $u_n \nearrow$, then $u_n v_n \to \infty \iff u_n(\min_{1 \leq i \leq n} v_i) \to \infty$.
However, in practice h_n is often decreasing, and $\underline{h}_n = h_n$.

14.3.5 The α-mixing case

Under strong mixing dependence, we lose a little on the mixing conditions and the rate of convergence. Here we just give sufficient conditions for a.s. convergence.

We retain the notation of the previous section and we now assume that K is a Lipschitzian function of parameter L with compact support.

Theorem 14.9 *Suppose that there exists $\gamma > 0$ such that $\sup_t E|X_t|^\gamma < \infty$ and $\limsup \underline{h}_n < \infty$. If there exists $p = p_n \leq n$ such that*

$$h_n^{-(d/2+1)d/2} a_{\triangleleft,1}(p_n) \searrow \text{ and } \sum_n \left[h_n^{-(d/2+1)d/2} a_{\triangleleft,1}(p_n)\right]^{2\gamma/(2\gamma+d)} < \infty,$$

and if

$$\frac{n\underline{h}_n^d}{p_n} \nearrow \text{ and } \frac{n\underline{h}_n^d}{p_n(\log p_n + \log\log n)} \to \infty,$$

then

$$\|\hat{f}_n - E\hat{f}_n\|_2 \longrightarrow 0 \quad \text{a.s.}$$

If in addition $\int K(x)\,dx = 1$ and $h_n \to 0$, then $\|\hat{f}_n - f\|_2 \to 0$ a.s. If the (X_t) are bounded, the mixing condition becomes

$$\sum_n h_n^{-(d/2+1)d/2} a_{\triangleleft,1}(p_n) < \infty.$$

Remark 14.15 In Rhomari (2002b), we prove L^1-a.s. convergence, under the condition $\sum_n \left[h_n^{-d/2} a_{\triangleleft,1}(p_n)\right]^{2\gamma/(2\gamma+d)} < \infty$. But when the r.v.'s $(X_i)_i$ are bounded, the reduced condition $\sum_n h_n^{-d/2} a_{\triangleleft,1}(p_n) < \infty$ suffices.

On the other hand, for absolutely regular mixing, a.s. convergence is obtained under $\sum_n \beta(p)/p < \infty$.

14.4 Proofs

We begin by proving the applications; proofs of the maximal inequalities are given at the end.

14.4.1 Proofs of applications

We first recall two simple properties of the convergence of series of positive terms. Take an increasing function $\varphi : \mathbb{N} \longrightarrow \mathbb{N}$ with $\varphi(n) \nearrow \infty$, and a decreasing sequence of positive numbers $u_n \geq 0$, then it is easy to see that

$$\sum_n (\varphi(n+1) - \varphi(n))u_{\varphi(n)} < \infty$$

$$\Rightarrow \sum_n u_n < \infty$$

$$\Rightarrow \sum_n (\varphi(n+1) - \varphi(n))u_{\varphi(n+1)} < \infty.$$

- If $\sum_n u_n < \infty$ then $\forall \varepsilon > 0, \exists N_\varepsilon > 0$ such that $\forall n \geq N_\varepsilon$: $0 \leq u_n \leq \varepsilon/n$.

Proof of Corollary 14.3 $\|Y_i\| \leq M$

Put $\Delta_n = \max_{1 \leq k \leq n} \left\|\sum_{t=1}^k Y_t\right\|$. Since Δ_n grows with respect to n, then in order that $\Delta_n/n \to 0$ it suffices that there exists a subsequence increasing to infinity, $n_k \to \infty$ as $k \to \infty$, such that $\limsup_{k \to \infty} n_{k+1}/n_k < \infty$ and $\Delta_{n_k}/n_k \to 0$. (Indeed $\forall n, \exists k$ such that $n_k \leq n \leq n_{k+1}$ and $0 \leq \Delta_n/n \leq (\Delta_{n_{k+1}}/n_{k+1})(n_{k+1}/n_k)$). We show that $\Delta_{n_k}/n_k \to 0$ for $n_k = 2^k$.

1. For a.s. convergence: applying Theorem 14.5, with $\varepsilon = A 2^k$, $A > 0$, by roughly bounding $\tilde{\sigma}_r^2(p) \leq pnM^2$ (using the notation of Theorem 14.5), we have

$$P\left(\Delta_{2^k} \geq A 2^k + \left[\frac{p_{2^k}}{2}\right]_e M\right)$$

$$\leq 4\exp\left(-\frac{A^2 2^k}{4M p_{2^k}(2M + A/3)}\right) + 2(2^k/p_{2^k})\beta(p_{2^k}). \tag{14.4}$$

The series of general terms on the right-hand side of (14.4) is summable if and only if

(a) $\sum_n \beta(p_n)/p_n < \infty$, and
(b) $\sum_n n^{-1} \exp\left(-\frac{A^2 n}{2M p_n(2M+A/3)}\right) < \infty.$

This holds because, for positive increasing $(u_n)_n$, we have $\sum_n u_n < \infty$ if and only if $\sum_n 2^k u_{2^k} < \infty$ and $(\beta(p_n)/p_n)_n$ is decreasing. Moreover, since $(n/p_n)_n$ grows, then

$$2^{-1} \sum_n \exp\left(-\frac{A^2 2^k}{4M p_{2^k}(2M + A/3)}\right) \leq \sum_n n^{-1} \exp\left(-\frac{A^2 n}{2M p_n(2M + A/3)}\right).$$

But (b) is convergent for all $A > 0$ if $p \log \log n/n \to 0$. By the Borell–Cantelli lemma we conclude that since $p_n/n \to 0$, $\Delta_{2^k}/2^k \to 0$ a.s.; hence $\Delta_n/n \to 0$ a.s.

2. Proposition (i) of Corollary 14.3 being a particular case of (ii), we give the bound in terms of c_0. Firstly, by the covariance inequality of Hilbertian r.v.'s (Dehling and Philipp 1982, Lemma 2.2), we have the upper bound:

$$E\|Y_{(i-1)p+1} + \cdots + Y_{ip}\|^2 \leq pM^2\left(1 + 10\sum_{i=1}^{p-1} \beta_{1,1}(i)\right) = pM^2\tilde{\beta}_{1,1}(p).$$

(ii) Let us show that for $n_k = [\rho^k]_e$, $\rho > 1$, and under (ii) of Corollary 14.3, we have $\lim_{k \to \infty} \frac{\tilde{\beta}_{1,1}(p_{n_k})}{\tilde{\beta}_{1,1}(p_{n_{k+1}})} = 1$; p_n is as in Corollary 14.3, i.e. $p_n \leq n/2$ grows to infinity, such that $\sum_n \beta(p_n)/p_n < \infty$. Indeed, as $(\beta(p_n)/p_n)_n$ is decreasing and $\sum_n \beta(p_n)/p_n < \infty$, then for n large enough, $\beta(p_n)/p_n \leq 1/n$ and therefore $\beta(p_{n_k}) \leq p_{n_k}/n_k$ for large k. We remark that

$$\rho(1 - \rho^{-k-1}) \leq \frac{n_{k+1}}{n_k} \leq \frac{\rho}{1 - \rho^{-k}} \leq 2\rho$$

for $k \geq \log 2/\log \rho$ and $\lim_{k \to \infty} n_{k+1}/n_k = \rho$. On the other hand, since $\beta_{1,1}(i) \leq \beta(i)$ and $(\beta_{1,1}(i))_i$ is decreasing, and

$$1 \leq \tilde{\beta}_{1,1}(p_{n_{k+1}}) = \tilde{\beta}_{1,1}(p_{n_k}) + 10 \sum_{i=p_{n_k}}^{p_{n_{k+1}}-1} \beta_{1,1}(i)$$

$$\leq \tilde{\beta}_{1,1}(p_{n_k}) + 10(p_{n_{k+1}} - p_{n_k})\beta_{1,1}(p_{n_k})$$

$$\leq \tilde{\beta}_{1,1}(p_{n_k}) + 10 p_{n_{k+1}} p_{n_k}/n_k,$$

then for k large enough,

$$1 \leq \frac{\tilde{\beta}_{1,1}(p_{n_k})}{\tilde{\beta}_{1,1}(p_{n_{k+1}})} + 10 \frac{p_{n_{k+1}}^2}{\tilde{\beta}_{1,1}(p_{n_{k+1}}) n_k}$$

$$\leq 1 + 20 \frac{\rho p_{n_{k+1}}^2}{\tilde{\beta}_{1,1}(p_{n_{k+1}}) n_{k+1}} = 1 + O\left(\frac{1}{\log_2 n_{k+1}}\right),$$

because under (ii) of Corollary 14.3,

$$\frac{p_{n_{k+1}}^2}{\tilde{\beta}_{1,1}(p_{n_{k+1}}) n_{k+1}} = O(1/\log_2 n_{k+1})$$

and this implies

$$\tilde{\beta}_{1,1}(p_{n_k})/\tilde{\beta}_{1,1}(p_{n_{k+1}}) \to 1 \text{ when } k \to \infty.$$

With the notation of Theorem 14.5, we have $2\tilde{\sigma}_r^2(p) \leq n(1 + 2p/n)M^2\tilde{\beta}_{1,1}(p)$, and this theorem allows, for all $A > 0$,

$$P\left(\Delta_{n_{k+1}} \geq A\sqrt{M^2\tilde{\beta}_{1,1}(p_{n_k})n_k \log_2 n_k} + \left[\frac{p_{n_{k+1}}}{2}\right]_e M\right)$$

$$\leq 4\exp\left(-\frac{A^2 n_k \tilde{\beta}_{1,1}(p_{n_k}) \log_2 n_k}{4n_{k+1}\tilde{\beta}_{1,1}(p_{n_{k+1}})\left[1 + \frac{2p_{n_{k+1}}}{n_{k+1}} + \frac{A}{3}\sqrt{\frac{n_k p_{n_{k+1}}^2 \log_2 n_k}{n_{k+1}^2 \tilde{\beta}_{1,1}(p_{n_{k+1}})}}\right]}\right)$$

$$+ \left(\frac{n_{k+1}}{p_{n_{k+1}}} + 2\right)\beta(p_{n_{k+1}}).$$

$=: (14.5\text{I}) + (14.5\text{II}).$ \hfill (14.5)

The last term (14.5II) of (14.5) is summable (see 1 above). And (14.5I) is summable, under (ii) of Corollary 14.3, if $\exp\left(-(A^2 \log_2 n_k)/\left(4\rho\left(1 + \frac{Ac_0}{3}\right)\right)\right)$ is also summable, which is the case if $A^2/\left(4\rho\left(1 + \frac{Ac_0}{3}\right)\right) > 1$, because if

$$\limsup_n \frac{p_n^2 \log_2 n}{n\widetilde{\beta}_{1,1}(p_n)} \leq c_0^2,$$

then $\frac{2p_{n_{k+1}}}{n_{k+1}} \to 0$ and

$$\limsup_k \frac{n_k p_{n_{k+1}}^2 \log_2 n_k}{n_{k+1}^2 \widetilde{\beta}_{1,1}(p_{n_{k+1}})} \leq \frac{p_{n_{k+1}}^2 \log_2 n_{k+1}}{n_{k+1} \widetilde{\beta}_{1,1}(p_{n_{k+1}})} \leq c_0^2.$$

Consequently, (14.5I) converges if $A^2 - 4\rho c_0 A/3 - 4\rho > 0$, i.e. $A > 2\rho c_0/3 + 2\sqrt{\rho^2 c_0^2/9 + \rho}$. Therefore we choose $A > 2c_0/3 + 2\sqrt{c_0^2/9 + 1}$, and ρ such that $1 < \rho < A^2/(4Ac_0/3 + 4)$.

(iii) But if $\frac{p_n^2 \log_2 n}{\widetilde{\beta}_{1,1}(p_n)n} \to \infty$, then

$$P\left(\Delta_{n_{k+1}} \geq Ap_{n_k} \log_2 n_k + \left[\frac{p_{n_{k+1}}}{2}\right]_e M\right)$$

$$\leq 4\exp\left(-\frac{3Ap_{n_k} \log_2 n_k}{4Mp_{n_{k+1}}\left[\frac{6M\widetilde{\beta}_{1,1}(p_{n_{k+1}})n_{k+1}}{Ap_{n_{k+1}} p_{n_k} \log_2 n_k} + 1\right]}\right) \quad (14.6)$$

$$+ (n_{k+1}/p_{n_{k+1}} + 2)\beta(p_{n_{k+1}}).$$

First, for $k \geq \log 2/\log \rho$, we have

$$\frac{\widetilde{\beta}_{1,1}(p_{n_{k+1}})n_{k+1}}{p_{n_{k+1}} p_{n_k} \log_2 n_k} \leq \frac{2\rho\widetilde{\beta}_{1,1}(p_{n_k})n_k}{p_{n_k}^2 \log_2 n_k} + \frac{10\sum_{i=p_{n_k}}^{p_{n_{k+1}}-1} \beta_{1,1}(i)n_{k+1}}{p_{n_{k+1}} p_{n_k} \log_2 n_k}$$

$$\leq \frac{2\rho\widetilde{\beta}_{1,1}(p_{n_k})n_k}{p_{n_k}^2 \log_2 n_k} + \frac{10(p_{n_{k+1}} - p_{n_k})\beta_{1,1}(p_{n_k})n_{k+1}}{p_{n_{k+1}} p_{n_k} \log_2 n_k}$$

$$\leq \frac{2\rho\widetilde{\beta}_{1,1}(p_{n_k})n_k}{p_{n_k}^2 \log_2 n_k} + \frac{10(p_{n_{k+1}} - p_{n_k})(p_{n_k}/n_k)n_{k+1}}{p_{n_{k+1}} p_{n_k} \log_2 n_k}$$

$$\leq \frac{2\rho\widetilde{\beta}_{1,1}(p_{n_k})n_k}{p_{n_k}^2 \log_2 n_k} + \frac{2\rho}{\log_2 n_k} \to 0.$$

(The monotonicity of $(n/p_n)_n$ makes easier the justification of the above convergence to 0). If $(n/p_n)_n \nearrow$ then

$$1 \leq \limsup_{k\to\infty} p_{n_{k+1}}/p_{n_k} \leq \limsup_{k\to\infty} n_{k+1}/n_k \leq \rho.$$

Thus the first term on the right of inequality (14.6) is summable if

$$\liminf_k \frac{3Ap_{n_k}}{4Mp_{n_{k+1}}} > 1;$$

that is, $3A/(4M\rho^{\theta_0}) > 1$.

To conclude the proof, we use the Borel–Cantelli Lemma once again, then we tend the constants A to their lower bounds, noting that under (ii),

$$\lim_k \frac{p_{n_{k+1}}^2}{\widetilde{\beta}_{1,1}(p_{n_k})n_k \log_2 n_k} = \lim_k \frac{\rho p_{n_{k+1}}^2}{\widetilde{\beta}_{1,1}(p_{n_{k+1}})n_{k+1} \log_2 n_{k+1}} = 0,$$

and under (iii), $\lim_k p_{n_{k+1}}/(p_{n_k} \log_2 n_k) = 0$; we then deduce the results announced in (i)–(iii).

3. We first note that under the mixing being considered,

$$0 < \widetilde{\beta}_{1,1}(p) \le \widetilde{\beta}_{1,1}(\infty) < \infty.$$

Let $\eta_n = 1/\log_2 n$ and $p_n = \eta_n \sqrt{n/\log_2 n}$; thus

$$\lim_n \frac{p_n^2 \log_2 n}{n \widetilde{\beta}_{1,1}(p_n)} = \lim_n \eta_n^2 = 0.$$

It is clear that the series $\sum_n \eta_n \sqrt{n \log n} \beta(\eta_n \sqrt{n/\log n}) < \infty$ under the mixing we are considering, with $(a > 1$ and $b, c \in \mathbb{R})$ or $(a = 1, b > 1,$ and $c \in \mathbb{R})$ or $(a = b = 1$ and $c > 2)$. We are therefore in case (i) of the corollary; whence

$$\limsup_n \frac{1}{\sqrt{n \log_2 n}} \max_{1 \le k \le n} \|S_k\| \le 2M \sqrt{1 + 10 \sum_i \beta_{1,1}(i)} \text{ a.s.}$$

If there exists $0 < \kappa < 1$ such that $\sum_{i=1}^{\infty} \beta_{1,1}^{\kappa}(i) < \infty$, we can give a more precise bound. We apply the covariance inequality for Hilbertian r.v.'s having moments (Dehling and Philipp 1982, Lemma 2.2) to get the bound above:

$$E\|Y_{(i-1)p+1} + \cdots + Y_{ip}\|^2$$

$$\le p \max_i E\|Y_i\|^2 + 15p \max_i (E\|Y_i\|^{2/(1-\kappa)})^{1-\kappa} \sum_{i=1}^{p-1} \beta_{1,1}^{\kappa}(i)$$

$$\le p \max_i (E\|Y_i\|^{2/(1-\kappa)})^{1-\kappa} \left(1 + 15 \sum_{i=1}^{p-1} \beta_{1,1}^{\kappa}(i)\right).$$

In this case, let $\widetilde{\beta}_{1,1}(p, \kappa) = 1 + 15 \sum_{i=1}^{p-1} \beta_{1,1}^{\kappa}(i)$, then

$$2\widetilde{\sigma}_r^2(p) \le n(1 + 2p/n) \max_i (E\|Y_i\|^{2/(1-\kappa)})^{1-\kappa} \widetilde{\beta}_{1,1}(p, \kappa).$$

We therefore obtain

$$\limsup_n \frac{1}{\sqrt{n \log_2 n}} \max_{1 \le k \le n} \left\| \sum_{i=1}^k Y_i \right\|$$

$$\le 2 \max_i (E \|Y_i\|^{2/(1-\kappa)})^{(1-\kappa)/2} \sqrt{1 + 15 \sum_i \beta_{1,1}^\kappa(i)} \text{ a.s.}$$

The desired bound follows.

Proof of Corollary 14.4
Here is a Banach space of type $1 < \tau \le 2$ and $\|Y_i\| \le M$ a.s.
Let $e^*(n) = \max \left(E \left\| \sum_{i=1}^r S_{2i}^* \right\|, E \left\| \sum_{i=1}^r S_{2i-1}^* \right\| \right)$, where $(S_i^*, i = 1, \ldots, 2r)$ are independent r.v.'s such that

$$S_i^* \stackrel{\text{law}}{=} Y_{(i-1)p+1} + \cdots + Y_{ip}, \quad r = \left[\frac{n}{2p}\right]_e + 1.$$

First evaluate $e^*(n)$. To do this, set

$$e^*(n) = \max \left(e_0^*(n), e_1^*(n) \right) \text{ and } e_l^*(n) = E \left\| \sum_{i=1}^r S_{2i-l}^* \right\|, l = 0, 1.$$

As the $(S_{2i}^*)_i$ are independent and centered in a Banach space of type τ, then there exists a constant c_τ, depending only on τ, such that we have (after an application of Hölder's inequality):

$$e_0^{*\tau}(n) \le E \left\| \sum_{i=1}^r S_{2i}^* \right\|^\tau \le c_\tau \sum_{i=1}^r E \|S_{2i}^*\|^\tau$$

$$\le c_\tau p^{\tau-1} \sum_{i=1}^r \sum_{j=1}^p E \|Y_{j+(i-1)p}\|^\tau$$

$$\le c_\tau \max_i E \|Y_i\|^\tau p^{\tau-1} r p$$

$$\le c_\tau \max_i E \|Y_i\|^\tau p^{\tau-1} n(1/2 + p/n).$$

Therefore $e^*(n) \le c_\tau^{1/\tau} \max_i (E \|Y_i\|^\tau)^{1/\tau} p^{1-1/\tau} n^{1/\tau} (1/2 + p/n)^{1/\tau}$.
Thus, $e^*(n)/n \le C(p/n)^{1-1/\tau} \to 0$ when $p/n \to 0$.
We proceed as in the proof of Corollary 14.3. Set $\Delta_n = \max_{1 \le k \le n} \left\| \sum_{t=1}^k Y_t \right\|$ and $n_k = [\rho^k]_e$, $\rho > 1$. Although the a.s. convergence is derived from the bound given, we will prove it separately.
  1. Theorem 14.6, applied with $\tilde{M}^2(p) = pn(1/2 + p/n)M^2$ and $z_i = 0$, gives

$$P\left(\Delta_{n_k} \geq \varepsilon n_k + 2e^*(n_k) + \left[\frac{p_{n_k}}{2}\right]_e M\right)$$

$$\leq 2\exp\left(-\frac{\varepsilon^2 n_k}{8M^2 p_{n_k}}\right) + 2(n_k/p_{n_k})\beta(p_{n_k}). \tag{14.7}$$

The right-hand side of (14.7) is summable for all $\varepsilon > 0$ if and only if $\sum_n \beta(p)/p < \infty$ and $\sum_n n^{-1}\exp\left(-\frac{\varepsilon^2 n}{8M^2 p_n}\right)$ are also summable, because $(\beta(p_n)/p_n)_n \searrow$ and $(n/p_n)_n \nearrow$. This latter series is finite for all $\varepsilon > 0$ if $p\log\log n/n \to 0$. Since $p/n \to 0$ and $e^*(n)/n \to 0$, the Borel–Cantelli Lemma allows us to conclude that $\Delta_{n_k}/n_k \to 0$ and hence $\Delta_n/n \to 0$.

2. For the bound, Theorem 14.6 once again with

$$\varepsilon = A\sqrt{p_{n_k} n_k \log_2 n_k} + 2e^*(n_{k+1})$$

implies

$$P\left(\Delta_{n_{k+1}} \geq A\sqrt{p_{n_k} n_k \log_2 n_k} + 2e^*(n_{k+1}) + \left[\frac{p_{n_{k+1}}}{2}\right]_e M\right)$$

$$\leq 2\exp\left(-\frac{A^2 n_k p_{n_k} \log_2 n_k}{4M^2\left(1 + \frac{2p_{n_{k+1}}}{n_{k+1}}\right) n_{k+1} p_{n_{k+1}}}\right)$$

$$+ \left(\frac{n_{k+1}}{p_{n_{k+1}}} + 2\right)\beta(p_{n_{k+1}}). \tag{14.8}$$

We have

$$\frac{n_k p_{n_k}}{p_{n_{k+1}} n_{k+1}} \geq \frac{n_k^2}{n_{k+1}^2} \geq \frac{(1-\rho^{-k})^2}{\rho^{-2}} \to \rho^{-2}$$

when $k \to \infty$. Then the first term of the RHS of inequality (14.8) is summable if $\sum_k \exp\left(-\frac{A^2 \log k}{4M^2\rho^2}\right)$ is finite; this converges if $A > 2M\rho$, or also if $A > 2M$ and $1 < \rho < A/(2M)$. The Borel–Cantelli Lemma implies, for large enough k,

$$\Delta_{n_{k+1}} \leq A\sqrt{p_{n_k} n_k \log_2 n_k} + 2e^*(n_{k+1}) + \frac{p_{n_{k+1}}}{2}M \text{ a.s.}$$

Now for $n_k \leq n < n_{k+1}$, we have $\frac{p_{n_{k+1}}}{n_k} \leq \frac{\rho}{(1-\rho^{-k})}\frac{p_n}{n} =: \rho_k \frac{p_n}{n}$; $\rho_k \to \rho$. Since for large enough k, $\rho_k \leq (1 + 2\rho_k p/n)^{1/\tau}\rho_k \leq A/(2M)$, with the above bound of $e^*(n)$ we finally get, for large n and $A > 2M$,

$$\left\|\sum_{i=1}^n Y_i/n\right\| \leq A\sqrt{\frac{p\log_2 n}{n}}$$

$$+ 2c_\tau^{1/\tau}\max_i(E\|Y_i\|^\tau)^{1/\tau}\left(\frac{1}{2} + \frac{\rho_k p}{n}\right)^{1/\tau}\rho_k\left(\frac{p}{n}\right)^{1-1/\tau} + \rho_k \frac{p_n}{2n}M$$

$$\leq A\sqrt{\frac{p\log_2 n}{n}}$$

$$+(1/2)^{1/\tau}(A/M)c_\tau^{1/\tau}\max_i(E\|Y_i\|^\tau)^{1/\tau}\left(\frac{p}{n}\right)^{1-1/\tau}+\frac{Ap_n}{4n}\quad\text{a.s.}$$

Remark 14.16 *This bound may be improved if we can finely approximate $e^*(n)$, using a good bound of $E\|Y_{(i-1)p+1}+\cdots+Y_{ip}\|^\tau$. This is possible, for example, in the case when we can approximate the Banach space by a Hilbert space (e.g. a subspace of L^s, $s\geq 1$) by using covariance inequalities for Hilbertian r.v.'s. Then in this case it is preferable to use Theorem 14.7 instead of Theorem 14.6, in order to obtain close bounds in the Hilbert-space case.*

Proof of Theorem 14.8 (Convergence of the recursive Kernel estimate in L^2: the β-mixing case).
Recall that

$$\hat{f}_n(x)=(1/n)\sum_{i=1}^n\left(1/h_i^d\right)K((x-X_i)/h_i)\quad\text{for }x\in\mathbb{R}^d.$$

Set $K_i:\mathbb{R}^d\longrightarrow L^2(\mathbb{R}^d,\lambda)$, $u\longmapsto K_i(u)\equiv\{x\longmapsto(1/h_i^d)K((x-u)/h_i)\}$, where λ is the Lebesgue measure. Put $Y_i=K_i(X_i)\in\mathcal{Y}=L^2(\mathbb{R}^d,\lambda)$. Thus, we can write $\hat{f}_n=(1/n)\sum_{i=1}^n Y_i$; $\hat{f}_n\in L^2$. On the other hand $\|Y_i\|_2^2=\|K\|_2^2/h_i^d$ and $\|Y_i-EY_i\|_2^2\leq 4\|K\|_2^2/h_i^d$. Whence

$$\|Y_i-EY_i\|_2\leq 2\|K\|_2/h_i^{d/2}\leq 2\|K\|_2/\underline{h}_n^{d/2},$$

where we have noted $\underline{h}_n=\min_{1\leq i\leq n}h_i$.

By the covariance inequality for Hilbertian r.v.'s (Dehling and Philipp 1982, Lemma 2.2), for $l\geq 0$ we obtain

$$E\|(Y_{l+1}-EY_{l+1})+\cdots+(Y_{l+p}-EY_{l+p})\|_2^2$$

$$\leq\sum_{i=1}^p\frac{\|K\|_2^2}{h_{l+i}^d}+10\sum_{k=1}^{p-1}\sum_{i=1}^{p-1-k}\beta_{1,1}(k)\frac{\|K\|_2^2}{h_{l+i}^{d/2}h_{l+k+i}^{d/2}}$$

$$\leq\frac{\|K\|_2^2}{\min_{1\leq j\leq p-1}h_{l+j}^{d/2}}\sum_{i=1}^p\frac{1}{h_{l+i}^{d/2}}\left(1+10\sum_{k=1}^{p-1}\beta_{1,1}(k)\right).$$

To simplify this last expression we bound it, for all $l\geq 0$, by

$$E\|(Y_{l+1}-EY_{l+1})+\cdots+(Y_{l+p}-EY_{l+p})\|_2^2$$

$$\leq\frac{p\|K\|_2^2}{\underline{h}_n^d}\left(1+10\sum_{k=1}^{p-1}\beta_{1,1}(k)\right)=\frac{p\|K\|_2^2}{\underline{h}_n^d}\widetilde{\beta}_{1,1}(p).$$

With the notation of Theorem 14.5, $2\tilde{\sigma}_r^2(p) \leq n(1+2p/n)\frac{\|K\|_2^2}{h_n^d}\tilde{\beta}_{1,1}(p)$ and $M(p) = 2p\frac{\|K\|_2}{h_n^{d/2}}$.

Let $\Delta_n = \max_{1 \leq k \leq n}\left\|\sum_{t=1}^k (Y_t - EY_t)\right\|_2$, so Δ_n is increasing with respect to n. In order that $\Delta_n/n \to 0$ it suffices that there exists an infinitely increasing subsequence, $n_k \to \infty$ when $k \to \infty$, such that $\limsup_{k\to\infty} n_{k+1}/n_k < \infty$ and $\Delta_{n_k}/n_k \to 0$. (Indeed $\forall n$, $\exists k$ such that $n_k \leq n \leq n_{k+1}$ and $0 \leq \Delta_n/n \leq (\Delta_{n_{k+1}}/n_{k+1})(n_{k+1}/n_k)$). We show then that $\Delta_{n_k}/n_k \to 0$ for $n_k = 2^k$.

Exactly as before, by Theorem 14.5, for all $\varepsilon > 0$ we obtain

$$P\left(\Delta_{n_k} \geq n_k\varepsilon + \frac{p_{n_k}}{h_{n_k}^{d/2}}\|K\|_2\right)$$

$$\leq 4\exp\left(-\frac{\varepsilon^2 n_k h_{n_k}^d}{8\|K\|_2(\|K\|_2\tilde{\beta}_{1,1}(p_{n_k}) + p_{n_k}h_{n_k}^{d/2}\varepsilon/3)}\right) + (n_k/p_{n_k} + 2)\beta(p_{n_k})$$

$$=: 4\exp(-U_{n_k}) + (n_k/p_{n_k} + 2)\beta(p_{n_k}).$$

The last term is summable (see above). The first term on the right of the above inequality is summable under the corollary's conditions:

(i) Let $C < 0$ such that $\limsup_n \frac{p_n h_n^{d/2}}{\tilde{\beta}_{1,1}(p_n)} \leq C$, thus for large k

$$U_{n_k} \geq \frac{\varepsilon^2 n_k h_{n_k}^d}{8\|K\|_2\tilde{\beta}_{1,1}(p_{n_k})(\|K\|_2 + 2C\varepsilon/3)},$$

and since $\frac{nh_n^d}{\tilde{\beta}_{1,1}(p_n)} \nearrow$ then, for example for $n_k = 2^k$, the series $\sum_k \exp(-U_{n_k})$ is finite for all $\varepsilon > 0$ when

$$\sum_n n^{-1}\exp\left(-\frac{\varepsilon^2 n h_n^d}{8\|K\|_2\tilde{\beta}_{1,1}(p_n)(\|K\|_2 + 2C\varepsilon/3)}\right)$$

is also finite (see the beginning of the proof of Corollary 14.3). And it is convergent for all $\varepsilon > 0$ if $\frac{nh_n^d}{\tilde{\beta}_{1,1}(p_n)\log\log n} \to \infty$.

(ii) If $\lim_n \frac{p_n h_n^{d/2}}{\tilde{\beta}_{1,1}(p_n)} \to \infty$ and since $\frac{nh_n^{d/2}}{p_n} \nearrow$, then $\sum_k \exp(-U_{n_k})$ is convergent for all $\varepsilon > 0$ if $\sum_n n^{-1}\exp\left(-\frac{\varepsilon n h_n^{d/2}}{8\|K\|_2 p_n}\right)$ is also summable. This latter series is finite for all $\varepsilon > 0$ if $\frac{nh_n^{d/2}}{p_n \log\log n} \to \infty$.

The Borel–Cantelli Lemma allows us to conclude, in both cases, that $\|\hat{f}_n - E\hat{f}_n\|_2 \to 0$ a.s. when $n \to \infty$, because $\frac{p_n}{nh_n^{d/2}} \to 0$ in both cases.

If in addition $\int K(x)\,dx = 1$ and $h_n \to 0$ then, since $\|\hat{f}_n - f\|_2 \le \|\hat{f}_n - E\hat{f}_n\|_2 + \|E\hat{f}_n - f\|_2$, it remains to show that the bias $E\hat{f}_n - f$ goes to 0 in L^2, i.e. $\|E\hat{f}_n - f\|_2 \to 0$.

First recall the following result: if $\int K(x)\,dx = 1$ and $g \in L^2$ then (see, for example, Wheeden and Zygmund 1977) $g * K_{h_t} \to g$ in L^2 when $h_t \to 0$.

Now, for $x \in \mathbb{R}^d$, $E\hat{f}_n(x) = \frac{1}{n}\sum_{t=1}^n f * K_{h_t}(x)$, where $*$ designates the convolution product, and $K_{h_t}(u) = (1/h_t^d) K(u/h_t)$. Thus

$$\|E\hat{f}_n - f\|_2 \le \frac{1}{n}\sum_{t=1}^n \|f * K_{h_t} - f\|_2.$$

But $\|f * K_{h_t} - f\|_2 \to 0$ when $t \to \infty$ (since $h_t \to 0$). We conclude by Toeplitz's Lemma that $\|E\hat{f}_n - f\|_2 \to 0$ as $n \to \infty$. Whence the result claimed.

Proof of Corollary 14.6 (bound recursive kernel estimate in L^2).

We keep the same notation as above.

1. Theorem 14.5 gives, for all $A > 0$,

$$P\left(\Delta_{n_{k+1}} \ge A\sqrt{\tilde{\beta}_{1,1}(p_{n_k})\frac{n_k}{\underline{h}_{n_k}^d}\log_2 n_k} + \frac{p_{n_{k+1}}}{\underline{h}_{n_{k+1}}^{d/2}}\|K\|_2\right)$$

$$\le 4\exp\left(-\frac{A^2 n_k \underline{h}_{n_{k+1}}^d \tilde{\beta}_{1,1}(p_{n_k})\log_2 n_k}{4\|K\|_2^2 n_{k+1}\underline{h}_{n_k}^d \tilde{\beta}_{1,1}(p_{n_{k+1}})\left[1 + \frac{2p_{n_{k+1}}}{n_{k+1}} + \frac{2A}{3}\sqrt{\frac{p_{n_{k+1}}^2 n_k \underline{h}_{n_{k+1}}^d \log_2 n_k}{n_{k+1}^2 \underline{h}_{n_k}^d \tilde{\beta}_{1,1}(p_{n_{k+1}})\|K\|_2^2}}\right]}\right)$$

$$+ \left(\frac{n_{k+1}}{p_{n_{k+1}}} + 2\right)\beta(p_{n_{k+1}})$$

$$=: V_{n_k} + \left(\frac{n_{k+1}}{p_{n_{k+1}}} + 2\right)\beta(p_{n_{k+1}}). \tag{14.9}$$

The last term is summable (see above). For the first term, and for $n_k = [\rho^k]_e$, $\rho > 1$, recall that $\tilde{\beta}_{1,1}(p_{n_k})/\tilde{\beta}_{1,1}(p_{n_{k+1}}) \to 1$ as $k \to \infty$ if $\limsup_n \frac{p_n^2 \log_2 n}{n\tilde{\beta}_{1,1}(p_n)} \le c_0^2$, (see the proof of 2. (i) of Corollary 14.3), $\frac{\underline{h}_{n_{k+1}}^d}{\underline{h}_{n_k}^d} \le 1$, and since $n\underline{h}_n^d \nearrow$, then

$$1 \ge \frac{n_k \underline{h}_{n_{k+1}}^d}{n_{k+1}\underline{h}_{n_k}^d} \ge \frac{n_k^2}{n_{k+1}^2} \ge \frac{(1-\rho^{-k})^2}{\rho^{-2}} \to \rho^{-2} \text{ when } k \to \infty.$$

If $\frac{p_n^2 \log_2 n}{n\tilde{\beta}_{1,1}(p_n)} \to 0$, then

$$\frac{p_{n_{k+1}}^2 n_k \underline{h}_{n_{k+1}}^d \log_2 n_k}{n_{k+1}^2 \underline{h}_{n_k}^d \tilde{\beta}_{1,1}(p_{n_{k+1}})\|K\|_2^2} \to 0$$

and thus $\sum_k V_{n_k}$ is finite once $\sum_k \exp\left(-\frac{A^2 \log_2 n_k}{4\|K\|_2^2 \rho^2}\right)$ is also finite, which converges for $n_k = [\rho^k]_e$, $\rho > 1$, if $A > 2\|K\|_2$ and $1 < \rho < A/(2\|K\|_2)$. We remark that $\frac{p_n^2}{n\tilde{\beta}_{1,1}(p_n)\log_2 n} \to 0$. By Borel–Cantelli's Lemma and letting the constant A go to its lower bound, we finally obtain

$$\limsup_n \sqrt{\frac{nh_n^d}{\tilde{\beta}_{1,1}(p_n)\log_2 n}} \|\hat{f}_n - E\hat{f}_n\|_2 \le 2\|K\|_2 \text{ a.s.}$$

On the other hand, if $\limsup_n \frac{p_n^2 \log_2 n}{n\tilde{\beta}_{1,1}(p_n)} \le c_0^2$, then

$$\frac{p_{n_{k+1}}^2 n_k \underline{h}_{n_{k+1}}^d \log_2 n_k}{n_{k+1}^2 \underline{h}_{n_k}^d \tilde{\beta}_{1,1}(p_{n_{k+1}}) \|K\|_2^2} \le c_0^2/\|K\|_2^2,$$

and hence $\sum_k V_{n_k} < \infty$ when $\sum_k \exp\left(-\frac{A^2 \log_2 n_k}{4\|K\|_2 \rho^2(\|K\|_2 + 2Ac_0/3)}\right)$ is also finite, which converges for $n_k = [\rho^k]_e$, $\rho > 1$, if

$$A > 2\|K\|_2 \rho \left(2\rho c_0/3 + \sqrt{4\rho^2 c_0^2/9 + 1}\right).$$

That is to say

$$A > 2\|K\|_2 \left(2c_0/3 + \sqrt{4c_0^2/9 + 1}\right)$$

and

$$1 < \rho < A^2(4\|K\|_2)^{-1}(\|K\|_2 + 2Ac_0/3)^{-1}.$$

And one still has $\frac{p_n^2}{n\tilde{\beta}_{1,1}(p_n)\log_2 n} \to 0$.

But if $\frac{p_n^2 \log_2 n}{n\tilde{\beta}_{1,1}(p_n)} \to \infty$, then

$$P\left(\Delta_{n_{k+1}} \ge A \frac{p_{n_k}}{\underline{h}_{n_k}^{d/2}} \log_2 n_k + \frac{p_{n_{k+1}}}{\underline{h}_{n_{k+1}}^{d/2}} \|K\|_2\right)$$

$$\le 4\exp\left(-\frac{3Ap_{n_k} \underline{h}_{n_{k+1}}^{d/2} \log_2 n_k}{8\|K\|_2 p_{n_{k+1}} \underline{h}_{n_k}^{d/2} \left[\frac{3\|K\|_2 n_{k+1} \underline{h}_{n_k}^{d/2} \tilde{\beta}_{1,1}(p_{n_{k+1}})}{Ap_{n_k} p_{n_{k+1}} \underline{h}_{n_{k+1}}^{d/2} \log_2 n_k} + 1\right]}\right)$$

$$+ \left(\frac{n_{k+1}}{p_{n_{k+1}}} + 2\right) \beta(p_{n_{k+1}})$$

$$=: V'_{n_k} + \left(\frac{n_{k+1}}{p_{n_{k+1}}} + 2\right) \beta(p_{n_{k+1}}). \quad (14.10)$$

However since $(n/p_n)_n$ and $(n\underline{h}_n^d)_n$ are increasing, for large k:

$$\frac{n_{k+1}\underline{h}_{n_k}^{d/2}\widetilde{\beta}_{1,1}(p_{n_{k+1}})}{p_{n_k}p_{n_{k+1}}\underline{h}_{n_{k+1}}^{d/2}\log_2 n_k} \leq (2\rho)^{3/2}\frac{n_{k+1}\widetilde{\beta}_{1,1}(p_{n_{k+1}})}{p_{n_{k+1}}^2 \log_2 n_k} \to 0 \text{ as } k \to \infty,$$

and

$$\frac{p_{n_k}\underline{h}_{n_{k+1}}^{d/2}}{p_{n_{k+1}}\underline{h}_{n_k}^{d/2}} \geq \left(\frac{n_k}{n_{k+1}}\right)^{3/2} \geq \left(\frac{1-\rho^{-k}}{\rho}\right)^{3/2} \to \rho^{-3/2} \text{ as } k \to \infty.$$

Therefore $\sum_k V'_{n_k}$ is finite when $\sum_k \exp\left(-\frac{3A\log_2 n_k}{8\|K\|_2\rho^{3/2}}\right)$ is also finite, which converges for $n_k = [\rho^k]_e, \rho > 1$ if $A > 8\|K\|_2\rho^{3/2}/3$. Note that

$$\frac{p_{n_{k+1}}\underline{h}_{n_k}^{d/2}}{p_{n_k}\underline{h}_{n_{k+1}}^{d/2}\log_2 n_k} \leq \frac{\rho^{3/2}(1-\rho^{-k})^{-3/2}}{\log_2 n_k} \to 0.$$

To conclude, we again use Borel–Cantelli's Lemma and let the constants go to their lower bounds.

We then get $\limsup_n \frac{\underline{h}_n^{d/2}}{p_n \log_2 n} \Delta_n \leq A$, that is to say,

$$\limsup_n \frac{n\underline{h}_n^{d/2}}{p_n \log_2 n}\|\hat{f}_n - E\hat{f}_n\|_2 \leq 8\|K\|_2/3 \text{ a.s.}$$

3. This is similar to 3. of Corollary 14.3.

Proof of Theorem 14.9 (Convergence of the recursive kernel estimate in L^2 under α-mixing).

Here K is Lipschitzian of parameter L and has compact support:

$$|K(x) - K(y)| \leq L|x-y| \quad x, y \in \mathbb{R}^d.$$

Let $K_i : \mathbb{R}^d \longrightarrow L^2(\mathbb{R}^d, \lambda)$ and $u \longmapsto K_i(u) \equiv \{x \longmapsto (1/h_i^d) K((x-u)/h_i)\}$, where λ is the Lebesgue measure. Set $Y_i = K_i(X_i) \in \mathcal{Y} = L^2(\mathbb{R}^d, \lambda)$. We verify that the (K_i) are Lipschitzian:

$$\|K_i(X) - K_i(Y)\|_2^2 = (1/h_i^d)\int |K((x-X)/h_i) - K((x-Y)/h_i)|^2 dx/h_i^d$$

$$= (1/h_i^d)\int |K(u) - K(u+(X-Y)/h_i)|^2 du$$

$$\leq (1/h_i^{d+2}) L^2|X-Y|^2 \lambda\Big(\text{supp}K \cup \{\text{supp}K - (X-Y)/h_i\}\Big)$$

$$\leq (2L^2\lambda(K)/h_i^{d+2})|X-Y|^2$$

$$\leq (2L^2\lambda(K)/\underline{h}_n^{d+2})|X-Y|^2,$$

where $\underline{h}_n = \min_{1\leq i\leq n} h_i$; here $\lambda(K)$ denotes the (Lebesgue) measure of the support of K. Therefore, with $L_n = L\sqrt{2\lambda(K)}/\underline{h}_n^{d/2+1}$,

$$\|K_i(X) - K_i(Y)\|_2 \leq L_n|X - Y|.$$

On the other side, $\|Y_i\|_2^2 = \|K\|_2^2/h_i^d$ and $\|Y_i - EY_i\|_2^2 \leq 4\|K\|_2^2/h_i^d$. Thus

$$\|Y_i - EY_i\|_2 \leq 2\|K\|_2/\underline{h}_n^{d/2}$$

and

$$E\|Y_i - EY_i\|_2^2 \leq E\|Y_i\|_2^2 \leq \|K\|_2/\underline{h}_n^d.$$

Let $\Delta_n = \max_{1\leq k\leq n}\left\|\sum_{i=1}^k(Y_i - EY_i)\right\|_2$; Δ_n is increasing with respect to n. In order that $\Delta_n/n \to 0$ it suffices that there exists an indefinitely increasing subsequence $n_k \to \infty$ as $k \to \infty$, such that $\limsup_{k\to\infty} n_{k+1}/n_k < \infty$ and $\Delta_{n_k}/n_k \to 0$. Hence let us show that $\Delta_{n_k}/n_k \to 0$ (e.g. for $n_k = 2^k$). With the notation of Theorem 14.2, $\sigma^2 = \|K\|_2^2/\underline{h}_n^d$ and $\tilde{K} = 2\|K\|_2/\underline{h}_n^{d/2}$.

Exactly as before, by Theorem 14.2, for all $\varepsilon > 0$ and $0 < \theta < 1$ we obtain

$$P\left(\Delta_{n_k} \geq n_k\varepsilon + \frac{p_{n_k}}{\underline{h}_{n_k}^{d/2}}\|K\|_2\right)$$

$$\leq 2\exp\left(-\frac{(1-\theta)^2\varepsilon^2 n_k \underline{h}_{n_k}^d}{4p_{n_k}(\|K\|_2^2 + (1-\theta)\|K\|_2\varepsilon \underline{h}_{n_k}^{d/2}/3)} + \log p_{n_k}\right)$$

$$+ C(n_k)n_k\left[\underline{h}_{n_k}^{-(d/2+1)d/2}\alpha_{\triangleleft,1}(p_{n_k})\right]^{2\gamma/(2\gamma+d)}$$

$$=: 2\exp\left(-U_{n_k}\right) + C(n_k)n_k\left[\underline{h}_{n_k}^{-(d/2+1)d/2}\alpha_{\triangleleft,1}(p_{n_k})\right]^{2\gamma/(2\gamma+d)},$$

where

$$C(n) = (2\sqrt{2}\ 3^{d/2} + 1_{\gamma<\infty})\left\{3\max\left(\frac{L\sqrt{2\lambda(K)}M_\gamma}{\theta\varepsilon}, \underline{h}_n^{d/2+1}\right)\right\}^{d\gamma/(2\gamma+d)}$$

and $M_\gamma^\gamma = \sup_t E|X_t|^\gamma$; $\limsup_n C_n < \infty$.

Since $\limsup \underline{h}_n \leq C < \infty$, and $\frac{p_n \log p_n}{n\underline{h}_n^d} \to 0$, then for large enough k

$$U_{n_k} \geq \frac{(1-\theta)^2\varepsilon^2 n_k\underline{h}_{n_k}^d}{8p_{n_k}(\|K\|_2^2 + (1-\theta)\|K\|_2\varepsilon C)}.$$

Because $\frac{n\underline{h}_n^d}{p_n} \nearrow$ then $\sum_k \exp\left(-U_{n_k}\right) < \infty$ for all $\varepsilon > 0$ if

$$\sum_n n^{-1}\exp\left(-\frac{(1-\theta)^2\varepsilon^2 n\underline{h}_n^d}{8p_n(\|K\|_2^2 + (1-\theta)\|K\|_2\varepsilon C)}\right) < \infty \text{ for all } \varepsilon > 0.$$

Then it is summable if $\frac{n\underline{h}_n^d}{p_n \log\log n} \to \infty$.

For the other term in $a_{\triangleleft,1}(\cdot)$, since $h_n^{-(d/2+1)d/2} a_{\triangleleft,1}(p_n) \searrow$, then it is summable if

$$\sum_n [h_n^{-(d/2+1)d/2} a_{\triangleleft,1}(p_n)]^{2\gamma/(2\gamma+d)} < \infty.$$

We conclude again using Borel–Cantelli's Lemma, and noting that $\frac{p_n}{n h_n^{d/2}} \to 0$.

14.4.2 Proofs of maximal inequalities

We proceed in the classic way, by approaching the dependent r.v.'s by means of independent ones and applying to these new random vectors the inequalities of the independent case; see Pinelis and Sakhanenko (1985) or Pinelis (1990). The proofs are thus in two stages and will be presented as a series of lemmas. We start by stating a lemma on the approximation of dependent random vectors by independent ones (Rhomari 2002a).

Lemma 14.1 *Let $X_1, \ldots, X_m$ be m random vectors from $\mathbb{R}^d$, then for all $C_i \in \mathbb{R}^d$, $\gamma > 0$ and $0 < q \le \min_{2 \le i \le m} \|X_i + C_i\|_\gamma$, there exists $X_1^* = X_1, X_2^*, \ldots, X_m^*$, independent random vectors, such that $X_i^* \stackrel{\mathcal{L}}{=} X_i$ and, for all $i = 2, \ldots, m$,*

$$P(|X_i - X_i^*| > q) \le (2\sqrt{2}\, 3^{d/2} + \mathbb{1}_{\gamma < \infty})(\|X_i + C_i\|_\gamma / q)^{d\gamma/(2\gamma+d)}$$
$$\times [\alpha(\sigma(X_1, \ldots, X_{i-1}), \sigma(X_i))]^{2\gamma/(2\gamma+d)}.$$

Given n random vectors $Z_1, \ldots, Z_n$, we establish the convention that $Z_m = 0$ when $m > n$ and for a positive integer p, $1 \le p \le n$, $1 \le j \le p$, and $q = \left[\frac{n}{p}\right]_e + 1$, where $[x]_e$ denotes the integer part of x, we denote $Z_{j,i} = Z_{j+(i-1)p}$ and

$$S_{2,j}^Z = \sum_{i=1}^q Z_{j,i} \quad \text{and} \quad S_n^Z := \sum_{t=1}^n Z_t = \sum_{j=1}^p S_{2,j}^Z. \tag{14.11}$$

And for $1 \le p \le n/2$ and $1 \le i \le 2r$, where $r = \left[\frac{n}{2p}\right]_e + 1$,

$$S_{1,i}^Z = \sum_{j=1}^p Z_{j,i} \quad \text{and} \quad S_n^Z = \sum_{t=1}^n Z_t = \sum_{i=1}^{2r} S_{1,i}^Z = \sum_{i=1}^r S_{1,2i}^Z + \sum_{i=1}^r S_{1,2i-1}^Z. \tag{14.12}$$

We first remark, if $\|Z\| \le M$ a.s., that on the one hand

$$\max_{1 \le k \le n} \|S_k^Z\| \le \sum_{j=1}^p \max_{1 \le l \le q} \left\| \sum_{i=1}^l Z_{j,i} \right\| + \left[\frac{p}{2}\right]_e M, \tag{14.13}$$

and secondly

$$\max_{1\le k\le n}\|S_k^Z\| \le \max_{1\le l\le r}\left\|\sum_{i=1}^{l} S_{1,2i}^Z\right\| + \max_{1\le l\le r}\left\|\sum_{i=1}^{l} S_{1,2i-1}^Z\right\| + \left[\frac{p}{2}\right]_e M. \qquad (14.14)$$

Indeed, for all $1 \le k \le n$ there exists l such that $(l-1)p \le k < lp$ and

$$\|S_k^Z\| \le \|S_{(l-1)p}^Z\| + \left\|\sum_{t=(l-1)p+1}^{k} Z_t\right\| \le \|S_{(l-1)p}^Z\| + [k-(l-1)p]M,$$

and one also has

$$\|S_k^Z\| \le \|S_{lp}^Z\| + \left\|\sum_{t=k+1}^{lp} Z_t\right\| \le \|S_{lp}^Z\| + (lp-k)M.$$

Since $\min\{k-(l-1)p, lp-k\} \le \left[\frac{p}{2}\right]_e$, we have

$$\|S_k^Z\| \le \max\left(\|S_{(l-1)p}^Z\|, \|S_{lp}^Z\|\right) + \left[\frac{p}{2}\right]_e M.$$

For $1 \le l < q$, because $S_{lp}^Z = \sum_{j=1}^{p}\sum_{i=1}^{l} Z_{j+(i-1)p}$, we have

$$\|S_k^Z\| \le \sum_{j=1}^{p} \max\left(\left\|\sum_{i=1}^{l-1} Z_{j,i}\right\|, \left\|\sum_{i=1}^{l} Z_{j,i}\right\|\right) + \left[\frac{p}{2}\right]_e M,$$

and from this one deduces (14.13). But if $1 \le l < 2r$,

$$\max_{1\le k\le n}\|S_k^Z\| \le \max_{1\le l\le 2r}\left\|\sum_{i=1}^{l} S_{1,i}^Z\right\| + \left[\frac{p}{2}\right]_e M,$$

and hence (14.14) follows from this last inequality.

Thus, with $Z_t = K_t(X_t)$ and $Z_t^* = K_t(X_t^*)$, we have

Lemma 14.2 *Under* (K), $\|K_t(X_t)\| \le \tilde{K}$ *and if* $\exists \gamma > 0$ *such that* $\|X_t\|_\gamma \le M_\gamma$, *then for all* $\varepsilon > 0$, $0 < \theta < 1$, $\delta > 1$, $1 \le p \le n$, *and* $j \in \{1,\ldots,p\}$, *there exists a sequence of independent random vectors* $\left(X_{j,i}^*, i=1,\ldots,q\right)$ *such that* $X_{j,i}^*$ *has the same law as* $X_{j,i}$ (*hence* $Z_{j,i}^* \stackrel{\mathcal{L}}{=} Z_{j,i}$) *and*

$$P\left(\max_{1\le k\le n}\left\|\sum_{t=1}^{k} K_t(X_t)\right\| \ge \varepsilon + \left[\frac{p}{2}\right]_e \tilde{K}\right)$$

$$\le \sum_{j=1}^{p} P\left(\max_{1\le l\le q}\left\|\sum_{i=1}^{l} K_{j,i}\left(X_{j,i}^*\right)\right\| \ge (1-\theta)\varepsilon/p\right)$$

$$+ C_K(\gamma) n \left[a_{\triangleleft,1}(p)\right]^{2\gamma/(2\gamma+d)},$$

where $C_K(\gamma)$ is that of Theorem 14.2.

Proof of Lemma 14.2

Let $Z_t = K_t(X_t)$. Lemma 14.1 ensures that there exists, for each fixed j in $\{1, \ldots, p\}$ and for all $C_{j,i}$ in $\mathbb{R}^d$ and $0 < \xi \leq \min_{j,1 \leq i \leq q} \|X_{j,i} + C_{j,i}\|_\gamma$, a sequence of independent r.v.'s $\left(X^*_{j,i}, i = 1, \ldots, q\right)$ such that $X^*_{j,i} \stackrel{\mathcal{L}}{=} X_{j,i}$ and

$$P\left(|X^*_{j,i} - X_{j,i}| \geq \xi\right)$$

$$\leq (2\sqrt{2}\, 3^{d/2} + \mathbb{1}_{\gamma<\infty}) \left(\frac{\|X_{j,i} + C_{j,i}\|_\gamma}{\xi}\right)^{d\gamma/(2\gamma+d)} \{a_{\triangleleft,1}(p)\}^{2\gamma/(2\gamma+d)}, \quad (14.15)$$

because $\max_{1 \leq j \leq p} \max_{1 \leq i \leq q} \alpha(\sigma(X_{j,1}, \ldots, X_{j,i-1}), \sigma(X_{j,i})) \leq a_{\triangleleft,1}(p)$. The inequality (14.13) and the hypothesis (**K**) implies

$$\max_{1 \leq k \leq n} \|S^Z_k\| \leq \sum_{j=1}^{p} \max_{1 \leq l \leq q} \left\|\sum_{i=1}^{l} K_{j,i}\left(X^*_{j,i}\right)\right\|$$

$$+ \sum_{j=1}^{p} \max_{1 \leq l \leq q} \sum_{i=1}^{l} L_{j,i} |X_{j,i} - X^*_{j,i}| + \left[\frac{p}{2}\right]_e \tilde{K}$$

$$= \sum_{j=1}^{p} \max_{1 \leq l \leq q} \left\|\sum_{i=1}^{l} K_{j,i}\left(X^*_{j,i}\right)\right\| + \sum_{t=1}^{n} L_t |X_t - X^*_t| + \left[\frac{p}{2}\right]_e \tilde{K}$$

Thus for all $0 < \theta < 1$,

$$P\left(\max_{1 \leq k \leq n} \left\|\sum_{t=1}^{k} K_t(X_t)\right\| \geq \varepsilon + \left[\frac{p}{2}\right]_e \tilde{K}\right)$$

$$\leq \sum_{j=1}^{p} P\left(\max_{1 \leq l \leq q} \left\|\sum_{i=1}^{l} K_{j,i}\left(X^*_{j,i}\right)\right\| \geq (1-\theta)\varepsilon/p\right)$$

$$+ \sum_{t=1}^{n} P\left(|X_t - X^*_t| > \frac{\theta\varepsilon}{L_t n}\right).$$

Define $L = L(n) = \max_{1 \leq t \leq n} L_t$, and let $\delta > 1$, $C_t = \delta M_\gamma$, and $\xi = \min\left(\frac{\theta\varepsilon}{Ln}, (\delta-1)M_\gamma\right)$, hence thanks to (14.15), the last sum is bounded above by

$$(2\sqrt{2}\, 3^{d/2} + \mathbb{1}_{\gamma<\infty}) \left((\delta+1)M_\gamma/\xi\right)^{d\gamma/(2\gamma+d)} n\, \{a_{\triangleleft,1}(p)\}^{2\gamma/(2\gamma+d)},$$

since $0 < \xi \leq (\delta-1)M_\gamma \leq \min_{1 \leq t \leq n} \|X_t + C_t\|_\gamma \leq \max_{1 \leq t \leq n} \|X_t + C_t\|_\gamma \leq (\delta+1)M_\gamma$. But

$$(\delta+1)\frac{M_\gamma}{\xi} = \frac{\delta+1}{\delta-1}\max\left(\frac{L(\delta-1)M_\gamma n}{\theta\varepsilon}, 1\right).$$

This ends the proof of lemma.

If the process is strictly stationary, we can choose $C_t = 0$ and $\xi = \min\left(\frac{\theta\varepsilon}{Ln}, M_\gamma\right)$; thus in this case

$$C_K(\gamma) = (2\sqrt{2}\, 3^{d/2} + \mathbb{1}_{\gamma<\infty})\left\{\max\left(\frac{LM_\gamma n}{\theta\varepsilon}, 1\right)\right\}^{d\gamma/(2\gamma+d)}.$$

The following lemma concerns the transformation (K_t) with values in a space of finite dimension. Without loss of generality we assume that the (K_t) are the identity of $\mathbb{R}^d$. With the decompositions (14.12), we have

Lemma 14.3 $(\mathbb{R}^d)$ *If for* $1 \leq p \leq n/2$, $|X_{(i-1)p+1} + \cdots + X_{ip}| \leq M(p)$ *a.s., then for all* $\varepsilon > 0$, $0 < \theta < 1$, $\delta > 1$ *there exist two sequences of independent random vectors* $(S^{X*}_{1,2i}, i = 1, \ldots, r)$ *and* $(S^{X*}_{1,2i-1}, i = 1, \ldots, r)$ *such that* $S^{X*}_{1,i}$ *has the same law as* $S^X_{1,i}$ *for* $i = 1, \ldots, 2r$, *and*

$$P\left(\max_{1\leq k\leq n}\left|\sum_{t=1}^{k} X_t\right| \geq \varepsilon + \left[\frac{p}{2}\right]_e M\right)$$

$$\leq P\left(\max_{1\leq l\leq r}\left|\sum_{i=1}^{l} S^{X*}_{1,2i}\right| \geq (1-\theta)\varepsilon/2\right)$$

$$+ P\left(\max_{1\leq l\leq r}\left|\sum_{i=1}^{l} S^{X*}_{1,2i-1}\right| \geq (1-\theta)\varepsilon/2\right) + C_b\, 2r\, \alpha(p),$$

with $C_b = 2\sqrt{2}\, 3^{d/2}\left\{\frac{\delta+1}{\delta-1}\max\left(\frac{2(\delta-1)rM(p)}{\theta\varepsilon}, 1\right)\right\}^{d/2}$ *and* $M = M(1)$.

Proof of Lemma 14.3

The proof is similar to the previous one. Using the notation (14.12), Lemma 14.1 ensures that, for all C_i in $\mathbb{R}^d$ and

$$0 < \xi \leq \min_{1\leq i\leq 2r}\left\|S^X_{1,i} + C_i\right\|_\infty$$

there are two sequences of independent random vectors $(S^{X*}_{1,2i}, i = 1, \ldots, r)$ and $(S^{X*}_{1,2i-1}, i = 1, \ldots, r)$, such that $S^{X*}_{1,i} \stackrel{\mathcal{L}}{=} S^X_{1,i}$ and

$$P\left(\left|S^{X*}_{1,i} - S^X_{1,i}\right| \geq \xi\right) \leq 2\sqrt{2}\, 3^{d/2}\left(\frac{\|S^X_{1,i} + C_i\|_\infty}{\xi}\right)^{d/2}\alpha(p), \qquad (14.16)$$

since $\max_{j=0,1} \max_{1 \le i \le r} \alpha\left(\sigma\left(S^X_{1,2-j}, \ldots, S^X_{1,2(i-1)-j}\right), \sigma\left(S^X_{1,2i-j}\right)\right) \le \alpha(p)$. On the other hand the inequality (14.14) gives

$$\max_{1 \le k \le n} |S^X_k| \le \max_{1 \le l \le r} \left|\sum_{i=1}^{l} S^{X*}_{1,2i}\right| + \max_{1 \le l \le r} \left|\sum_{i=1}^{l} S^{X*}_{1,2i-1}\right|$$
$$+ \sum_{i=1}^{2r} |S^{X*}_{1,i} - S^X_{1,i}| + \left[\frac{p}{2}\right]_e M.$$

Therefore for all $0 < \theta < 1$

$$P\left(\max_{1 \le k \le n} \left|\sum_{t=1}^{k} X_t\right| \ge \varepsilon + \left[\frac{p}{2}\right]_e M\right)$$
$$\le \sum_{j=0}^{1} P\left(\max_{1 \le l \le r} \left|\sum_{i=1}^{l} S^{X*}_{1,2i-j}\right| \ge (1-\theta)\varepsilon/2\right) + \sum_{i=1}^{2r} P\left(|S^{X*}_{1,i} - S^X_{1,i}| > \frac{\theta\varepsilon}{2r}\right).$$

For $\delta > 1$, $C_i = \delta M(p)$, and $\xi = \min\left(\frac{\theta\varepsilon}{2r}, (\delta-1)M(p)\right)$, thanks to (14.16) the last sum is bounded above by

$$2\sqrt{2}\, 3^{d/2} ((\delta+1)M(p)/\xi)^{d/2}\, 2r\, \alpha(p),$$

because

$$0 < \xi \le (\delta-1)M(p) \le \min_{1 \le i \le 2r} \|S^X_{1,i} + C_i\|_\infty$$
$$\le \max_{1 \le i \le 2r} \|S^X_{1,i} + C_i\|_\infty \le (\delta+1)M(p).$$

But

$$(\delta+1)\frac{M(p)}{\xi} = \frac{\delta+1}{\delta-1}\max\left(\frac{(\delta-1)2r\,M(p)}{\theta\varepsilon}, 1\right),$$

That finishes the proof.

If the process is strictly stationary, we choose $C_t = 0$ and $\xi = \min\left(\frac{\theta\varepsilon}{2r}, M(p)\right)$; hence we have $C_b = 2\sqrt{2}\, 3^{d/2} \left\{\max\left(\frac{2r\,M(p)}{\theta\varepsilon}, 1\right)\right\}^{d/2}$.

We now consider $Y_1, \ldots, Y_n$ to be n random vectors in a separable Banach space $\mathcal{Y}$.

Lemma 14.4 *If $\|Y_t\| \le M$, a.s., then for all $\varepsilon > 0$, $1 \le p \le n$, and $j \in \{1, \ldots, p\}$ there exist two sequences of independent random vectors $\left(Y^*_{j,i}, i = 1, \ldots, q\right)$ such that $Y^*_{j,i} \stackrel{\mathcal{L}}{=} Y_{j,i}$ and*

$$P\left(\max_{1\leq k\leq n}\left\|\sum_{t=1}^{k}Y_{t}\right\|\geq\varepsilon+\left[\frac{p}{2}\right]_{e}M\right)$$

$$\leq\sum_{j=1}^{p}P\left(\max_{1\leq l\leq q}\left\|\sum_{i=1}^{l}Y_{j,i}^{*}\right\|\geq\varepsilon/p\right)+n\,\beta_{\triangleleft,1}(p).$$

Lemma 14.5 *If* $\|Y_{t}\|\leq M$ *a.s., then for all* $\varepsilon>0$ *and* $1\leq p\leq n/2$, *there are two sequences of independent random vectors* $(S_{1,2i}^{Y*},i=1,\ldots,r)$ *and* $(S_{1,2i-1}^{Y*},i=1,\ldots,r)$ *such that* $S_{1,i}^{Y*}$ *have the same law as* $S_{1,i}^{Y}$ *for* $i=1,\ldots,2r$, *and*

$$P\left(\max_{1\leq k\leq n}\left\|\sum_{t=1}^{k}Y_{t}\right\|\geq\varepsilon+\left[\frac{p}{2}\right]_{e}M\right)$$

$$\leq\sum_{j=0,1}P\left(\max_{1\leq l\leq r}\left\|\sum_{i=1}^{l}S_{1,2i-j}^{Y*}\right\|\geq\varepsilon/2\right)+2r\,\beta(p).$$

The proofs of Lemmas 14.4 and 14.5 are analogous to those of Lemmas 14.2 and 14.3 using (instead of Lemma 14.1) the following lemma which is a corollary of Berbee (1979)'s theorem.

Lemma 14.6 *Let* $Y_{1},\ldots,Y_{m}$ *be m random vectors in a separable Banach space* S. *Then* $\exists Y_{1}^{*}=Y_{1},Y_{2}^{*},\ldots,Y_{m}^{*}$ *(m independent random vectors) such that* $Y_{i}^{*}\stackrel{\mathcal{L}}{=}Y_{i}$ *and, for all* $i=2,\ldots,m$,

$$P\left(Y_{i}\neq Y_{i}^{*}\right)=\beta(\sigma(Y_{1},\ldots,Y_{i-1}),\sigma(Y_{i})).$$

The proof of this lemma is omitted because it is similar to that of Lemma 14.1 and uses (instead of Theorem 1 in Rhomari (2002a)), the following theorem:

Theorem A. (Berbee 1979) *Let* X *and* Y *be two r.v.'s with values in two Polish spaces* S *and* S', *respectively, and let* U *be a random variable uniformly distributed on* $[0,1]$ *and independent of* (X,Y). *There exists a random vector* Y^{*} *in* S', $\sigma(X,Y,U)$-*measurable, satisfying:*

(i) Y^{*} *is independent of* X,
(ii) Y^{*} *and* Y *have the same probability distribution, and*
(iii) $P(Y\neq Y^{*})=\beta(X,Y)$.

Proofs of Lemmas 14.4–14.5
Just slightly modify those of Lemmas 14.2–14.3 by using Lemma 14.6 above instead of Lemma 14.1.

Proofs of theorems
The proofs of the theorems are similar and can be derived from: the previous lemmas, the exponential inequalities due to Pinelis and Sakhanenko (1985) and Pinelis (1990) obtained for independent r.v.'s, and remark 3 on page 145 of Pinelis

and Sakhanenko (1985) on deducing maximal inequalities (with the same bounds) through Doob's submartingale inequalities. We simply cite the lemmas needed in each proof.

Proof of Theorem 14.1
Lemma 14.3, with $\gamma = \infty$, Corollary 2 on pages 144–5 and Remark 3 on page 145 of Pinelis and Sakhanenko (1985).

Proof of Theorem 14.2
Lemma 14.2, Corollary 2 and Remark 3 on page 145 of Pinelis and Sakhanenko (1985).

Proof of Theorem 14.3
Lemma 14.2, Theorem 1.2, on page 605 of Pinelis (1990) and Remark 3 on page 145 of Pinelis and Sakhanenko (1985).

Proof of Theorem 14.4
Lemma 14.2, Corollary 2 and Remark 3 on page 145 of Pinelis and Sakhanenko (1985).

Proof of Theorem 14.5
Lemma 14.5, Corollary 2 and Remark 3 on page 145 of Pinelis and Sakhanenko (1985).

Proof of Theorem 14.6
Lemma 14.5, Theorem 1.2 of Pinelis (1990) and Remark 3 on page 145 of Pinelis and Sakhanenko (1985).

Proof of Theorem 14.7
Lemma 14.5, Corollary 2 and Remark 3 on page 145 of Pinelis and Sakhanenko (1985).

APPENDIX

Definition Let $1 \leq s \leq 2$ and $(\xi_i)_i$ be a Rademacher sequence (i.e. (ξ_i) i.i.d. and $\xi_i = \pm 1$ with probability $1/2$). A Banach space $\mathcal{Y}$ is of type s if there exists a constant C_s such that, for all integers n and all $y_1, \ldots, y_n$ in $\mathcal{Y}$,

$$E \left\| \sum_{i=1}^{n} \xi_i y_i \right\|^s \leq C_s \sum_{i=1}^{n} \|y_i\|^s.$$

From Hoffmann-Jørgensen and Pisier (1976), $\mathcal{Y}$ is a Banach space of type s iff there exists a constant C'_s such that, for all independent random variables $Y_1, \ldots, Y_n$ in $\mathcal{Y}$ with mean 0 and finite sth moment,

$$E\left\|\sum_{i=1}^n Y_i\right\|^s \leq C'_s \sum_{i=1}^n E\|Y_i\|^s.$$

Example: Let (Δ, Σ, ν) be a measurable space with σ-finite measure ν. Then for $1 \leq p < \infty$, $L^p_{\mathbb{R}}(\Delta, \Sigma, \nu)$ is of type $s = \min(p, 2)$.

Lemma: (Hoffmann-Jørgensen 1974) Let $Y_1, \ldots, Y_n$ be independent r.v.'s taking values in a separable Banach space with mean 0 and pth moment ($Y_i \in L^p$), $p \geq 1$. Then there exists a constant $C = C(p) > 0$, depending only upon p, such that

$$\left\{E\left[\left\|\sum_{i=1}^n Y_i\right\|^p\right]\right\}^{1/p} \leq C\left\{E\left[\max_{1\leq i\leq n}\|Y_i\|^p\right]\right\}^{1/p} + CE\left[\left\|\sum_{i=1}^n Y_i\right\|\right].$$

(See also Ledoux and Talagrand (1991), Proposition 6.8 or 6.20.)

The following result is a Bernstein type inequality for partial sums of bounded Banach-valued independent random variables.

Lemma: (Pinelis and Sakhanenko (1985) and the remark on page 605 of Pinelis (1990)) Let $Y_1, \ldots, Y_n$ be independent r.v.'s taking values in a separable Banach space $\mathcal{Y}$, and let $y_1, \ldots, y_n$ be n vectors in $\mathcal{Y}$. Assume that $\|Y_i - y_i\| \leq M$ a.s., then for all $\varepsilon > 0$

$$P\left(\left\|\sum_{i=1}^n Y_i\right\| - E\left\|\sum_{i=1}^n Y_i\right\| > \varepsilon\right) \leq \exp\left(-\frac{\varepsilon^2}{2\sum_{i=1}^n E\|Y_i - y_i\|^2 + 2M\varepsilon/3}\right).$$

But if $\|Y_i - y_i\| \leq M_i$ a.s., then (Pinelis 1990) we have, for all $\varepsilon > 0$,

$$P\left(\left\|\sum_{i=1}^n Y_i\right\| - E\left\|\sum_{i=1}^n Y_i\right\| > \varepsilon\right) \leq \exp\left(-\frac{\varepsilon^2}{2\sum_{i=1}^n M_i^2}\right).$$

References

ALLAM, A., MOURID, T. (2001). Propriétés de mélanges des processus autorégressifs banachiques. *C. R. Acad. Sci. Paris, Ser. I*, **333**, 363–8.

BERBEE, H.C.P. (1979). *Random Walks with Stationary Increments and Renewal Theory.* Math. Centre Tracts, Amsterdam.

BERBEE, H.C.P. (1987). Convergence rates in the strong law for bounded mixing sequences. *Probab. Th. Relat. Fields*, **74**, 255–70.

BOSQ, D. (2000). *Linear Processes in Function Spaces. Theory and Applications.* Springer, New York.

BRADLEY, R. (1986). Basic properties of strong mixing conditions. In *Dependence in Probability and Statistics* (E. Eberlein, M.S. Taqqu, eds), 165–92. Birkhäuser, Basel.

BRADLEY, R. (2005). Basic properties of strong mixing conditions. A survey and some open questions. *Probability Surveys*, **2**, 107–44.

DEHLING, H., PHILIPP, W. (1982). Almost sure invariance principles for weakly dependent vector-valued random variables. *Ann. Probab.*, **10**, 689–701.

DEVROYE, L., GYÖRFI, L. (1985). *Nonparametric Density Estimation: The L^1 View*. John Wiley & Sons, New York.

DOUKHAN, P. (1994). *Mixing: Properties and Examples*. Springer, New York.

HOFFMANN-JØRGENSEN, J. (1974). Sums of independent Banach space valued random variables. *Studia Math.*, **52**, 159–86.

HOFFMANN-JØRGENSEN, J., PISIER, G. (1976). The law of large numbers and the central limit theorem in Banach spaces. *Ann. Probab.*, **4**, 587–99.

LEDOUX, M., TALAGRAND, M. (1991). *Probability in Banach Spaces*. Springer, New York.

PINELIS, I.F. (1990). Inequalities for distribution of sums of independent random vectors and their application to estimating density. *Theory Probab. Appl.*, **35**, 605–7.

PINELIS, I.F., SAKHANENKO, A.I. (1985). Remarks on inequalities for large deviation probabilities. *Theory Probab. Appl.*, **30**, 143–8.

RHOMARI, N. (2002a). Approximation et inégalités exponentielles pour les sommes de vecteurs aléatoires dépendants. *C. R. Acad. Sci. Paris, Ser. I*, **334**, 149–54.

RHOMARI, N. (2002b). Contribution à l'estimation fonctionnelle dans les espaces de Banach. Thèse de Doctorat d'Žtat, University Mohamed I, Morocco.

RIO, E. (1995a). A maximal inequality and dependent Marcinkiewicz-Zygmund strong laws. *Ann. Probab.*, **23**, 918–37.

RIO, E. (1995b). The functional law of the iterated logarithm for stationary strongly mixing sequences. *Ann. Probab.*, **23**, 1188–203.

RIO, E. (2000). *Théorie Asymptotique des Processus Aléatoires Faiblement Dépendants*. Springer, New York.

ROSENBLATT, M. (1956). A central limit theorem and a strong mixing condition. *Proc. Nat. Acad. Sci. U.S.A.*, **42**, 43–7.

ROZANOV, YU.A., VOLKONSKII, V.A. (1959). Some limit theorems for random functions I. *Theory Probab. Appl.*, **4**, 178–97.

SHAO, Q.M. (1993). Complete convergence for α-mixing sequences. *Stat. Prob. Lett.*, **16**, 279–87.

SHAO, Q.M. (1995). Maximal inequalities for partial sums of ρ-mixing sequences. *Ann. Probab.*, **23**, 948–65.

VAKHANIA, N.N., TARIELADZE, V.I., CHOBANYAN, S.A. (1987). *Probability Distributions on Banach Spaces*. Reidel, Dordrecht.

WEGMAN, E.J., DAVIES, H.I. (1979). Remarks on some recursive estimators of a probability density. *Ann. Statist.*, **7**, 302–27.

WHEEDEN, R.L., ZYGMUND, A. (1977). *Measure and Integral*. Marcel Dekker, New York.

WOLVERTON, C.T., WAGNER, T.J. (1969). Asymptotically optimal discriminant functions for pattern classification. *IEEE Trans. Inform. Theory*, **IT-15**, 258–65.

YAMATO, H. (1971). Sequential estimation of a continuous probability density function and the mode. *Bull. Math. Statist.*, **14**, 1–12.

YURINSKII, V.V. (1976). Exponetial inequalities for sums of random vectors. *J. Multi. Anal.*, **6**, 473–99.

CHAPTER 15

ON PRODUCT MEASURES ASSOCIATED WITH STATIONARY PROCESSES

ALAIN BOUDOU

YVES ROMAIN

15.1 INTRODUCTION

THE study of stationary processes is an important branch of probability theory and statistics. We can associate different notions with a stationary process: with relation to time, the well-known shift operator and, with relation to frequency, the (projection-valued) spectral measure and the random measure. The connections between these fundamental tools have not, to our knowledge, been much explored in the classical literature; the first motivation for our work has therefore been to clarify the properties of these notions and to explore their connections in depth. The second motivating factor concerns stationary processes which can be written as the product of two independent components. For these processes, a natural question arises: what are the corresponding elements associated with such processes? In other words, can we define the product of two (projection-valued) spectral measures and of two random measures? Can we establish the properties of the shift operator which is associated with such products? These questions admit positive answers and as a consequence we have also been led to set out the convolution products of two spectral measures and of two random measures.

The field of applications of these results is of course large. For example, we present some interpolation problems of one stationary process by another, and the solutions of some functional equations where the unknown quantities are measures.

This chapter is clearly reader-oriented and the three sections which follow are presented in increasing level of complexity. So the Section 15.2 contains historical facts and terminology, basic concepts, our motivation, and our mathematical approach in eight points. Section 15.3 presents the main results and their first potential applications. Without explicit proofs, the reader may find the answers to the questions posed above, and the consequences of these answers. The final section is more technical and presents mathematical developments of the previous results (31 properties are given, along with elements of proofs or hints for the main cases). Finally, we conclude with some remarks on connected, recent, and future work.

15.2 GENERALITIES

15.2.1 Terminology and general background

A *stochastic process* is a family of random variables (r.v.) which is indexed by a set denoted G. To each element g of G, an r.v. may be associated, so we may use the term *random function* instead of process. When G is the set $\mathbb{Z}$ of integers, we speak about *a random series* or *a discrete-time process*. When G is the set $\mathbb{R}$ of real numbers, the process is called *a continuous-time process*. Other cases are well known in the literature as *spatial processes* when G is $\mathbb{Z}^k$ or $\mathbb{R}^k$. In a particular context (for example, geodesic studies), G may be $\Pi = [-\pi, \pi[$ (because of a natural one-to-one map onto the meridian lines). Finally *mixed cases* may also be encountered, for example, when G is $\mathbb{R} \times \mathbb{Z}$.

A random process is called *stationary* when some stability properties are inherent to its behaviour and, more precisely, when some of its characteristics are *translation* invariant. Different definitions of stationarity exist: *strict-sense stationarity* is associated with invariance properties of the probability distribution function of the process, whereas *second-order stationarity* (or *wide-sense stationarity*) is expressed by characteristics such as mathematical expectation or the covariance function (or operator) of the process. In this chapter, we consider the latter notion of stationarity and we will see that the mathematical background of Hilbert spaces is particularly well adapted to this case.

Stationary processes are used in several concrete situations; one of their major mathematical properties is that a stationary process is the *Fourier transform* of a measure. At this stage, the terminology may be made more precise. Instead of the Fourier transform, Brillinger (2001) speaks about the *Cramer representation* and Priestley (1981) uses the term *spectral representation*. The previous measure is called *a random measure* (r.m.) by some authors, namely Rozanov (1967) and Azencott and Dacunha-Castelle (1984); this is justified by the fact that it is defined on subsets with values in random variable sets. Sometimes we may also encounter the term "a process with orthogonal increments."

To say that a process is the Fourier transform of a measure implies that each of its elements may be expressed as an integral and, as the integral is with respect to an r.m., we may speak about the *stochastic integral*. From the r.m. we can usually obtain the "spectral measure" (and the "spectral density," very well known by the statistics community). In fact, in this chapter we use a second possible *spectral measure* whose values are projection operators; this projection-valued measure (PVSM) is also well known and is generally associated with a *unitary operator* (often also called the *shift operator*). This PVSM is also called a *spectral family* by some authors (Riesz and Nagy 1990) and, in some particular cases, the *resolution of the identity* (Dunford and Schwartz 1963). The Fourier transform establishes a duality between the time and the frequency domains. Let us recall that, for example, a stationary series may be considered as the "sum" of noncorrelated periodic processes.

Of course, the terminology "time and frequency domains" is appropriate when G is $\mathbb{Z}$ or $\mathbb{R}$. Nevertheless, when G is $\mathbb{Z}^k$ or $\mathbb{R}^k$ these terms will be maintained.

In the frequency domain, the resolution of problems is sometimes easier. Let us follow Brillinger (2001) and consider this example: take a p-dimensional vector-valued stationary series and try to replace it "as well as possible" by another lower-dimensional stationary series (of course, the criteria of "optimality" has to be specified). Such an analysis is provided by a *principal component analysis* (PCA) of each of its spectral components, which are vectors in a p-dimensional space. The fact of working in the frequency domain also allows various potential problems to be solved: for example, the interpolation of one process by another or the definition of a random field in a given direction (these examples are developed later in this text). In addition, in Brillinger (2001) we may find other good reasons for working in the frequency domain.

15.2.2 Basic concepts and motivation

Let us now present the origins and the development of this work. Although the approach given here is personal, it may be interesting to examine it from different points of view.

At the beginning of our research, we wanted to solve an interpolation problem expressed in this way: from a given stationary series $(Y_n)_{n\in\mathbb{Z}}$ we want to define all the stationary series $(X_n)_{n\in\mathbb{Z}}$ such that $X_{2n} = Y_n$ for all n in $\mathbb{Z}$ (naturally the integer 2 may be replaced by any integer $k \geq 2$). Let us consider the shift operator V (it is a unitary operator) which is associated with the series $(Y_n)_{n\in\mathbb{Z}}$. Thus the above problem is equivalent to that of finding all unitary operators U such that $U^2 = V$. It is easy to verify that, if U is one solution, then $U \circ W$, where W is another unitary operator commuting with U and such that $W^2 = I$, is also a solution. So if we have one solution, we may bring in all the cases. After this we naturally asked another question: given three unitaries U, W, and $U \circ W$, can we find the relationship between their PVSM's which we denote by $\mathcal{E}_U$, $\mathcal{E}_W$, and $\mathcal{E}_{U \circ W}$, respectively? The answer is made easier by the fact that $\mathcal{E}_W$ is concentrated at 0 and $-\pi$ (because $W^2 = I$), and the expression of $\mathcal{E}_{U \circ W}$ in terms of $\mathcal{E}_U$ and $\mathcal{E}_W$ suggests a convolution product. Following this observation, we have been led to determine the (still unknown) convolution product of two PVSM's defined on the Borel σ-field $\mathcal{B}_\Pi$ of Π.

These last results were first obtained in a series framework. For a synthetic and more general presentation, we then considered the case where the index domain G is a group, and particularly a locally compact Abelian group (abbreviated l.c.a.g. in the rest of the chapter). From this point of view, we studied and established the convolution product of two PVSM's defined on $\mathcal{B}_{\widehat{G}}$, the Borel σ-field of $\widehat{G}$, where $\widehat{G}$ denotes the dual group of G.

Therefore the next step is natural: after determining the above convolution product, we were interested in the convolution product of two r.m.'s and we have proved that the Fourier transform of this last convolution product is (up to isometry) the product of two independent stationary random functions.

15.2.3 The mathematical approach in eight points

While trying to answer our various questions, we have been led to develop some new mathematical tools. Here we will present our main results and the corresponding tools in some sort of pedagogical order. We highlight the following eight points:

(i) A set of r.m.'s with some properties in common is named a *stationarily correlated random measure family* (see later for the exact definition) and will be abbreviated RMF. A first tool that will be useful in this chapter is the notion of a *PVSM associated* with an *RMF* (see Section 15.4.4).
(ii) This tool permits us to clarify the link between shift operators and PVSM's which is not so evident in the literature.
(iii) Then, from an r.m. Z, we can build an RMF and, with the previous notion, we are able to associate a PVSM to the r.m. Z.

(iv) So we can propose a first "toolbox" for the case of a stationary series, that presents the relationships between the classical elements which are naturally associated. We summarize this tool box by the following diagram:

Time domain		Frequency domain
$(X_n)_{n\in\mathbb{Z}}$	$\Longleftrightarrow$	Z
the stationary series		the random measure (r.m.)
$\Downarrow$		$\Downarrow$
U	$\Longleftrightarrow$	$\mathcal{E}$
the unitary operator		the spectral measure (PVSM)

Let us notice that Z is the r.m. whose Fourier transform is the series $(X_n)_{n\in\mathbb{Z}}$; U is the shift operator associated with the series $(X_n)_{n\in\mathbb{Z}}$, and $\mathcal{E}$ is the PVSM associated to U and to Z. Let us note that a two-way arrow symbolizes a one-to-one correspondence (as between $(X_n)_{n\in\mathbb{Z}}$ and Z—this is well known—but also between U and the PVSM $\mathcal{E}$). Furthermore, a one-way arrow also gives another kind of information: from a stationary series $(X_n)_{n\in\mathbb{Z}}$ only one shift operator U may be defined, but conversely from this shift (or unitary) operator we may build an infinity of stationary series. The other arrow tells us that to a given r.m. Z only one PVSM may be associated, but conversely this same PVSM may be associated with several r.m.'s.

Another new result here may be discovered by looking at the associations between the PVSM $\mathcal{E}$ and the unitary U and the r.m.: the PVSM $\mathcal{E}$ may be defined from the r.m. whose Fourier transform is the series $(X_n)_{n\in\mathbb{Z}}$.

A final remark about the above diagram would seem interesting: the right part of the schema is in the frequency domain and the left part is in the time domain. So the double horizontal arrows underline the duality between these two well-known domains.

(v) Now consider two PVSMs $\mathcal{E}_1$ and $\mathcal{E}_2$ (with some commutativity properties to be specified). We can build an RMF and then define a third PVSM (using the previous link given between a PVSM and a RMF) which is denoted by $\mathcal{E}_1 \otimes \mathcal{E}_2$. If U_1 and U_2 are the unitary operators whose PVSM's are respectively $\mathcal{E}_1$ and $\mathcal{E}_2$, we show that $\mathcal{E}_1 * \mathcal{E}_2$—the image of $\mathcal{E}_1 \otimes \mathcal{E}_2$ under a sum mapping—is the PVSM associated to the unitary operator $U_1 \circ U_2$. In this way, the convolution product is a kind of transport in the frequency domain of the composition of the two unitaries U_1 and U_2. Furthermore, this product admits remarkable algebraic properties such as the following: the image of a convolution product under a homomorphism is the convolution product of the images.

(vi) Various applications may be considered as "corollaries" of these theoretical results. For instance, the previous result gives the solutions of some equations where the unknown quantities are r.m.s; then some interpolation problems of stationary functions can be tackled. Here the work in the frequency domain appears to be useful although the problem arises in the time domain.

(vii) The different properties of the convolution product permit us to associate a family of unitaries to a PVSM. So the links between a stationary process and its spectral elements are clearer and may be presented in a general framework. Therefore we are able to provide a second toolbox where the index set G is now a group:

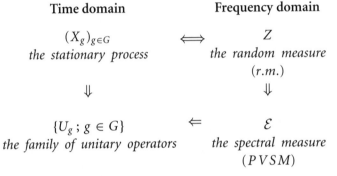

In this diagram Z is the r.m. whose Fourier transform is the stationary continuous random function $(X_g)_{g \in G}$, and $\{U_g \,;\, g \in G\}$ is the family of shift operators, that is, the group of unitaries derived from the PVSM $\mathcal{E}$ which is associated to the r.m. Z. Note that the arrows have the same significances as for the first box (with a slight difference between the family of shift operators and the PVSM). Then, in this diagram, we may keep in mind that the family $\{U_g \,;\, g \in G\}$ of shift operators is derived from the PVSM which is itself derived from the r.m. Z, the r.m. whose Fourier transform is precisely the stationary process $(X_g)_{g \in G}$.

(viii) Finally, using parallel work on *ampliation operators* (see Romain (2002)), we are able to define the product of two PVSM's as studied by Broszkiewicz (1987) and Birman and Solomjak (1996). From this new product, the convolution product of two r.m.'s may be derived. This convolution product is also an r.m. whose Fourier transform is (under certain conditions) a stationary process which may be written in a multiplicative form using two independent stationary processes.

The eight points detailed above give new tools, provide explanatory and original results, and enlarge the fields of application. In the following section, we develop the mathematical framework and the main results in more detail.

15.3 Main results and some applications

15.3.1 The mathematical framework

In this section the mathematical environment is specified, as well as some choices we have been led to make.

The scalar field is the complex plane $\mathbb{C}$ (here the Fourier transform plays an important role) and let us recall that even though the elements of the stationary process are real r.v.'s, their spectral components may be complex r.v.'s.

As we focus on stationary processes, the set of r.v.'s are of $L^2(\Omega, \mathcal{A}, P)$-type because the existence of the second-order moment is imposed by the stationarity expressed in the covariances of the r.v.'s. In fact, the Hilbert geometry of $L^2(\Omega, \mathcal{A}, P)$ is very important and useful, and almost all the spectral theory results for stationary processes may be presented in the extended framework where $L^2(\Omega, \mathcal{A}, P)$ is replaced by a $\mathbb{C}$-Hilbert space H. This is the reason why the definition of the r.m. we have adopted has values in H. This set function is generally defined on a σ-field $\mathcal{F}$ (as is usual for a "classic" measure); nevertheless this σ-algebra is in most cases the Borel σ-field $\mathcal{B}_{\widehat{G}}$ of $\widehat{G}$ (where $\widehat{G}$ is the dual group of G). The r.m. Z is a vector-valued measure satisfying the σ-summability property and such that, for two disjoints elements A and B of $\mathcal{F}$, the images $Z(A)$ and $Z(B)$ are orthogonal (note that this last point has given rise to the terminology "orthogonal increment process"). Then we may affirm that the set function μ_Z defined by $\mu_Z(A) = \|Z(A)\|^2$ is a finite measure. This measure is called the *spectral measure* of the process (not to be confused with the PVSM) and its derivative (in the Radon–Nikodym sense with regard to the Lebesgue measure, if it exists) is classically called the *spectral density* of the process.

Let us specify now which kind of stochastic integral will be used in this chapter. Here, an isometry is a one-to-one linear mapping between two $\mathbb{C}$-Hilbert spaces which leaves the scalar product invariant. Let us consider the space $\overline{\text{span}}\,\{1_A\,;\,A \in \mathcal{F}\}$ (this is in fact $L^2(\mu_Z)$ with the well-known property of density of the set of indicator mappings) and take H_Z to be the closed subspace defined by $\overline{\text{span}}\,\{Z(A)\,;\,A \in \mathcal{F}\}$. Then the stochastic integral is defined as the unique isometry between $L^2(\mu_Z)$ and H_Z which associates $Z(A)$ to 1_A, for each A in $\mathcal{F}$. Let us remark that the random element in the integral is the differential term: this fact has to be compared with other kinds of stochastic integrals. In the Ito integral, the part to be integrated and the differential term are both random; if, in a third kind of integral, only the part to be integrated is random, we can speak of the integral of random variables.

At this stage, the reader has to keep in mind the importance of the orthogonality property (given in the definition of a r.m.) when building the stochastic integral. Furthermore, the r.m. is a vector-valued measure and in general we are able to give the integral with respect to such a measure when it has finite variation (see Dinculeanu (2000)). In fact, this technique is not available for our stochastic integral because in general an r.m. may not be of finite-variation type (see, for instance, the r.m. associated with white noise).

After defining the stochastic integral, we may ask some questions concerning the image of an r.m. under a measurable mapping and concerning the existence of a transport-type theorem. In both cases we establish the answers: the image of an r.m. under a measurable mapping is also an r.m., and using a suitable transport property we may give the stochastic integral of such an image measure.

Let us now specify the choice of the index domain G. In the literature, studies of series are presented first and, if possible, extended to random functions. If we want to give a synthetic presentation of all the (major) possible cases, we are led to consider G to be an Abelian group. In fact, with a topology on G under appropriate conditions, we obtain a topological group. In all this chapter we assume that G is a l.c.a.g. such that its dual $\widehat{G}$ admits a countable basis: this last restriction is useful technically because on the one hand (see Métivier (1968)) we know that each measure defined on $\mathcal{B}_{\widehat{G}}$, the Borel σ-field of $\widehat{G}$, is regular, and on the other hand, that $\mathcal{B}_{\widehat{G}} \otimes \mathcal{B}_{\widehat{G}}$ is the Borel σ-field of $\widehat{G} \times \widehat{G}$. We may say that this restriction is very light because all the classical cases, $\mathbb{Z}$, $\mathbb{R}$, $\mathbb{Z}^k$, $\mathbb{R}^k$, $\Pi = [-\pi, \pi[$, Π^k, and $\mathbb{Z}\times\mathbb{R}$ are contained in our choice. Of course, as usual $\widehat{G} \times \widehat{G}$ (respectively, $\mathbb{R}$, Π, $\mathbb{Z}$) is identified with the dual of $G \times G$ (respectively, $\mathbb{R}$, $\mathbb{Z}$, Π). Rudin (1967) provides a study of the relationship between the topology on a group and the topology on its dual group. Finally here, let us note that Π admits a group structure for which the corresponding sum law, denoted by $\oplus$, is the sum modulo 2π.

Another notion will play a particular role in what follows: the transpose of a continuous homomorphism between groups. Recall that a homomorphism is a map preserving the group structures and, as it is continuous, it is also measurable. The transpose, as defined in Berg and Forst (1975), is also a continuous homomorphism and is called the *dual homomorphism*. We have preferred the term "transpose" because of the analogy of its definition with that of the transpose of a linear mapping. We may note that the algebraic properties of the transpose of a homomorphism are similar to the corresponding ones for a linear mapping. To conclude the presentation of our mathematical environment, let us now present the development of some notions associated with the stationary process.

Usually, the set of r.v.'s is $L^2(P)$ and, as the process mathematics only rests on the Hilbertian structure, for the sake of generalization in this chapter a $\mathbb{C}$-Hilbert space H is substituted for $L^2(P)$. We consider the notion of a *continuous stationary random function* (CSRF), which is a family $(X_g)_{g \in G}$ of elements of H such that the mapping $g \longmapsto X_g$ is continuous and such that

$$\langle X_g, X_{g'} \rangle = \langle X_{g-g'}, X_0 \rangle$$

for all pairs (g, g') of G.

When H is $L^2(P)$-type and when each X_g is centered, this gives the classical notion of wide-sense stationarity.

We show that a unique r.m. Z exists (called the *r.m. associated to the CSRF* $(X_g)_{g \in G}$) such that, for all $g \in G$:

$$X_g = \int (\cdot, g)_{\widehat{G}G} \, dZ,$$

and we say that $(X_g)_{g \in G}$ is the Fourier transform of the r.m. Z.

This last result is well known when $H = L^2(P)$. Its proof in the general case is based on Bochner's Theorem (see Rudin (1967)), which allows us to affirm the existence of the measure μ_Z whenever the map $g \longmapsto \langle X_g, X_0 \rangle$ is positive.

Now consider two CSRF's $(X_g)_{g \in G}$ and $(X'_g)_{g \in G}$. They are said to be *stationarily correlated* when

$$\langle X_g, X'_{g'} \rangle = \langle X_{g-g'}, X'_0 \rangle$$

for all pairs $(g, g') \in G$. In the spectral domain, this definition implies that, for all disjoint Borel subsets A and B,

$$\langle Z(A), Z'(B) \rangle = 0,$$

where Z and Z' are the r.m.'s respectively associated with the CSRF's $(X_g)_{g \in G}$ and $(X'_g)_{g \in G}$. In this case, Z and Z' are also said to be *stationarily correlated*.

A last point concerns the PVSM's: they are defined with values in the set $\mathcal{P}(H)$ of orthogonal projectors of H. They satisfy the property of additivity and consequently, for a PVSM $\mathcal{E}$ and for all pairs (A, B) of $\mathcal{F}$, we have the equality

$$(\mathcal{E}(A)) \circ (\mathcal{E}(B)) = \mathcal{E}(A \cap B).$$

Some final notation is given by the following definition: $\mathcal{E}(\cdot)(X)$ is an r.m. and we put $Z_{\mathcal{E}}^X(\cdot) = \mathcal{E}(\cdot)(X)$.

15.3.2 The PVSM associated with a unitary operator

Let us recall first that the shift operators (i.e. the linear mappings $X_n \longmapsto X_{n+1}$ for a stationary random series) are unitary operators (i.e. $U^* \circ U = U \circ U^* = I$, where U^* is the adjoint of U). The link which leads us from the unitary operator to a PVSM is well known (see, for example, Dunford and Schwartz (1963) or Riesz and Nagy (1990)). Therefore to each PVSM $\mathcal{E}$ for H on $\mathcal{B}_\Pi$, we can associate a unitary operator U: this unitary operator may be expressed in two ways. For an element X of H we have

$$U(X) = \int e^{i \cdot 1} \, dZ_{\mathcal{E}}^X(\cdot),$$

or we also have

$$U = \lim_n \sum_{k=0}^{k=n-1} e^{i(-\pi + k\frac{2\pi}{n})} \mathcal{E}\left(\left[-\pi + k\frac{2\pi}{n}, -\pi + (k+1)\frac{2\pi}{n}\right]\right),$$

which may remind us of a Riemann sum.

Conversely, to a unitary operator U we may associate a PVSM $\mathcal{E}$, which is particular because of the previous notion of a PVSM associated with an RMF (here, this is the family of r.m. Z^X associated with the different stationary series $(U^n(X))_{n\in\mathbb{Z}}$ which are pairwise stationarily correlated). Then the PVSM $\mathcal{E}$ is defined by

$$(\mathcal{E}(A))(X) = Z^X(A)$$

for all (A, X) of $\mathcal{B}_\Pi \times H$.

So when a unitary operator V is such that $V^2 = I$ then, for each $X \in H$, $(V^n(X))_{n\in\mathbb{Z}}$ is a stationary series. As it is also periodic (the period is 2), the r.m. of which it is the Fourier transform may be written

$$Z^X(\cdot) = \delta_0(\cdot)\frac{1}{2}(X + V(X)) + \delta_{-\pi}(\cdot)\frac{1}{2}(X - V(X)),$$

where δ_λ is the Dirac measure at λ.

With the previous properties, the PVSM $\mathcal{E}_V$ which is associated with V is such that

$$(\mathcal{E}_V(A))(X) = Z^X(A),$$

and then

$$\mathcal{E}_V(\cdot) = \delta_0(\cdot)\frac{1}{2}(I + V) + \delta_{-\pi}(\cdot)\frac{1}{2}(I - V),$$

where we remark that $\frac{1}{2}(I + V)$ and $\frac{1}{2}(I - V)$ are orthogonal projectors whose product is zero.

The expression of V relative to the PVSM $\mathcal{E}_V$ is then

$$V = \mathcal{E}_V(\{0\}) + e^{-i\pi}\mathcal{E}_V(\{-\pi\}),$$

because $V(X) = \int e^{i\cdot 1}\, dZ^X$.

15.3.3 The convolution product of two PVSM's

Consider two unitaries U and V; then $U \circ V$ is also a unitary operator. We want to determine the relationship between the PVSM of $U \circ V$ and those of U and V. When U and V commute, the answer is clear.

Let us note first the particular case of the previous section where

$$V = \mathcal{E}_V(\{0\}) + e^{-i\pi}\mathcal{E}_V(\{-\pi\}).$$

In this case, it is easy to establish that

$$\mathcal{E}_{U \circ V}(A) = (\mathcal{E}_V(\{0\})) \circ (\mathcal{E}_U(A \ominus 0)) + (\mathcal{E}_V(\{-\pi\}))) \circ (\mathcal{E}_U(A \ominus (-\pi))).$$

This equality suggests a convolution product.

Having in mind the convolution product for probability measures ($P_1 * P_2 = S(P_1 \otimes P_2)$, where S is the sum mapping), we have first to define the PVSM associated with the product $\mathcal{E}_V \otimes \mathcal{E}_U$. This is the PVSM associated with the RMF whose elements are associated to the stationary series $((V^m \circ U^n)(X))_{(n,m) \in \mathbb{Z} \times \mathbb{Z}}$ (stationarity here comes from the commutativity of U and V).

For each (A, B) of $\mathcal{B}_\Pi \times \mathcal{B}_\Pi$, we have

$$(\mathcal{E}_U \otimes \mathcal{E}_V)(A \times B) = (\mathcal{E}_U(A)) \circ (\mathcal{E}_V(B))$$

and the notation $\otimes$ seems appropriate: let us remark that, as $U \circ V = V \circ U$, $\mathcal{E}_U(A)$, and $\mathcal{E}_V(B)$ are commuting projectors, then $(\mathcal{E}_U(A)) \circ (\mathcal{E}_V(B))$ is an orthogonal projector.

Then consider the measurable sum mapping

$$S_\Pi : (\lambda_1, \lambda_2) \in \Pi \times \Pi \longmapsto \lambda_1 \oplus \lambda_2 \in \Pi.$$

By this mapping, the PVSM image of $\mathcal{E}_U \otimes \mathcal{E}_V$ is naturally called the *convolution product* of $\mathcal{E}_U$ and $\mathcal{E}_V$ and is denoted by $\mathcal{E}_U * \mathcal{E}_V$. Furthermore, we can verify that this is the PVSM associated with the unitary operator $U \circ V$.

15.3.4 The unitary operator group generated by a PVSM

Since we can define a unitary operator from a PVSM on $\mathcal{B}_\Pi$ for H, we are able to define a family of unitaries from a PVSM $\mathcal{E}$ on $\mathcal{B}_{\widehat{G}}$ for H. It is possible to give a group structure to this family, which justifies the terminology "the unitary operators group deduced from the PVSM $\mathcal{E}$". To this end, we proceed in the following way.

For $g \in G$, let us denote by ${}^t h_g$ the transpose homomorphism of the mapping

$$n \in \mathbb{Z} \longmapsto ng \in G,$$

which is measurable from $\widehat{G}$ in Π (itself identified with the dual group of $\mathbb{Z}$). Then, ${}^t h_g(\mathcal{E})$, the image of the PVSM $\mathcal{E}$ under ${}^t h_g$, is a PVSM on $\mathcal{B}_\Pi$ for H and it may be associated with the unitary operator U_g in H.

Using some algebraic properties of ${}^t h_g$ and some results on the convolution product of PVSM on $\mathcal{B}_\Pi$ for H, we can write:

$${}^t h_{g_1+g_2}(\mathcal{E}) = ({}^t h_{g_1}(\mathcal{E})) * ({}^t h_{g_2}(\mathcal{E})),$$

and then

$$U_{g_1+g_2} = U_{g_1} \circ U_{g_2}.$$

That is the reason why the family $\{U_g\,;\,g \in G\}$ admits a group structure and is called the *unitary operator group* in H deduced from the PVSM $\mathcal{E}$.

Furthermore, the homomorphism family $\{{}^t h_g\,;\,g \in G\}$ has the following property: for two PVSM's $\mathcal{E}_1$ and $\mathcal{E}_2$ on $\mathcal{B}_{\widehat{G}}$ for H such that ${}^t h_g(\mathcal{E}_1) = {}^t h_g(\mathcal{E}_2)$ for all $g \in G$, then $\mathcal{E}_1 = \mathcal{E}_2$.

In the particular case where $\mathcal{E}$ is the PVSM associated with the r.m. Z, defined on $\mathcal{B}_{\widehat{G}}$, then U_g is the unique unitary operator in H_Z such that, for all $g' \in G$,

$$U_g\left(\int (\cdot, g')_{\widehat{G}G}\,dZ\right) = \int (\cdot, g+g')_{\widehat{G}G}\,dZ.$$

This last equation sets out the structure of the shift operator family which is associated with a CSRF.

15.3.5 Extension of the convolution product of two PVSM's

It is natural to try to extend the definition of the convolution product of two PVSM's $\mathcal{E}_1$ and $\mathcal{E}_2$ to the case when they are on $\mathcal{B}_{\widehat{G}}$ for H. Following a similar approach to that used previously, we obtain this extension by using the PVSM associated with an RMF. So we can show the existence of a unique PVSM on $\mathcal{B}_{\widehat{G}} \otimes \mathcal{B}_{\widehat{G}}$ for H which is denoted by $\mathcal{E}_1 \otimes \mathcal{E}_2$ and satisfies, for all (A, B) of $\mathcal{B}_{\widehat{G}} \times \mathcal{B}_{\widehat{G}}$,

$$(\mathcal{E}_1 \otimes \mathcal{E}_2)(A \times B) = (\mathcal{E}_1(A)) \circ (\mathcal{E}_2(B)).$$

The image of the PVSM $\mathcal{E}_1 \otimes \mathcal{E}_2$ under the sum mapping

$$S_{\widehat{G}} : (\gamma_1, \gamma_2) \in \widehat{G} \times \widehat{G} \longmapsto \gamma_1 + \gamma_2 \in \widehat{G}$$

is called the *convolution product of the PVSM's $\mathcal{E}_1$ and $\mathcal{E}_2$* and is denoted $\mathcal{E}_1 * \mathcal{E}_2$.

This product admits some remarkable algebraic properties: the PVSM $\mathcal{E}_{\widehat{G}} = \delta_0(\cdot)I$ is the identity for the convolution product, each PVSM has an inverse, and the convolution product is associative and commutative. Nevertheless, we cannot speak about a group structure because a convolution product is not possible when there is no commutativity. Finally, we have a "distributivity" property: the image under a homomorphism of a convolution product is the convolution product of the images.

Using this last result, we can prove that if $\{U_g^1\,;\,g \in G\}$ and $\{U_g^2\,;\,g \in G\}$ are unitary operator groups in H deduced from the PVSM's $\mathcal{E}_1$ and $\mathcal{E}_2$, respectively (which have to commute), then $\{U_g^1 \circ U_g^2\,;\,g \in G\}$ is the unitary operator group in H deduced from the PVSM $\mathcal{E}_1 * \mathcal{E}_2$.

15.3.6 An equation where the unknown quantity is a PVSM

Let us solve an equation where the unknown quantity is a PVSM. More precisely, we define all the PVSM's whose images under a homomorphism are a given PVSM. When the problem of defining all the unitaries U such that $U^2 = V$ (with V given and unitary) is transposed into the frequency domain, then we are led to such an equation.

To solve it, we use the distributivity property of the convolution product (given in the previous section). Then the general solution combines a particular solution and a solution of a homogeneous equation: that is, a PVSM whose image under the homomorphism being considered gives the identity element for the convolution product. Furthermore, we are able to solve equations where the unknown quantity is now an r.m. instead of a PVSM.

15.3.7 An interpolation problem

Let us present two apparently different problems (which are solved in Boudou (2003) and in Boudou (2007)).

Problem 15.1 *From a given stationary series* $(Y_n)_{n\in\mathbb{Z}}$, *can we define all the stationary series* $(X_n)_{n\in\mathbb{Z}}$ *such that* $(X_{nq})_{n\in\mathbb{Z}} = (Y_n)_{n\in\mathbb{Z}}$?

Problem 15.2 *From a given CSRF* $(Y_t)_{t\in\mathbb{R}}$, *can we define all the CSRF's* $(X_{(t_1,t_2)})_{(t_1,t_2)\in\mathbb{R}\times\mathbb{R}}$ *such that* $(X_{(at,bt)})_{t\in\mathbb{R}} = (Y_t)_{t\in\mathbb{R}}$ *with* $(a,b) \in \mathbb{R}\times\mathbb{R}$?

These two questions seem different but we can assert that they are particular cases of a more general problem which can be presented in the following form:

General problem *From a given CSRF defined on the l.c.a.g.* G_1, $(Y_{g_1})_{g_1\in G_1}$, *and with a given continuous homomorphism* h *from* G_1 *into* G_2, *a second l.c.a.g., can we determine all the CSRF's* $(X_{g_2})_{g_2\in G_2}$ *such that* $(X_{h(g_1)})_{g_1\in G_1} = (Y_{g_1})_{g_1\in G_1}$?

This problem may be approached by solving the equation

$$X \circ h = Y,$$

where Y is the mapping

$$g_1 \in G_1 \longmapsto Y_{g_1} \in H$$

and X is the mapping

$$g_2 \in G_2 \longmapsto X_{g_2} \in H.$$

Then we use the one-to-one correspondence between the r.m. and the CSRF, and the problem is transposed into the frequency domain as follows:

From a given Z_Y defined on $\mathcal{B}_{\widehat{G_1}}$ with values in H, can we determine all the r.m.'s Z_X defined on $\mathcal{B}_{\widehat{G_2}}$ with values in H such that ${}^t h(Z_X) = Z_Y$?

Here the reader is referred to the previous section which gives the solution of such an equation.

15.3.8 The convolution product of two r.m.'s

After defining the convolution product for two PVSM's, our interest turns to the convolution product for two r.m.'s. A particular tool will play an interesting role here: it is called the *functional tensor product*, denoted by $\overset{l}{\otimes}$, and defined in the following way (see Romain (2002) for more details). Let A_1 (respectively, A_2) be a continuous endomorphism of the separable $\mathbb{C}$-Hibert space H_1 (respectively, H_2), then the functional tensor product of A_1 and A_2, denoted by $A_1 \overset{l}{\otimes} A_2$, is defined by the mapping:

$$K \in \sigma_2(H_1, H_2) \longmapsto A_2 \circ K \circ A_1^* \in \sigma_2(H_1, H_2),$$

where $\sigma_2(H_1, H_2)$ is the space of Hilbert–Schmidt operators from H_1 into H_2.

When $\mathcal{E}$ is a PVSM on $\mathcal{B}_{\widehat{G}}$ for H_2 (respectively, H_1), then the mapping

$$E^1_{\mathcal{E}}(\cdot) = I_{H_1} \overset{l}{\otimes} \mathcal{E}(\cdot)$$

(respectively, $E^2_{\mathcal{E}}(\cdot) = \mathcal{E}(\cdot) \overset{l}{\otimes} I_{H_2}$) is a PVSM on $\mathcal{B}_{\widehat{G}}$ for $\sigma_2(H_1, H_2)$ called *the left ampliation of $\mathcal{E}$ in regard to H_1* (respectively, *the right ampliation of $\mathcal{E}$ in regard to H_2*).

Now consider $\mathcal{E}_1$ (respectively, $\mathcal{E}_2$), a PVSM on $\mathcal{B}_{\widehat{G}}$ for H_1 (respectively, H_2), then the PVSM's $E^2_{\mathcal{E}_1}$ and $E^1_{\mathcal{E}_2}$ are commuting and so we can take their product $E^2_{\mathcal{E}_1} \otimes E^1_{\mathcal{E}_2}$, which is a PVSM on $\mathcal{B}_{\widehat{G}} \otimes \mathcal{B}_{\widehat{G}}$ for $\sigma_2(H_1, H_2)$ and which is such that

$$E^2_{\mathcal{E}_1} \otimes E^1_{\mathcal{E}_2}(A \times B) = (\mathcal{E}_1(A)) \overset{l}{\otimes} (\mathcal{E}_2(B))$$

for all pairs (A, B) of elements of $\mathcal{B}_{\widehat{G}}$.

Finally, when $\mathcal{E}_1$ (respectively, $\mathcal{E}_2$) is the PVSM associated with the r.m. Z_1 (respectively, Z_2) which is defined on $\mathcal{B}_{\widehat{G}}$ with values in H_1 (respectively, H_2), we denote by $Z_1 \otimes Z_2$ the *product of the r.m.'s Z_1 and Z_2*, and the r.m. $Z^{Z_1(\widehat{G}) \otimes Z_2(\widehat{G})}_{E^2_{\mathcal{E}_1} \otimes E^1_{\mathcal{E}_2}}$ is defined on $\mathcal{B}_{\widehat{G}} \otimes \mathcal{B}_{\widehat{G}}$ with values in $\sigma_2(H_1, H_2)$. The image of $Z_1 \otimes Z_2$ under the mapping $S_{\widehat{G}}$ is then denoted by $Z_1 * Z_2$ and is called the *convolution product of the r.m.'s Z_1 and Z_2*. Furthermore, we can establish that its Fourier transform is the CSRF

$$\left(\left(\int (\cdot, -g)_{\widehat{G}G} \, dZ_1 \right) \otimes \left(\int (\cdot, g)_{\widehat{G}G} \, dZ_2 \right) \right)_{g \in G}.$$

15.4 Mathematical developments

In this section, we make the previous definitions and the mathematical tools and properties we have used or developed more precise. The proofs are often shortened and some difficulties (particularly to do with continuity problems) are avoided. The reader may refer back to the different steps in the previous sections. We provide this section in order to present this chapter in an almost self-contained form.

Let us recall that $\mathcal{B}_\Pi$ is the Borel σ-algebra of $\Pi = [-\pi, \pi[$, G (respectively, G_1, G_2) is an Abelian group, the dual group $\widehat{G}$ (respectively, $\widehat{G_1}$, $\widehat{G_2}$) of which admits a countable basis, and its Borel σ-algebra is denoted by $\mathcal{B}_{\widehat{G}}$ (respectively, $\mathcal{B}_{\widehat{G_1}}$, $\mathcal{B}_{\widehat{G_2}}$); finally, we denote by $\mathcal{P}(H)$ the orthogonal projector set in the $\mathbb{C}$-Hilbert space H.

15.4.1 Random measure (r.m.) and spectral measure (PVSM)

In this section $(F, \mathcal{F})$ is a measurable space. The first notion to be examined is what we call an r.m.

Definition 15.1 *An r.m. Z defined on $\mathcal{F}$ with values in H is a mapping from $\mathcal{F}$ into H such that:*

(i) *for all pairs (A, B) of disjoint elements of $\mathcal{F}$, $Z(A \cup B) = Z(A) + Z(B)$ and $\langle Z(A), Z(B) \rangle = 0$;*

(ii) *for all decreasing sequences $(A_n)_{n \in \mathbb{N}}$ of elements in $\mathcal{F}$ which converge to $\emptyset$ then, $\lim_n Z(A_n) = 0$.*

From this condition we can prove the following:

Property 15.1 *If Z is an r.m. defined on $\mathcal{F}$ with values in H, the mapping $\mu_Z : A \in \mathcal{F} \longmapsto \|Z(A)\|^2 \in \mathbb{R}_+$ is a finite positive measure.*

The second important notion to be specified is the stochastic integral.

Property 15.2 *If Z is an r.m. defined on $\mathcal{F}$ with values in H, a unique isometry exists from $L^2(F, \mathcal{F}, \mu_Z)$ onto $H_Z = \overline{\text{span}}\{Z(A), A \in \mathcal{F}\}$ which associates $Z(A)$ to 1_A for all $A \in \mathcal{F}$. The image under this isometry of an element φ of $L^2(F, \mathcal{F}, \mu_Z)$ is called the integral of φ with respect to the r.m. Z and is denoted by $\int \varphi \, dZ$.*

Consequently, we can also define the measure image of an r.m.:

Property 15.3 *If Z is an r.m. defined on $\mathcal{F}$ with values in H, if $(F', \mathcal{F}')$ is a second measurable space, and $\mathcal{L}$ is a measurable mapping from F into F', then:*

(i) *the mapping $\mathcal{L}(Z) : A' \in \mathcal{F}' \longmapsto Z(\mathcal{L}^{-1} A') \in H$ is an r.m. called the r.m. image of Z under $\mathcal{L}$;*

(ii) *$\mathcal{L}(\mu_Z) = \mu_{\mathcal{L}(Z)}$;*

(iii) if φ is in $L^2(F', \mathcal{F}', \mu_{\mathcal{L}(Z)})$, then $\varphi \circ \mathcal{L}$ belongs to $L^2(F, \mathcal{F}, \mu_Z)$ and $\int \varphi \, d\mathcal{L}(Z) = \int \varphi \circ \mathcal{L} \, dZ$.

Let us now give a useful result for proving uniqueness properties.

Property 15.4 *If Z_1 and Z_2 are r.m.'s defined on $\mathcal{F} \otimes \mathcal{F}$ with values in H such that $Z_1(A \times B) = Z_2(A \times B)$ for each pair (A, B) of elements of $\mathcal{F}$, then they are equal.*

Now let us consider a spectral measure taking its values in the set of orthogonal projectors $\mathcal{P}(H)$ of H; such a measure is denoted by PVSM.

Definition 15.2 *A PVSM $\mathcal{E}$ on $\mathcal{F}$ for H is a mapping from $\mathcal{F}$ into $\mathcal{P}(H)$ such that:*

(i) *for each pair (A, B) of disjoint elements of $\mathcal{F}$, $\mathcal{E}(A \cup B) = \mathcal{E}(A) + \mathcal{E}(B)$;*
(ii) *when $(A_n)_{n \in \mathbb{N}}$ is a decreasing sequence of elements of $\mathcal{F}$ which converges to $\emptyset$ then, for each X in H, $\lim_n \mathcal{E}(A_n)X = 0$.*

Using the previous results it is not difficult to obtain those that follow.

Property 15.5 *If $\mathcal{E}$ is a PVSM defined on $\mathcal{F}$ for H, then:*

(i) *for each pair (A, B) of elements of $\mathcal{F}$, one has $\mathcal{E}(A \cap B) = \mathcal{E}(A) \circ \mathcal{E}(B)$;*
(ii) $\mathcal{E}(\emptyset) = 0$;
(iii) *for each X in H, the mapping $Z_\mathcal{E}^X : A \in \mathcal{F} \longmapsto \mathcal{E}(A)X \in H$ is a r.m.*

Let us now consider the image of a PVSM under a mapping.

Property 15.6 *If $\mathcal{E}$ is a PVSM defined on $\mathcal{F}$ for H, $(F', \mathcal{F}')$ a second measurable space, and $\mathcal{L}$ a measurable mapping from F into F', then the mapping $\mathcal{L}(\mathcal{E}) : A' \in \mathcal{F}' \longmapsto \mathcal{E}(\mathcal{L}^{-1}A') \in \mathcal{P}(H)$ is a PVSM called the PVSM image of $\mathcal{E}$ under $\mathcal{L}$.*

15.4.2 The transpose mapping of a homomorphism

Let us recall that a homomorphism h from G_1 into G_2 is a mapping such that $h(g + g') = h(g) + h(g')$ for each pair (g, g') of elements of G_1. As in the case of the transpose of a linear mapping, the transpose of a homomorphism takes into account the dual spaces:

Property 15.7 *If h is a continuous homomorphism from G_1 into G_2 then:*

(i) *for each γ_2 of $\widehat{G_2}$, $\gamma_2 \circ h$ belongs to $\widehat{G_1}$;*
(ii) *the mapping ${}^th : \gamma_2 \in \widehat{G_2} \longmapsto \gamma_2 \circ h \in \widehat{G_1}$ is a continuous homomorphism from $\widehat{G_2}$ into $\widehat{G_1}$ called the transpose homomorphism of h;*
(iii) *one has $(\gamma_2, h(g_1))_{\widehat{G_2}G_2} = ({}^th(\gamma_2), g_1)_{\widehat{G_1}G_1}$ for each (γ_2, g_1) of $\widehat{G_2} \times G_1$.*

Furthermore, with obvious notation it easy to verify the formulae:

$${}^t(h_1 \circ h_2) = {}^t h_2 \circ {}^t h_1 \text{ and } {}^t(h_1 + h_2) = {}^t h_1 + {}^t h_2.$$

15.4.3 Stationarily correlated random measure families (RMF's)

With the same notation as above, let us recall that two r.m.'s Z_1 and Z_2 defined on $\mathcal{F}$ with values in H are said to be stationarily correlated when $\langle Z_1(A), Z_2(B)\rangle = 0$ for each pair (A, B) of disjoint elements of $\mathcal{F}$.

Definition 15.3 *A set $\{Z^X ; X \in H\}$ of r.m.'s defined on $\mathcal{F}$ with values in H constitutes a stationarily correlated random measure family, denoted by RMF, when the r.m.'s are pairwise stationarily correlated and when $Z^X(F) = X$ for each X in H.*

To such a family we can associate a PVSM in the following manner (see Boudou (2000)):

Property 15.8 *If $\{Z^X ; X \in H\}$ is an RMF defined on $\mathcal{F}$ with values in H, then:*
  (i) *for each $A \in \mathcal{F}$, the mapping $\mathcal{E}(A) : X \in H \longmapsto Z^X(A) \in H$ is an orthogonal projector;*
  (ii) *the mapping $\mathcal{E} : A \in \mathcal{F} \longmapsto \mathcal{E}(A) \in \mathcal{P}(H)$ is a PVSM on $\mathcal{F}$ for H, called the PVSM associated to the RMF $\{Z^X ; X \in H\}$.*

15.4.4 The PVSM associated to a r.m.

The one-to-one mapping set out in the definition of the stochastic integral with respect to an r.m. enables us to give the following result:

Property 15.9 *If Z is a random measure defined on $\mathcal{F}$ with values in H, we can affirm that:*
  (i) *for each X in H_Z, the mapping $Z^X : A \in \mathcal{F} \longmapsto \int 1_A \varphi \, dZ \in H_Z$, where φ is the unique element of $L^2(\mu_Z)$ such that $\int \varphi \, dZ = X$, is an r.m.;*
  (ii) *$\{Z^X ; X \in H\}$ is an RMF and the PVSM $\mathcal{E}$ associated to Z is called the PVSM associated to the family;*
  (iii) *for each $A \in \mathcal{F}$, one has $\mathcal{E}(A)(Z(F)) = Z(A)$.*

15.4.5 The PVSM associated with a unitary operator

Let U be a given unitary operator in H. It is easy to verify that $\{(U^n(X))_{n\in\mathbb{Z}} ; X \in H\}$ constitutes a family of stationary series that are pairwise correlated. Then if, for each X of H, we denote by Z^X the r.m. associated with the stationary series $(U^n(X))_{n\in\mathbb{Z}}$, we can assert the

Property 15.10 *When U is a unitary operator in H, then $\{Z^X ; X \in H\}$ constitutes an RMF; the PVSM on $\mathcal{B}_\Pi$ for H, which is associated to $\{Z^X ; X \in H\}$, is called the PVSM associated with the unitary operator U.*

Conversely, from a PVSM we can define a unitary operator:

Property 15.11 *Let $\mathcal{E}$ be a PVSM on $\mathcal{B}_\Pi$ for H, then the mapping $X \in H \longmapsto \int e^{i \cdot 1} dZ_{\mathcal{E}}^X \in H$ is a unitary operator whose associated PVSM is $\mathcal{E}$.*

15.4.6 The product of two PVSM's

Two unitaries U_1 and U_2 are commuting operators if and only if their respectively associated PVSM's $\mathcal{E}_1$ and $\mathcal{E}_2$ commute (i.e. if and only if $\mathcal{E}_1(A) \circ \mathcal{E}_2(B) = \mathcal{E}_2(B) \circ \mathcal{E}_1(A)$ for each pair (A, B) of elements of $\mathcal{B}_\Pi$). In this case, it is easy to verify that

$$\left\{ \left((U_1^n \circ U_2^m)(X) \right)_{(n,m) \in \mathbb{Z} \times \mathbb{Z}} \, ; \, X \in H \right\}$$

is a set of stationary series that are pairwise correlated. So if Z^X is the r.m. defined on $\mathcal{B}_\Pi \otimes \mathcal{B}_\Pi$ with values in H which is associated to the stationary series $\left((U_1^n \circ U_2^m)(X) \right)_{(n,m) \in \mathbb{Z} \times \mathbb{Z}}$, then $\{Z^X ; X \in H\}$ constitutes an RMF. Let us denote by $\mathcal{E}_1 \otimes \mathcal{E}_2$ its associated PVSM; then this is a PVSM on $\mathcal{B}_\Pi \otimes \mathcal{B}_\Pi$ for H.

If P is the measurable mapping $(\lambda_1, \lambda_2) \in \Pi \times \Pi \longmapsto \lambda_1 \in \Pi$, taking into account the integration rules with respect to a measure image we can verify that

$$\int e^{i \cdot n} dP(Z^X) = U_n^1(X).$$

Then $P(Z^X) = Z_{\mathcal{E}_1}^X$ and also $(\mathcal{E}_1 \otimes \mathcal{E}_2)(A \times \Pi) = \mathcal{E}_1(A)$ for each A in $\mathcal{B}_\Pi$.

An analogous proof would give $(\mathcal{E}_1 \otimes \mathcal{E}_2)(\Pi \times B) = \mathcal{E}_2(B)$ for each B of $\mathcal{B}_\Pi$. So that implies the following result:

Property 15.12 *If $\mathcal{E}_1$ and $\mathcal{E}_2$ are commuting PVSM's on $\mathcal{B}_\Pi$ for H, then a unique PVSM on $\mathcal{B}_\Pi \otimes \mathcal{B}_\Pi$ for H, denoted by $\mathcal{E}_1 \otimes \mathcal{E}_2$ and called the product of the two PVSM's $\mathcal{E}_1$ and $\mathcal{E}_2$, exists such that*

$$(\mathcal{E}_1 \otimes \mathcal{E}_2)(A \times B) = (\mathcal{E}_1(A)) \circ (\mathcal{E}_2(B))$$

for each pair (A, B) of elements of $\mathcal{B}_\Pi$.

Uniqueness is proved because of Property 15.4.

15.4.7 The convolution product of two PVSM's

Let us recall that S_Π denotes the (continuous) mapping

$$(\lambda_1, \lambda_2) \in \Pi \times \Pi \longmapsto \lambda_1 \oplus \lambda_2 \in \Pi.$$

We propose the following definition:

Definition 15.4 *Let $\mathcal{E}_1$ and $\mathcal{E}_2$ be two commuting PVSM's on $\mathcal{B}_\Pi$ for H. The PVSM on $\mathcal{B}_\Pi$ for H defined as the image of $\mathcal{E}_1 \otimes \mathcal{E}_2$ under the mapping S_Π is called the* convolution product *of the two PVSM's $\mathcal{E}_1$ and $\mathcal{E}_2$, and is denoted by $\mathcal{E}_1 * \mathcal{E}_2$.*

Let V be the unitary operator whose associated PVSM is $\mathcal{E}_1 * \mathcal{E}_2$, then one has:

$$\begin{aligned} V(X) &= \int e^{i.1}\, dZ^X_{\mathcal{E}_1 * \mathcal{E}_2} = \int e^{i.1}\, dZ^X_{S_\Pi(\mathcal{E}_1 \otimes \mathcal{E}_2)} \\ &= \int e^{i.1}\, dS_\Pi(Z^X_{\mathcal{E}_1 \otimes \mathcal{E}_2}) = \int e^{i.1} \circ S_\Pi\, d(Z^X_{\mathcal{E}_1 \otimes \mathcal{E}_2}) \\ &= \int e^{i(.1+.1)}\, d(Z^X_{\mathcal{E}_1 \otimes \mathcal{E}_2}) = (U_1 \circ U_2)(X). \end{aligned}$$

Consequently we can give

Property 15.13 *If $\mathcal{E}_1$ and $\mathcal{E}_2$ are two PVSM's respectively associated with the two commuting unitaries U_1 and U_2, then $\mathcal{E}_1 * \mathcal{E}_2$ is the PVSM associated with the unitary operator $U_1 \circ U_2$.*

Finally, it is easy to prove the following distributivity result:

Property 15.14 *If $\mathcal{E}$ is a PVSM on $\mathcal{B}_\Pi$ for H and if h_1 and h_2 are two measurable mappings defined from $\widehat{G}$ into Π, we can affirm that:*

(i) *the mapping $h_1 + h_2 : \gamma \in \widehat{G} \longmapsto h_1(\gamma) \oplus h_2(\gamma) \in \Pi$ is measurable;*
(ii) *the PVSM's $h_1(\mathcal{E})$ and $h_2(\mathcal{E})$, both on $\mathcal{B}_\Pi$ for H, are commuting and $(h_1 + h_2)(\mathcal{E}) = (h_1(\mathcal{E})) * (h_2(\mathcal{E}))$.*

15.4.8 The unitary operator group generated by a PVSM

For any element g of G, let us denote by h_g the continuous homomorphism $n \in \mathbb{Z} \longmapsto ng \in G$. Its transpose is such that $e^{i\,{}^t h_g(\gamma)} = (\gamma, g)_{\widehat{G}G}$ for each (γ, g) of $\widehat{G} \times G$ (of course, in the case where $\widehat{\mathbb{Z}}$ is identified with Π this is true because of the mapping $\lambda \in \Pi \longmapsto e^{i.\lambda} \in \widehat{\mathbb{Z}}$).

Since $h_{g_1+g_2} = h_{g_1} + h_{g_2}$ holds, then the previous property allows us to give

Property 15.15 *When $\mathcal{E}$ is a PVSM on $\mathcal{B}_{\widehat{G}}$ for H, for each pair (g_1, g_2) of elements of G, one has ${}^t h_{g_1+g_2}(\mathcal{E}) = ({}^t h_{g_1}(\mathcal{E})) * ({}^t h_{g_2}(\mathcal{E}))$.*

In fact, we are now able to establish a new notion.

Definition 15.5 *The family $\{U_g\,;\, g \in G\}$, where U_g is the unitary operator on H whose associated PVSM is ${}^t h_g(\mathcal{E})$, is called the* unitary operator group on H generated by the PVSM $\mathcal{E}$ on $\mathcal{B}_{\widehat{G}}$ for H.

Furthermore, combining Properties 15.13 and 15.15, we can affirm

Property 15.16 *If $\{U_g\,;\,g\in G\}$ is the unitary operator group on H generated by the PVSM $\mathcal{E}$ on $\mathcal{B}_{\widehat{G}}$ for H, then for each pair (g_1, g_2) of elements of G, we have $U_{g_1+g_2} = U_{g_1} \circ U_{g_2}$.*

This last result permits us to declare that the family $\{U_g\,;\,g\in G\}$ is a group. Then, applying the integral rules with respect to the image of an r.m., we can write:

$$U_g(X) = \int e^{i.1}\,dZ^X_{{}^t h_g(\mathcal{E})} = \int e^{i.1}\,d^t h_g\left(Z^X_{\mathcal{E}}\right)$$
$$= \int e^{i.1}\circ {}^t h_g\,dZ^X_{\mathcal{E}} = \int (\cdot,g)_{\widehat{G}G}\,dZ^X_{\mathcal{E}},$$

which implies

Property 15.17 *If $\{U_g\,;\,g\in G\}$ is the unitary operator group on H generated by the PVSM $\mathcal{E}$ on $\mathcal{B}_{\widehat{G}}$ for H, then for each X in H, $(U_g(X))_{g\in G}$ is a CSRF whose associated r.m. is $Z^X_{\mathcal{E}}$.*

From this last result, we can deduce the following fact:

Property 15.18 *Two PVSM's $\mathcal{E}_1$ and $\mathcal{E}_2$ on $\mathcal{B}_{\widehat{G}}$ for H, such that the equality ${}^t h_g(\mathcal{E}_1) = {}^t h_g(\mathcal{E}_2)$ holds for each $g\in G$, are equal.*

In fact, from ${}^t h_g(\mathcal{E}_1) = {}^t h_g(\mathcal{E}_2)$, we deduce that $U_g^1 = U_g^2$. Then for each X of H, the equality between the two CSRF's $\left(U_g^1(X)\right)_{g\in G}$ and $\left(U_g^2(X)\right)_{g\in G}$ holds. This last point and Property 15.17 allow us to write $Z^X_{\mathcal{E}_1} = Z^X_{\mathcal{E}_2}$. Then it can be deduced that $\mathcal{E}_1 = \mathcal{E}_2$. We remark that the results which are presented in this section are exposed in a more complete way in Boudou (2007).

15.4.9 Extension of the product of two PVSM's

Let $\mathcal{E}_1$ and $\mathcal{E}_2$ be two commuting PVSM's on $\mathcal{B}_{\widehat{G}}$ for H and let $\{U_g^1\,;\,g\in G\}$ and $\{U_g^2\,;\,g\in G\}$ be the respective unitary operator groups on H generated by with the PVSM's $\mathcal{E}_1$ and $\mathcal{E}_2$. From the commutativity of the PVSM's $\mathcal{E}_1$ and $\mathcal{E}_2$ we deduce the commutativity of the PVSM's ${}^t h_g(\mathcal{E}_1)$ and ${}^t h_g(\mathcal{E}_2)$. Consequently, $U_g^1 \circ U_g^2 = U_g^2 \circ U_g^1$ and

$$\left\langle \left(U_{g_1}^1 \circ U_{g_2}^2\right)(X),\left(U_{g_1'}^1 \circ U_{g_2'}^2\right)(X')\right\rangle = \left\langle \left(U_{g_1-g_1'}^1 \circ U_{g_2-g_2'}^2\right)(X),X'\right\rangle.$$

As the mapping $(g_1, g_2)\in G\times G \longmapsto \left(U_{g_1}^1 \circ U_{g_2}^2\right)(X)\in H$ is continuous (indeed $\left(U_g^1(X)\right)_{g\in G}$ and $\left(U_g^2(X)\right)_{g\in G}$ are two CSFR's), it is clear that

$$\left\{\left(U_{g_1}^1 \circ U_{g_2}^2\right)(X))_{(g_1,g_2)\in G\times G}\,;\,X\in H\right\}$$

is a set of pairwise correlated CSFR's. So denoting by Z^X the r.m. associated to the CSRF $\left(U_{g_1}^1 \circ U_{g_2}^2\right)(X))_{(g_1,g_2)\in G\times G}$, we can affirm that $\{Z^X\,;\,X\in H\}$ is an RMF.

If $\mathcal{E}_1 \otimes \mathcal{E}_2$ denotes the PVSM associated to this last RMF, following an analogous approach to that given in the previous section we can prove that $(\mathcal{E}_1 \otimes \mathcal{E}_2)(A \times \widehat{G}) = \mathcal{E}_1(A)$ and that $(\mathcal{E}_1 \otimes \mathcal{E}_2)(\widehat{G} \times B) = \mathcal{E}_2(B)$ for each pair (A, B) of elements of $\mathcal{B}_{\widehat{G}}$.

So we have

Property 15.19 *From two commuting PVSM's $\mathcal{E}_1$ and $\mathcal{E}_2$ on $\mathcal{B}_{\widehat{G}}$ for H, a unique PVSM on $\mathcal{B}_{\widehat{G}} \otimes \mathcal{B}_{\widehat{G}}$ for H exists which is denoted by $\mathcal{E}_1 \otimes \mathcal{E}_2$ and called the product of the PVSM's $\mathcal{E}_1$ and $\mathcal{E}_2$. This product is such that for each pair (A, B) of elements of $\mathcal{B}_{\widehat{G}}$,*

$$(\mathcal{E}_1 \otimes \mathcal{E}_2)(A \times B) = (\mathcal{E}_1(A)) \circ (\mathcal{E}_2(B)).$$

We note that here the uniqueness is again a consequence of Property 15.4. Let us remark that in Boudou and Romain (2002), we define the product $\mathcal{E}_1 \otimes \mathcal{E}_2$ where $\mathcal{E}_1$ is a PVSM on $\mathcal{B}_{\widehat{G_1}}$ for H and $\mathcal{E}_2$ is a PVSM on $\mathcal{B}_{\widehat{G_2}}$ for H.

15.4.10 The extension of the convolution product of two PVSM's

Let $S_{\widehat{G}}$ be the continuous mapping $(\gamma_1, \gamma_2) \in \widehat{G} \times \widehat{G} \longmapsto \gamma_1 + \gamma_2 \in \widehat{G}$.

Definition 15.5 *For two commuting PVSM's $\mathcal{E}_1$ and $\mathcal{E}_2$ on $\mathcal{B}_{\widehat{G}}$ for H, the image of $\mathcal{E}_1 \otimes \mathcal{E}_2$ under $S_{\widehat{G}}$ is called the* convolution product *of $\mathcal{E}_1$ and $\mathcal{E}_2$ and is denoted by $\mathcal{E}_1 * \mathcal{E}_2$.*

Let us denote by $\mathcal{E}_{\widehat{G}}$ the PVSM $A \in \mathcal{B}_{\widehat{G}} \longmapsto \delta_0(A) I \in \mathcal{P}(H)$, where δ_0 is the Dirac measure at 0. Then we have

Property 15.20 *The PVSM $\mathcal{E}_{\widehat{G}}$ commutes with every PVSM $\mathcal{E}$ on $\mathcal{B}_{\widehat{G}}$ for H, and is such that $\mathcal{E}_{\widehat{G}} * \mathcal{E} = \mathcal{E}$.*

With obvious notation, we also obtain

Property 15.21 *The mapping $O : \gamma \in \widehat{G} \longmapsto 0_{\widehat{G_1}} \in \widehat{G_1}$ is measurable and for each PVSM $\mathcal{E}$ on $\mathcal{B}_{\widehat{G}}$ for H, one has $O(\mathcal{E}) = \mathcal{E}_{\widehat{G_1}}$.*

The following fact is also given without proof:

Property 15.22 *If $\mathcal{E}$ is a PVSM on $\mathcal{B}_{\widehat{G_1}}$ for H, and h_1 and h_2 are two measurable mappings from $\widehat{G_1}$ in $\widehat{G_2}$, then:*

(i) *the mapping $h_1 + h_2 : \gamma \in \widehat{G_1} \longmapsto h_1(\gamma) + h_2(\gamma) \in \widehat{G_2}$ is measurable;*
(ii) *the PVSM's $h_1(\mathcal{E})$ and $h_2(\mathcal{E})$ on $\mathcal{B}_{\widehat{G_2}}$ for H commute and $(h_1 + h_2)(\mathcal{E}) = (h_1(\mathcal{E})) * (h_2(\mathcal{E}))$.*

Finally, using the two previous results and denoting by w the continuous mapping $\gamma \in \widehat{G} \longmapsto -\gamma \in \widehat{G}$, we can give

Property 15.23 *For each PVSM $\mathcal{E}$ on $\mathcal{B}_{\widehat{G}}$ for H, we have $(w(\mathcal{E})) * \mathcal{E} = \mathcal{E}_{\widehat{G}}$.*

15.4.11 Equations for PVSM's

Let us consider the following problem:

For a given PVSM $\mathcal{E}_Y$ on $\mathcal{B}_{\widehat{G_2}}$ for H and for a continuous homomorphism h of $\widehat{G_1}$ into $\widehat{G_2}$, we want to define all the PVSM's $\mathcal{E}_X$, on $\mathcal{B}_{\widehat{G_1}}$ for H, such that $h(\mathcal{E}_X) = \mathcal{E}_Y$. For this purpose, we will assume the existence of a mapping u of $\widehat{G_2}$ into $\widehat{G_1}$ such that $h \circ u = i_{\widehat{G_2}}$.

Then we obtain

Property 15.24 *If $\mathcal{E}$ is a PVSM on $\mathcal{B}_{\widehat{G_1}}$ for H, which commutes with $u(\mathcal{E}_Y)$ and is such that $h(\mathcal{E}) = \mathcal{E}_{\widehat{G_2}}$, then $\mathcal{E} * (u(\mathcal{E}_Y))$ is one solution and all the solutions are of this type.*

The direct part of the proof is deduced from

$$h(\mathcal{E} * (u(\mathcal{E}_Y))) = (h(\mathcal{E})) * (h(u(\mathcal{E}_Y))) = \mathcal{E}_{\widehat{G_2}} * \mathcal{E}_Y = \mathcal{E}_Y.$$

Let us have a look at the inverse statement. Let us denote one solution by $\mathcal{E}_X$. The PVSM $\mathcal{E} = ((w \circ u \circ h) + i_{\widehat{G_1}})(\mathcal{E}_X)$ is such that

$$h(\mathcal{E}) = ((h \circ w \circ u \circ h) + h)(\mathcal{E}_X) = O(\mathcal{E}_X) = \mathcal{E}_{\widehat{G_2}}.$$

This PVSM commutes with $u(\mathcal{E}_Y) = u(h(\mathcal{E}_X))$ because they are both images of $\mathcal{E}_X$. Furthermore,

$$\begin{aligned}\mathcal{E} * (u(\mathcal{E}_Y)) &= (((w \circ u \circ h) + i_{\widehat{G_1}})(\mathcal{E}_X)) * ((u \circ h)(\mathcal{E}_X)) \\ &= ((w \circ u \circ h) + i_{\widehat{G_1}} + (u \circ h))(\mathcal{E}_X) = \mathcal{E}_X.\end{aligned}$$

15.4.12 Equations for random measures

Let us now examine a similar problem to the previous one:

For a given r.m. Z_Y defined on $\mathcal{B}_{\widehat{G_2}}$ with values in H and for a continuous homomorphism h from $\widehat{G_1}$ into $\widehat{G_2}$, we want to define all the r.m.'s Z_X, defined on $\mathcal{B}_{\widehat{G_1}}$ with values in H, such that $h(Z_X) = Z_Y$ and such that $H_{Z_X} = H_{Z_Y}$.

In Boudou (2007) a more general problem is solved. The condition $H_{Z_X} = H_{Z_Y}$ (that is assumed here) enables us to give a simpler proof by using the association between the r.m. and the PVSM. As previously, we assume that there exists a mapping u from $\widehat{G_2}$ in $\widehat{G_1}$ such that $h \circ u = i_{\widehat{G_2}}$.

Let $\mathcal{E}_Y$ be the PVSM on $\mathcal{B}_{\widehat{G_2}}$ for H_{Z_Y} which is associated with the r.m. Z_Y; we can state

Property 15.25 *If $\mathcal{E}$ is a PVSM on $\mathcal{B}_{\widehat{G_1}}$ for H_{Z_Y} which commutes with $u(\mathcal{E}_Y)$ and is such that $h(\mathcal{E}) = \mathcal{E}_{\widehat{G_2}}$, then $Z^{Z_Y(\widehat{G_2})}_{\mathcal{E}*(u(\mathcal{E}_Y))}$ is one solution and all the solutions are of this type.*

If $\mathcal{E}$ is a PVSM satisfying the hypotheses of Property 15.25, we can write

$$h\left(Z^{Z_Y(\widehat{G_2})}_{\mathcal{E}*(u(\mathcal{E}_Y))}\right) = Z^{Z_Y(\widehat{G_2})}_{h(\mathcal{E}*(u(\mathcal{E}_Y)))} = Z^{Z_Y(\widehat{G_2})}_{\mathcal{E}_Y} = Z_Y.$$

So the direct part is obtained. Conversely, let us assume that Z_X is one solution. Let $\mathcal{E}_X$ denote the PVSM on $\mathcal{B}_{\widehat{G_1}}$ for $H_{Z_X}(= H_{Z_Y})$ which is associated to it. Now consider an element (A, X) of $\mathcal{B}_{\widehat{G_2}} \times H_{Z_Y}$. Then there exists an element φ of $L^2(\mu_{Z_Y})$ such that $X = \int \varphi \, dZ_Y = \int \varphi \circ h \, dZ_X$. So we can write

$$\begin{aligned}(\mathcal{E}_X(h^{-1}A))(X) &= (\mathcal{E}_X(h^{-1}A))(\int \varphi \circ h \, dZ_X) \\ &= \int (1_A \varphi) \circ h \, dZ_X = \int 1_A \varphi \, dZ_Y. \\ &= (\mathcal{E}_Y(A))(X),\end{aligned}$$

This means that $h(\mathcal{E}_X) = \mathcal{E}_Y$ and so, using Property 15.24, we can affirm that there exists a PVSM $\mathcal{E}$ on $\mathcal{B}_{\widehat{G_1}}$ for H_{Z_Y} which commutes with $u(\mathcal{E}_Y)$ and such that $\mathcal{E} * (u(\mathcal{E}_Y)) = \mathcal{E}_X$. Then we conclude the proof with the equality

$$Z_X = Z^{Z_X(\widehat{G_1})}_{\mathcal{E}_X} = Z^{Z_Y(\widehat{G_2})}_{\mathcal{E}*(u(\mathcal{E}_Y))}.$$

15.4.13 Ampliations of PVSM's

In this section and those that follow, the chosen index domain is $\mathbb{Z}$ for a "light" mathematical framework. All the results can be generalized to an l.c.a.g index domain whose dual group admits a countable basis. Here H_1 and H_2 are both separable $\mathbb{C}$-Hilbert spaces and σ_2 is the set of Hilbert–Schmidt operators from H_1 into H_2. Let us recall (see Section 15.3.8) that when A_1 and A_2 are, respectively, bounded endomorphisms in H_1 and in H_2, then the mapping $A_1 \overset{l}{\otimes} A_2$ is defined by $K \in \sigma_2 \longmapsto A_2 \circ K \circ A_1^* \in \sigma_2$.

The following result is then obtained:

Property 15.26 *If $\mathcal{E}$ is a PVSM on $\mathcal{B}_\Pi$ for H_2 then*

(i) *for each A of $\mathcal{B}_\Pi$, $E(A) = I_{H_1} \overset{l}{\otimes} (\mathcal{E}(A))$ is an orthogonal projector;*
(ii) *the mapping $E : A \in \mathcal{B}_\Pi \longmapsto E(A) \in \mathcal{P}(\sigma_2)$ is a PVSM on $\mathcal{B}_\Pi$ for σ_2 called the left functional ampliation of $\mathcal{E}$ with respect to H_1.*

It now seems natural to want to find the unitary operator V of σ_2 whose associated PVSM is E. Let us denote by U the unitary operator whose associated PVSM is $\mathcal{E}$, then we have:

$$V = \lim_n \sum_{nk=0}^{k=n-1} e^{i(-\pi+k\frac{2\pi}{n})} E\left(\left[-\pi+k\frac{2\pi}{n}, -\pi+(k+1)\frac{2\pi}{n}\right)\right)$$

$$= \lim_n \sum_{k=0}^{k=n-1} e^{i(-\pi+k\frac{2\pi}{n})} I_{H_1} \overset{l}{\otimes} \left(\mathcal{E}\left(\left[-\pi+k\frac{2\pi}{n}, -\pi+(k+1)\frac{2\pi}{n}\right)\right)\right)$$

$$= \lim_n I_{H_1} \overset{l}{\otimes} \left(\sum_{k=0}^{k=n-1} e^{i(-\pi+k\frac{2\pi}{n})} \mathcal{E}\left(\left[-\pi+k\frac{2\pi}{n}, -\pi+(k+1)\frac{2\pi}{n}\right)\right)\right)$$

$$= I_{H_1} \overset{l}{\otimes} \left(\lim_n \sum_{k=0}^{k=n-1} e^{i(-\pi+k\frac{2\pi}{n})} \mathcal{E}\left(\left[-\pi+k\frac{2\pi}{n}, -\pi+(k+1)\frac{2\pi}{n}\right)\right)\right)$$

$$= I_{H_1} \overset{l}{\otimes} U$$

So we can give

Property 15.27 *If U is a unitary operator in H_2 whose PVSM is $\mathcal{E}$, then $I_{H_1} \overset{l}{\otimes} U$ is a unitary operator in σ_2 whose PVSM is the left functional ampliation of $\mathcal{E}$ with respect to H_1.*

As previously, we can also state

Property 15.28 *If U is a unitary operator in H_1 whose PVSM is $\mathcal{E}$, then:*

(i) *for each A of $\mathcal{B}_\Pi$, $E(A) = (\mathcal{E}(A)) \overset{l}{\otimes} I_{H_2}$ is an orthogonal projector;*
(ii) *the mapping $E : A \in \mathcal{B}_\Pi \longmapsto E(A) \in \mathcal{P}(\sigma_2)$ is a PVSM on $\mathcal{B}_\Pi$ for σ_2 called the right functional ampliation of $\mathcal{E}$ with respect to H_2;*
(iii) $U^* \overset{l}{\otimes} I_{H_2}$ *is a unitary operator in σ_2 whose associated PVSM is E.*

15.4.14 The convolution product of two r.m.'s

Let Z_1 and Z_2 be two r.m.'s defined on $\mathcal{B}_\Pi$ with values in H_1 and H_2, respectively. Let us denote by $(X_n^1)_{n\in\mathbb{Z}}$ (respectively, $(X_n^2)_{n\in\mathbb{Z}}$) the Fourier transform of the r.m. Z_1 (respectively, Z_2), by U_1 (respectively, U_2) the shift operator $X_n^1 \longmapsto X_{n+1}^1$ (respectively, $X_n^2 \longmapsto X_{n+1}^2$), and by $\mathcal{E}_1$ (respectively, $\mathcal{E}_2$) the PVSM on $\mathcal{B}_\Pi$ for H_{Z_1} (respectively, H_{Z_2}) which is associated with the r.m. Z_1 (respectively, Z_2), and which is also the PVSM associated with the unitary operator U_1 of H_{Z_1} (respectively, U_2 of H_{Z_2}). Finally, let E_1 and E_2 be the respective right ampliation of $\mathcal{E}_1$ with respect to H_{Z_2} and the left ampliation of $\mathcal{E}_2$ with respect to H_{Z_1}.

It is easy to verify that the PVSM's E_1 and E_2 commute and so their product $E_1 \otimes E_2$, which is a PVSM on $\mathcal{B}_\Pi \otimes \mathcal{B}_\Pi$ for σ_2 (the set of Hilbert–Schmidt operators from H_{Z_1} to H_{Z_2}) may be considered. If we put $Z_1 \otimes Z_2 = Z_{E_1 \otimes E_2}^{X_0^1 \otimes X_0^2}$, for each (A_1, A_2) of

$\mathcal{B}_\Pi \times \mathcal{B}_\Pi$ we have:

$$Z_1 \otimes Z_2(A_1 \times A_2) = ((E_1 \otimes E_2)(A_1 \times A_2))\left(X_0^1 \otimes X_0^2\right)$$
$$= ((\mathcal{E}_1(A_1)) \overset{l}{\otimes} (\mathcal{E}_2(A_2)))\left(X_0^1 \otimes X_0^2\right)$$
$$= ((\mathcal{E}_1(A_1))(X_0^1)) \otimes ((\mathcal{E}_2(A_2))(X_0^2))$$
$$= (Z_1(A_1)) \otimes (Z_2(A_2)).$$

So the new result is:

Property 15.29 *Let Z_1 and Z_2 be two r.m.'s defined on $\mathcal{B}_\Pi$ with values in H_1 and H_2, respectively. Then a unique r.m., denoted by $Z_1 \otimes Z_2$, exists and is called the product of the two r.m.'s Z_1 and Z_2. This product is such that*

$$Z_1 \otimes Z_2(A_1 \times A_2) = (Z_1(A_1)) \otimes (Z_2(A_2))$$

for each (A_1, A_2) of $\mathcal{B}_\Pi \times \mathcal{B}_\Pi$.

Consequently we introduce

Definition 15.6 *Let Z_1 and Z_2 be two r.m.'s defined on $\mathcal{B}_\Pi$ with values in H_1 and H_2, respectively. The image of the product $Z_1 \otimes Z_2$ under S_Π is called the* convolution product *of the two r.m.'s Z_1 and Z_2 and is denoted by $Z_1 * Z_2$.*

Since

$$Z_1 * Z_2 = S_\Pi(Z_1 \otimes Z_2)$$
$$= S_\Pi\left(Z_{E_1 \otimes E_2}^{X_0^1 \otimes X_0^2}\right)$$
$$= Z_{S_\Pi(E_1 \otimes E_2)}^{X_0^1 \otimes X_0^2}$$
$$= Z_{E_1 * E_2}^{X_0^1 \otimes X_0^2},$$

and since $E_1 * E_2$ is the PVSM associated with the unitary operator

$$\left(U_1^* \overset{l}{\otimes} I_{H_{Z_2}}\right) \circ (I_{H_{Z_1}} \overset{l}{\otimes} U_2) = U_1^* \overset{l}{\otimes} U_2,$$

we can affirm that $Z_1 * Z_2$ is the r.m. associated to the stationary series $((U_1^* \overset{l}{\otimes} U_2)^n(X_0^1 \otimes X_0^2))_{n \in \mathbb{Z}} = (X_{-n}^1 \otimes X_n^2)_{n \in \mathbb{Z}}$.
Then we have

Property 15.30 *Let Z_1 and Z_2 be two r.m.'s defined on $\mathcal{B}_\Pi$ with values in H_1 and H_2, respectively. Then for all n in $\mathbb{Z}$ one has*

$$X_{-n}^1 \otimes X_n^2 = \int e^{i.n}\, dZ_1 * Z_2.$$

15.4.15 The product of two independent stationary series

Let us recall that the composition product of an isometry and an r.m. is an r.m., and that the integral with respect to the latter is its integral image. More precisely and with obvious notation, we may write $\int \varphi \, d\mathcal{I} \circ Z = \mathcal{I}(\int \varphi \, dZ)$, which is not surprising as the integral is a sum and an isometry is linear. This property will permit us to define the r.m. associated with the product of two stationary series.

Let us consider the probability space $(\Omega, \mathcal{A}, P)$ and two independent sub-σ-algebras $\mathcal{B}_1$ and $\mathcal{B}_2$. Let $\mathcal{U}$ be the σ-algebra generated by the family $\{B_1 \cap B_2 ; (B_1, B_2) \in \mathcal{B}_1 \times \mathcal{B}_2\}$. We can demonstrate (see Boudou (1986)) the existence of a unique isometry $\mathcal{I}$ from $\sigma_2(L^2(\mathcal{B}_1), L^2(\mathcal{B}_2))$ (the set of Hilbert–Schmidt operators from $L^2(\mathcal{B}_1)$ to $L^2(\mathcal{B}_2)$) on $L^2(\mathcal{U})$ which associates $X_1 \cdot X_2$ to $\overline{X_1} \otimes X_2$ for each (X_1, X_2) in $L^2(\mathcal{B}_1) \times L^2(\mathcal{B}_2)$ (of course $\bar{x}$ is the conjugate of the complex number x).

Combining the different previous results we are able to give

Property 15.31 *If $\left(X_n^1\right)_{n \in \mathbb{Z}}$ (respectively, $\left(X_n^2\right)_{n \in \mathbb{Z}}$) is a stationary series of elements belonging to $L^2(\mathcal{B}_1)$ (respectively, $L^2(\mathcal{B}_2)$) with associated r.m. Z_1(respectively, Z_2), then $\left(\overline{X_{-n}^1} X_n^2\right)_{n \in \mathbb{Z}}$ is a stationary series of elements belonging to $L^2(\mathcal{U})$ with associated r.m. $\mathcal{I} \circ (Z_1 * Z_2)$.*

Indeed, we have

$$\int e^{i \cdot n} \, d\mathcal{I} \circ (Z_1 * Z_2) = \mathcal{I}\left(\int e^{i \cdot n} \, dZ_1 * Z_2\right)$$
$$= \mathcal{I}\left(\overline{X_{-n}^1} \otimes X_n^2\right)$$
$$= \overline{X_{-n}^1} X_n^2.$$

The last two sections are studied in a more general mathematical framework in Boudou and Romain (2010).

15.5 Conclusion

In this chapter, we have demonstrated that the commutativity assumption (of two unitaries and of two PVSM's) is very important and allows for elegant proofs. This hypothesis may appear restrictive to some extent, and it leaves some open problems.

Nevertheless, the commutativity assumption would permit us to enlarge the notion of stationary correlation. Indeed, two stationary series with two respective commuting shift operators are stationarily correlated, but the converse is false. This is an interesting topic for investigation in the future.

Another tool that plays an important role in this chapter is the notion of the transpose of a homomorphism: although this notion completes the study of the group structure, it is not present in the literature. It permits us to transport some properties expressed in the time domain over to the frequency domain, for example, transformations of the index using the homomorphism will be expressed for spectral elements via the transpose of this homomorphism (see Boudou (2007) for more details).

We do not need the commutativity assumption when we use the so-called left and right ampliation operators (because the ampliations always commute) to define the convolution product of two r.m.'s: this product is associated with a tensor product of processes and we are looking for general practical results. But if we add an independence hypothesis for the two processes, then the results are easier to apply because of an isometry between the spaces of type σ_2 and L^2. To a certain extent, stochastic independence replaces the commutativity hypothesis (for more details see Boudou and Romain (2002)). So the main result is that the r.m. associated with the product of two independent processes is the convolution product of their respective r.m.'s: this reminds us of the well-known property of the distribution of the sum of two independent r.v.'s which is the convolution product of their respective laws.

In recent work, we have obtained complementary results. In Boudou and Romain (2002) one can find a Fubini-type formula which allows us to give the integral with respect to a product of two r.m.'s and in Boudou and Romain (2010) we complete the earlier paper by giving an intrinsic form of the previous Fubini-type integral.

Finally, let us give a last extension of the results presented here. All the stationary processes are Hilbert-space valued (and even finite-dimensional for some particular cases). The generalization to Banach spaces where the orthogonality notion is abandoned is another challenge for investigation. In Benchikh et al. (2007), a first step in this direction has been proposed.

References

Arveson, W. (2001). *A Short Course on Spectral Theory*. Springer, New York.
Azencott, R., Dacunha-Castelle, D. (1984). *Séries d'Observations Irrégulières. Modélisation et Prévision*. Masson, Paris.
Benchikh, T., Boudou, A., Romain, Y. (2007). Mesures aléatoires opératorielles et banachiques. Applications aux séries stationnaires. *C. R. Math. Acad. Sci. Paris, Série I*, 345(6), 345–8.

BERG, C., FORST, G. (1975). *Potential Theory on Locally Compact Abelian Groups.* Spinger, Berlin.

BIRMAN, M., SOLOMJAK, M. (1996). Tensor product of a finite number of spectral measures is always a spectral measure. *Integr. Equat. Oper. Th.*, **24**, 179–87.

BOSQ, D. (2000). *Linear Processes in Function Spaces: Theory and Applications.* Springer, New York.

BOUDOU, A. (1986). Analyse en composantes principales de données aléatoires. *Cahiers du Centre d'Etudes de Recherche Opérationnelle*, **28**(4), 265–81.

BOUDOU, A. (2000). Produits de mesures et produits de convolution de mesures spectrales. *Publi. Labo. Stat. Proba. Toulouse*, **14-00**, 1–19.

BOUDOU, A. (2003). Interpolation de processus stationnaire. *C. R. Acad. Sci. Paris, Série I*, **336**, 1021–4.

BOUDOU, A. (2006). Approximation of the principal component analysis of a stationary function. *Statistics & Probability Letters.*, **76**(6), 571–8.

BOUDOU, A. (2007). Groupe d'opérateurs unitaires déduit d'une mesure spectrale–une application. *C. R. Acad. Sci. Paris Série I*, **344**, 791–4.

BOUDOU, A., DAUXOIS, J. (1994). Principal component analysis for stationary random function defined on locally compact Abelian group. *J. Mult. Anal.*, **51**, 1–516.

BOUDOU, A., ROMAIN, Y. (2001). Processus hilbertien associé à la convolée de deux mesures aléatoires, *C. R. Acad. Sci. Paris, Série* 1, **332**, 1–4.

BOUDOU, A., ROMAIN, Y. (2002). On spectral and random measures associated to discrete and continuous time processes. *Statistics & Probability Letters.*, **59**(2), 145–57.

BOUDOU, A., ROMAIN, Y. (2010). On the integral with respect to the tensor product of two random measures. *J. Multivariate Anal.* **101**(2), 385–394.

BOUDOU, A., VIGUIER-PLA, S. (2006). On proximity between PCA in the frequency domain and usual PCA. *Statistics*, **40**, 447–64.

BRILLINGER, D.R. (2001). *Time Series Data Analysis and Theory.* Society for Industrial Applied Mathematics (SIAM), Philadelphia, PA.

BROSZKIEWICZ, I. (1987). A note on tensor products of semigroups of contractions. *Univ. Lagel. Acta Math.*, **26**, 191–8.

DACUNHA-CASTELLE, D., DUFLO, M. (1983*). Probabilités et Statistiques. Problèmes à Temps Mobile.* Masson, Paris.

DAUXOIS, J., ROMAIN, Y., VIGUIER, S. (1994). Tensor products and Statistics. *Lin. Alg. Appl.*, **210**, 59–88.

DEHAY, D. (1991). On the product of two harmonizable time series. *Stochastic Process. Appl.*, **38**, 347–58.

DINCULEANU, N. (2000). *Vector Integration and Stochastic Integration in Banach spaces.* Wiley, New York.

DUNFORD, N., SCHWARTZ, J. (1963). *Linear Operators.* Wiley, New York.

MÉTIVIER, M. (1968). *Notions Fondamentales de la Théorie des Probabilités.* Dunod, Paris.

PRIESTLEY, M.B. (1981). *Spectral Analysis and Times Series.* Academic Press, London.

RIESZ, F., NAGY B. SZ. (1990). *Functional Analysis.* Dover, Mineola, NY.

ROMAIN, Y. (2002). Perturbation of functional tensors with applications to covariance operators. *Statistics & Probability Letters.*, **58**(3), 253–64.

ROSENBERG, M. (1974). Operators as spectral integrals of operator-valued functions from the study of multivariate stationary stochastic processes. *J. Mult. Anal.*, **4**, 166–209.

ROZANOV, Y.A. (1967). *Stationary Random Processes.* Holden-Day, San Francisco, CA.

RUDIN, W. (1967). *Fourier Analysis On Groups*. Wiley, New York.
SCHAEFER, H.H. (1974). *Banach Lattices and Positive Operators*. Springer, Berlin.
SHUMWAY, R.H., STOFFER, D.S. (2000). *Time Series Analysis and its Applications*. Springer, New York.
WIENER, N., MASANI, P. (1957). The prediction theory of multivariate stochastic processes. I. The regularity condition. *Acta Math.*, **98**, 111–50.

CHAPTER 16

AN INVITATION TO OPERATOR-BASED STATISTICS

YVES ROMAIN

16.1 INTRODUCTION

READERS might wonder what we mean by operator-based statistics, and ask themselves what the most important advantages of such an approach might potentially be. In response, we will pose five questions, and attempt to answer them from different practitioner perspectives. Historical and pedagogical aspects of the subject are considered first, followed by practical and theoretical motivations; finally we look at synthetic and conceptual arguments. This chapter offers a progressive way forward for a multivariate data analyst confronted with the problem of high or infinite dimension. Other related fields are identified, as is the current move towards a unification of classical and quantum statistics, an endeavour which will necessarily demand an operator-based approach.

Operators are used in statistics (as in other branches of mathematics) for two principal reasons:

1. The first reason concerns "intrinsic coordinates," equivalent to a "coordinate-free" approach: in the basic formulation of a linear mapping g from one vector space E to another vector space F, we are interested in the intrinsic definition of g and not in any possible representation of it (i.e. without choosing any basis in E and F). In fact, we pay attention to all the methods, concepts, and

properties that are basis free (think about the trace or the determinant, for instance). So this first motivation for the use of operators holds in finite as well as in infinite dimensions, and in this sense the operator-based approach is a "functional" version of the matrix-based approach.

2. The second reason for using operators is related to the problem of the infinite dimension of the mathematical framework: in many contexts the application of operator theory, and more generally of functional analysis, is determined by the nature of the empirical data to be analyzed. Let us take the example of classical factor multivariate methods, such as principal component analysis (PCA), canonical analysis (CA), discriminant analysis (DA), and factor correspondence analysis (FCA). From the early synthetic work of Dauxois and Pousse (1975), we know that all these methods (and a majority of related ones) are essentially linked through two operator-based formulations—the PCA of an operator and the CA of two operators—and it can be shown that respective analyses can be carried out through the *spectral analysis of a positive selfadjoint operator*.

The operator-based approach therefore immediately offers at least two advantages. If applied in finite-dimensional situations, it anticipates (with a slight need for attention in the most frequently encountered cases) the infinite-dimensional case (think here about the natural transition from the Euclidean-space to the Hilbert-space formulation). When applied in an infinite context, each choice of specific bases will give different matrix representations, but the intrinsic nature of the method is the same.

This chapter is divided into five sections. The first part focuses on some generalities (among them, the problem of checking the statistical results hold in infinite dimensions, an important issue that is not often addressed). In the second part, we develop the operator-based approach for factor multivariate methods (and for their asymptotic studies), before offering several examples that show the value of operators in statistics. Thus we speak of *covariance operators* (directly linked to PCA), *Hankel and Toeplitz operators* (linked here to CA), *regression operators* (essentially linked to "functional data analysis"), *measure-associated operators* (with studies in Banach spaces), *tensor operators* (which lead naturally to noncommutative statistics), and some other important categories (which are more fully developed by other authors in this volume).

In the third part, we venture into a connected but less well-known domain, looking at noncommutative or *quantum statistics*. Even in a very brief overview the mathematical environment of this topic (which furnishes models for the microscopic world) can clearly be identified as purely tensor-based and operator-based. Thus, if we need to transfer knowledge from one version to another the natural way to do this is to use the operator-based approach. Finally, in the fourth part we discuss some conceptual and synthetic questions where the notions of projection operators and the algebra they generate (called a von Neuman algebra) are fundamental.

By now readers might have the impression that the statistics with which we are principally concerned here is that in infinite dimensions and using a nonparametric approach (in the sense that few stochastic distribution assumptions are made). For mathematical statistics (only a few relevant works are mentioned here), the infinite-dimensional case generally concerns problems on abstract spaces, and this is not the main focus of this chapter (a first reference in the same "spirit" would be Small and McLeish (1994)).

Finally, we note that this overview of operator theory in statistics is not intended to be comprehensive. The infinite-dimensional case is very close to abstract formulations and consequently very close to deeper reflections: we end this chapter (and this book) with some remarks on this topic.

16.2 Generalities

16.2.1 Statistics: a tree with various branches

Statistics is an area of mathematics with a wide range of domains of application. It has different and complementary attributes: descriptive and inferential, parametric and nonparametric, asymptotic and nonasymptotic, univariate and multivariate, Bayesian and nonBayesian, ... From concrete exploratory cases to elaborate models, statistics is an attractive approach to various concepts often linked with stochastic versions of important theories, such as measure and integration theory and functional analysis. Associated with the calculus of probability, we often have to use tools from (linear and general) algebra, (analytic and differential) geometry, and (real and complex) analysis.

The impact of statistics-based research is quasi-universal in a very broad sense (for example, it is used in econometrics, medicine, engineering, biometrics, psychometrics, ...) and, when associated with physics, thermodynamics, and mechanics, it gives rise to authentic branches of science.

The "microscopic world" (as opposed to the traditional "macroscopic world") is receiving growing attention, and one result of this is new quantum versions of physics, chemistry, mechanics, and, more recently, probability and statistics. The term "noncommutative" is interchangeable with "quantum", despite some historical differences. This derives from the fact that while the product of two simple "variables" (or functions) is commutative, the product of two "observables" (the quantum analogy to "variable") is generally not, since each observable is an operator. The mathematical background of this branch of mathematics therefore *naturally requires* operator theory.

At this point, we need only to acknowledge that our "old friend," a random variable defined on a measurable space, is now replaced by our "new friend," a selfadjoint operator whose spectrum is also its set of values. Moreover, the old concept is embedded in the new, since each random variable can be associated with a multiplication operator whose spectrum is analogous to the set of values of the random variable. This correspondence is relevant in the context of the aspiration to a "unified version" of statistics containing the "macro" and "micro" versions (see the last section of this chapter on this topic).

16.2.2 Low, high, and infinite dimension

In exploratory statistics, practitioners display data in tables. A statistical variable or a "question" X may be real and vector-valued in a (countable or uncountable) set, but might also be categorical with values in an (ordered or unordered) generally finite set of possibilities (or "words", or "answers"). The units (or individuals), often denoted by ω, are chosen from a subset of a population, Ω say, with finite cardinality denoted by n (the "sample size"). Several variates are studied—let us denote by p their number (possibly a combination of quantitative or polychotomous variates). These variates may be observed on the same units at different dates (or conditions) and here the index domain is generally denoted by T (often finite), where t is one possible element (or "date").

So we may obtain a cubic data table with p, n, and k as dimensional parameters where, for example, we may consider a table as the observation of a variable $X = (X_1, \ldots, X_p) \in \mathbb{R}^p$ or H on $\{\omega_1, \ldots, \omega_n\} \in \Omega$ at $\{t_1, \ldots, t_k\} \in T$. This is a *low-dimensional context* with $\dim H = p$, $\operatorname{card} \Omega = n$, and $\operatorname{card} T = k$.

In fact *a high- or infinite-dimensional context* may be encountered in four possible situations:

1. First, when the sample size n grows infinitely: this is the usual case in sample-based asymptotic behaviour studies (notice here that, even with a finite population, we may have infinite observations if the sampling is with replacement).
2. The second case concerns the dimension of H and is the most usual context when the statistician speaks of methods in "infinite dimension": from the vector space $\mathbb{R}^p$ we may consider the Euclidean space H and more generally the Hilbert space H of infinite dimension (other cases, such as Banach spaces, may also be treated).
3. The third possibility is associated with the index domain T: the classic countable context is for a finite subset of $\mathbb{N}$. As soon as k tends to infinity (or when we consider an uncountable subset of $\mathbb{R}$), then we have another

way of understanding the term "infinite dimension" (a general framework is obtained here by considering T as a locally Abelian compact group).

4. Finally, there exists a more analytical way to encounter this problem of dimensionality. For example, when we want to explore potential links between variables we may choose different levels of connection, where the first step is the potential linearity between two variables. Then we might generalize the study to two or more subsets of variables, and we might also be interested in nonlinear transformations. In this last case the notion of (vector) subspace is naturally replaced by the notion of σ-algebra (i.e. $L^2(X^{-1}(\mathbb{B}_{\mathbb{R}'}))$ in a simple, well-known example). We can say here that the possibly infinite dimension is brought in by the *complexity level parameter* of the study.

In conclusion we have *high-* (sometimes also called *complex*) or *infinite-dimensional data analysis* studies or methods in the case when one of the four parameters above is large or tends to infinity.

Note that, except for one special case, the term "operator" implies "*linear operator*" throughout this chapter.

16.2.3 An important remark?

The work of a statistician may be briefly divided into three steps:

1. First, data must be described and explored in order to gather the most salient information about the target phenomenon, and there must be explicit recognition of the fact that the results obtained are only valid for the given "sample" (in a broad sense).
2. Second, a (mathematical or formal) model in a deterministic and/or stochastic setting might be proposed, for forecasting unobserved values.
3. Third, the proposed model must be evaluated using different theoretical or at least empirical methods.

These three steps are naturally sequential, but at the end of the third step it is possible (and sometimes preferable) to return to the original (perhaps now expanded) dataset, to iterate the procedure if the model proves inadequate.

To illustrate what we mean by "at least empirical methods" in the evaluation of the model, we give the following example to illustrate this fact for *the lowest-dimensional case*. We take the simplest case where n is finite and small, where $p = 2$ (that is, when two variables are observed) only at one date ($t = 1$). There exists a very pertinent example where "the same calculations and results may be obtained by very different data". In Lebart *et al.* (1984), we find a remarkable case of five studies of simple linear regression between pairs of variates (X^j, Y^j) for each $j \in \{1, 5\}$ on $n = 16$ units. If, for each j, we enter these five different data $\left(x_i^j, y_i^j\right)_{i=1,16}$ separately into regression software, or even process them using a classroom calculator with a "black-box approach," then we observe that all the respective empirical means and variances are equal, all the empirical covariances are equal, all the determination coefficients are equal (near 0.7), and consequently all the ("satisfactory") regression-line equations are identical.

The question that the practitioner might ask is "What is the validity of this simple linear model," whereas the statistician might ask "How can we check the relevance of the results?" So, in our example, we have a very simple graphical method: for each $j \in \{1, 5\}$ we can display the data for the 16 pair units on a bi-plot and observe the results. The surprising outcome is that only one case among the five can be retained—the other four are immediately rejected for different visible reasons (such as outlier presence, only local fit, nonadaptability to the linearity, etc.). The *exploratory empirical graphical method* is a good validation technique here.

The title of this section is a question. The new question is: "Are we able, in high-dimensional cases, to propose relevant validation techniques and, furthermore, how relevant is this question when we have to work in infinite dimensions (for curves or in a nonlinear context for instance)?"

We highlight this problem of validation because a great part of the expanding literature on "functional data analysis methods" is presented within such a "black-box" procedural framework.

16.2.4 First arguments for the operator-based approach

The arguments for an operator-based approach will emerge more naturally as different application examples are discussed in the following sections. But even at this level of generality it might nevertheless be useful to identify some important directions of interest.

(i) Firstly, the operator-based approach has become more and more *natural* in evolutionary history. After much research into the matrix-based approach, the need for an abstract formulation or language (evidently associated with the development of probability theory in abstract spaces) is now evident. If, thirty years ago, the question "why consider infinite dimensions in data analysis?" was often asked, this question is no longer posed.

(ii) Secondly, the approach is clearly *pedagogic* from several perspectives. It clarifies some of the notions of the matrix-based version (take the simple example of the adjoint A^* and the transpose $^t A$ of an operator A whose matrix representations are the same although they admit different spaces of definition). Furthermore, it offers a *natural extension* of the Euclidean case to the (infinite-dimensional) Hilbert-space case.

(iii) Thirdly, the approach permits a *synthetic background* for multivariate analysis. Take the example of PCA, many variations of which exist in the literature (not centered, centered, reduced, or in finite and infinite dimensions), and other related methods (FCA, for instance). All these cases can be presented in the general context of the study of an operator (see the next section for this development).

(iv) Finally, we will see (in the last section) the importance of a possible "unified version" of statistics encompassing the macro and micro contexts. From this perspective, only the operator-based (associated with a tensor-based) approach will permit the required degree of generality. In the same vein, we can say that the operator-based approach is being *embraced by other branches of mathematics*.

16.3 OPERATOR-BASED STATISTICS

16.3.1 Factor multivariate analysis

We start from a data analysis point of view and a first question may be the following:

Question 16.1 *Does a synthetic formulation exist in infinite dimensions that allows us to present the (most familiar) factor multivariate methods in a unitary framework and to study their asymptotic behavior by sampling?*

The answers here are affirmative and require operator theory and functional analysis to build such a framework. We can present the main results on two levels:

(i) Each (classical) factor multivariate analysis can be obtained by the *spectral analysis of a positive compact selfadjoint operator:* this is perhaps the most important "message" given by the earlier work of Dauxois and Pousse (1975, 1976) (and simultaneous papers by Kleffe (1973) and Deville (1974)).

(ii) As the sample size increases indefinitely, each approximation *under an i.i.d. (independent and identically distributed) sampling design* of all the (classical factor) multivariate analysis methods is *consistent*, and we know its asymptotic behavior as well as those of its sample spectral elements (i.e. its associated sample eigenvalues, eigenprojectors, and eigenvectors): see Dauxois et al. (1982) for PCA and Fine (2003) for CA for these two fundamental methods (and Anderson (1963) for PCA and Anderson (1999) and Eaton and Tyler (1994) for CA are corresponding references for the matrix-based approach).

This presentation is very general and has many different properties and applications.

Concerning (i), we can see that different specific areas may be envisaged, such as the linear, mixed, or nonlinear cases, the finite- or infinite-dimensional context, the real or complex domain, ... Furthermore, we can make explicit the relationships between different methods themselves, and on this particular topic the operator-based PCA and CA can be considered as fundamental.

For (ii), we may emphasize some particular results. Here, by asymptotic behavior we mean the convergence of the sample estimator sequences of the operators whose spectral analysis gives the method under consideration, and also the convergence of all its eigenelement sequences. Furthermore, by applying an adapted perturbation technique, the asymptotic expansions are obtained without any distribution assumption, and limiting laws are given in explicit forms within a general class of distributions: the family of elliptic distributions which contains the Gaussian case.

Now let us focus (in a very shortened presentation) on operator-based PCA and CA (let us notice again that several methods may be considered as "particular cases" of these two). Let E be a Euclidean space and Φ a (linear) operator defined from the dual space E^* with values in a second Euclidean space F. Let us provide E (respectively, F) with the metrics M (respectively, D): this also means that M (respectively, D) may denote the Riesz canonical isometric mapping between E and its dual space E^* (respectively, F and F^*). Then we define the covariance operator $V = {}^t\Phi D \Phi$ and $W = \Phi M {}^t\Phi$.

What is the PCA of order q (with $q < \min(\dim E, \dim F)$) of the operator Φ?

This PCA is defined by searching for the unit element u of E^* such that $\|\Phi u\|_F^2$ is a maximum and then by iterations under orthonormality constraints. So it is well known that this PCA is obtained, for instance, by the spectral analysis of the M-symmetric operator VM whose eigenvalues are the *principal values* and whose images of the eigenvectors under the mapping Φ are the *principal components* (and only the q greatest eigenvalues and associated components may be retained).

If $(\Omega, \mathcal{A}, P)$ is a probability space and $L_E^2(P)$ denotes the set of random variables defined on $(\Omega, \mathcal{A}, P)$ with values in E with square norm P-integrable, let us consider the random variable (r.v.) X belonging to $L_E^2(P)$ and the (canonically associated) operator Φ_X defined by

$$\Phi_X : f \in E^* \longmapsto \langle X, f \rangle_E \in L^2(P),$$

and then the PCA of the r.v. X is exactly the PCA of the operator Φ_X.

This intermediate-level presentation admits, on the one hand, well-known particular cases as, for example, the original PCA of Hotelling (of a set of real r.v.'s) and the PCA in descriptive statistics (of the initial tabular $[X]$ of the observed data) and, on the other hand, potential important generalizations. For the latter cases, perhaps the most illustrative example is the PCA of a vector random function.

Let us develop this context a little bit. Let us introduce a second measure space (T, E, μ) and consider a *random function* X defined on $(\Omega \times T, \mathcal{A} \otimes E, P \otimes \mu)$ with values in E and with square norm integrable. Then with

$$\Phi_X : f \in E^* \longmapsto \langle X, f \rangle_E \in L_E^2(P \otimes \mu),$$

we can again define *the PCA of the random function X by the PCA of the operator* Φ_X.

So the operator-based approach to PCA is a general framework containing several particular cases. (Furthermore, we do not develop here all the well-known links with FCA and DA that may be presented as a particular version of PCA but also as a particular version of CA.)

We will also give a short presentation on the CA of two operators. Consider Φ and Ψ to be two operators with the same context as previously. The CA of Φ and Ψ is defined as the CA of the closure of their respective ranges and, knowing the general CA of two subspaces (see Dauxois and Pousse (1976)), then the CA of Φ and Ψ can be obtained, for example, by the spectral analysis of the product of their respective projection operators. Indeed, let P (respectively, Q) be the orthogonal projector whose range closure is $\overline{R(\Phi)}$ (respectively, $\overline{R(\Psi)}$), then the CA of Φ and Ψ is obtained, for example, by the spectral analysis of the symmetric operator PQP.

This operator-based approach (in an analogous manner to that for PCA) contains many "avatars," among them linear and nonlinear CA and also other important factor multivariate methods such as DA and FCA. Let us give some illustrative examples.

In particular, when P and Q are the respective conditional expectations relative to the sub-σ-algebras B and C, then this is the CA of the two tribes B and C. If $X = (X_1, \ldots, X_p)$ and $Y = (Y_1, \ldots, Y_q)$ are two vector-valued r.v.'s, then P (respectively, Q) is the projector on the (vector) subspace $\text{span}(X_i)_{i=1,p}$ (respectively, $\text{span}(Y_j)_{j=1,q}$) and this is the original Hotelling's CA of two sets of r.v.'s. The list of particular cases is very large and also admits recent developments as, for example, Dauxois and Nkiet (2000, 2002), and Dauxois *et al.* (2007).

Therefore the operator-based approach to PCA and CA provides fundamental methods, including many varieties of connected techniques.

When we are interested in studying asymptotic behavior using sampling of these multivariate methods (a sample replaces Ω via a given design and, in general, the well-known i.i.d case is considered), then a second question arises:

Question 16.2 *Are the matrix-based and operator-based languages used in multivariate statistics (and in particular in the asymptotics of the previous factor methods) comparable?*

To answer this question precisely, let us illustrate this problem with PCA asymptotic studies under the elliptical distribution assumption (let us recall that this distribution family contains the Gaussian case, which is obtained by replacing the "kurtosis" parameter by the null value).

Let V (respectively, V_n) be the population covariance operator (respectively, the classical empirical estimator for a sample of size n): we know that the spectral analysis of these operators gives the theoretical and the estimated PCA's (in many cases we can consider L^2-type spaces for E and F, with M = Identity). Let v

(respectively, v_n) denote the population covariance matrix (respectively, the classical empirical estimator). Working with V (respectively, V_n) can be equally adapted to the infinite- and the finite-dimensional cases: working with v (respectively, v_n) means the work is in a finite-dimensional context.

Under an i.i.d. sampling design and when n increases infinitely, in the literature (see mostly the same references as for Question 16.1) we may find three kinds of expression for the limiting distribution of the sequences of the empirical estimators V_n and v_n as $n \to +\infty$:

1. The "pure" operator-based approach gives:

$$\sqrt{n}\,(V_n - V) \xrightarrow{d} N(0, (1+\kappa)\eta + \kappa\xi),$$

   where η is the asymptotic covariance operator of $\sqrt{n}\,(V_n - V)$ under a Gaussian assumption, ξ is a tensor product of V by itself (see later), and d stands for the usual notion of convergence in distribution (in an appropriate space).

2. A semi operator-based and basis-dependent language gives a more explicit form. Let $V = \sum_i \lambda_i\, e_i \otimes e_i$ be the Schmidt decomposition of V (i.e. the operator-based version of the singular value decomposition, where λ_i are the eigenvalues and e_i the associated eigenvectors). Then the previous operator η may be written:

$$\sum_{i \neq j} \lambda_i \lambda_j\, \varphi\left(e_i \otimes e_j\right) \tilde{\otimes} \varphi\left(e_i \otimes e_j\right) + 2 \sum_i \lambda_i^2 \left((e_i \otimes e_i) \tilde{\otimes} (e_i \otimes e_i)\right),$$

   where $\otimes$ and $\tilde{\otimes}$ are the well-known respective tensor products (for vectors $u \otimes v(\cdot) = \langle \cdot, u \rangle v$ and for operators $f \tilde{\otimes} g(\cdot) = \langle \cdot, f \rangle g$), and where φ is the mapping which associates $A + A^*$ with the operator A (and A^* denotes the adjoint of A).

3. finally a third, "purely" matrix-based version is presented in a slightly different way:

$$\sqrt{n}\,\text{vec}\,(V_n - V) \xrightarrow{\mathcal{L}} Y,$$

   where Y is the Gaussian $N\left(0, (1+\kappa)(I+C)\left(V \otimes^K V\right) + \kappa\,(\text{vec}\,V)^t(\text{vec}\,V)\right)$. Here vec denotes the usual vec transformation and $\otimes^K$ is the classical Kronecker product, C denotes the commutation matrix, and I the identity.

At this stage, we can see that there is a *language problem* and a translation between the different terminologies (from matrix-based to operator-based language, but also conversely) is necessary. For that purpose, a *dictionary* has been proposed by Dauxois et al. (1994) that gives a translation of each form into the other one. Such a dictionary offers many advantages (the most important one seems to be that of avoiding some evident redundancy); perhaps some researchers underestimate the scale of this problem of communication between themselves.

Sometimes a translation is not possible: it is interesting in such cases to ask "why?" We will speak about this remark again in Section 16.5.1.

16.3.2 Progressive enlargement to other operator-based areas in statistics

After the previous case of factor multivariate methods where the operator-based approach is important from a synthetic and methodological point of view, a third question may arise of giving other possibilities for using operators in statistics.

Question 16.3 *Starting from PCA and CA methods, are there many other operator-based areas in statistics?*

In the last section, we have seen that the operator-based approach works in finite as well as in infinite dimensions. In this section, we give some other examples. In fact, one may be interested in the classification of statistical methods according to those which absolutely require the operator-based approach (and here the criterion of dimensionality is a determining factor) and other methods where it is not essential.

We now illustrate some of the different domains in which various classes of operators are often used and studied.

Starting from PCA: covariance operators

A first area of study, because of the natural link with the previous section on PCA, concerns the class of *covariance operators*. These are positive symmetric operators but their compactness is not in general so obvious (see Tarieladze and Chobanja (1973), for example): nevertheless, for the principal cases studied multivariate statistics, they may be assumed compact. This class generalizes the well-known notion of the covariance matrix and is directly linked with reproducing kernel theory: as parameters, they may also appear in the Gaussian distribution on abstract (Hilbert or Banach) spaces and as concerns applications, they are useful, for example, in communication and information theory. For general references see, for instance, Grenander (1981) and Fortet (1995) (with an extensive bibliography in each book).

For more statistical-oriented purposes, we give various areas where this class of operators appears. For pointwise estimation, Bosq (1989, 1999a, 1999b, 2000, 2002a) studies the consistency of the empirical estimator with φ-mixing conditions, generalizing previous work of Dauxois and Pousse (1976) and Arconte et al. (1980). In Grenander (1981) or Antoniadis and Beder (1989) "sieve" estimators are used. For domain-asymptotic estimation we may see Dippon (1993). In Vakhania (1993), a canonical factorization is given for the Gaussian case. The problem of comparison (i.e. of partial or total simultaneous spectral analysis) of two covariance operators has been the focus of many works such as Kadota (1967), Baker (1969), and Dauxois *et al.* (1993) where the operators are estimated on different samples. Studies on this topic are also directly linked with functional limit theorems: see Mas (2006, 2008) for recent developments.

Starting from CA : Hankel and Toeplitz operators

A second area of study, but now because of a statistical link with CA, concerns the class of *Hankel and Toeplitz operators*. In operator theory, a Toeplitz operator is the compression of a multiplication operator on the circle to the Hardy space: a bounded operator on this space is Toeplitz if and only if its matrix representation has constant diagonals.

In the context of CA, Jewell and Bloomfield (1983) use the theory of Hankel and Toeplitz operators to characterize the structure of some series (such as ARMA, strongly mixing, absolutely regular), and Pavon (1984) classifies linear stochastic systems according to information transfer given in terms of the system Hankel operator.

The forecasting problem in the statistics of stochastic processes has always needed specific tools and methods (see, for example, Gelfand and Yaglom (1957) and Yaglom (1965) in the nonlinear prediction case). Studies on the past and the future give rise to important developments; see Hida (1960) for the stationary case, Ibragimov and Rozanov (1978) for the discrete Gaussian context, and Vaninskii and Yaglom (1990) for the continuous case. In Grenander (1981), we may find the introduction of Hankel matrices (see also Lord (1954)) where the author remarks that the value of the maximum correlation coefficient is equal to the norm of such a matrix. Some results on the past and the future are obtained under the condition that the value of the maximum correlation coefficient is strictly less than one: this observation (and other considerations) have motivated a particular piece of research directly linked with the uncertainty principle (see Koudou and Romain (2000) in an unpublished paper).

Some regularity conditions enable us to classify different processes via Hankel and Toeplitz operators, as in Peller and Krushchev (1982). In an important paper, Peller (1990) gives the extensions of previous work to the completely regular multivariate stationary case. Finally, basic references for Hankel and Toeplitz operators may be found in Power (1980), Page (1970), Clark (1968), and Peller (2003).

From factor multivariate analysis to modeling statistics: regression operators

It is necessary to work in function spaces and in infinite dimensions when the data are curves. With a few hypotheses (such as second-order integrability, for example) the Hilbert geometry is available for L^2-type spaces. In 1997 (see 2002 for the second edition) Ramsay and Silverman wrote a well-known book on *Functional Data Analysis*. This book has been the starting point for many developments: for instance, Bosq (2000) and later Ferraty and Vieu (2006) are important references for extensions of the initial work. The *regression operators R* appear naturally in the model $Y = R(X) + \varepsilon$ where, even if Y may be a real (or Hilbertian) answer, X may be in an infinite-dimensional space and the operator R may be nonlinear. Here a model is proposed by introducing the "residual" ε and must be "evaluated" in the best way.

For approximation problems, we know that spline functions are useful tools (see, for example, Besse and Thomas-Agnan (1991) for a first review), but other research is also interesting: on the topic of optimal (not necessarily linear) transformations in multivariate regression see Breiman and Friedman (1985) and de Leurgans *et al.* (1993) for CA. Since the 1990s (see, for example, the work of Cardot (1997) and Cardot *et al.* (1998, 1999)), the literature in this domain has grown considerably: the reader can find several more recent developments in Dabo-Niang and Ferraty (2008). Finally, in Chapter 1 of this volume, we can see how nonparametric statistics and the operator-based approach nowadays give important areas of intersection.

From L^2-type Hilbert spaces to general (Hilbert and Banach) spaces: measure-associated operators

This section is devoted to a specific recent piece of work on links between measure theory and some useful operators. We focus on another domain where Banach spaces take the place of Hilbert spaces, as proposed by Benchick *et al.* (2007). We are interested in a countable family of operators (indexed by $\mathbb{Z}$) whose elements belong to the set $K(H, E)$ of compact operators defined from the $\mathbb{C}$-Hilbert space H with values in the $\mathbb{C}$-Banach space E. The stationarity of this family is defined by using the notion of the "quasi-transpose" operator of a compact operator (when E is Hilbert-type, this transpose form may be compared to the adjoint transformation). Then we show that this family is the Fourier transform of a so-called operator-based measure which is defined on the Borel tribe of $[-\Pi, \Pi[$ with values in $K(H, E)$. An example of such a family (of compact operators) may be found in studies of Banach-valued stationary series; this allows us to express the stationarity of a Banach-valued series, as proposed by Bosq (2000, 2002a, 2002b). These results may then be applied to give approximations of the Banach-valued series by the Fourier transform of a so-called Banach-valued measure.

Of course many other studies reflect the importance of enlarging the usual Hilbertian framework to a different or a more general context (see Ledoux and Tallagrand (1991) for a general framework, or Hall and Hosseini-Nasab (2009) and see Rhomari's Chapter 14 in this volume for more specific examples).

From the single to the multiplicative case: tensor operators

Nowadays, tensor algebra and functional analysis hold a very important place in statistics. We now give particular attention to three other operators (see Dauxois et al. (1994) and Romain (2000a, 2000b)).

Let E and F be two real Hilbert spaces, let 1_E and 1_F be their respective identity mappings, and let $\mathcal{L}(E, F)$ denote the set of homomorphisms from E into F (and $\mathcal{L}(E)$ is $\mathcal{L}(E, E)$). When $\mathcal{L}(E, F)$ is equipped with the Hilbert–Schmidt inner product, it is the Hilbert space denoted by $\sigma_2(E, F)$. If a belongs to $\mathcal{L}(E)$ and b to $\mathcal{L}(F)$, there exists one way to embed these endomorphisms into a common space by an "ampliation technique". For u in $\mathcal{L}(E, F)$, consider the elementary operators $a \overset{l}{\otimes} 1_F$ and $1_E \overset{l}{\otimes} b$ defined by $\left(a \overset{l}{\otimes} 1_F\right)(u) = ua^*$ and $\left(1_E \overset{l}{\otimes} b\right)(u) = bu$, where a^* denotes the adjoint of a. These operators are called the *functional ampliations of a and b in $\mathcal{L}(\mathcal{L}(E, F))$*. Then the composition product (respectively, the sum, the difference) of these functional ampliations defines the *functional tensor product* (respectively, the *sum*, the *difference*) in $\mathcal{L}(\mathcal{L}(E, F))$. So

$$a \overset{l}{\otimes} b = \left(a \overset{l}{\otimes} 1_F\right) \circ \left(1_E \overset{l}{\otimes} b\right), \quad a \overset{l}{\oplus} b = a \overset{l}{\otimes} 1_F + 1_E \overset{l}{\otimes} b$$

and

$$a \overset{l}{\ominus} b = a \overset{l}{\otimes} 1_F - 1_E \overset{l}{\otimes} b.$$

These operators have many remarkable properties because of the natural commutativity of the ampliations. There are many operations where they can be used "as easily as real numbers." Furthermore, in Boudou and Romain (2001, 2010), some examples are given where functional ampliations are useful in establishing new results. Here we recall a tensor perturbation rule for positive symmetric operators and its applications in multivariate statistics (see Romain (2002)).

The operators $a \in \mathcal{L}(E)$ and $b \in \mathcal{L}(F)$ are to be perturbed under the same conditions, that is, $a(\epsilon) = a + \epsilon u(\epsilon)$ (respectively, $b(\epsilon) = b + \epsilon v(\epsilon)$) by adding the selfadjoint operator $\epsilon u(\epsilon)$ (respectively, $\epsilon v(\epsilon)$), with $\epsilon \in\ ]0, 1[$.

Then a "perturbation rule for tensor operators" may be given in the following form:

The eigenvalues of the perturbed functional tensor sum (respectively, difference, product) are defined in $\mathbb{R}$ by the sum (respectively, difference, product) of the perturbed eigenvalues of each operator. The eigenvectors of the perturbed functional tensor sum, difference, and product are defined in $\mathcal{L}(E, F)$ by the tensor product of the perturbed eigenvectors of each operator. The eigenprojectors of the perturbed functional tensor sum, difference, and product are defined in $\mathcal{L}(\sigma_2(E, F))$ by the functional tensor product of the perturbed eigenprojectors of each operator.

From a multivariate statistical point of view, the positive symmetric operators a of $\mathcal{L}(E)$ and b of $\mathcal{L}(F)$ may be considered as covariance operators, and the random perturbation is obtained by classical i.i.d. sampling approximations. Let us consider the following: if λ_a is a simple eigenvalue and π_{λ_a} is the associated eigenprojector, then $a = \sum_{\lambda_a} \lambda_a \pi_{\lambda_a}$, where h_{λ_a} denotes a unit eigenvector associated with λ_a and then $\pi_{\lambda_a} = h_{\lambda_a} \otimes h_{\lambda_a}$.

The conditions for the application of the perturbation theory are then satisfied, and the previous results can be applied to obtain the following properties:

(i) $\sqrt{n}\left(a_n \overset{l}{\otimes} b_n - a \overset{l}{\otimes} b\right) = \left(a \overset{l}{\otimes} v_n + u_n \overset{l}{\otimes} b\right) + O\left(\frac{2\log\log n}{\sqrt{n}}\right)$,

(ii) $\sqrt{n}(\lambda_{a_n}\lambda_{b_n} - \lambda_a \lambda_b) = (\lambda_a \, tr \, \pi_{\lambda_b} v_n + \lambda_b \, tr \, \pi_{\lambda_a} u_n) + O\left(\frac{2\log\log n}{\sqrt{n}}\right)$,

(iii) $\sqrt{n}\left(h_{\lambda_{a_n}} \otimes h_{\lambda_{b_n}} - h_{\lambda_a} \otimes h_{\lambda_b}\right) = (S_{\lambda_a} u_n \overset{l}{\oplus} S_{\lambda_b} v_n)\left(h_{\lambda_a} \otimes h_{\lambda_b}\right)$
$+ O\left(\frac{2\log\log n}{\sqrt{n}}\right)$,

(iv) $\sqrt{n}\left(\pi_{\lambda_{a_n}} \overset{l}{\otimes} \pi_{\lambda_{b_n}} - \pi_{\lambda_a} \overset{l}{\otimes} \pi_{\lambda_b}\right) = \left(\pi_{\lambda_a} \overset{l}{\otimes} \varphi(\pi_{\lambda_b} v_n S_{\lambda_b}) + \varphi(\pi_{\lambda_a} u_n S_{\lambda_a}) \overset{l}{\otimes} \pi_{\lambda_b}\right)$
$+ O\left(\frac{2\log\log n}{\sqrt{n}}\right)$.

These perturbation properties may be considered as statistical single sampling design corollaries for tensor covariance operators. Consequently, previous studies on the asymptotic behavior of such operators (see Bosq (2000, 2008)) or of multivariate methods (such as PCA for instance; see Dauxois et al. (1982)) may be extended to the tensor case and consequently to the multiplicative case.

Some other cases considered in this volume

Before presenting the operator-based quantum version of statistics, we just want to quickly emphasize other categories of interesting operators in statistics. The reader may find them easily in this volume.

- *Trigonometric operators*: see the remarkable research presented in detail in Chapter 13 by K. Gustafson (and the large corresponding list of references).
- *Integration and measure*: in Chapter 12, N. Dinculeanu develops an original measure theory and integration where, for particular purposes, the measure or the integral part is possibly operator-valued.
- *Shift operators* (or *unitary operators*): in Chapter 15, the reader's attention is particularly drawn to the results on shift operators associated with the product and the convolution product of spectral and random measures (themselves associated with stationary processes).

Of course, we do not pretend in any way to be comprehensive on this topic; other types of operators (namely differential, integral, nonbounded, ...) may be encountered and prove to be useful.

16.4 OPERATOR-BASED QUANTUM STATISTICS

Before giving some unifying favourable arguments for the use of the operator-based approach to statistics set out in the last section, we propose here to look at an area which is less well-known in the statistics community. We simply want to emphasize the importance of operator-based and tensor-based formulations, which is fundamental in the noncommutative version of probability and statistics. Therefore we continue our presentation by posing a new question.

Question 16.4 *Do we need a deep commitment to understand the links between classical and the quantum statistics?*

The terms "quantum" and "noncommutative" may be considered equivalent for a first approach, even if there is some chronology in the terminology. Let us give some basic notions of noncommutative probability just to see how the classical case may be embedded in the non-classical one.

Historically, particle physics (i.e. the study of the microscopic world) has been rapidly developed in the last century, with early work of Bohr, Sommerfeld, Heisenberg, Schrödinger, and others. The formulation of quantum mechanics (see Birkhoff and von Neumann (1936)) is based on quantum logic: propositions are represented by orthogonal projectors in a separable Hilbert space H. On the set of these projectors denoted by $\mathcal{P}(H)$ (or equivalently on the set of their range-subspaces), the classic laws of Boolean algebra no longer work in general. For example, for subspaces there is no distributivity between $\cap$ and $+$; in fact, "and $= \cap$", "or $= \cup$", "not = complement" are respectively replaced by the infimum $\wedge$, the supremum $\vee$, and the orthocomplement. Then $\mathcal{P}(H)$ is a complete orthomodular lattice which is nondistributive in general (see Holland (1970) and Kalmbach (1983)): furthermore, we can consider the notion of a "state" on $\mathcal{P}(H)$ which admits similar properties to those of a probability measure. Gleason's theorem (1957) is central because it characterizes all the states ρ on this family by proving that $\rho(\pi) = \text{tr}\,(\nu\pi)$, where π belongs to $\mathcal{P}(H)$ and ν is a positive trace class operator with $tr\,\nu = 1$ and is called the density operator of the state ρ.

For systems in a given state, the measures are made with the notion of one "observable": this is the microscopic version of the classic random variable. An observable is a selfadjoint operator; we can then understand noncommutative probability by studying its correspondences with the notion of random variable. Before developing this topic, let us give some introductory references in quantum mechanics.

The fundamental framework for this approach may be found in Cohen-Tannoudji *et al.* (1977) and Prugovecki (1981). For logical and geometric aspects, see Varadarajan (1985) and for algebraic concepts see Brattelli and Robinson (1981). Finally, more specific research may be found in Aragon and Couot (1976); this is a rare example (to our knowledge) of the intersection between multivariate statistics and quantum mechanics.

For *quantum* or *noncommutative probability*, we refer here to the (first pages of) the books of Meyer (1993), Parthasarathy (1992), and Biane (1995). A recent survey may be found in Rédei and Summers (2007).

We are now interested (even in the finite-dimensional case) to investigate the possible correspondences between a random variable and an observable. Let X be a random variable defined on $(\Omega, \mathcal{P}(\Omega), P)$: a selfadjoint operator T_X belonging to $\mathcal{L}\left(L^2_{\mathbb{C}}(\Omega, \mathcal{P}(\Omega), P)\right)$ may be associated with X by the equation $T_X f = Xf$. If we consider the orthonormal basis $\left(1/\sqrt{P(\{\omega\})}\, \mathbb{1}_{\{\omega\}}\right)_{\omega \in \Omega}$ of $L^2_{\mathbb{C}}(\Omega, \mathcal{P}(\Omega), P)$, then the operator T_X is diagonal and its eigenvalues are the values taken by X as:

$$T_X\left(\frac{1}{\sqrt{P(\{\omega\})}}\, \mathbb{1}_{\{\omega\}}\right) = \frac{X(\omega)}{\sqrt{P(\{\omega\})}}\, \mathbb{1}_{\{\omega\}}.$$

Furthermore, its mathematical expectation may be written $\mathbb{E}(X) = \langle X \mathbb{1}_\Omega, \mathbb{1}_\Omega \rangle$.

Conversely, let $A = \sum_{i \in I} \lambda_i \pi_i$ be the spectral Schmidt expansion of the selfadjoint operator A on a Hilbert space E. Let u be a unit vector and P_u the orthogonal projector in the direction of u. So, considering $\mu(i) = \text{tr}\,(P_u \pi_i)$, we define a probability measure on $(I, \mathcal{P}(I))$ and determine an isometry J from $\ell^2(I, \mathcal{P}(I), \mu)$ on E by associating the element $\pi_i(u)$ with $\mathbb{1}_{\{i\}}$. Let X denote the random variable defined on $(I, \mathcal{P}(I))$ by $X(i) = \lambda_i$, then the operator $A = J \circ T_X \circ J^*$ may be interpreted as multiplication by X and the expectation may be expressed by

$$\mathbb{E}(X) = \langle A(u), u \rangle = \text{tr}\,(P_u A).$$

The correspondence is thus well established, and we can go further. Let us consider a positive operator S with unit trace for generalizing P_u, let $(u_k)_k$ be a family of unit vectors, and S be a convex combination $\sum_k p_k \pi_{u_k}$ (with $\sum_k p_k = 1$). Then a *quantum probability space* is a pair (E, S) and a state corresponding to S is a linear functional on $\mathcal{L}(E)$ which associates the real $\text{tr}\,(A S)$ with each A. A *quantum (or noncommutative) random variable* on this space is a selfadjoint operator on E whose distribution may be defined in the following way. Let sp(A) be the spectrum of the quantum random variable A and f be a real function on sp(A). Let us consider the selfadjoint operator $f(A)$ defined by $\sum_{i \in I} f(\lambda_i) \pi_i$, then the mapping $\mu_A : f \mapsto \text{tr}\, f(A) S$ is a positive linear functional and, by Riesz's theorem, this defines a probability measure on sp(A) (with its Borel σ-field) which is called the *distribution of the quantum random variable* A (and for $f := \mathbb{1}$ constant and equal to 1, one has $\mu_A(\mathbb{1}) = 1$, and for each λ_i of sp(A) one has $\mu(\{\lambda_i\}) = \text{tr}\,(\pi_i S)$).

We immediately see that the product of two quantum r.v.'s is not commutative in general and so the joint distribution of two quantum r.v.'s is possible only under the assumption of commutativity of the two operators (this is the reason why theoretical results on this property are important: see Samoilenko (1991) for a general reference).

Of course a complete presentation of quantum probability is not our purpose here. Just let us note that *quantum events are projectors* (or subspaces), and some particular laws are also well studied and known (such as the quantum Bernoulli distribution, for instance).

For quantum statistics, we have to emphasize that:

- for independent replicated experiments, the statistical model is based on the tensor product $E^{\otimes n}$ (instead of the Cartesian product Ω^n in the classic case);
- the tensor product of operators is then fundamental because it is the basic notion of the observables (or quantum variables) that is defined on that product.

Here we have to recall that this tensor product of operators $(A_i)_{i=1,n}$ is defined for the tensor element $\otimes_{i=1}^n x_i$ of $E^{\otimes n}$ by

$$\otimes_{i=1}^n A_i \left(\otimes_{i=1}^n x_i \right) = \otimes_{i=1}^n A_i(x_i).$$

Remark 16.1 *Let us notice that for $u = x \otimes y$ the previous functional product $A \overset{l}{\otimes} B$ will give $A \overset{l}{\otimes} B(u) = A(x) \otimes B(y)$ and this then generalizes (in this sense) the tensor product of operators: this is also one of our motivations for working on ampliations and functional tensors, as seen before in Section 16.3.6.*

We can now present a statistical model for quantum statistics. The state set $\mathcal{E}$ may be indexed by a set Θ to obtain the parametric model $\{E, \mathcal{E}_\theta, \theta \in \Theta\}$. Then a decision theory in quantum statistics is available with analogous notions to those of (Bayesian or not) estimation theory, the Cramer–Rao inequality, sufficiency, maximum likelihood,...Nevertheless, let us remark that some notions need different backgrounds: for example, conditional expectation is no longer defined with respect to a sub-σ-field, but to a von Neumann subalgebra (see Takesaki, 1972. This formulation of quantum decisional statistics may be found in Holevo (1973, 1982) and Helstrom (1976). Recent developments may be found in an important review paper of Barndorff-Nielsen *et al.* (2003), with discussion by other authors working on this stimulating topic.

One possible answer to Question 16.4 is that the operator-based and tensor-based mathematical framework is *necessary* for the statistics of quantum phenomena. So each piece of research that establishes links between the two (classical and non-commutative) versions will allow these two branches to be embedded in a general formulation (see Holevo (1973) and the following section on this topic). Finally, we give the reference which allows us, following my personal approach, to provide motivation for the study of the links between the "classical" theory and the "new one": see the paper by Malley and Hornstein (1993).

16.5 Arguments for operator-based statistics

Following this review of examples and domains where the operator-based approach, or more simply where some operator categories, appear in statistics, we are now interested in setting out some positive arguments to invite the reader to follow such a direction.

16.5.1 More arguments

In this section we conclude our previous comments on this topic (see Section 16.2.4). The different examples we have presented show that a common language is often necessary in order to study the intersections between several areas. This observation is obvious in mathematics and science: for example in sofware engineering, the lowest-level machine language is often reserved for people working in this speciality, rather than other more elaborate languages which allow a majority of researchers to "speak" to each other. The matrix-based language is well known and historically well adapted to concrete cases. Statistical software and languages have also allowed the same evolution to a possible higher level. New stages in potential applications have recently been correlated with the present calculating capacity (such as for simulation, optimization, complex tabular, bootstrap methods, stochastic algorithms, etc.). So functional, symbolic, and operator-based languages will certainly be developed more and more in the future.

We have seen that a "coordinate-free" or "invariant" property is often implicit when working with an operator-based approach. We have many other examples to show that this way of presenting a method in an intrinsic way is well adapted to a synthetic presentation (and sometimes it is more rigorous). In a kind of counterexample (to "confirm the rule" ...), this property may be used as a default position for some specific proofs (we think here of Halmos (1982) in his third problem, where the author prefers a nonfunctional demonstration of the Riesz Representation Theorem, even if the manner of proof is less elegant).

Finally, reflecting on the languages used in statistics permits us to classify the tools for which an intrinsic form does exist and those for which such a form does not exist: consequently, we may have an *extension in infinite dimensions* of this notion. For example, it is well known that the trace of a matrix is independent of the choice of basis, and so a class of operators with finite trace has been considered (the trace class or nuclear operators). Another example is given by the functional tensor product denoted $a \overset{1}{\otimes} b$ and presented in Section 16.3.2: a matrix-based representation of this product is the Kronecker product, and an operator-based definition on Hilbert spaces of tensor products is the (tensor) product of operators. In fact, a "good" question arises (with no answer in some cases): does each matrix-based tool have an intrinsic form? Take the Hadamard product (also called the pointwise product of matrices): to our knowledge, this product has no operator-based (or intrinsic) form, and this question is a conjecture because we do not "know" the algebraic structure or the functional expression which are involved in this notion.

16.5.2 Conceptual and synthetic arguments: projectors in noncommutative statistics

Let us now ask a final question. By a "unified statistics," we mean the potential union of traditional and quantum statistics in a Hilbert-space context.

Question 16.5 *What are the key notions and formulations of a "unified statistics"?*

Mathematical statistics is a (parametrized or not) version of probability and of measure theory. This last theory admits a general formulation (and consequently the same is true for statistics and probability), which is the notion of the von Neumann algebra. This algebra is generated by (orthogonal) projectors and so the answers to Question 16.5 are orthogonal projection for the key notion and the von Neumann algebra for the formulation. Let us now develop these two elements in more detail.

First of all, a von Neumann algebra is a subalgebra of the set of (bounded) operators $\mathcal{L}(H)$ (on a Hilbert space H); this admits two presentations. The first one is topological: a von Neumann algebra is a subalgebra that is closed in the weak operator topology and contains the identity operator (this is a particular case of a C^*-algebra which is closed in the norm topology). The second presentation is algebraic: denoting by A' the commutant of A in $\mathcal{L}(H)$ (i.e. the set of all elements in H commuting with all elements of A), then a von Neumann algebra $\mathcal{U}$ is equal to its bicommutant, which means $\mathcal{U} = (\mathcal{U}')' = \mathcal{U}''$ holds.

An important result of Murray and von Neumann (1937) shows that these two ways of defining a von Neumann algebra are equivalent. Therefore we can understand this general formulation better when we read A. Connes (1990): "in the commutative case, the von Neumann algebra theory is equivalent to the Lebesgue measure theory and to the spectral theorem of self-adjoint operators." This result could give deep developments, but we may see it as a continuation of the correspondence we have seen (cf. Section 16.4) between a random variable and a selfadjoint multiplication operator.

So we can keep in mind that statistics, probability, and measure theory may be considered as particular (commutative) cases of a more general background theory.

The second pointer concerns functional (or symbolic) calculus in relation to the spectral theorem: some properties here may illuminate the "structure" of a von

Neumann algebra $\mathcal{U}$. If we consider a Borel measurable bounded function f on the spectrum of the operator T, then we can give a meaning to $f(T)$ which still belongs to $\mathcal{U}$. Thus it can be deduced that $\mathcal{U}$ is generated by the orthogonal projectors which belong to it. We may illustrate this fact in the case of countable dimension by writing, for T symmetric, $T = \sum_i \lambda_i P_i$, where λ_i is an eingenvalue associated with the eigenprojector P_i and then $f(T) = \sum_i f(\lambda_i) P_i$.

Consequently, the notion of orthogonal projection is fundamental. And another notion also appears important in von Neumann algebras: it is that of a factor. A factor is a von Neumann algebra whose center (i.e. $\mathcal{U} \cap \mathcal{U}'$) is reduced to scalar operators; we may discover that every von Neumann algebra can be built as countable or noncountable "sums" or "direct integrals" of factors.

So the study of a von Neumann algebra $\mathcal{U}$ is equivalent to the study of different types of factors. Because of the lattice structure of the set of its projectors, three types of factors have emerged in the literature (see Jones and Sunder (1996)).

Then the two known versions of statistics admit a general formulation using the von Neumann algebra notion, and in which orthogonal projection plays a key role in the noncommutative version. In the next section we will give other arguments to highlight the fact that this key notion is also very important in "traditional" statistics. For introductory studies on von Neumann algebras, we may look at Dixmier (1957), Sunder (1987), Connes (1990), and Jones and Sunder (1996).

16.5.3 A particular point of view: projectors in (classical) statistics

Perhaps "a projector-based approach to statistics" could be the title of a modern course in statistics, taking account of the importance of this notion. First of all, let us remark that we now have a large spectrum of knowledge of the mathematical properties of the set of projectors in $\mathcal{L}(H)$, or equivalently on the set of the closed subspaces in H.

(i) From an *algebraic point of view*, the unifying notion is the lattice structure of closed subspaces (see Nordström and von Rosen (1987) for a reference in the finite-dimensional case with applications to statistics, and see Holland (1970) and Kalmbach (1983) for general properties). For example, the lattice is modular if and only if H is finite dimensional (it is orthomodular in infinite dimensions) and furthermore, the orthomodularity property characterizes the complete spaces between the inner product spaces: for three subspaces E, F (containing E), and G of H, the modular law implies that $E \cap (F \cup G) = F \cup (E \cap G)$ but the orthomodular law implies that $E \vee (E^\perp \wedge F) = F$.

(ii) From the *geometrical point of view*, the study of the relative position of two subspaces has given rise to important research in complementary directions. The initial research here is that of Jordan (1875) in $\mathbb{R}^n$ and then, for example, Dixmier (1948), Davis (1958), Halmos (1969): for the case of more than two subspaces see Sunder (1988). In an operator-based context, see Lenard (1972) for the joint numerical range, Bhatia (1997) for the links between CS decomposition and the Sylvester equation, Sano and Watatani (1994) on the angles between two type II_1-factors.

Of course, this area concerns the CA method in multivariate statistics: the most important references here are Hotelling (1936), Gelfand and Yaglom (1957), Hannan (1961), and Dauxois and Pousse (1975). See also recent developments proposed by Cupidon *et al.* (2008) and Eubank and Hsing (2008).

(iii) From a *stochastic point of view*, we have seen (in the work of Gleason (1957)) that any finite and σ-additive measure μ on $\mathcal{L}(H)$ can be expressed as $\mu_T(\pi) = \text{tr}(\pi T)$, where T is a trace class operator and π is a projector in $\mathcal{L}(H)$. Furthermore, it is important to highlight the important role of conditional expectation as a projector in statistics. Of course, each projector is not a conditional expectation (see estimating functions, for example), but the converse is true and many applications of this result are more-or-less known (see, for example, Small and McLeish (1994) for the case of incomplete data studied using conditional likelihood).

Regarding asymptotic studies (under an i.i.d sampling design), Dossou-Gbete and Pousse (1991) present the properties of the sequence of eigen-projectors of estimators of compact positive selfadjoint operators: these general results give rise to various applications (Dauxois *et al.* (1982) and Fine (2003)) and also to data analyses such as that of Jeffries *et al.* (1985) and Paulraj *et al.* (1993) in signal processing.

For another topic (this also relates to point (i)), see Paszkiewicz (1986) on the definitions of various types of stochastic convergence in operator algebras (and therefore including von Neumann algebras).

(iv) Finally, we are interested in completing the spectrum of above properties from a *methodological point of view*. In statistics, we may emphasize that in various situations we have to obtain decompositions of a space into orthogonal components (or subspaces), and consequently to consider the corresponding orthogonal projectors. Let us illustrate this fact by some examples:

- in parametric estimation, when we try to remove nuisance parameters;
- in the Cramer–Rao context, when we want to reduce the variance of some estimators;
- in a conditional inference strategy, when we want to project onto the subspace (or subalgebra) spanned by a statistical random variable;

- in a multivariate context (such as a linear model, signal processing, ...), when we want to decompose an initial space into a target subspace and an orthogonal residual subspace;
- in the approximation problem of an operator (or a tabular) by a linear combination of projectors;
- ... in fact, each reader may complete this list from his own experience.

Orthogonal projection is a key notion which allows us to simplify various statistical problems in a geometrical and functional manner. Furthermore, in many situations it allows us to extend to infinite dimensions and to various other extensions (such as the noncommutative domain).

16.6 Conclusion

In this chapter, we have tried to lead the reader from well-known descriptive multivariate methods into different directions. Even if a "unified course" on statistics seems to be not mature enough nowadays for a complete abstract theoretical treatment (from von Neumann algebras to descriptive statistics), the operator-based approach is a natural setting from which to open the classical fields of application to new ones. Beginning with a translation problem between the two languages used in asymptotics for multivariate factor methods, we have seen that the dimensionality problem is large and includes several components: sample size, abstract space structure, and the exploration of complexity. Furthermore, the quantum version of statistics is less well known but its natural formulation needs operator theory in infinite dimensions in general. If we explore the study of noncommutative subspaces a little more deeply, we see that a notion of uncountable (or non-integer) dimension exists, taking its values in $[0, 1]$ in some cases. Even if we may have met such non-integer dimensions in the fractal context (see Mandelbrot (1995) for example), this draws our attention to the new direction of "continuous geometry".

But this last point does not seem the most important in statistics: a more crucial point is that both *traditional* and *quantum* versions of statistics are naturally concerned with the concept of infinity. Will the power of the methods and their theoretical potential imply a loss of contact with applications? For us, it seems that the opposite tendency may be observed, but we have to take care to control the techniques being used, as previously remarked.

This concept of infinity is an old problem in science. Some scientists (and nonscientists) have provided reflections that go outside the strict domain of their research activity in order to give enlightenment on various philosophical points of view: see Houzel *et al.* (1976), Apery *et al.* (1982), and Guichard (2000) for examples in mathematics, and Atkins (2004) for another example. Nowadays, the statistics community may also contribute to this conceptual debate.

References

AKHIEZER, N.I., GLAZMAN, I.M. (1961). *Theory of Linear Operators in Hilbert Space*, Vols 1 and 2. Frederik Ungar, New York.

AMREIN, W.O., SINHA, K.B. (1994). On pairs of projections in a Hilbert space. *Linear Algebra Appl.*, 208/209, 425–35.

ANDERSON, T.W. (1958). *An Introduction to Multivariate Analysis*. Wiley, New York.

ANDERSON, T.W. (1963). Asymptotic theory for principal component analysis. *Ann. Math. Statist.*, 34, 122–48.

ANDERSON, T.W. (1984). Estimating linear relationships. *Ann. Statist.*, 12, 1–45.

ANDERSON, T.W. (1993). Nonnormal multivariate distributions. Inference based on elliptically contoured distributions. In *Multivariate Analysis: Future Directions* (C.R. Rao, ed.), 1–24. North-Holland, Amsterdam.

ANDERSON, T.W. (1999). Asymptotic theory for the canonical analysis. *J. Multivariate Anal.*, 61, 1–16.

ANTONIADIS, A., BEDER, J.H. (1989). Joint estimation of the mean and the covariance of a Banach valued Gaussian vector. *Statistics*, 20, 77–93.

APERY, R., CAVERING, M., DIEUDONNÉ, J. et al. (1982). *Penser les Mathématiques*. Seuil, Paris.

ARAGON, Y., COUOT, J. (1976). Une définition de l'opérateur d'Escoufier. *C. R. Acad. Sci. Paris Sér.*, A-B 283 no. 11, 867–869.

ARCONTE, A., DAUXOIS, J., POUSSE, A., ROMAIN, Y. (1980). Etude asymptotique de l'analyse en composantes principales linéaires d'une fonction aléatoire à valeurs dans un espace de Hilbert. *C. R. Acad. Paris Sér. A Math.*, 291, 319–22.

ATKINS, P. (2004). *Le Doigt de Galilée–Dix Grandes Idées pour Comprendre la Science*. Dunod, Paris. (Translation of *Galileo's Finger. The Ten Great Ideas of Science*, Oxford University Press, 2003.)

BAHADUR, R.R. (1955). Measurable subspaces and subalgebra. *Proc. Am. Math. Soc.*, 6, 565–70.

BAKER, C.R. (1969). Complete simultaneous reduction of covariance operators. *SIAM J. Appl. Math.*, 17, 972–83

BARNDORFF-NIELSEN, O., GILL, R., JUPP, P. (2003). On quantum statistical inference. With discussion and reply by the authors. *J. R. Stat. Soc. Ser. B Stat. Methodol.*, 65(4), 775–816.

BEAUZAMY, B. (1988). *Introduction to Operator Theory and Invariant Subspaces*. North-Holland, Amsterdam.

BEDER, J.M. (1988). A sieve estimator for the covariance of a Gaussian process. *Ann. Statist.*, 16, 648–60.

BENCHICK, T., BOUDOU, A., ROMAIN, Y. (2007). Mesures aléatoires opératorielle et banachique. Application aux séries stationnaires. *C. R. Math. Acad. Sci. Paris*, 345(6), 345–8.

BESSE, P., THOMAS-AGNAN, C. (1991). Le lissage par fonctions splines en statistique: revue-bibliographique. *Statist. Anal. Données*, 14, 55–83.

BHATIA, R. (1997). *Matrix Analysis*. Springer, New York.

BHATIA, R., ROSENTHAL, P. (1997). How and why to solve the operator equation $ax - xb = y$. *Bull. Lond. Math. Soc.*, **29**, 1–12.

BIANE, P. (1995). Calcul stochastique non-commutatif. In *Lectures on Probability Theory* (P. Bernard, ed.), Vol. 1668, 1–96. Ecole d'Eté de Probabilités de Saint-Flour, No. 23–1993

BILLINGSLEY, P. (1968). *Convergence of Probability Measures*. Wiley, New York.

BIRKHOFF, G., VON NEUMANN, J. (1936). The logic of quantum mechanics. *Ann. Math.*, **37**, 823–42.

BIRMAN, M., SOLOMJAK, M. (1996). Tensor product of a finite number of spectral measures is always a spectral measure. *Integr. Equat. Oper. Th.*, **24**, 179–87.

BOSQ, D. (1989). Propriétés des opérateurs de covariance empiriques d'un processus stationnaire hilbertien. *Note C. R. Acad. Sci. Paris, I*, **309**, 873–5.

BOSQ, D. (1999a). Autoregressive Hilbertian processes. *Ann. I. U. S. P.*, **43**(2–3), 25–55.

BOSQ, D. (1999b). Représentation autorégressive de l'opérateur de covariance empirique d'un ARH(1). Applications. *C. R. Acad. Paris Sér. I. Math.*, **329**(6), 531–4.

BOSQ, D. (2000). *Linear Processes in Function Spaces. Theory and Applications*. Lecture Notes in statistics, **149**. Springer, New York.

BOSQ, D. (2002a). Estimation of mean and covariance operator of autoregressive processes in Banach spaces. *Stat. Inference Stoch. Process.*, **5**(3), 287–306.

BOSQ, D. (2002b). Estimation of the autocorrelation operator and prediction for infinite-dimensional autoregressive processes. *Math. Methods Statist.*, **11**(4), 381–401.

BOSQ, D. (2008). On tensorial products of Hilbertian linear processes. in *Functional and Operatorial Statistics* (Dabo Niang, S., Ferraty, F., eds), 71–6. Springer, New York.

BOUDOU, A., DAUXOIS, J. (1994). Principal component analysis for a stationary random function defined on a locally compact Abelian group. *J. Multivariate Anal.*, **51**, 1–16.

BOUDOU, A., ROMAIN, Y. (2001). Processus hilbertien associé à la convolée de deux mesures aléatoires. *C. R. Acad. Sci. Paris Sér. I Math.*, **332**(4), 361–4

BOUDOU, A., ROMAIN, Y. (2002). On spectral and random measures associated to discrete and continuous-time processes. *Statist. Probab. Lett.*, **59**(2), 145–57.

BOUDOU, A., ROMAIN, Y. (2010). On the integral with respect to the tensor product of two random measures. *J. Multivariate Anal.*, **101**(2), 385–94.

BOURBAKI, N. (1962). *Eléments de Mathématiques. Algèbre. Chapitre 2, Algèbre Linéaire*. Hermann, Paris.

BRATTELLI, O., ROBINSON, D.W. (1981). *Operator Algebras and Quantum Statistical Mechanics 1* (second edition). Springer, New York.

BREIMAN, L., FRIEDMAN, J.H. (1985). Estimating optimal transformations of multiple regression and correlation. *J. Amer. Statist. Ass.*, **80**, 580–98.

BUJA, A. (1990). Remarks on functional canonical variates, alternating least squares methods and ACE. *Ann. Statist.*, **18**, 1032–69.

CARDOT, H. (1997). Contribution à l'estimation et à la prévision statistique de données fonctionnelles. PhD Thesis, Université Paul Sabatier, Toulouse.

CARDOT, H., FERRATY, F., SARDA, P. (1998). Modèle linéaire fonctionnel. *Publi. Labo. Stat. Proba.*, **4–98**, 1–18. Univ. Paul Sabatier, Toulouse.

CARDOT, H., FERRATY, F., SARDA, P. (1999). Functional Linear Model. *Stat. Prob. Letters*, **45**, 11–22.

CARDOT, H., MAS, A., SARDA, P., (2007). CLT in functional linear regression models. *Probab. Theory Related Fields*, **138**, 325–61.

CHAKAK, A., DOSSOU-GBETE, S., POUSSE, A. (1986). Une formalisation de l'analyse en composantes principales de variables aléatoires complexes. *Statist. Anal. Données*, 11, 1–18.

CHEN, K.H., ROBINSON, J. (1989). Comparison of factor spaces of two related populations. *J. Multivariate Anal.*, 28, 190–203.

CLARK, D.N. (1968). On the spectra of bounded hermitian Hankel matrices. *Am. J. Math.*, 90, 627–56.

COHEN-TANNOUDJI, C., DIU, B., LALOE, F. (1977). *Mécanique Quantique*. Herman, Paris.

CONNES, A. (1990). *Géométrie Non Commutative*. Interéditions, Paris.

CUPIDON, J., EUBANK, R., GILLIAMA, D., RUYMGAART, F. (2008). Some properties of canonical correlations and variates in infinite dimensions. *J. Mult. Analysis*, 99(6), 1083–104.

DABO-NIANG, S., FERRATY, F., eds. (2008). *Functional and Operatorial Statistics*. IWFOS'2008 Toulouse. Springer, New York.

DACUNHA-CASTELLE, D. (1981). Inversion des opérateurs de Toeplitz et statistiques des champs aléatoires gaussiens. In *Colloque International du CNRS*, Vol. 307, 231–41. CNRS, Paris.

DAUXOIS, J., FERRE, L., YAO, A.F. (2001). Un modèle semi-paramétrique pour variables aléatoires hilbertiennes. *C. R. Acad. Sci. Paris Sér. I Math.*, 333(10), 947–52.

DAUXOIS, J., NKIET, G.M. (1997). Canonical analysis of two Euclidean subspaces and its applications. *Linear Algebra. Appl.*, 264, 355–88.

DAUXOIS, J., NKIET, G.M. (1998). Nonlinear canonical analysis and independence tests. *Ann. Statist.*, 26(4), 1254–78.

DAUXOIS, J., NKIET, G.M. (2000). Lois asymptotiques de fonctions des valeurs propres d'une suite d'opérateurs aléatoires auto-adjoints. *C. R. Acad. Sci. Paris Sér. I Math.*, 330(7), 601–4.

DAUXOIS, J., NKIET, G.M. (2002). Measures of association for Hilbertian subspaces and some applications. *J. Multivariate Anal.*, 82(2), 263–98.

DAUXOIS, J., NKIET, G.M., ROMAIN, Y. (2001). Projecteurs orthogonaux, opérateurs associés et statistique multidimensionnelle. *Revue Annales ISUP*, 45(1), 31–54.

DAUXOIS, J., NKIET, G.M., ROMAIN, Y. (2004). Canonical analysis relative to a closed subspace. *Linear Algebra Appl.*, 388, 119–45.

DAUXOIS, J., NKIET, G.M., ROMAIN, Y. (2004). Linear relative canonical analysis of Euclidean random variables, asymptotic study and some applications. *Ann. Inst. Statist. Math.*, 56(2), 279–304.

DAUXOIS, J., NKIET, G.M., ROMAIN, Y. (2007). On invariance in canonical analysis with applications to covariate discriminant analysis. *Test*, 16(2), 314–32.

DAUXOIS, J., POUSSE, A. (1975). Une extension de l'analyse canonique. Quelques applications. *Ann. Inst. Henri Poincaré*, 11, 355–79.

DAUXOIS, J., POUSSE, A. (1976). Les analyses factorielles en calcul des probabilités et en statistique: essai d'étude synthétique. Thèse de Doctorat d'Żtat Université Paul Sabatier, Toulouse.

DAUXOIS, J., POUSSE, A., ROMAIN, Y. (1982). Asymptotic theory for the principal component analysis of a vector random function: some applications to statistical inference. *J. Multivariate Anal.*, 12, 136–54.

DAUXOIS, J., ROMAIN, Y., VIGUIER, S. (1993). Comparison of two factor subspaces. *J. Multivariate Anal.*, 44, 160–78.

DAUXOIS, J., ROMAIN, Y., VIGUIER, S. (1994). Tensor products and statistics. *Linear Algebra Appl.*, 210, 59–88.

DAVIS, C. (1958). Separation of two linear subspaces. *Acta. Sci. Math. Szeged*, **19**, 172–87.

DE LEURGANS et al. (1993).

DEVILLE, J.C. (1974). Méthodes statistiques et numériques de l'analyse harmonique. *Ann. Insee*, **15**, 3–101.

DEVILLE, J.C. (1981). Décompositions des processus aléatoires. Une comparaison entre les méthodes factorielles et l'analyse spectrale. *Statist. Anal. Données*, **3**, 5–33.

DIPPON, J. (1993). Asymptotic confidence spheres in certain Banach spaces via covariance operator. *J. Multivariate Anal.*, **47**, 48–58.

DIXMIER, J. (1948). Position relative de deux variétés linéaires fermées d'un espace de Hilbert. *Revue Scientifique*, **86**, 387–99.

DIXMIER, J. (1957). *Les Algèbres d'Opérateurs dans l'Espace Hilbertien (Algèbre de von Neumann)*. Gauthier-Villars, Paris.

DOSSOU-GBETE, S., POUSSE, A. (1991). Asymptotic study of eigenelements of a sequence of random self adjoint operators. *Statistics*, **22**, 479–91.

EATON, M.L. (1983). *Multivariate Statistics. A Vector Space Approach*. Wiley, New York.

EATON, M.L., TYLER, D.E. (1994). The asymptotic distribution of singular values with applications to canonical correlations and correspondence analysis. *J. Multivariate Anal.*, **50**, 238–64.

EUBANK, R., HSING, T. (2008). Canonical correlation for stochastic processes. *Stochastic Process. Appl.*, **118**(9), 1634–61.

FAUSETT, D., FULTON, C. (1994). Large least squares problems involving Kronecker products. *SIAM J. Matrix Anal. Appl.*, **15**, 219–27.

FERRATY, F., VIEU, P. (2006). *Nonparametric Functional Data Analysis. Theory and Practice*. Springer, New York.

FERRATY, F., VIEU, P., VIGUIER-PLA, S. (2007). Factor-based comparison of groups of curves. *Comput. Statist. Data Anal.*, **51**(10), 4903–10.

FINE, J. (1987). On the validity of the perturbation method in asymptotic theory. *Statistics*, **18**, 401–14.

FINE, J. (1994). Asymptotic study of the multivariate functional model in the case of a random number of observations for each mean. *Statistics*, **25**, 285–306.

FINE, J. (2000). Etude asymptotique de l'analyse canonique. *Ann. I.S.U.P.*, **44**(2–3), 21–72.

FINE, J. (2003). Asymptotic study of canonical correlation analysis: from matrix and analytic approach to operator and tensor approach. *Stat. Oper. Research*, **27**(2), 165–74.

FINE, J., POUSSE, A. (1992). Asymptotic study of the multivariate functional model. Appplication to the metric choice in principal component analysis. *Statistics*, **23**, 62–83.

FINE, J., ROMAIN, Y. (1984). Reduced principal component analysis. *Statistics*, **15**, 493–512.

FORTET, R.M. (1995). *Vecteurs, Fonctions et Distributions Aléatoires dans les Espaces de Hilbert*. Hermès, Paris.

GELFAND, J.M., YAGLOM, A.M. (1957). Calculation of the amount of information about a random function contained in another such function. (In Russian.) *Usp. Mat. Nauk.*, **12**(1), 3–52. English translation: *Ann. Math. Soc. Transl.*, **12**, 192–246 (1959).

GLEASON, M.A. (1957). Measures on the closed subspaces of a Hilbert space. *J. Math. Mech.*, **6**, 885–93.

GLYNN, W., MUIRHEAD, R.J. (1978). Inference in canonical correlation analysis. *J. Multivariate. Anal.*, **8**, 468–78.

GOHBERG, I.C., KREJN, M.G. (1971). *Introduction à la Théorie des Opérateurs Linéaires Non-autoadjoints dans un Espace Hilbertien*. Dunod, Paris.

GRENANDER, U. (1963). *Probabilities on Algebraic Structures*. Almquist et Wiksell, Stockholm.
GRENANDER, U. (198.). *Abstract Inference*. Wiley, New York.
GUICHARD, J. (2000). *L'infini au Carrefour de la Philosophie et des Mathématiques*, Ellipses, Paris.
HALL, P., HOSSEINI-NASAB, M. (2009). Theory for high-order bounds in functional principal components analysis. *Math. Proc. of the Cambridge Phil. Soc.*, **146**, 225–56.
HALMOS, P.R. (1969). Two subspaces. *Trans. Amer. Math. Soc.*, **144**, 381–9.
HALMOS, P.R. (1982). *A Hilbert Space Problems Book* (second edition). Springer, Berlin.
HANNAN, E.J. (1961). The general theory of canonical correlations and its relation to functional analysis. *J. Austral. Math. Soc.*, **2**, 229–42.
HELSTROM, C.W. (1976). *Quantum Detection and Estimation Theory*. Academic Press, New York.
HIDA, T. (1960). Canonical representation of Gaussian processes and their applications. *Mem. Coll. Sci. Uni. Kyoto*, **33**, 109–55.
HOLEVO, A.S. (1973). Statistical decision for quantum systems. *J. Multivariate Anal.*, **3**, 262–75.
HOLEVO, A.S. (1982). *Probabilistic and Statistical Aspects of Quantum Theory*. North-Holland, Amsterdam.
HOLLAND, S.M. (1970). The current interest in orthomodular lattices. In *Trends in Lattice Theory* (J.C. Abbott ed.), 41–126. Van Nostrand, New York.
HOTELLING, H. (1933). Analysis of a complex of statistical variables into principal components. *J. Educ. Psychol.*, **24**, 417–41.
HOTELLING, H. (1936). Relations between two sets of variables. *Biometrika*, **28**, 321–77.
HOUZEL, C., OVAERT, J.-L., RAYMOND, P., SANSUC, J.J. (1976). *Philosophie et Calcul de l'Infini*. Maspéro, Paris.
IBRAGIMOV, J.A., ROZANOV, Y.A. (1978). *Gaussian Random Processes*. Springer, New York.
JEFFRIES, D.J., FARRIER, D.R. (1985). Asymptotic results for eigenvectors methods. *IEEE Proceedings*, **132**, 589–94.
JEWELL, N.P., BLOOMFIELD, P. (1983). Canonical correlation of past and future for time series: definition and theory. *Ann. Statist.*, **11**, 837–47.
JONES, V., SUNDER, V.S. (1996). *Introduction to Subfactors*. Lecture Note Series, **236**. London Math. Soc., Cambridge Univ. Press, Cambridge.
JORDAN, C. (1875). Essai sur la géométrie à n dimensions. *Bull. Soc. Math. France*, **3**, 103–174.
KADOTA, T.T. (1967). Simultaneous diagonalization of two covariance kernels and application to second order stochastic processes. *SIAM Appl. Math.*, **15**, 1470–80.
KALMBACH, G. (1983). *Orthomodular Lattices*. Academic Press, London.
KARHUNEN, K. (1947). Zur Spectraltheorie stochasticher Prozesse. *Ann. Ac. Sci. Fennicae*, **37**, 1–79.
KATO, T. (1980). *Perturbation Theory for Linear Operators*. Springer, New York.
KHATRI, C.G., BHAVSAR, C.D. (1990). Some asymptotic inferential problems connected with complex elliptical distribution. *J. Multivariate. Anal.*, **35**, 66–85.
KLEFFE, J. (1973). Principal component of random variables with values in a separable Hilbert space. *Math. Oper. Forch. v. Statist.*, **4**, 391–406.
KOUDOU, A.E., ROMAIN, Y. (2000). Principes d'incertitude et statistique: une première approche. *Publ. Lab. Statist. Probab.*, **17-00**, 1–24. Univ. P. Sabatier, Toulouse.
LEDOUX, M., TALLAGRAND, M. (1991). *Probability in Banach Spaces*. Springer, Berlin.

LEBART, L., MORINEAU, A., FENELON, J.P. (1984). *Traitement des Données Statistiques.* Dunod, Paris.

LEIBOVICI, D., EL MAACHE, H. (1997). Une décomposition en valeurs singulières d'un élément d'un produit tensoriel de k espaces de Hilbert séparables. *Note C. R. Acad. Sci. Paris Série 1*, **325**, 779–82.

LENARD, A. (1972). The numerical range of a pair of projections. *J. Funct. Anal.*, **10**, 410–23.

LEURGANS, S.E., MOYEED, R.A., SILVERMAN, B.W. Canonical correlation analysis when the data are curves. *J. Roy. Statist. Soc. Ser. B* **55**(1993), no. 3, 725–740.

LOEVE, M. (1948). Fonctions aléatoires du second ordre. Supplement to: P. Lévy, *Processus Stochastiques et Mouvement Brownien*, 299–352. Gauthier-Villars, Paris.

LORD, R.D. (1954). The use of Hankel transformation in statistics. *Biometrika*, **41**, 44–55.

MALLEY, J.D., HORNSTEIN, J. (1993). Quantum statistical inference. *Statist. Sci.* 8, 4, 433–57.

MANDELBROT, B. (1995). Les Objects Fractals. Champs, 301. Flammarion, Paris.

MANTEIGA, W.G., VIEU, P. (2007). Statistics for functional data. *Comput. Statist. Data Anal.*, **51**(10), 4788–92.

MAS, A. (2006). A sufficient condition for the CLT in the space of nuclear operators, application to covariance of random functions. *Statist. Probab. Lett.*, **76**(14), 1503–9.

MAS, A. (2008) Local functional principal component analysis. *Complex Analysis and Operator Theory*, **2**(1), 135–67.

MAS, A., MENNETEAU, L. (2003). Perturbation approach applied to the asymptotic study of random operators. In *High Dimensional Probability III* (Sandjberg, Denmark, 2002), 127–34. Progr. Probab., **55**. Birkhäuser, Basel.

MENNETEAU, L. (2005). Some laws of the iterated logarithm in Hilbertian autoregressive models. *J. Multivariate Anal.*, **92**(2), 405–25.

MEYER, P.A. (1993). *Quantum Probability for Probabilists*. Springer, New York.

MURRAY, F.J., VON NEUMANN, J. (1937). On rings of operators II. *Trans. Amer. Math. Soc.*, **41**, 208–48.

NORDSTRÖM, K., VON ROSEN, D. (1987). Algebra of subspaces with applications to problems in statistics. *Proc. 2nd Inter. Tamper. Conf. Statis.* (T. Pukhila, S. Puntanen, eds), 603–14. Univ. Tampere, Finland.

OHYA, M. (1998). Foundation of entropy, complexity and fractals in quantum systems. In *Probability towards 2000* (L. Accardi, C.C. Heyde, eds), 263–86. Lecture Notes in Statistics, **128**. Springer, New York.

PAGE, L.B. (1970). Bounded and compact vectorial Hankel operators. *Trans. Am. Math. Soc.*, **150**, 529–39.

PAIGE, C.C., WEI, M. (1994). History and generality of the CS decomposition. *Linear Algebra Appl.*, **208/209**, 303–26.

PARTHASARATHY, K.R. (1967). *Probability Measures on Metric Spaces*. Academic Press, New York.

PARTHASARATHY, K.R. (1992). *An Introduction to Quantum Stochastics Calculus*. Birkhäuser, Basel.

PASZKIEWICZ, A. (1986). Convergence in W^*-algebra. *J. Func. Anal.*, **69**, 143–54.

PAVON, M. (1984). Canonical correlations of past inputs and future outputs for linear stochastic systems. *Systems Control Lett.*, **4**, 209–15.

PAULRAJ, A., OTTERSTEN, B., ROY, R., SWINDLEHURST, A., XU, G., KAILATH, T. (1993). Subspace methods for directions-of-arrival estimation. In *Handbook of Statistics*, Vol 10 (C.R. Rao, N.K. Bose, eds), 693–739. Elsevier, Amsterdam.

PELLER, V.V. (1990). Hankel operators and multivariate stationary processes. *Proc. Symp. in Pure Math.*, **51**, 357–71.

PELLER, V. (2003). *Hankel Operators and Their Applications*. Springer, New York.

PELLER, V.V., KRUSHCHEV, S.V. (1982). Hankel operators, best approximations and stationary Gaussian processes. *Russian Math. Surveys*, **37**(1), 61–144.

POUSSE, A. (1992). Etudes asymptotiques. In *Modèles pour l'Analyse des Données Multidimensionnelles* (J.J. Droesbeke et al., eds) *Economica*, Paris. 83–128.

POWER, S.C. (1980). Hankel operators on Hilbert spaces. *Bull. London Math. Soc.*, **12**, 422–42.

PRUGOVECKI, E. (1981). *Quantum Mechanics in Hilbert Spaces* (second edition). Academic Press, New York.

RAMSAY, J.O., DALZELL, C.J. (1991). Some tools for functional data analysis. *J. Roy. Statist. Soc. Ser. B*, **53**, 539–72.

RAMSAY, J.O., SILVERMAN, B. (2002). *Functional Data Analysis*, (second edition). Springer, New York.

RAO, M.M. (1993). An approach to stochastic integration (a generalized unified treatment). In *Multivariate Analysis: Future Directions* (C.R. Rao, ed.), 347–74. North-Holland/Elsevier, Amsterdam.

RÉDEI, M., SUMMERS, S.J. (2007). Quantum probability theory. *Stud. Hist. Philos. Sci. B Stud. Hist. Philos. Modern Phys.*, **38**(2), 390–417.

ROMAIN, Y. (2000a). Etude des somme, différence et produit tensoriels fonctionnels de deux endomorphismes. *Pub. Lab. Stat. Proba.*, 02-00, 1–65. Université Paul Sabatier, Toulouse.

ROMAIN, Y. (2000b). Eléments tensoriels fonctionnels généralisés. *Publ. Lab. Statist. Probab.*, 18-00, 1–31. Université Paul Sabatier, Toulouse.

ROMAIN, Y. (2002). Perturbation of functional tensors with applications to covariance operators. *Statist. Probab. Lett.*, **58**(3), 253–64.

SAMOILENKO, Y.S. (1991). *Spectral Theory of Families of Selfadjoint Operators*. Mathematics and its Applications, **57**. Kluwer, Dordrecht.

SANO, T., WATATANI, Y. (1994). Angles between two subfactors. *J. Oper. Theory*, **32**, 209–41.

SILVERMAN, B.W. (1985). Some aspects of the spline smoothing approach to non-parametric regression curve fitting (with discussion). *J. R. Statist. Soc. B*, **47**, 1–52.

SMALL, C.G., McLEISH, D.L. (1994). *Hilbert Space Methods in Probability and Statistical Inference*. Wiley, New York.

SUBRAMANYAM, A., NAIK-NIMBALKAR, U.V. (1990). Optimal unbiased statistical estimating functions for Hilbert space valued parameters. *J. Statist. Plann. Inference*, **24**, 95–105.

SUNDER, V.S. (1987). *An Invitation to von Neumann Algebras*. Springer, Paris.

SUNDER, V.S. (1988). N subspaces. *Canad. J. Math.*, **11**(1), 38–54.

TAKESAKI, M. (1972). Conditional expectation in von Neumann algebras. *J. Funct. Anal*, **9**, 306–21.

TARIELADZE, V.I., CHOBANJA, S.A. (1973). The complete continuity of the covariance operator. *Sakharth. SSR Mecn. Akad. Moambe*, **70**, 273–6.

URBANIK, K. (1972). Levy's probability measures on Euclidean spaces. *Studia Math.*, **44**, 119–48.

VAKHANIA, N.N. (1981). *Probability Distribution on Linear Spaces*. North-Holland, New York.

VAKHANIA, N.N. (1993). Canonical factorization of Gaussian covariance operators and some of its application. *Theory Probab. Appl.*, **38**, 498–505.

VARADARAJAN, V.S. (1985). *Geometry of Quantum Theory*. Springer, New York.
VANINSKII, K.L., YAGLOM, A.M. (1990). Stationary processes with a finite number of non-zero canonical correlations between future and past. *J. Time Series Analysis*, 11, 361–75.
YAGLOM, A.M. (1965). Stationary Gaussian processes satisfying the strong mixing condition and the best predictable functionals. In *Bernoulli, Bayes and Laplace Anniversary Volume* (L. LeCam, J. Neyman, eds), 241–52. Springer, Berlin.

Index

Note: page numbers in *italics* refer to Figures.

absolutely regular mixing (β-mixing)
  definition of β coefficient 385
  maximal inequalities
    Banachian framework 392–3
    covariance operator estimation of random vectors 396–8
    Hilbertian framework 391
    m-dependence case 393–4
    strong law of large numbers and bounded LIL 394–6
  recursive kernel density estimates in L^2 398–400
  proof 408–10
adapted functions 346
additive functional model (AFM) 9–10, 107–8
  asymptotic results 110–12
  comments 112
  general assumptions 109–10
  generalized 11
  interactive 11–12
  kernel estimation 108–9
additivity, testing procedure 114, 116
almost-sure convergence, kernel smoothing methods for α-mixing variables 146–9
α-mixing coefficients 137–8
α-mixing processes 138–9
  see also strong mixing (weak dependence)
α-mixing sequences 137
α-mixing variables 131, 137–8
  kernel smoothing methods 142–3, 158–60
    asymptotic properties 145–57
    functional kernel estimators 143–5
    useful tools 139–41
α-trimming functions 281
ampliation operators 428, 465
ampliations of PVSMs 436, 445–6, 449
amplitude mean square error 244
amplitude modulation function 257
amplitude variation 236–7, *238*

angle-between-subspaces theory, relation to operator trigonometry 378–9
antieigenvalues 357, 364, 365
Antieigenvector Reconstruction Lemma 368–70
antieigenvectors 358–9, 364–5, 365, 370–1
ARHD process 65, 68
  application to El Niño sea surface temperature data 67
ARHX(1) model 64–5
arithmetic α-mixing coefficients 137
association measures, application of operator trigonometry 377–8
asymptotic behaviour 459
asymptotic mean square error (AMSE), functional coefficient models 172
asymptotic normality, kernel estimator for α-mixing variables 149–53
asymptotic optimality property of functional index 100–1
asymptotic results
  additive functional model (AFM) 110–12
  autoregressive processes 58–9
  functional linear regression 31
    L^2 estimator 37–8
    mean square prediction error 33–7
    types 32
  general linear processes 51–4
  kernel functional regression
    L^2 expansion for the kernel estimate 82–3
    L^p expansion for the kernel estimate 84–5
    pointwise asymptotic normality for the kernel estimate 83–4
    uniform results 85–8
  kernel smoothing methods for α-mixing variables 145–57
  partial linear functional model 105–6
  single functional index model 99–100
asymptotic studies, assumptions 32
autocorrelation operator, convergence results 62–3
  hypothesis testing 63–4

autoregressive moving average (ARMA)
    models 205
autoregressive processes 55–6
    asymptotics for the mean and
        covariance 58–9
    confidence bands 193–4
    convergence results for autocorrelation
        operator and predictor 62–4
    extensions 64–5
    identifiability 59
    inverse problem 59–61
    numerical predictions for functional
        data 65–7
    representation of stochastic processes 56–7
    statistical problems 68

Banach spaces 328–9, 420–1
    covariance operator estimation of random
        vectors 396–8
    maximal inequalities
        α-mixing case 389–90
        β-mixing case 392–3
Banach-valued series 464
bandwidth selection
    functional coefficient models 172–3
    kernel functional regression 81
bandwidths, kernel functional regression 88–90
basis functions 216, 220, 223, 300–1
"Bayes rule" (optimal classification rule) 262
Berkeley Growth study 284
    amplitude and phase variation 236–7
    landmark registration 247–8
    time-warping functions 238–40
        deformation function 240–1
        functional inverse warping function 241–2
        parametric and nonparametric families of
            functions 242–3
best linear unbiased estimator (BLUE) 356
beta mixing see absolutely regular mixing
beta-coefficients
    definition 385
    functional coefficient models 178–82
    market model 168–9
bias correction, nonparametric regression 194
binary classification, partial least squares (PLS)
        functional classifiers 270
binomial distributions, generalized functional
        linear model 38–41
BLUP (Best Linear Unbiased Prediction)
        formula 31
Bochner integrability 332–3
Bochner integral 327, 328, 331, 333–5
bootstraps 190, 199, 207
    applications

fMRI data 205
language classification 206–7
sulfur dioxide concentrations 205–6
confidence bands for density estimation 192–3
confidence bands in nonparametric regression
    and autoregression 193–4
for empirical processes 203–5
for functional statistics 200–1
    simulation studies 201–3
kernel functional regression 90–1
in nonparametric functional regression 195
testing procedures 118–19
wild bootstrap sampling scheme, functional
    coefficient models 174
bounded law of the iterated logarithm
    (LIL) 394–6
    proofs 401–8

cadlags 346
canonical analysis (CA) 473
    Hankel and Toeplitz operators 463
    operator-based 460
canonical correlations 371–5
    past and future of stochastic processes 374–6
Capital Asset Pricing Model (CAPM) 168–9, 179,
    180
cell data 285
Central Limit Theorem (CLT) 48, 201, 204
    application to α-mixing variables 141
    application to functional linear
        regression 36–7
chemometrics, functional linear regression 22
classification methods 259–61
    general references 261
    proofs of results 289–93
    real-data examples 284–7
    simulation studies 287–9
    see also clustering; supervised functional
        classification; unsupervised classification
classification rules
    empirical risk and empirical minimization
        rules 262–3
    kernel rules 267–9
    k-NN rules 265–7
    linear discrimination rules 263–4
    optimal classifier and plug-in rules 262
clock time
    Berkeley Growth study 236
    time-warping functions 238–40
cluster analysis 275–6
    choice of number of clusters 276
    functional clustering model 307–8
        curve estimation 314–15
        discriminant functions 313–14

extensions to functional
classification 315–16
K-means clustering 277–8
for data in a Hilbert space 278–9
impartial trimming 280–3
low-dimensional graphical
representations 308–13
simulation studies 288–9
see also unsupervised classification
commutativity
PVSMs 434–5, 442–3
stationary processes 448–9
conditional cumulative distribution
function 143
kernel estimator 153
conditional density functions 144
kernel estimation 122–3
conditional distribution functions, kernel
estimation 122
conditional hazard functions 144, 145
kernel estimation 123, 153
conditional M- estimator 144, 145
conditional mean 143, 145
conditional median 41
conditional mode 144, 145
conditional quantiles 41, 144, 145
kernel estimator 153
conditional Z- estimator 144, 145, 153
confidence intervals
in density estimation 192–3
in nonparametric regression and
autoregression 193–4
consistency of classification rules
kernel rules 268
k-NN rules 266–7
continuity-type functional model 8
continuous registration methods
curve alignment 248–52, *250*
model-fitting procedure 252–4, 255
continuous stationary random functions
(CSRFs) 430–1
interpolation problems 435–6
continuous-time processes 48, 424
contraction semigroups 363–4
convergence rates, functional regression
optimality 120
topological effects on pointwise results
119–20
topological effects on uniform results 120
convergence theorems 349–50
convex linear operations 256–7
convolution products
PVSMs 426, 427–8, 432–3, 440–1
extension 434, 443–4

random measures 436, 446–7
coordinate-free approach, operator-based
statistics 452–3, 470
covariance inequalities 139–40
covariance matrices 356, 358, 361
covariance operators 462–3
of autoregressive processes 58–9
estimation of Hilbert- or Banach-valued
random vectors 396–8
functional linear regression 24
of general linear processes 52–3
covariate-adjusted regression, functional
coefficient models 168
Cramer representations 425
crisp k-means clustering algorithms 288
cross-validation, principal component analysis
(PCA) 220, 221–2
curve alignment
continuous registration methods 248–52,
250
model-fitting procedure 252–4, 255
landmark registration 247–8
curve features 236–*8*
curve registration 235
amplitude and phase variation 236–7
amplitude/phase decomposition 243–6
model-based approaches 254, 256–7
software resources 257
time-warping functions 238–40
deformation function 240–1
functional inverse warping function
241–2
parametric and nonparametric families of
functions 242–3

datasets 287
Berkeley Growth study 284
cell data 285
ECG data 284
food industry 287
medflies 286
phonemes 285–6
Tecator 285
DBWK (Durbin, Bloomfield, Watson, Knott)
lower bound 356, 358, 371
deformation functions 240–1, 254, 255
density estimation, confidence bands 192–3
density of functional data 121, 226–8
density surrogate, functional data 228–9
dependence structure 131
depth of data 272
depth-based classification 272–4
dimension reduction, PCA 210–11, 217–20
interpretation 220–4

dimensionality-reduction nonparametric functional models 6–7
  additive functional model (AFM) 9–10
    extensions 11–12
  varying-coefficient functional model 14–15
discrete-time processes 424
discretization, functional linear regression 30
discriminant functions 313–14
Donsker classes 204–5
dual homomorphism 430

education, marginal returns, functional coefficient instrumental-variable models 175, 177–8
eigenvalue ordering methods, principal components 222–3
El Niño sea surface temperature data 132–3
  application of ARH models 66–7
  functional modelization 134, 135
electricity demand modeling 169
electrocardiogram (ECG) data 284
empirical α-trimming functions 281
empirical objective function, K-means clustering 277–8
empirical processes
  bootstrap 203–5
  subsampling 199
empirical risk, supervised classification 262–3
endogenous variables, functional coefficient models 167, 175–6, 177
envelope functions 204
estimation, functional linear regression 27
  asymptotic behaviour 31–8
  discretization 30
  penalized least squares estimators 29–30
  projection-based estimators 27–9
  sparse data cases 30–1
Euler equation 366, 371
evanescent processes 346
evanescent spaces 345
Expectation Maximization (EM) algorithm 302, 305
exponential family of distributions, generalized functional linear model 39
exponential inequalities 140

factor multivariate analysis, operator-based approach 458–62
Fatou's Lemma 343
filtering, functional linear discrimination 264
financial models 168–70
  functional coefficient beta models 178–82

functional coefficient instrumental-variable models 175–8
  parameter instability 169–70
Fisher's linear classifier 263
  simulation studies 288
FLiRTI (Functional Linear Regression That's Interpretable) 320–1
fMRI data, resampling approach 205
food industry data 287
Fourier transforms 425
functional clustering model 307–8
  curve estimation 314–15
  discriminant functions 313–14
  extensions to functional classification 315–16
  low-dimensional graphical representations 308–13
functional coefficient models 166–7
  applications 168–71
    beta models 178–82
    instrumental-variable models 175–8
  bandwidth selection 172–3
  misspecification testing 173–4
  nonparametric estimation of functional coefficients 171–2
functional data
  definition v
  density 121, 226–8
  log-density 229–30
  surrogate for density 228–9
functional delta method 201
functional discrimination 94, 263–4
functional index model (FIM) 10–11
functional inverse warping function 241–2
functional linear discriminants 310–11
functional linear regression (FLR) 7–8, 23–6
  asymptotic results 31–2
    L^2 estimator 37–8
    mean square prediction error 33–7
  bibliography 42–3
  estimation 27
    discretization 30
    projection-based 27–9
    sparse data cases 30–1
  extensions 43
  generalized model 38–41
  introduction 21–2
  principal component analysis (PCA) 218–20
    interpretation of choice of dimension 220–4
    weighted least squares 224–6
  on quantiles 41–2
  smoothing parameters 33–4, 43

Functional Linear Regression That's
    Interpretable (FLiRTI) 320–1
functional logistic regression 40
functional principal component regression
    (FPCR) 27–8
functional principal oscillation pattern
    decomposition, autocorrelation
    operator 62–3
functional quasi-likelihood models 39
functional regression models
  bibliography 16
  conclusions 16–17
  density of functional variable 121
  links with multivariate models 121–2
  nonparametric
    additive functional model (AFM) 9–10,
      11–12
    continuity-type functional model 8
    definition 6
  parametric
    definition 6
    functional linear model 7–8
  partial linear varying-coefficient functional
    model (PLVCFM) 15
  practical issues 122
  rates of convergence
    optimality 120
    topological effects on pointwise
      results 119–20
    topological effects on uniform results 120
  with scalar response 74
  semiparametric
    definition 7
    functional index model 10–11
    partial linear functional model 13–14
    single functional index model 10, 12–13
  for sparse data cases 316–18
  testing procedures 113–15
    comments 118–19
    kernel-based test statistics 115–18
  transformed regression 15
  varying-coefficient functional model 14–15
  see also functional linear regression (FLR)
functional regression operator 5, 6
functional regression problems, definition 5
functional tensor product 436, 465, 470–1
functionality of models, testing procedure 115,
  116
fuzzy k-means clustering algorithms 288

G classes 144
G operator, time series 134–5
gene expression data 286–7

general linear processes 50
  asymptotics 51–4
  covariance operators 52–3
  invertibility 50–1
  mean 51–2
  perspectives and trends 54–5
General Two-Component Lemma, operator
    trigonometry 366–7, 370
generalized additive functional model
    (GAFM) 11
generalized cross-validation (GCV), in
    functional linear regression 43
generalized functional linear model 38–41
geometric α-mixing coefficients 137
Grenander's theory of sieves 62
grids v
growth time, Berkeley Growth study 236

h-modal depth 273–4
Hadmard product 471
Hankel operators 463
heteroscedasticity, functional coefficient
    models 168
hierarchical clustering methods 275–6
high-dimensional contexts 455–6
Hilbert spaces 48–9
  covariance operator estimation of random
    vectors 396–8
  general linear processes 50
    covariance operators 52–3
    invertibility 50–1
    mean 51–2
    perspectives and trends 54–5
  maximal inequalities
    α-mixing case 388–9
    β-mixing case 391
Hilbert–Schmidt operators 49
Hoffman–Jørgensen lemma 421
Hotelling correlation coefficient 372–3

identifiability, autoregressive processes 59
identification condition, functional coefficient
    models 167
impartial trimmed K-means (ITKM) 280–3
index models 179
  single functional index model 95–101
indistinguishable processes 347
inequalities
  application of operator trigonometry 371–5
  see also maximal inequalities
inference, resampling methods 190, 192, 207
  bootstrap 199–201
  subsampling 195–8, 197

infinite-dimensional contexts 455–6
infinity concept 474
integrable semivariation 328, 331, 341, 344–5, 351
   measurability with respect to a vector measure 341–2
   seminorm in $m_{F,G}(f)$ 342–3
   space $F_D(m_{F,G})$ 343–4
   summability of processes 352
integrable variation 331, 351
   summability of processes 352
integral of step functions 330–1
interactive additive functional model (IAFM) 11–12
interest rates, term structure 169
"interestingness" in dimension reduction 211
International Workshop on Functional and Operatorial Statistics (IWFOS) vi
invariant approach, operator-based statistics 452–3, 470
inverse problem, autoregressive processes 59–61
invertibility, general linear processes 50–1
Iris data set 284
I-splines 243
I_X 347

Kantorovich inequality 374
Karhunen–Loève expansion 22, 24, 214, 216, 302–3
   slope function estimator, functional linear regression 40
k-centres functional clustering methodology 288
kernel classifiers 267–9
   simulation studies 288
kernel estimation 143–5
   additive functional model (AFM) 108–9
      asymptotics 110–12
      general assumptions 109–10
   partial linear functional model 103
      asymptotics 106
      general assumptions 104–5
   single functional index model 96–7
      asymptotics 99–100
      general assumptions 97
      modeling conditions 98, 98–9
   uses 122–3
kernel functional regression 73
   bibliography 93–4
   comments on the hypothesis 92–3
   L^2 expansion for the kernel estimate 82–3
   L^p expansion for the kernel estimate 84–5
   models and estimates 74–5
   pointwise almost-sure consistency
      general conditions 75–6

      kernel estimation 76
      kNN estimation 80–1
      local linear smoothing 78–80
   pointwise asymptotic normality for the kernel estimate 83–4
   pointwise asymptotic results 82
   uniform asymptotic results 85, 92
      application to bandwidth choice 88–90
      application to bootstrapping 90–1
      Kolmogorov entropy 85–6
kernel ideas 72–3
"kernel mode" estimator 202
kernel regression estimate, rates of uniform consistency 86–8
kernel smoothing estimators 139
kernel smoothing methods
   functional coefficient models 171–2
   α-mixing variables 158–60
   asymptotic properties 145–57
kernel-based test statistics 115–17, 159–60
   theoretical advantages 117–18
kernels, reproducing 270–1
   applications to supervised classification 271–2
K-means clustering 275, 277–8
   for data in a Hilbert space 278–9
k nearest neighbour (k-NN) method
   functional regression estimation 80–1
   bibliography 94
   supervised functional classification 265–7
   simulation studies 287–8
Kolmogorov entropy 85–6
Kronecker delta 217
K-sample subsampling 197–8

L^2 estimator
   functional linear regression 37–8
   kernel functional regression 82–3
   single functional index model 99–100
land use prediction, multilogit functional linear model 40
landmark registration 247–8
language classification, functional bootstrap approach 206–7
language problems 470
   factor multivariate analysis 460–1
Lasso method 320
lattice structure of closed spaces 472
Lebesgue's theorem 345
linear closed spaces (LCS) 54
linear discriminant plots 309, 310–11, 312
linear discrimination rules 263–4
linear processes 47–9

general 50
  asymptotics 51–4
  covariance operators 52–3
  invertibility 50–1
  mean 51–2
  perspectives and trends 54–5
  statistical problems 68
  see also autoregressive processes
linearity, testing procedures 114, 116
local linear estimator 193
  functional coefficient models 171, 177
local linear functional regression 78–80
  bibliography 93–4
local smoothing techniques 302–3
locally-stationary time-series model 169
log-density 229–30
logistic regression models 40
long memory functional processes, kernel estimator 153
low-dimensional statistics 455
L^p convergence, kernel estimators 157
L^p expansion for the kernel estimate 84–5
L^q convergence, kernel estimators 153–7

market model, finance 168–9
Markov chains, α-mixing 138
martingales 352–3
MATLAB, functions for curve registration 257
matrix-based languages 470
  multivariate statistics 460–1
maximal inequalities 384
  α-mixing case
    Banach-space case 389–90
    finite-dimensional framework and Hilbertian transformations 386–9
  β-mixing case
    Banachian framework 392–3
    covariance operator estimation of random vectors 396–8
    Hilbertian framework 391
    m–dependence case 393–4
    recursive kernel density estimates in L^2 398–400
    strong law of large numbers and bounded LIL 394–6
  proofs 414–20
m-dependence, maximal inequalities 393–4
mean
  of autoregressive processes 58
  of general linear processes 51–2
mean square prediction error, functional linear regression 32, 33–7

measurability 331–2
measurable functions 329–30
measure-associated operators 464
measures with finite variation 335–6
  integration with respect to 336–8
medflies data 286
median, functional linear regression 41–2
Mercer's theorem 214
M-estimation 42
Min–Max theorem, operator trigonometry 368–9, 370
MISE, single functional index model 100, 101
MIT-BIH Arrhythmia Database 284
mitochondrial calcium overload (MCO) data 285
mixed-effects models 301–2, 307–8
mixing coefficients, notation and definitions 385–6
m-negligible/measurable sets 337, 342
modification of processes 346
modular law 472
Monotone Convergence Theory 343
Moore–Penrose pseudo-inverse 60
most predictive design points 319
moving average (MA) processes 68
moving window rule 267
  consistency 268
MSFT price analysis, functional coefficient beta model 180–2
multilogit functional linear model, land use prediction 40
multinomial distributions, generalized functional linear model 38–41
multiple functional index regression model (MFIM) 12–13
multivariate models, links with functional regression models 121–2
multivariate prediction problems 136

Nadaraya–Watson convolution kernel estimate 76–8, 193
naive bootstrap 91
neural networks 275
noncommutative probability 454, 467, 468–9, 474
non-effect, testing procedure 114, 116
nonparametric classes, definition 5
nonparametric functional regression 17
  consistency of bootstrap 195
  continuity-type functional model 8
  definition 6
  see also functional coefficient models

nonparametric smoothing 191–2
  confidence bands
    in density estimation 192–3
    in regression and autoregression 193–4
non-stationary sequences, kernel estimator 153
norming spaces, semivariation 339–40
nuclear (trace class) operators 49
nullity of slope function, functional linear regression 42–3

observables 467
operator-based languages, multivariate statistics 460–1
operator-based quantum statistics 467–9
operator-based statistics
  advantages 452–3, 457–8, 470–1
  covariance operators 462–3
  factor multivariate analysis 458–62
  Hankel and Toeplitz operators 463
  measure-associated operators 464
  projection operators 472–4
  regression operators 463–4
  shift operators (unitary operators) 466
  tensor operators 465–6
operator theory 49
operator trigonometry 355, 379–80, 466
  Antieigenvector Reconstruction Lemma 368–70
  application to statistical efficiency 355–62
  canonical correlations 371–5
    stochastic processes 377–8
  essentials 365–6, 370–1
  General Two-Component Lemma 366–7
  Min–Max theorem 368–9
  origins 363–5
  relation to angle-between-subspaces theory 378–9
operator turning angles 357, 358, 364
optimal α-trimming functions 281
optimal classification rule 262
optimality property of functional index 100–1
optional processes 346
Ornstein–Uhlenbeck (O–U) process 56, 57
orthogonal projection 472, 473–4
orthogonality property, random measures 425, 429
orthomodular law 472
ozone pollution forecasting 170–1

PACE (Principal Analysis by Conditional Estimation) 302–3
parameter instability, financial models 169–70
parametric bootstrap, simulation studies 202–3

parametric classes, definition 5
parametric regression models 17
  definition 6
partial least squares (PLS) functional classifiers 269–70
  simulation studies 287, 288
partial linear functional model 13–14, 102–3
  asymptotic results 105–6
  comments 106–7
  general assumptions 104–5
  kernel estimation 103
partial linear varying-coefficient functional model (PLVCFM) 15
penalized least squares estimators, functional linear regression 29–30, 40
perturbation rule for tensor operators 465–6
Pettis integral 328
phase mean square error 244
phase variation 235, 236–7, 238
  deformation functions 254, 255
phoneme data 285–6
Picard condition 26
Pinelis–Sakhanenko lemma 421
pivots 192
plug-in methodology, supervised classification 262
pointwise almost-sure convergence, kernel smoothing methods for α-mixing variables 146–9
pointwise product of matrices (Hadmard product) 471
Poisson distributions, generalized functional linear model 38–41
predictable processes 346
prediction problems, principal component analysis (PCA) 218–20
prediction theory, application of operator trigonometry 375–8
Principal Analysis by Conditional Estimation (PACE) 302–3
principal component analysis (PCA) 210–12, 304–7, 425
  applications 212
  curve registration 252–4, 255
  and "density" concept 228–30
  dimension reduction 217–20
    interpretation of choice of dimension 220–4
  finite-dimensional case 213–15
  for functional data 215–17
  functional linear regression, weighted least squares 224–6
  operator-based 457–8, 459–60

principal component scores, finite-dimensional case 213
principal components, eigenvalue ordering methods 222–3
probe functions 248
Procrustes iteration 249, 252, 253
projection-based estimation, functional linear regression 27–9
projection estimators 28–9
projection operators 472–4
projection pursuit 211
projection-pursuit functional model (PPFM) 13
projection-valued spectral measure (PVSM) 425, 427
  ampliations 436, 445–6
  associated random measures 439
  associated unitary operators 431–2, 439–40
  convolution products 426, 427–8, 432–3, 440–1
    extension 434, 443–4
  definition and properties 438
  equations 444
  generation of unitary operator groups 433–4, 441–2
  products 440
    extension 442–3
  properties 431
  as unknown quantities 435

quantiles, functional linear regression 41–2
quantum mechanics 467
quantum probability spaces 468
quantum random variables 468
quantum statistics 454–5, 468–9, 474

R, functions for curve registration 257
random functions 424
random measures 423–4, 425, 429
  associated PVSMs 439
  convolution products 436, 446–7
  definition and properties 437–8
  equations 444–5
  image under measurable mapping 430
  stationarily correlated random measure families 426, 439
random projections methodology 287, 288
random series 424
Rayleigh quotients 374
recursive kernel density estimates in L^2
  α-mixing case 400–1
  β-mixing case 398–400
  proofs 408–14
registered (synchronised) curves 239

regression operators 463–4
  testing procedures 113–15
regression relation 4–5
regularity, regression operator 8, 9
regularity of predictive curve, functional linear regression 32
regularization, functional linear discrimination 264
reproducing kernels 270–1
  applications to supervised classification 271–2
resampling 190
  background and notation 190–1
  bootstrap 199–201
    applications 205–7
    for empirical processes 203–5
    simulation studies 201–3
  confidence intervals
    in density estimation 192–3
    in nonparametric regression and autoregression 193–4
  subsampling 195–6
    empirical processes 199
    K-sample subsampling 197–8
    parameter estimation 196–7
residual sum of squares (RSS), functional coefficient models 174
resolution of the identity 425
roots, probabilistic distribution 195–6

SARIMA model, application to El Niño sea surface temperature data 67
SASDA (sequential algorithm for selecting design automatically) 319–20
"satisfactory" integration theories 328
second-order (wide-sense) stationarity 424
self-modeling non-linear regression (SEMOR) 256
semi-metrics, choice in functional kernel methods 158–9, 268–9
semiparametric functional regression models viii, 7, 17, 95
  asymptotic results 99–100
  functional index model 10–11
  partial linear functional model 13–14, 102–3
    asymptotic results 105–6
    comments 106–7
    general assumptions 104–5
    kernel estimation 103
  single functional index model 10, 95–6
    comments 100–1
    extensions 12–13
    kernel estimation 96–7
    modeling conditions 98–9

semiparametric functional regression
    models (*cont.*)
  testing procedure 114, 116
semivariation 338–9
  and norming spaces 339–40
  of σ-additive measures 340–1
SEMOR (self-modeling non-linear
    regression) 256
sequential algorithm for selecting design
    automatically (SASDA) 319–20
shift operators (unitary operators) 423–4, 425,
    466
σ-additive measures, semivariation 340–1
σ-algebras 346
simulation studies, classification methods 287–9
single functional index model 10, 95–6
  asymptotic results 99–100
  comments 100–1
  extensions 12–13
  kernel estimation 96–7
  modeling conditions 98–9
slope function, functional linear regression
    24–7
  asymptotic results 31–8
  estimation of derivatives 38
  nullity 42–3
  penalized least squares estimators 29–30
  projection-based estimation 27–9
  projection-based estimators
  truncated Karhunen–Loève estimator 40
small-ball probability function 227–8
  conditions 92–3
  kernel functional regression 76
  uniform version 85–6
smoothed bootstrap, simulation studies 202–3
smoothing methods 300–1, 302–3
smoothing parameters, functional linear
    regression 33–4, 43
smoothness conditions, functional linear
    regression 32
Sobolev spaces 49
software
  for curve registration 257
  for functional regression 122
sonority of speech, application of
    bootstrap 206–7
sparse data cases 298–9
  basis functions approach 300–1
  functional clustering model 307–8
    curve estimation 314–15
    discriminant functions 313–14
    extensions to functional
        classification 315–16
  low-dimensional graphical
      representations 308–13
  functional principal component analysis
      (FPCA) 304–7
  functional regression 30–1, 316–18
  local smoothing 302–3
  mixed-effects methods 301–2
spatial processes 424
spectral density 429
spectral families 425
spectral measures 423–4, 425, 429
spectral representations 425
speech-recognition data 285–6
spline estimators, functional linear
    regression 29–30, 34
spline functions 464
spline smoothing, application to El Niño sea
    surface temperature data 67
square integrable stochastic processes 379–80
standard bootstrap, simulation studies 202–3
stationarily correlated random measure families
    (RMFs) 426
  definition and properties 439
stationary correlation, CSRFs 431
stationary processes 423–4, 424–5
  interpolation problems 426, 435–6
    convolution product of two PVSMs 432–3,
        434
    mathematical approach 426–8
    mathematical framework 429–31
    PVSM associated with a unitary
        operator 431–2
    PVSMs as unknown quantities 435
    unitary operator group generated by a
        PVSM 433–4
stationary series, products 448
statistical efficiency, application of operator
    trigonometry 355–62
step functions, integral 330–1
stochastic integrals 348–9, 425, 429–30
  convergence theorems 349–50
  local summability and local integrability
      350
  martingales 352–3
stochastic intervals 346
stochastic processes 379–80, 424
  canonical correlations 374–6
  notation and definitions 345–7
stock price analysis, functional coefficient beta
    model 180–2
stopped processes 346
stopping times 346
strict-sense stationarity 424

strong law of large numbers 394–6
  proofs 401–8
strong mixing coefficient (α), definition 385
strong mixing (weak dependence) 131
  maximal inequalities
    Banach-space case 389–90
    finite-dimensional framework and Hilbertian transformations 386–9
  recursive kernel density estimates in L^2 400–1
    proof 412–14
  see also α-mixing variables
strong mixing modelizations 142
structural-change models 169–70, 179, 179–80
structural intensities 248
structural mean 249
studentization 207
subsampling 190, 195–6, 207
  empirical processes 199
  K-sample subsampling 197–8
  parameter estimation 196–7
sulfur dioxide concentrations, functional bootstrap approach 205–6
summable processes 347, 349, 350, 352
supervised functional classification 260–1
  depth-based 272–4
  empirical risk and empirical minimization rules 262–3
  kernel rules 267–9
  k–NN rules 265–7
  linear discrimination rules 263–4
  neural networks 275
  optimal classifier and plug-in rules 262
  partial least squares (PLS) functional classifiers 269–70
  reproducing kernels 271–2
  simulation studies 287–8
  support vector machines (SVMs) 274–5
synchronised (registered) curves 239
system time, transformation to clock time 238–40

Tecator dataset 285
tensor operators 465–6, 470–1
testing procedures
  functional coefficient models 173–4
  functional regression models 113–15
    comments 118–19
    kernel-based test statistics 115–18, 159–60
  see also validation techniques
threshold models 170, 176
  stock price analysis 180–2
thresholded projection estimator, functional linear regression 28–9, 35

time series
  functional modelization 134–6, 159
  locally-stationary model 169
  standard discretized model 132–4
  trending time-varying coefficient model 168
time-shift parameters 239
time-warping functions 235, 238–40, 256
  deformation function 240–1
  functional inverse warping function 241–2
  parametric and nonparametric families of functions 242–3
  software 257
Toeplitz operators 463
topology
  effects on pointwise convergence 119–20
  effects on uniform results 120
total mean square error 243
T-periodic shapes 134
trace-class (nuclear) operators 49
training samples 260
transformation regression models 15
transpose homomorphisms 430, 438, 449
trending time-varying coefficient model 168
triangular arrays, α-mixing coefficients 140–1
trigonometric operators 466
trimming functions 280–3
truncated Karhunen–Loève estimator, slope function 40

under-smoothing 194
unified statistics 471–2
unitary operator groups, generation by PVSMs 433–4, 441–2
unitary operators (shift operators) 425, 466
  associated PVSMs 431–2, 439–40, 445–6
unsupervised classification 259–60, 275–6
  choice of number of clusters 276
  K-means clustering 277–8
    for data in a Hilbert space 278–9
    impartial trimming 280–3
  simulation studies 288–9
  see also cluster analysis

validation techniques 456–7
  see also testing procedures
variable selection methods 299
  FLiRTI 320–1
  principal component analysis (PCA) 218
  SASDA 319–20
varying-coefficient functional model (VCFM) 14–15
  functional linear regression 43

vector integration 327–8
vector measures 341–2
very high-dimensional data, application of FDA 211
volatity trajectory modeling 171
von Neumann algebras 471–2
Voronoi cells 278

Watson factorized efficiencies 359
Wavelet II, application to El Niño sea surface temperature data 67
weak convergence 191
weak dependence (strong mixing) 131
    maximal inequalities
        Banach-space case 389–90
        finite-dimensional framework and Hilbertian transformations 386–9

recursive kernel density estimates in L^2 400–1
    proof 412–14
see also α-mixing variables
weak dependence (strong mixing) modelizations 142
weight construction, kernel functional regression 75–6
weighted least squares, functional linear regression 224–6
weighted predictors, principal component analysis (PCA) 223–4
wide-sense (second-order) stationarity 424
wild bootstrap 91
    functional coefficient models 174
Wong process 56, 57

Yule–Walker equation 59